# Biomarkers in
# Cardiovascular Diseases

T0179293

# Biomarkers in Cardiovascular Diseases

*Editors*

**Dimitris Tousoulis MD PhD FESC FACC**

First Cardiology Department
Athens University Medical School
Hippokration Hospital
Greece

**Christodoulos Stefanadis MD FACC FESC**

First Cardiology Department
Athens University Medical School
Hippokration Hospital
Greece

**CRC Press**
Taylor & Francis Group
Boca Raton London New York

CRC Press is an imprint of the
Taylor & Francis Group, an **informa** business

A SCIENCE PUBLISHERS BOOK

CRC Press
Taylor & Francis Group
6000 Broken Sound Parkway NW, Suite 300
Boca Raton, FL 33487-2742

First issued in paperback 2019

ISBN-13: 978-1-4665-8714-4 (hbk)
ISBN-13: 978-0-367-37970-4 (pbk)

Library of Congress Cataloging-in-Publication Data

Biomarkers in cardiovascular diseases / editors, Dimitris Tousoulis, Christodoulos Stefanadis.
    p. ; cm.
  Includes bibliographical references and index.
  ISBN 978-1-4665-8714-4 (hardback : alk. paper)
  I. Tousoulis, Dimitris. II. Stefanadis, Christodoulos.
  [DNLM: 1. Cardiovascular Diseases--diagnosis. 2. Cardiovascular Diseases--prevention & control. 3. Biological Markers. 4. Risk Factors.  WG 141]

  RC670
  616.1'075--dc23
                                                        2013010316

Visit the Taylor & Francis Web site at
http://www.taylorandfrancis.com

and the CRC Press Web site at
http://www.crcpress.com

# Preface

It is well known that atherosclerosis begins from an early age and evolves during the years, via altered underlying processes, leading to its manifestations the most significant being cardiovascular disease. The latter is the leading cause of morbidity and mortality in Western societies and several risk factors have been associated with development of heart diseases. Thus, several "tools" have been investigated in order to monitor not only heart diseases, but also atherogenesis. Current research has identified several biomarkers that have been found to possess a crucial role. Many years have passed since the first time when the term "biomarker" was introduced as "a measurable and quantifiable biological parameter". Numerous biomarkers have been strongly linked to the underlying processes such as inflammation, thrombosis and increased oxidative stress. For several years now classical biomarkers, including C-reactive protein, fibrinogen, pro-inflammatory cytokines, adhesion molecules and others, have been used as a diagnostic tool to determine the condition of cardiovascular diseases patients.

However, further questions led to the discovery of new biomarkers such as endothelial progenitor cells, markers of the extracellular matrix and others or the most recently debated microRNAs. Moreover, all these biomarkers are still under investigation and till now there is not one single marker that includes both specificity and cost-effectiveness. Research is continuing for the ideal biomarker that will offer reliability, accessibility and easy interpretation by clinicians. Considering the fact that no individual biomarker exhibits all the necessary requirements, perhaps the use of a combination of biomarkers appears the most likely and effective alternative, but this also has to be proven through large scale studies.

Therefore, in the present book, we focus on the presentation and evaluation of the most promising biomarkers used in all conditions(stable or unstable) of cardiovascular diseases and we review their potential contribution in the pathophysiology of atherosclerosis.

In the first chapter Tousoulis et al. reviews the role of biomarkers in early atherosclerosis, while in chapters two-four Androulakis et al., Tousoulis et al. and Toutouzas et al. discuss the role of biomarkers in hypertension as screening strategies and they comment on the role of circulating biomarkers in states of hyperlipidemia and acute coronary syndromes.

In the next three chapters Andreoli et al., Brili et al. and Aggeli et al. elaborate on the role of biomarkers in congestive heart failure, pulmonary embolism and hypertension, as well as the role of cardiac biomarkers in coronary artery by-pass grafting respectively.

In the following three chapters Charakida et al. discuss the role of biomarkers in peripheral vascular disease, Tsioufis et al. reviews the role of biomarkers in the cardiorenal syndrome, while Tousoulis et al. highlight the implication of C-reactive protein in cardiovascular disease.

In the second part of this book the authors focus on novel biomarkers. Tousoulis et al. comment on the role of novel biomarkers in stable and unstable coronary artery disease, Oikonomou et al. discuss the role of novel cardiac biomarkers and their application to heart failure, while Antonopoulos et al. expand on the role of genetics/genomics as cardiac biomarkers. In the final three chapters Papageorgiou et al. highlight the role of microRNAs in cardiovascular disease, Dilaveris et al. discuss the role of circulating biomarkers in cardiac arrhythmias, while Zaromitidou et al. report the available data on the potential role of biomarkers in the antiplatelet treatment.

## Acknowledgements

We would like to thank Dr. Nikolaos Papageorgiou for his important technical work on this book.

*Editors*
**Dimitris Tousoulis**
**Christodoulos Stefanadis**

# Contents

## Part II: Novel Biomarkers

# Part I
# Classical Biomarkers

# CHAPTER 1

# C-Reactive Protein in Cardiovascular Disease

Dimitris Tousoulis,* Emmanuel Androulakis, Nikolaos
Papageorgiou and Christodoulos Stefanadis

## Introduction

Atherosclerosis is a chronic disease affecting the entire arterial tree and involves complex processes leading to the formation of atherosclerotic plaque. It represents in large part, an inflammatory response in the vessels (Gutstein et al. 1999; Tousoulis et al. 2012). Of note, it is now considered an interplay of genetics and inflammation which leads to a greater understanding of the pathophysiology of atherosclerosis (Tousoulis et al. 2008a). Furthermore, endothelial injury, resulting in endothelial dysfunction (ED), is now regarded an important early event in the development of atherosclerosis (Tousoulis et al. 2012), while classic and novel risk factors for atherosclerosis represent crucial parameters associated with ED (Khot et al. 2003; Greenland et al. 2003).

Various novel inflammatory biomarkers have been used to predict cardiovascular risk (Kampoli et al. 2009; Tousoulis et al. 2008b). Importantly, studies have indicated that these molecules may predict cardiovascular events. Several circulating inflammatory biomarkers have been proposed in the last few years to contribute to inflammation in atherosclerosis and may predict cardiovascular events (Kampoli et al. 2012). However, further studies are required to evaluate the role of inflammatory biomarkers in states of atherosclerosis.

1st Department of Cardiology, 'Hippokration' Hospital, University of Athens Medical School, Athens, Greece.
*Corresponding author: drtousoulis@hotmail.com

C-reactive protein (CRP), which is an acute phase reactant and a potent inflammatory marker, has raised a debate over its role in the pathogenesis of atherothrombosis. It seems to be rather a mediator than marker and may contribute directly to atherogenesis. For example, it has been demonstrated that CRP may be detected in the atherosclerotic tissue (Torzewski et al. 1998), while mRNAs encoding CRP may be increased in atherosclerotic plaque (Yasojima et al. 2001).

In the present article, we will review important aspects of the role of inflammation in cardiovascular disease (CVDs) and will discuss the role of CRP in that state, both as a marker and as a mediator.

## Inflammation: A Crucial Process in Cardiovascular Disease

Endothelium is a semi-permeable barrier between blood and the underlying tissue that lines the entire cardiovascular system (Sandoo et al. 2010). Endothelial cells (ECs) protect vascular homeostasis by releasing several molecules including nitric oxide (NO), prostacyclin, antithrombotic substances (tissue type plasminogen activator-(tPA)), growth factors, cytokines and other mediators (Kopfstein et al. 2007). NO is a potent endothelial molecule, an endogenous anti-atherogenic molecule which exhibits vasodilatory properties, as well as anti-thrombotic, anti-inflammatory and anti-oxidant activities (Kanner et al. 1991). Classic risk factors downregulate endothelial nitric oxide synthase (eNOS) activity and stimulate reactive oxygen species (ROS) formation. Thus, reduced NO bioavailability predisposes to atherosclerosis, through up-regulation of adhesion molecules (intercellular adhesion molecule-1-ICAM-1, vascular cell adhesion protein-1-VCAM-1) and the enhanced circulation of inflammatory mediators (interleukins-IL, tumor necrosis factor-a-TNF-a) (Tousoulis et al. 2006a; Tousoulis et al. 2011a). Furthermore, NO regulates the secretion of plasminogen activator inhibitor-1 (PAI-1), TNF-a and IL-1 which favor leukocyte migration and platelet aggregation in states of ED (Tousoulis et al. 2011a). Also, ED at both macro- and micro-vascular level may contribute to limit blood flow. This state can be observed in microvascular coronary dysfunction and may contribute to limit blood flow (Agarwal et al. 2010). Patients with angina pectoris and microvascular dysfunction have been shown to present with high levels of inflammatory biomarkers, such as CRP, suggesting an underlying inflammatory process (Agarwal et al. 2010).

Atherosclerosis is a disease of large and medium sized arteries characterized by ED, vascular inflammation, and the build-up of lipids, cholesterol and cellular debris within the intima of the vessel wall. Subsequently, this lesion results in vascular remodeling, acute and chronic luminal obstruction and plaque formation with abnormalities of blood flow, and diminished oxygen supply to target organs (Briasoulis et al. 2012; Tousoulis et al. 2006b). It is well established that vascular smooth muscle cells (VSMCs) under the influence of growth factors migrate and proliferate within the intima, while circulating immune cells (leukocytes, monocytes, lymphocytes) attach to the vessel (Libby et al. 2002). Moreover, isolated from blood antioxidants, low density lipoprotein (LDL) is modified to oxidized forms (ox-LDL) contributing to the formation of atheromatous plaque via triggering systemic inflammatory responses and transforming macrophages into foam cells. Apparently, triggered by multiple risk factors, inflammatory and oxidative mechanisms are dominant in the pathophysiology of atherosclerosis (Rajavashisth et al. 1990). In turn, the foam cells' interactions with other cells, such as T lymphocytes and VSMCs contribute to the development of the sub-endothelial lipid core, a crucial event of progression to vulnerable plaque and plaque-related clinical events (Hansson 2005). Of note, a growing body of evidence has identified several processes leading to the atherosclerotic plaques formation in which subclinical inflammation seems to play crucial role (Li 2011). Moreover, macrophages also participate at the eventual rupture of a thin fibrous cap, given reports which observed rupture at sites of increased macrophage content (Drakopoulou et al. 2011). Increased density of vasa vasorum in lesions with inflammatory activation and intraplaque hemorrhage also contribute to plaque progression (Drakopoulou et al. 2011).

In diabetes mellitus, impaired vasodilation is present due to reduced NO production and increased NO inactivation. In that state, it has been observed that enhanced leukocytes-endothelial cells aggregation and increased vascular permeability are responsible for ED, mediated by mechanisms such as activation of protein kinase C, decrease of cellular NADPH, insulin resistance and ROS (Varughese 2007). Furthermore, insulin resistance has been shown to activate the mitogen-activated protein kinase stimulating pro-inflammatory adhesion molecules enhanced expression of the vasoconstrictor entdothein-1 (ET-1) (Vicent et al. 2003).

Hyperlipidemia, major risk factor for atherosclerosis and CAD, is known to mobilize inflammatory processes and to reduce NO bioavailability (Hacknman et al. 1996). In addition, ox-LDL stimulates various inflammatory processes in the vasculature, thus promoting the local production of ROS favoring mechanisms related to the progression of atherosclerosis (Stehouwer et al. 1997). Besides, hypercholesterolemia promotes leukocyte and platelet adhesion as it enhances the circulation of pro-inflammatory adhesion molecules (Sampietro et al. 1997). Moreover, recent data have proposed that inflammation may play a significant role in the pathophysiology of increased blood pressure (Savoia and Schiffrin 2006). More specifically, this is provoked through a complex set of interactions among inflammatory cells, leading to enhanced expression of mediators, such as growth factors, adhesion molecules, cytokines and chemokines. Such markers have been found to be useful tools in the diagnosis of hypertension in the future, while inflammatory process appears to be a potential therapeutic target, given reports suggesting that hypertension-associated inflammation may be associated to increased cardiovascular risk (Androulakis et al. 2009; Androulakis et al. 2011).

## CRP in Cardiovascular Disease

### *Properties*

CRP is the prototype of acute-phase protein with a molecular weight of 115 kDa, which is produced in the liver in response to a variety of inflammatory cytokines. It has a homopentameric structure and Ca-binding specificity for phosphocholine. It is phylogenetically ancient and belongs to proteins named as pentraxin, composed of five equal sub-groups, each with 206 amino acids that are combined in a three-dimensional structure (Srinivasan et al. 1994). Of note, the crystal structure, as well as its topology and chemical composition has been determined. Levels of CRP can increase rapidly up to 1,000-fold after the onset of inflammation and decrease with the resolution of aggression. Its expression is mainly regulated by IL-6 during the acute phase and it may require dissociation from a pentameric to a monomeric form in the aim of exerting its pro-atherosclerotic effects (Volanakis 2001). Interestingly, CRP binds phosphocholine which is present in bacterial membranes, cell membrane and lipoproteins and potentially recognizes the nuclear constituent in damaged cells. Also, it may activate C3 convertase through the classical pathway generating opsonic complement fragments, while interactions with Fc receptors lead to the productions of cytokines and ROS (Volanakis 2001).

CRP is a distinctive and sensitive marker for the detection of sub-clinical inflammation and has been suggested to be implicated in the pathophysiology of atherosclerosis. It participates in several related processes, such as the binding of LDL-cholesterol from macrophages to form foam cells and the development of vulnerable plaques (Wilson et al. 2006). Moreover, CRP stimulates the secretion of pro-inflammatory cytokines and tissue factor and may contribute to vascular remodeling via its effects on eNOS. Specifically, it has been indicated that CRP contributes to uncoupling of eNOS, which increases superoxide production and decreases NO bioavailability (Hein et al. 2009) (Fig. 1).

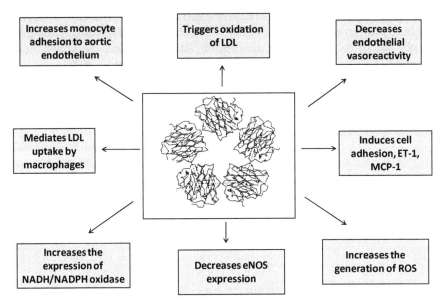

**Figure 1.** Potential mechanisms of effect for C-reactive protein.

**Abbreviations:** ROS: reactive oxygen species, LDL: low density lipoprotein, eNOS: endothelial nitric oxide synthase, ET-1: endothelin-1, MCP-1: monocyte chemotactic protein-1 (Image adapted from Srinivasan et al. 1994).

## CRP and endothelial dysfunction

CRP levels have been inversely correlated with ED while contributing to the prediction of future or recurrent cardiovascular events. They induce increased expression of several molecules, such as ICAM, VCAM and E-selectin (Fichtlscherer et al. 2000). The proposed pro-atherogenetic role of CRP is largely dependent on the ability of CRP to interact with vascular ECs. Previously, it has been demonstrated that CRP may be able to up-regulate adhesions of the selectin family and the immunoglobulin superfamily in cultured vascular ECs (Pasceri et al. 2001). Thus, it seems possible that CRP induces a significant impairment of endothelial-dependent vasorelaxation in various conditions which could lead to an increased cardiovascular risk profile. According to experimental data, CRP has been demonstrated to decrease NOS expression and bioactivity in aortic ECs, as a result of reduced mRNA and protein for endothelial NOS (Ikeda et al. 2003). Reasonably, plasma CRP concentrations have been inversely correlated with endothelium-dependent vasorelaxation (Ferri et al. 2007). In accordance, Venugopal et al. tested the effect of CRP on eNOS expression and bioactivity in cultured human aortic ECs. They demonstrated that CRP decreased eNOS mRNA, protein abundance, and enzyme activity in these cells, while pre-incubation of cells with CRP also significantly increased the adhesion of monocytes to cultured human aortic ECs (Venugopal et al. 2002).

Furthermore, increased CRP plasma levels across the coronary circulation have been associated with a profound impairment in coronary endothelial-dependent function, highlighting a novel pathophysiological mechanism linking CRP to ACS (Tousoulis et al. 2011b). Moreover, CRP has increased oxidative stress status in coronary arterioles thus inhibiting NO-dependent dilation (Qamirani et al. 2005). In line with this thinking, it has been elsewhere suggested that increases in coronary blood flow during infusion of substance P and of adenosine were lower in patients with elevated than in those with normal CRP levels (Tomai et al. 2005). However, even though data have noted the detrimental effects of the inflammatory process on coronary endothelium in the ground of NO bioavailability, CRP was not useful to evaluate such conclusions. Thus, CRP has not been related to the severity and extent of coronary atherosclerosis, although it seems to be related to the degree of local inflammation (Tousoulis et al. 2011b). Similarly, coronary sinus-aorta differences in CRP were similar among groups with different extent of left coronary arterial atherosclerosis (Gensini score), revealing a relationship between endothelial cell injury and inflammation in atherosclerotic coronary arteries, independent of CRP (Mezaki et al. 2003).

## CRP and atherothrombosis

A large amount of evidence supports a role for CRP in all phases of atherosclerosis. For example, it has been hypothesized that CRP may have pro-coagulant effects related to its ability to enhance expression of tissue factor (Cermak et al. 1993). Furthermore, according to significant evidence CRP and CRP peptides may be involved in processes related to shedding of some cellular adhesion molecules (Zouki et al. 1997), while it has also been hypothesized that sub-clinical inflammation detected by CRP reflects evidence of chronic infection. Given experimental data, CRP can be found within endothelial vessel walls (Reynolds and Vance 1987), and can induce complement activation (Wolbink et al. 1996).

A growing amount of data has proposed that CRP may promote atherothrombosis. In transgenic CRP-overexpressing mice, Danenberg et al. showed that compared with mice, in which CRP levels are undetectable, transluminal wire injury to the femoral artery was followed by thrombotic occlusion in the femoral artery in a significant higher proportion at 28 days (Danenberg et al. 2003). Similarly, photochemical injury to the carotids decreased clot formation time in the human transgenic CRP mice compared with wild type, even though questions have been raised regarding the validity of the mouse model (Verma et al. 2006). Moreover, CRP has been suggested to activate the complement in rats, increase inducible NOS (iNOS) expression in inflammation-related cells and stimulate nuclear factor-κB (NF-κB) in rat vascular smooth muscle cells (Devaraj et al. 2009). Of note, in seven male volunteers, it has been shown that an infusion of recombinant human CRP on two occasions resulted in an increase of inflammatory and coagulation biomarkers as evidenced by increased serum IL-6, IL-8, serum amyloid-A (SAA), prothrombin 1 and 2, and D-dimer (Bisoendial et al. 2007). Moreover, several lines of evidence have indicated that CRP may induce superoxide anion release and tissue factor activity *in vivo* along with promoting ox-LDL uptake, matrix metalloproteinase (MMP) release, and myeloperoxidase from macrophages (Devaraj et al. 2009). Also it has been shown that adult rats which underwent cerebral artery occlusion and then received human CRP similarly developed significantly larger cerebral infarcts compared with control subjects receiving human serum albumin (Gill et al. 2004).

## *CRP and vasculature of the coronaries*

It has been shown that CRP plasma levels were significantly higher across the coronary circulation in ACS patients, while in patients with normal coronary arteries and stable angina, no differences between coronary sinus and aorta CRP plasma levels have been observed (Forte et al. 2011). Furthermore, novel data have suggested that IL-6, a major inducer of CRP, increased in the coronary sinus after percutaneous coronary intervention (PCI) and were found to be higher in patients presenting with late restenosis compared to those without (Hojo et al. 2000). Similarly, Deliargyris et al. have examined the source of elevated IL-6 levels in patients with ACS and congestive heart failure. Samples were obtained from the femoral artery, femoral vein, left main coronary artery, and coronary sinus in 57 patients during cardiac catheterization. Notably, an elevated transcardiac IL-6 gradient was present in patients with ACS compared with controls (Deliargyris et al. 2000). Moreover, it is important to note that IL-6 and SAA are increased at the site of ruptured plaque, whereas CRP is decreased, in patients with acute myocardial infarction (AMI) (Maier et al. 2005). CRP is produced and released within the coronary circulation of patients with ACS; this is associated with impairment of endothelial function, suggesting a new pathophysiological link between CRP and ACS (Forte et al. 2011). Other data support that aside from the liver, CRP is produced in atherosclerotic lesions and various other cells. Given that several pro-atherogenic effects of CRP have been documented in ECs, human aortic ECs have been examined for CRP production. It has been suggested that stimulated synthesis and secretion of CRP in the atherosclerotic lesion could result in local concentrations of CRP and could contribute to pro-inflammatory and pro-atherogenic effects (Venugopal et al. 2005). Also, CRP mRNA has been detected in coronary plaque and has been suggested that CRP within coronary plaque might contribute to enhanced CRP levels across coronary circulation in a state of ACS (Ishikawa et al. 2004). However, other data contradicts the aforementioned notions, given that Sun et al. could not demonstrate CRP mRNA in the atherosclerotic lesion therefore suggesting that CRP is not synthesized in the atheroma (Sun et al. 2005). Of note, in another experimental study, CRP appeared to decelerate atherogenesis, possibly through proteasome-mediated protein degradation (Kovacs et al. 2007). Furthermore, genetic data have questioned the causal role of CRP in atherogenesis as well as its contribution to aortic calcification (Tousoulis et al. 2011b). However, there are several reports highlighting artifacts, such as contamination of CRP with endotoxin and azide. It is worth mentioning that in some studies using recombinant CRP, endotoxin and azide may have explained the effects seen with CRP (Devaraj et al. 2009).

## *The role in prognosis of CVD*

A growing body of data have focused on the predictive value of CRP in subjects both with and without CVD (Liuzzo et al. 1994). At this point, it is important to mention that careful interpretation of CRP values are always necessary, given that there are several conditions which affect CRP values (Table 1). Of note, according to the Cardiovascular Health Study, which included 5,888 subjects followed up for as long as 12 years for CVD incidence, elevations of CRP levels (> 3 mg/L) were associated with increased risk for CVD death (72%) and all-cause mortality (52%). Reasonably, CRP has been recently proposed as a risk factor independently of traditional risk factors such as total dyslipidaemia, age, smoking, body mass index, and

**Table 1.** Conditions associated with increased levels of C-reactive protein.

Age
Elevated body mass index
Cigarette smoking
Hypertension
Metabolic syndrome
Diabetes mellitus
Insulin resistance
Low HDL/high triglycerides
Progestogen hormone use
Chronic infections (bronchitis)
Chronic inflammation (rheumatoid arthritis, systemic lupus)
Genetic factors
Mistakenly used techniques/Inappropriate sample

blood pressure (Cao et al. 2007). Also, increased levels of high sensitivity CRP (hsCRP) have independently predicted advanced carotid plaques and early-onset carotid atherosclerosis in dyslipidemic subjects, as indicated by increased intima-media thickness (Lorenz et al. 2007; Blackburn et al. 2001; Magyar et al. 2003). Furthermore, several studies have emerged focusing on the predictive role of CRP in subjects without prior CVD. More specifically, CRP levels have shown to be predictive of future CVD events. Men in the highest levels of CRP had a threefold increased risk of AMI and twofold risk of ischemic stroke compared with men in the lowest quartile (Ridker et al. 1997). Based on prospective data derived from the Framingham Offspring Study (Rutter et al. 2004), CRP is considered as an independent risk factor for CVDs. In addition, hsCRP was the strongest univariate predictor of the risk of cardiovascular events; more specifically, the relative risk of events for women in the highest as compared with the lowest quartile for this marker was 4.4 (Ridker et al. 2000). High CRP levels have been associated with an increased risk of coronary artery disease (CAD) in both men (participating in the Health Professionals Follow-up Study) and women (participating in the Nurses' Health Study) (Pai et al. 2004).

Moreover, in essential hypertension with metabolic syndrome, elevated CRP levels were associated with a greater incidence of left ventricular hypertrophy and diastolic dysfunction. Cardiovascular disorders seem to be more frequent in that state and this is significantly related to the increased levels of inflammation (Sciarretta et al. 2007; Stefanadi et al. 2010). Also, according to recent data, microalbuminuria, accompanied by increased levels of CRP, is a strong correlate of metabolic abnormalities in essential hypertension and identifies a patient subset at very high cardiovascular risk (Pedrinelli et al. 2004). Moreover, according to retrospective data of 302 autopsies, circulating concentrations of hsCRP in the highest quartile were associated with a marked increment in the relative risk of symptomatic atherosclerosis (Burke et al. 2002). Furthermore, significant data have demonstrated that CRP levels could be a useful tool in long-term risk assessment in a state of established CAD. For example, after adjustment for other risk factors in patients suspected for CAD (ECAT study), the relative risk of an event was about two times greater in the fifth quintile of CRP concentration compared to the first four quintiles (Haverkate et al. 1997). Also, approximately one third of the CAD events which occurred within the two-year follow-up were among patients with baseline CRP concentrations of more than 3.6 mg/L (Haverkate et al. 1997). In accordance, Lindahl et al. have indicated that high levels of CRP were associated with coronary heart disease. In particular, the rates of death from cardiac causes were 5.7% among the 314 patients with blood CRP levels of less than 2 mg per liter, 7.8% among the 294 with levels of 2 to 10 mg per liter, and 16.5% among the 309 with levels of more than 10 mg per liter (Lindahl et al. 2000). Regarding unstable angina, other studies have evaluated the predictive value of CRP levels. Specifically, it has been suggested that CRP might be associated with increased risk of death in patients with unstable angina; the probabilities of death were 2.2%, 3.6%, and 7.5% after stratification of patient data by tertiles of CRP levels (< 2, 2 to 10, and > 10 mg/L) (Toss et al. 1997).

The hypothesis has been supported that CRP might be a predictor of sudden cardiac death in both atherosclerotic patients (Burke et al. 2002) and healthy subjects (Albert et al. 2002). Moreover, it has been demonstrated that the majority of patients with sudden cardiac death had no personal history of coronary heart disease. In particular, healthy male subjects, who were followed up for 17 years, were in the highest quartile of hsCRP and manifested with a 2.7-fold increased risk of sudden cardiac death (Albert et al. 2002). Furthermore, CRP may play a crucial role as a predictor of recurrent events (Biasucci et al. 1999). A small proportion of patients with discharge levels of CRP <3 mg/L, but almost 70% of those with elevated CRP were readmitted because of recurrence of instability or new AMI. However, based on the THROMBO study (Harb et al. 2002), there is no convincing evidence supporting the association between CRP and recurrent coronary events in patients with AMI.

Furthermore, the available data are confusing regarding the role of CRP levels in predicting restenosis after PCI. In the light of this consideration, Rittersma et al. have demonstrated a lack of association between preprocedural plasma CRP concentrations and angiographic coronary in-stent restenosis or clinically driven target lesion revascularization (Rittersma et al. 2004). Moreover, one month after bare-metal coronary stenting and within a follow-up time after PCI of 16.6 months, it has been suggested that patients who experienced an event had higher hsCRP levels 24 hours and 30 days after stenting. Also, low hsCRP levels have been associated with better survival rates (Fournier et al. 2008). Interestingly, data from the JUPITER study (Ridker et al. 2008) have attributed an extremely important role to CRP in the field of primary prevention. More specifically, this trial was conducted among 17,802 men and women with low density lipoprotein-C < 130 mg/dL and hsCRP ≥ 2 mg/L, and has provided the only available data on rosuvastatin for primary prevention of CVD. In apparently healthy subjects without hyperlipidemia with elevated hsCRP levels, rosuvastatin significantly reduced the incidence of major cardiovascular events. However, other data focusing on JUPITER's results have raised several controversial questions. Saely et al. have evaluated serum CRP in a consecutive series of 703 statin-naïve Caucasian patients with angiographically proven stable CAD. Of note, only 69.2% met the ≥ 2.0 mg/l serum CRP inclusion criterion of the JUPITER trial, while median CRP in their patients was 3.3, which was significantly lower than the median CRP in JUPITER, thus suggesting considerable subclinical atherosclerosis in the patients studied in JUPITER (Saely et al. 2009).

Consequently, in spite of previous evidence, scepticism still surrounds the role of CRP as an additional cardiovascular risk factor. For example, it is worth noting the controversial results regarding the predictive value as regards clinical outcomes and atherogenesis, as well as arterial restenosis in patients undergoing PCI (Tousoulis 2011b). It is relevant to mention that there are several confounding factors in CRP measurement. More specifically, this is nicely exemplified with factors related to blood samplings, such as the optimal time of collection. It has been suggested, despite the absence of clear data on the matter, that blood samplings for CRP assay should be usually performed on the morning after an overnight fast, as is routinely made for the more common laboratory measurements. Indeed, various experimental models, modifications or mistakenly used techniques, along with inadequate samples of subjects seem to play a significant role in this discrepancy in the available data (Ferri et al. 2007).

## Conclusion

Recent evidence suggests that vascular inflammation plays an important role in the pathogenesis and the clinical evolution of atherosclerosis, while several circulating inflammatory biomarkers, such as CRP, have been proposed to play a crucial role in the progression of atherosclerosis. Emerging data have suggested the production and release of CRP within the coronary circulation in patients with ACS and this is correlated with impaired coronary endothelial function. Although most reports support this statement, there are still controversial studies complicating the potential pathophysiological link between CRP and ACS. Moreover, a number of studies have focused on the positive predictive role of CRP in populations with or without prior CVD. It has been shown that CRP could be not only a risk factor independent of traditional risk factors, but also a useful predictor for cardiovascular events. However, uncertainty still exists regarding the role of CRP as an additional cardiovascular risk factor. Several parameters, including mistakenly used techniques and small samples of subjects for such evaluation seem to play a significant role in this discrepancy of the available data. Therefore, better designed studies, especially in humans, are required to investigate its potential role in CVD risk stratification.

# References

Agarwal, M., P.K. Mehta, and C.N. Bairey Merz. 2010. Nonacute coronary syndrome anginal chest pain. Med. Clin. North Am. 94: 201–216.

Albert, C.M., J. Ma, N. Rifai, M.J. Stampfer, and P.M. Ridker. 2002. Prospective study of C-reactive protein, homocysteine, and plasma lipid levels as predictors of sudden cardiac death. Circulation 105: 2595–2599.

Androulakis, E., D. Tousoulis, N. Papageorgiou, G. Latsios, G. Siasos, C. Tsioufis, A. Giolis, and C. Stefanadis. 2011. Inflammation in hypertension: current therapeutic approaches. Curr. Pharm. Des. 17: 4121–4131.

Androulakis, E.S., D. Tousoulis, N. Papageorgiou, C. Tsioufis, I. Kallikazaros, and C. Stefanadis. 2009. Essential hypertension: is there a role for inflammatory mechanisms? Cardiol. Rev. 17: 216–221.

Biasucci, L.M., G. Liuzzo, R.L. Grillo, G. Caligiuri, A.G. Rebuzzi, A. Buffon, F. Summaria, F. Ginnetti, G. Fadda, and A. Maseri. 1999. Elevated levels of C-reactive protein at discharge in patients with unstable angina predict recurrent instability. Circulation 99: 855–860.

Bisoendial, R.J., J.J. Kastelein, S.L. Peters, J.H. Levels, R. Brijmohun, J.I. Rotmans, D. Hartman, J.C. Meijers, M. Levi, and E.S. Stroes. 2007. Effects of CRP infusion on endothelial function and coagulation in normocholesterolemic and hypercholesterolemic subjects. J. Lipid Res. 48: 952– 960.

Blackburn, R., P. Giral, E. Bruckert, J.M. André, S. Gonbert, M. Bernard, M.J. Chapman, and G. Turpin. 2001. Elevated C-reactive protein constitutes an independent predictor of advanced carotid plaques in dyslipidemic subjects. Arterioscler. Thromb. Vasc. Biol. 21: 1962–1968.

Briasoulis, A., D. Tousoulis, E.S. Androulakis, N. Papageorgiou, G. Latsios, and C. Stefanadis. 2012. Endothelial dysfunction and atherosclerosis: focus on novel therapeutic approaches. Recent Pat. Cardiovasc. Drug Discov. 7: 21–32.

Burke, A.P., R.P. Tracy, F. Kolodgie, G.T. Malcom, A. Zieske, R. Kutys, J. Pestaner, J. Smialek, and R. Virmani. 2002. Elevated C-reactive protein values and atherosclerosis in sudden coronary death: association with different pathologies. Circulation 55: 2019–2023.

Cao, J.J., A.M. Arnold, T.A. Manolio, J.F. Polak, B.M. Psaty, C.H. Hirsch, L.H. Kuller, and M. Cushman. 2007. Association of carotid artery intima-media thickness, plaques, and C-reactive protein with future cardiovascular disease and all-cause mortality: The Cardiovascular Health Study. Circulation 116: 32–38.

Cermak, J., N.S. Key, R.R. Bach, J. Bhalla, H.S. Jacob, and G.M. Vercellotti. 1993. C-reactive protein induces human peripheral blood monocytes to synthesize tissue factor. Blood 82: 513–520.

Danenberg, H.D., A.J. Szalai, R.V. Swaminathan, L. Peng, Z. Chen, P. Seifert, W.P. Fay, D.I. Simon, and E.R. Edelman. 2003. Increased thrombosis after arterial injury in human C-reactive protein-transgenic mice. Circulation 108: 512–515.

Deliargyris, E.N., R.J. Raymond, T.C. Theoharides, W.S. Boucher, D.A. Tate, and G.J. Dehmer. 2000. Sites of interleukin-6 release in patients with acute coronary syndromes and in patients with congestive heart failure. Am. J. Cardiol. 86: 913–918.

Devaraj, S., U. Singh, and I. Jialal. 2009. The evolving role of C-reactive protein in atherothrombosis. Clin. Chem. 55: 229–238.

Drakopoulou, M., K. Toutouzas, A. Michelongona, D. Tousoulis, and C. Stefanadis. 2011. Vulnerable plaque and inflammation: potential clinical strategies. Curr. Pharm. Des. 17: 4190–4209.

Ferri, C., G. Croce , V. Cofini, G. De Berardinis, D. Grassi, R. Casale, G. Properzi, and G. Desideri. 2007. C-reactive protein: interaction with the vascular endothelium and possible role in human atherosclerosis. Curr. Pharm. Des. 13: 1631–1645.

Fichtlscherer, S., G. Rosenberger, D.H. Walter, S. Breuer, S. Dimmeler, and A.M. Zeiher. 2000. Elevated C-reactive protein levels and impaired endothelial vasoreactivity in patients with coronary artery disease. Circulation 102: 1000–1006.

Forte, L., G. Cimmino, F. Loffredo, R. De Palma, G. Abbate, P. Calabrò, D. Ingrosso, P. Galletti, C. Carangio, B. Casillo, R. Calabrò, and P. Golino. 2011. C-reactive protein is released in the coronary circulation and causes endothelial dysfunction in patients with acute coronary syndromes. Int. J. Cardiol. 152: 7–12.

Fournier, J.A., C. Delgado-Pecellín, A. Cayuela, S. Cabezón, and M.D. Mendoza. 2008. The high-sensitivity C-reactive protein level one month after bare-metal coronary stenting may predict late adverse events. Rev. Esp. Cardiol. 61: 313–316.

Gill, R., J.A. Kemp, C. Sabin, and M.B. Pepys. 2004. Human C-reactive protein increases cerebral infarct size after middle cerebral artery occlusion in adult rats. J. Cereb. Blood Flow Metab. 24: 1214–1218.

Greenland, P., M.D. Knoll, J. Stamler, J.D. Neaton, A.R. Dyer, D.B. Garside, and P.W. Wilson. 2003. Major risk factors as antecedents of fatal and nonfatal coronary heart disease events. JAMA 290: 891–897.

Gutstein, D.E. and V. Fuster. 1999. Pathophysiology and clinical significance of atherosclerotic plaque rupture. Cardiovasc. Res. 41: 323–333.

Hackman, A., Y. Abe, W. Jr. Insull, H. Pownall, L. Smith, K. Dunn, A.M. Gotto Jr., and C.M. Ballantyne. 1996. Levels of soluble adhesion molecules in patients with dyslipidemia. Circulation 93: 1334–1338.

Hansson, G.K. 2005. Inflammation, atherosclerosis, and coronary artery disease. N. Engl. J. Med. 352: 1685–1695.

Harb, T.S., W. Zareba, A.J. Moss, P.M. Ridker, V.J. Marder, N. Rifai, L.F. Watelet, R. Arora, M.W. Brown, R.B. Case, E.M. Dwyer Jr., J.A. Gillespie, R.E. Goldstein, H. Greenberg, J. Hochman, R.J. Krone, C.S. Liang, E. Lichstein, W. Little, F.I. Marcus, D. Oakes, C.E. Sparks, and L. VanVoorhees. 2002. Association of C-reactive protein and serum amyloid A with recurrent coronary events in stable patients after healing of acute myocardial infarction. Am. J. Cardiol. 89: 216–221.

Haverkate, F., S.G. Thompson, and S.D.M. Pyke. 1997. Production of C-reactive protein and risk of coronary events in stable and unstable angina. Lancet 449: 462–466.

Hein, T.W., U. Singh, J. Vasquez-Vivar, S. Devaraj, L. Kuo, and I. Jialal. 2009. Human C-reactive protein induces endothelial dysfunction and uncoupling of eNOS *in vivo*. Atherosclerosis 206: 61–68.

Hojo, Y., U. Ikeda, T. Katsuki, O. Mizuno, H. Fukazawa, K. Kurosaki, H. Fujikawa, and K. Shimada. 2000. Interleukin 6 expression in coronary circulation after coronary angioplasty as a risk factor for restenosis. Heart 84: 83–87.

Ikeda, U., M. Takahashi, and K. Shimada. 2003. C-reactive protein directly inhibits nitric oxide production by cytokine-stimulated vascular smooth muscle cells. J. Cardiovasc. Pharmacol. 42: 607–611.

Ishikawa, T., T. Imamura, K. Hatakeyama, H. Date, T. Nagoshi, R. Kawamoto, A. Matsuyama, Y. Asada, and T. Eto. 2004. Possible contribution of C-reactive protein within coronary plaque to increasing its own plasma levels across coronary circulation. Am. J. Cardiol. 93: 611–614.

Kampoli, A.M., D. Tousoulis, C. Antoniades, G. Siasos, and C. Stefanadis. 2009. Biomarkers of premature atherosclerosis. Trends Mol. Med. 15: 323–332.

Kampoli, A.M., D. Tousoulis, N. Papageorgiou, Z. Pallatza, G. Vogiatzi, A. Briasoulis, E. Androulakis, C. Toutouzas, P. Stougianos, C. Tentolouris, and C. Stefanadis. 2012. Clinical utility of biomarkers in premature atherosclerosis. Curr. Med. Chem. 19: 2521–2533.

Kanner, J., S. Harel, and R. Granit. 1991. Nitric oxide as an antioxidant. Arch. Biochem. Biophys. 289: 130–136.

Khot, U.N., M.B. Khot, C.T. Bajzer, S.K. Sapp, E.M. Ohman, S.J. Brener, S.G. Ellis, A.M. Lincoff, and E.J. Topol. 2003. Prevalence of conventional risk factors in patients with coronary heart disease. JAMA 290: 898–904.

Kopfstein, L., T. Veikkola, V.G. Djonov, V. Baeriswyl, T. Schomber, K. Strittmatter, S.A. Stacker, M.G. Achen, K. Alitalo, and G. Christofori. 2007. Distinct roles of vascular endothelial growth factor-D in lymphangiogenesis and metastasis. Am. J. Pathol. 170: 1348–1361.

Kovacs, A., P. Tornvall, R. Nilsson, J. Tegnér, A. Hamsten, and J. Björkegren. 2007. Human C-reactive protein slows atherosclerosis development in a mouse model with human-like hypercholesterolemia. Proc. Natl. Acad. Sci. USA 104: 13768–13773.

Li, J.J. 2011. Inflammation in coronary artery diseases. Chin. Med. J. (Engl.) 124: 3568–3575.

Libby, P., P.M. Ridker, and A. Maseri. 2002. Inflammation and atherosclerosis. Circulation 105: 1135–1143.

Lindahl, B., H. Toss, A. Siegbahn, P. Venge, and L. Wallentin. 2000. Markers of myocardial damage and inflammation in relation to long-term mortality in unstable coronary artery disease. N. Engl. J. Med. 343: 1139–1147.

Liuzzo, G., L.M. Biasucci, J.R. Gallimore, R.L. Grillo, A.G. Rebuzzi, M.B. Pepys, and A. Maseri. 1994. The pronostic value of C-reactive protein and serum amyloid A protein in severe unstable angina. N. Engl. J. Med. 331: 417–424.

Lorenz, M.W., P. Karbstein, H.S. Markus, and M. Sitzer. 2007. High-sensitivity C-reactive protein is not associated with carotid intima-media progression: the carotid atherosclerosis progression study. Stroke 38: 1774–1779.

Magyar, M.T., Z. Szikszai, J. Balla, A. Valikovics, J. Kappelmayer, S. Imre, G. Balla, V. Jeney, L. Csiba, and D. Bereczki. 2003. Early-onset carotid atherosclerosis is associated with increased intima-media thickness and elevated serum levels of inflammatory markers. Stroke 34: 58–63.

Maier, W., L.A. Altwegg, R. Corti, S. Gay, M. Hersberger, F.E. Maly, G. Sütsch, M. Roffi, M. Neidhart, F.R. Eberli, F.C. Tanner, S. Gobbi, A. von Eckardstein, and T.F. Lüscher. 2005. Inflammatory markers at the site of ruptured plaque in acute myocardial infarction: locally increased interleukin-6 and serum amyloid A but decreased C-reactive protein. Circulation 111: 1355–1361.

Mezaki, T., T. Matsubara, T. Hori, K. Higuchi, A. Nakamura, I. Nakagawa, S. Imai, K. Ozaki, K. Tsuchida, A. Nasuno, T. Tanaka, K. Kubota, M. Nakano, T. Miida, and Y. Aizawa. 2003. Plasma levels of soluble thrombomodulin, C-reactive protein, and serum amyloid A protein in the atherosclerotic coronary circulation. Jpn. Heart J. 44: 601–612.

Pai, J.K., T. Pischon, J. Ma, J.E. Manson, S.E. Hankinson, K. Joshipura, G.C. Curhan, N. Rifai, C.C. Cannuscio, M.J. Stampfer, and E.B. Rimm. 2004. Inflammatory markers and the risk of coronary heart disease in men and women. N. Engl. J. Med. 351: 2599–2610.

Pasceri, V., J.S. Cheng, J.T. Willerson, and E.T. Yeh. 2001. Modulation of C-reactive protein-mediated monocyte chemoattractant protein-1 induction in human endothelial cells by anti-atherosclerosis drugs. Circulation 103: 2531–2534.

Pedrinelli, R., G. Dell'Omo, V. Di Bello, G. Pellegrini, L. Pucci, S. Del Prato, and G. Penno. 2004. Low-grade inflammation and microalbuminuria in hypertension. Arterioscler. Thromb. Vasc. Biol. 24: 2414–2419.

Qamirani, E., Y. Ren, L. Kuo, and T.W. Hein. 2005. C-reactive protein inhibits endothelium-dependent NO-mediated dilation in coronary arterioles by activating p38 kinase and NAD(P)H oxidase. Arterioscler. Thromb. Vasc. Biol. 25: 995–1001.

Rajavashisth, T.B., A. Andalibi, M.C. Territo, J.A. Berliner, M. Navab, A.M. Fogelman, and A.J. Lusis. 1990. Induction of endothelial cell expression of granulocyte and macrophage colony-stimulating factors by modified low-density lipoproteins. Nature 344: 254–257.

Reynolds, G.D. and R.P. Vance. 1987. C-reactive protein immunohistochemical localization in normal and atherosclerotic human aortas. Arch. Pathol. Lab. Med. 111: 265–269.

Ridker, P.M., C.H. Hennekens, J.E. Buring, and N. Rifai. 2000. C-reactive protein and other markers of inflammation in the prediction of cardiovascular disease in women. N. Engl. J. Med. 342: 836–843.

Ridker, P.M., E. Danielson, F.A. Fonseca, J. Genest, A.M. Jr. Gotto, J.J. Kastelein, W. Koenig, P. Libby, A.J. Lorenzatti, J.G. MacFadyen, B.G. Nordestgaard, J. Shepherd, J.T. Willerson, and R.J. Glynn. JUPITER Study Group. 2008. Rosuvastatin to prevent vascular events in men and women with elevated C-reactive protein. N. Engl. J. Med. 359: 2195–2207.

Ridker, P.M., M. Cushman, M.J. Stampfer, R.P. Tracy, and C.H. Hennekens. 1997. Inflammation, aspirin, and the risk of cardiovascular disease in apparently healthy men. N. Engl. J. Med. 336: 973–979.

Rittersma, S.Z.H., R.J. de Winter, K.T. Koch, C.E. Schotborgh, M. Bax, G.S. Heyde, J.P. van Straalen, K.L. Mulder, J.G. Tijssen, G.T. Sanders, and J.J. Piek. 2004. Preprocedural C-reactive protein is not associated with angiographic restenosis or target lesion revascularization after coronary stent placement. Clin. Chem. 50: 1589–1596.

Rutter, M.K., J.B. Meigs, L.M. Sullivan, R.B. D'Agostino Sr., and P.W. Wilson. 2004. C-reactive protein, the metabolic syndrome, and prediction of cardiovascular events in the Framingham Offspring Study. Circulation 110: 380–385.

Saely, C.H., P. Rein, A. Vonbank, and H. Drexel. 2009. Serum levels of C-reactive protein in patients with stable coronary artery disease: JUPITER in perspective. Int. J. Cardiol. 144: 448–449.

Sampietro, T., M. Tuoni, M. Ferdeghini, A. Ciardi, P. Marraccini, C. Prontera, G. Sassi, M. Taddei, and A. Bionda. 1997. Plasma cholesterol regulates soluble cell adhesion molecule expression in familiar hypercholesterolemia. Circulation 96: 1381–1385.

Sandoo, A., J.J. van Zanten, G.S. Metsios, D. Carroll, and G.D. Kitas. 2010. The endothelium and its role in regulating vascular tone. Open Cardiovasc. Med. J. 4: 302–312.

Savoia, C. and E.L. Schiffrin. 2006. Inflammation in hypertension. Curr. Opin. Nephrol. Hypertens. 15: 152–158.

Sciarretta, S., A. Ferrucci, G.M. Ciavarella, P. De Paolis, V. Venturelli, G. Tocci, L. De Biase, S. Rubattu, and M. Volpe. 2007. Markers of inflammation and fibrosis are related to cardiovascular damage in hypertensive patients with metabolic syndrome. Am. J. Hypertens. 20: 784–791.

Srinivasan, N., H.E. White, and J. Emsley. 1994. Comparative analyses of pentraxins: implications for promoter assembly and ligand binding. Structure 2: 1017–1027.

Stefanadi, E., D. Tousoulis, E.S. Androulakis, N. Papageorgiou, M. Charakida, G. Siasos, C. Tsioufis, and C. Stefanadis. 2010. Inflammatory markers in essential hypertension: potential clinical implications. Curr. Vasc. Pharmacol. 8: 509–516.

Stehouwer, C.D., J. Lambert, A.J. Donker, and V.W. van Hinsbergh. 1997. Endothelial dysfunction and pathogenesis of diabetic angiopathy. Cardiovasc. Res. 34: 55–68.

Sun, H., T. Koike, T. Ichikawa, K. Hatakeyama, M. Shiomi, B. Zhang, S. Kitajima, M. Morimoto, T. Watanabe, Y. Asada, Y.E. Chen, and J. Fan. 2005. C-reactive protein in atherosclerotic lesions: its origin and pathophysiological significance. Am. J. Pathol. 167: 1139–1148.

Tomai, F., F. Ribichini, A.S. Ghini, V. Ferrero, G. Andò, C. Vassanelli, F. Romeo, F. Crea, and L. Chiariello. 2005. Elevated C-reactive protein levels and coronary microvascular dysfunction in patients with coronary artery disease. Eur. Heart J. 26: 2099–2105.

Torzewski, J., M. Torzewski, D.E. Bowyer, M. Fröhlich, W. Koenig, J. Waltenberger, C. Fitzsimmons, and V. Hombach. 1998. C-reactive protein frequently colocalizes with the terminal complement complex in the intima of early atherosclerotic lesions of human coronary arteries. Arterioscler. Thromb. Vasc. Biol. 18: 1386–1392.

Toss, H., B. Lindahl, A. Siegbahn, and L. Wallentin. 1997. Prognostic influence of increased fibrinogen and C-reactive protein levels in unstable coronary artery disease. Circulation 96: 4204–4210.

Tousoulis, D., A.M. Kampoli, C. Tentolouris, N. Papageorgiou, and C. Stefanadis. 2012. The role of nitric oxide on endothelial function. Curr. Vasc. Pharmacol. 10: 4–18.

Tousoulis, D., M. Koutsogiannis, N. Papageorgiou, G. Siasos, C. Antoniades, E. Tsiamis, and C. Stefanadis. 2010. Endothelial dysfunction: potential clinical implications. Minerva Med. 101: 271–284.

Tousoulis, D., A.M. Kampoli, E. Stefanadi, C. Antoniades, G. Siasos, A.G. Papavassiliou, and C. Stefanadis. 2008b. New biochemical markers in acute coronary syndromes. Curr. Med. Chem. 15: 1288–1296.

Tousoulis, D., C. Antoniades, and C. Stefanadis. 2006a. Nitric oxide in coronary artery disease: effects of antioxidants. Eur. J. Clin. Pharmacol. 62: 101–107.

Tousoulis, D., I. Andreou, C. Antoniades, C. Tentolouris, and C. Stefanadis. 2008a. Role of inflammation and oxidative stress in endothelial progenitor cell function and mobilization: therapeutic implications for cardiovascular diseases. Atherosclerosis 201: 236–247.

Tousoulis, D., M. Charakida, and C. Stefanadis. 2006b. Endothelial function and inflammation in coronary artery disease. Heart 92: 441–444.

Tousoulis, D., A.M. Kampoli, N. Papageorgiou, E. Androulakis, C. Antoniades, K. Toutouzas, and C. Stefanadis. 2011a. Pathophysiology of atherosclerosis: the role of inflammation. Curr. Pharm. Des. 17: 4089–4110.

Tousoulis, D., N. Papageorgiou, G. Latsios, G. Siasos, C. Antoniades, and C. Stefanadis. 2011b. C-reactive protein and endothelial dysfunction: gazing at the coronaries. Int. J. Cardiol. 152: 1–3.

Varughese, G.I. 2007. The impact of diabetes mellitus on endothelial dysfunction. South. Med. J. 100: 128–129.

Venugopal, S.K., S. Devaraj, and I. Jialal. 2005. Macrophage conditioned medium induces the expression of C-reactive protein in human aortic endothelial cells: potential for paracrine/autocrine effects. Am. J. Pathol. 166: 1265–1271.

Venugopal, S.K., S. Devaraj, I. Yuhanna, P. Shaul, and I. Jialal. 2002. Demonstration that C-reactive protein decreases eNOS expression and bioactivity in human aortic endothelial cells. Circulation 106: 1439–1441.

Verma, S., S. Devaraj, and I. Jialal. 2006. Is C-reactive protein an innocent bystander or proatherogenic culprit? C-reactive protein promotes atherothrombosis. Circulation 113: 2135–2150.

Vicent, D., J. Ilany, T. Kondo, K. Naruse, S.J. Fisher, Y.Y. Kisanuki, S. Bursell, M. Yanagisawa, G.L. King, and C.R. Kahn. 2003. The role of endothelial insulin signaling in the regulation of vascular tone and insulin resistance. J. Clin. Invest. 111: 1373–1380.

Volanakis, J.E. 2001. Human C-reactive protein: expression, structure, and function. Mol. Immunol. 38: 189–197.

Wilson, A.M., M.C. Ryan, and A.J. Boyle. 2006. The novel role of C-reactive protein in cardiovascular disease: risk marker or pathogen. Int. J. Cardiol. 106: 291–297.

Wolbink, G.J., M.C. Brouwer, S. Buysmann, I.J. ten Berge and C.E. Hack. 1996. C-reactive protein mediated activation of complement *in vivo*: assessment by measuring circulating complement-C-reactive protein complexes. J. Immunol. 157: 473–479.

Yasojima, K., C. Schwab, E.G. McGeer, and P.L. McGeer. 2001. Generation of C-reactive protein and complement components in atherosclerotic plaques. Am. J. Pathol. 158: 1039–1051.

Zouki, C., M. Beauchamp, C. Baron, and J.G. Filep. 1997. Prevention of *in vitro* neutrophil adhesion to endothelial cells through shedding of L-selectin by C-reactive protein and peptides derived from C-reactive protein. J. Clin. Invest. 100: 522–529.

# CHAPTER 2

# Biomarkers in Early Atherosclerosis

Dimitris Tousoulis,* Anna-Maria Kampoli, George Latsios,
Gerasimos Siasos and Christodoulos Stefanadis

## Introduction

Atherosclerosis is a chronic, progressive, inflammatory, and very complex disease of large and medium-sized muscular arteries with a long asymptomatic phase. Evidence indicates that atherosclerosis begins in childhood with the accumulation of lipid in the intima of arteries to form fatty streaks (McGill 1988). Nearly all children have at least some degree of aortic fatty streaks by 3 years of age (Napoli et al. 2005), and these fatty streaks increase after 8 years of age (Strong et al. 1962), with atherosclerotic plaques present in the coronary arteries during adolescence (Stary et al. 1992). This atherosclerotic process results in changes in the structure and function of the arterial tree (Ross 1999). Disease progression can lead eventually to the occurrence of acute cardiovascular events such as myocardial infarction, unstable angina pectoris and sudden cardiac death. Cardiovascular disease is the leading cause of death in Western societies. Statistics show CAD to be the leading cause of death among both men and women in the United States and in Europe. For example, approximately 12,800,000 Americans suffer from CAD and nearly 500,000 Americans die from heart attacks caused by CAD. Over 12 million Americans have a history of myocardial infarction or angina or both (Roger et al. 2011). Similarly, two million Europeans die from CAD each year. Death rates from CAD are higher in Northern, Central and Eastern Europe and lower in Southern and Western Europe (European cardiovascular disease statistics 2008). Based on the aforementioned data,

1st Department of Cardiology, 'Hippokration' Hospital, University of Athens Medical School, Athens, Greece.
*Corresponding author: drtousoulis@hotmail.com

it is obvious that the diagnosis of atherosclerosis in the early stages of the disease can be the cornerstone in the prevention of future cardiovascular disease. In order to be able to achieve this goal, the establishment of precise and reliable biomarker tests for the early stages of atherosclerosis is of great importance. The ideal biomarker should have the following characteristics: highly sensitive, specific, reliable, accessible, standardized, dependable, cost effective, and easily interpretable by clinicians. In this chapter, we will focus on the presentation and evaluation of the most promising biochemical markers used in risk assessment of premature atherosclerosis.

## Atherosclerotic Process: Stages and Predisposing Factors

The initial lesion in atherosclerosis involves the intima of the artery and begins in childhood with the development of fatty streaks. The first obvious change, observed at the initiation of plaque formation, is the accumulation and aggregation of lipoprotein particles, at the preferential lesion sites in the intima (Chi and Melendez 2007). More precisely, accumulation of low density lipoproteins (LDL) in the endothelial layer is the primary event of atherosclerosis (Chi and Melendez 2007; Aldons 2000; Ross 1993). LDL can diffuse into the subendothelial matrix through cell junctions and one major constituent of LDL, apolipoprotein B (apo B), interacts with the matrix proteoglycans in the endothelial monolayer (Borén et al. 1998). It is shown that only modified LDL can be taken by macrophages to form foam cells (Goldstein et al. 1979). It is believed that accumulated LDL undergoes oxidation, lipolysis, proteolysis or aggregation and that oxidation is the most significant modification for lesion initiation. As the inflammatory process continues, cell adhesion is followed by leukocyte transmigration through the endothelial layer into the intima (Chi and Melendez 2007; Varki 1994; Colling 2000). Accumulated oxidized LDL (oxLDL) can activate the endothelial cells to express adhesion molecules, which are indispensable to blood cell recruitment (Chi and Melendez 2007; Cybulsky and Gimbrone 1991). The first step of 'adhesion' is the tethering and rolling of blood cells along the surface of the endothelial monolayer, which is mediated by the selectin family of adhesion proteins (Chi and Melendez 2007; Springer 1990). P- and E-selectin are expressed on the surface of the activated endothelium and they bind to carbohydrate ligands on leukocytes (Varki 1994). The firm attachment of monocytes and T cells to the endothelium is mediated by the interaction of intercellular adhesion molecule1 (ICAM1) or vascular cell adhesion molecule1 (VCAM1) and integrin VLA-4 on the endothelium and the monocytes, respectively (Chi and Melendez 2007; Shih 1999a). Monocytes will differentiate into macrophages after their entrapment in the intima. Macrophage colony stimulating factor (M-CSF) is known to play a key role in stimulating the differentiation of monocytes into macrophages

(Rosenfeld et al. 1992). It is proposed that the initial round of T cell activation may occur in the regional lymph nodes because the antigen presenting cells may traffic from the plaque to the lymph nodes (Angeli et al. 2004). After that, the activated T cells bind to adhesion molecules to enter the plaque, where macrophages might present antigens to these T cells, leading to their further rounds of activation (Goran and Peter 2006).

In the meanwhile, within the arterial wall, macrophages contribute to the formation of fibrous plaque by metabolizing various subendothelial components (Itabe 2003). It has been noticed that in the early stage of atherosclerosis, the exit of some foam cells from plaque to the bloodstream may occur, but it is seldom seen in the late stage, when the fibrous cap forms (Chi and Melendez 2007; Daugherty et al. 2005). The transformation of macrophages to foam cells might be influenced by various factors other than just the ingestion of oxLDLs: it is suggested that immune complexes, as well as the peroxisome proliferator activated receptor, play a role in foam cell formation (Duval et al. 2002). After their formation, foam cells are subject to intensive aggregation that leads to the development of the necrotic core (containing lipids and the dead cell debris). In the center of the core, the foam cells die and the debris accumulates. This is accompanied by the accumulation of extracellular lipids. Cell death may be due to the DNA damage caused by ox-LDL (Chi and Melendez 2007; Hegyi et al. 2001). The necrotic core is covered by a fibrous cap forming the fibrous plaque. The fibrous cap is covered by a monolayer of endothelial cells, which sequesters peripheral blood from the subendothelial lesion before it ruptures. Death of the endothelial cells (ECs) may result from the cytokines produced by T cells, mast cells, a few B cells or probably even nature killer T cells (Bobryshev and Lord 1995). The degradation of matrix within the fibrous cap by various proteases such as collagenases, gelatinases and stromolysin will make the plaque more vulnerable and more susceptible to rupture (Libby 1999a). Tissue factor in the lesion lipid core is also a key player in thrombosis initiation. It is produced by activated ECs and macrophages, enhanced by oxLDL or ligation of CD40 on CD40L (Ait-Oufella et al. 2006). It is found that rupture often occurs in lesion edges, where foam cells are found in large numbers, so it is concluded that inflammatory factors may also play a role in thrombosis (Chi and Melendez 2007).

As is obvious from the aforementioned, multiple factors contribute to the pathogenesis of atherosclerosis including endothelial dysfunction, dyslipidemia, inflammatory and immunologic factors, diabetes mellitus, and smoking. Endothelial dysfunction plays a key role in atherosclerosis and especially in the initial steps of the process, mainly by causing an apparent decrease in the production of the vasodilator autacoid nitric oxide (NO) (Busse and Fleming 1996) and by increasing the production of superoxide anions and reactive oxygen species (ROS)—their excessive production is the main characteristic of oxidative stress (Victor et al. 2009). Dyslipidemia is the most important risk factor for atherosclerosis. LDL, very low density lipoprotein (VLDL) remnants, chylomicron remnants, small dense LDL (sdLDL), lipoproteins a (Lp(a)), and oxidized LDL are pro-atherogenic molecules that are increased in dyslipidemic patients, while on the other hand, high density lipoproteins (HDLs), which possess anti-atherogenic properties are decreased in these individuals (Koba and Hirano 2011). Recent studies have shown that infections and systemic inflammation play a critical role in the beginning and development of cardiovascular disease (Prasad et al. 2002; Gabrielli et al. 2002; Mayr et al. 2000). A number of different mechanisms have been accused for the inflammatory response in the vasculature. It has been found that viral and bacterial infections contribute to the development and progress of atherosclerosis inducing the expression of cytokines and adhesion molecules, migration and differentiation of smooth muscle cells, while inhibiting the apoptosis and increasing lipid accumulation (Jacob et al. 1992; Chiu et al. 1997). Active cigarette smoking is a well-known risk factor for atherosclerosis. It is associated with dose-dependent injury of the endothelium dependent dilation in healthy subjects (Celermajer et al. 1993; Tanriverdi et al. 2006; Noma et al. 2005) mainly by provoking increased asymmetric dimethylarginine (ADMA) levels, increased oxidative stress, platelet activation and the direct detrimental effects of smoke compound on endothelial cells. Finally, in diabetic subjects, hyperglycemia, elevated free fatty acids, and insulin resistance act in concert to target endothelial cells, resulting in oxidative stress and endothelial dysfunction (Kampoli et al. 2011; Tousoulis et al. 2009). Moreover the reduced nitric oxide production, the elevated levels of reactive oxygen species in combination with the increased endothelial production of endothelin-1 and other endothelium-derived vasoconstrictors further promote vasospasm and increase arterial stiffness (Kampoli et al. 2009a) (Fig. 1).

## Biomarkers of Early Atherosclerosis

The first step in determining who could be at risk for the development of atherosclerotic lesions is the identification of traditional cardiovascular risk factors such as hyperlipidemia, hypertension, smoking, diabetes mellitus, and physical inactivity. Their synergistic combination with cautiously selected biochemical markers is expected to lead to a more accurate risk stratification and treatment selection, fact indicative of the importance of the usage of markers in the clinical setting. Furthermore, the improvement of biomarker targeting, especially in the diagnosis of premature atherosclerosis and especially in individuals without known risk factors, can change the current perspectives of identifying patients with atherosclerotic risk. The

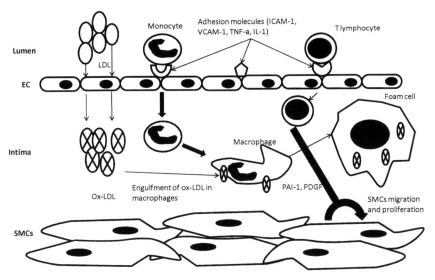

**Figure 1.** The process of atherosclerosis is illustrated. In response to adhesion molecules (ICAM-1, VCAM-1, TNF-a, IL-1), monocytes and T-lymphocytes are recruited. Monocytes differentiate into macrophages and endocytose (through interactions with scavenger receptors) into the vessel wall. These macrophages then take up oxidized low-density-lipoprotein cholesterol and become foam cells, which subsequently produce growth factors and cytokines (with the help of T lymphocytes that also produce inflammatory cytokines like PAI-1 and PDGF) that lead to the proliferation of vascular-smooth-muscle cells and the development of plaques.

LDL = low density lipoprotein, Ox-LDL = oxidized low density lipoprotein, EC = endothelial cell, SMCs = smooth muscle cells, ICAM-1 = inter-cellular adhesion molecule 1, VCAM-1 = vascular cell adhesion molecule-1, PDGF = platelet-derived growth factor, PAI-1 = plasminogen activator inhibitor-1, TNF-$\alpha$ = tumor necrosis factor alpha, IL-1 = interleukin 1.

biomarkers that will be described include classical, already established biomarkers and novel, promising ones.

The classical and novel biomarkers of atherosclerosis have been summarized in Table 1.

## Classical Biomarkers

### C-reactive protein (CRP)

The most extensively studied biomarker of inflammation in cardiovascular disease is CRP; that is the reason for its being currently the most commonly used marker of inflammation (Kampoli et al. 2009b). C-reactive protein represents the classical acute phase protein and is a circulating pentraxin produced in the liver in response to inflammatory stimuli. Plasma levels

**Table 1.** Classical and Novel Biomarkers of Premature Atherosclerosis.

| BIOMARKERS OF PREMATURE ATHEROSCLEROSIS | |
|---|---|
| **Classical** | **Novel** |
| C-Reactive Protein | Lipoprotein-associated phospholipase A2 |
| Interleukins | Asymmetric dimethylarginine |
| Tumor Necrosis Factor alpha | Myeloperoxidase |
| Apolipoproteins | Cathepsins and Cystatin C |
| Fibrinogen and fibrin | Microparticles |
| Homocysteine | |

of high-sensitive C-reactive protein (hs-CRP) provide a sensitive marker of increased inflammatory activity in the arterial wall (Pfützner et al. 2010; Pfützner and Forst 2006). It has been proved that chronic low level increase of the acute-phase protein turns out to be predictive of the risk of future cardiovascular events (Pfützner et al. 2010; Pfützner and Forst 2006). Data suggests that C-reactive protein, beyond its role as a cardiovascular risk marker, may also participate directly in atherogenesis (Blake and Ridker 2003; Ridker et al. 2005), in multiple ways including the binding to the phosphocholine of oxidized low-density lipoprotein, the inhibition of endothelial nitric oxide synthase expression in endothelial cells, the increase in plasminogen activator inhibitor-1 expression and the activity and upregulation of the expression of adhesion molecules in endothelial cells that will attract monocytes to the site of injury (Pfützner et al. 2010). Recently, high concentrations of CRP mRNA have been demonstrated to be present in atherosclerotic plaques (Yasojima et al. 2001). Two research groups revealed independently that CRP is produced by human artery smooth muscle cells of atherosclerotic lesions in response to inflammatory cytokines (Jabs et al. 2003; Calabro et al. 2003).

Despite a lack of specificity for the cause of inflammation, data from more than 30 epidemiologic studies have shown a significant association between elevated serum or plasma concentrations of CRP and the prevalence of underlying atherosclerosis, the risk of recurrent cardiovascular events among patients with established disease, and the incidence of first cardiovascular events among individuals at risk of atherosclerosis (Kampoli et al. 2012; Pearson et al. 2003). Based on the previous findings, the 2002 AHA/CDC Consensus Report (Kampoli et al. 2012; Pearson et al. 2003) recommends the measurement of CRP in asymptomatic subjects at intermediate risk for future coronary events. The most important studies, that have shown that the addition of the CRP measurement to global risk assessment by the Framingham score leads to an improved prediction of CAD, are MONICA (Koenig et al. 2004), CHS (Cushman et al. 2005), and WHS (Ridker et al. 2002). Low level increases in C-reactive protein appear to be a strong independent predictor of future cardiovascular events, including myocardial infarction, ischemic stroke, peripheral vascular disease, and sudden cardiac death among individuals with and without prior evidence of cardiovascular disease (Ridker et al. 2003; Ridker et al. 2004a; Ridker and Cook 2004b; Ridker 2004c; Pai et al. 2004; Danesh et al. 2004; Ridker 2007). Several studies have evaluated the prognostic role of CRP in premature atherosclerosis especially in children (Kampoli et al. 2012). One study aimed to evaluate the relationship between serum C-reactive protein and intima-media thickness (IMT) and functions of the common carotid artery (CCA) in children and adolescents with type 1 diabetes (Kampoli et al. 2012; Atabek et al. 2006). Using a multivariate regression model, the study included CRP and metabolic and anthropometric parameters such as duration of diabetes, HbA1c, BMI, waist to hip ratio, age, and systolic and diastolic blood pressure, as independent variables in the model for CCA structure and functions. CRP emerged as an independent correlation for mean IMT (beta = 0.51, $p < 0.001$) and diastolic wall stress (beta = 0.61, $p < 0.001$) (Atabek et al. 2006). A more recent similar study examined the role of inflammatory process in the initiation of atherosclerosis early during childhood in children with a positive family history (PFH) of CAD (Fragakis et al. 2010). The study concluded that in individuals with PFH of CAD the inflammatory process of atheromatosis appears to begin early in childhood and correlated the inflammatory markers with premature atherosclerosis. Lately, an interesting study revealed the direct role of CRP in the development and/or progression of atherosclerosis by directly quenching (at concentrations known to predict cardiovascular event) the expression of eNOS and by diminishing NO production, procedures that may serve to impair endothelial progenitor cell (EPC) function and the promotion of EPC apoptosis through the receptor for advanced glycation end-products (RAGE) (Chen et al. 2012).

## Interleukins

Interleukins are the cytokines that act specifically as mediators between leucocytes. The majority of interleukins are synthesized by helper CD4+ T lymphocytes, as well as through monocytes, macrophages, and endothelial cells. They promote the development and differentiation of T, B, and hematopoietic cells (Kampoli et al. 2012). Interleukins are considered to be key players in the chronic vascular inflammatory response that is typical of atherosclerosis. Thus, the expression of proinflammatory interleukins and their receptors has been demonstrated in atheromatous tissue, and the serum levels of several of these cytokines have been found to be positively correlated with (coronary) arterial disease and its sequelae (von der Thüsen et al. 2003). Thus far, more than 30 major members of the interleukin family have been identified, and the majority of these have been shown to play a role in atherogenesis. As applies to cytokines in general, it is possible to subdivide the interleukins into families according to the homology of their amino acid sequences or the homology of the receptor complexes to which they bind (von der Thüsen et al. 2003). Of these subgroups, the gp130 receptor family (IL-6) comprises principally proatherogenic interleukins, but most other families have both anti- and proatherogenic members. Among the numerous interleukins, IL-1, IL-6 and IL-18 have been more extensively studied for their relationship with premature atherosclerosis.

## Interleukin-1 family

The IL-1 family comprises four proteins that share considerable sequence homology and contain a β-pleated sheet structure: IL-1α, IL-1β, IL-1 receptor antagonist (IL-1Ra), and IL-18 (Dinarello 1997). Release of mature IL-1α requires extracellular calpain-mediated cleavage of a pro-IL-1α, whereas mature IL-1β is derived proteolytically from pro-IL-1β by intracellular IL-1β-converting enzyme (ICE or caspase-1) activity. Upon binding of IL-1α or IL-1β to the IL-1 receptor type I (IL-1RI), IL-1R accessory protein (IL-1RIAcP) is recruited by the receptor complex, and intracellular signal transduction is triggered through a p38 mitogen-activated protein kinase (MAPK)-activated phosphorylation cascade (von der Thüsen et al. 2003; Dinarello 1997). The signaling cascade culminates in the nuclear translocation of the transcription factors nuclear factor kappa B (NF-κB) and activating protein-1 (AP-1) and the ensuing transcription of a variety of proinflammatory genes, including autocrine amplification of IL-1 production (Suzuki et al. 1989). Il-1 has been proved to have proinflammatory effects on endothelial cells (Stanford et al. 2000) and macrophages (Sica et al. 1990), to be produced at the site of atherosclerotic lesions (Galea et al. 1996), to increase leukocyte adhesion to endothelial cells (Wang et al. 1995) and mediate leukocyte transmigration (Moser et al. 1989). Furthermore, locally produced IL-1 may serve to maintain an inflammatory milieu by autocrine and paracrine stimulation of cytokine (Garcia et al. 2000) and adhesion molecule expression (Collins and Cybulsky 2001). In the advanced plaque, IL-1-induced upregulation of matrix metalloproteinases may destabilize the proteinaceous scaffold of the cap and thereby have a hand in plaque rupture (Galis et al. 1995). For all the aforementioned, IL-1 was one of the first cytokines to be considered instrumental in the propagation of vessel wall inflammation in atherosclerosis. Lately a very interesting study conducted in obese children has found that cytokines IL-1Ra and MCP-1 were most significantly increased and could reliably predict the development of diabetes and atherosclerosis (Stoppa-Vaucher et al. 2012).

As far as IL-18 is concerned, due to its chemical similarity to IL-1, it has also been found to upregulate the expression of intercellular adhesion molecule 1 (ICAM-1) and cytokines by monocytes, including IL-1β, IL-6, and IL-8 (Dinarello 1999), and the production of vascular cell adhesion molecule-1 (VCAM-1) by endothelial cells (Vidal-Vanaclocha et al. 2000). Its pro-atherogenic properties (Mallat et al. 1999) have been demonstrated in atherosclerotic plaques in human carotids, which is primarily localized to macrophages. More specifically, it was noticed that the corresponding receptor, IL-18R, was expressed on endothelial cells and macrophages and barely present on SMCs. These findings have subsequently been confirmed histologically and *in vitro* by Gerdes et al. who also demonstrated the functionality of the IL-18 receptor on these cells through IL-18-mediated induction of pro-atherogenic factors, including IL-6, IL-8, ICAM-1, and matrix metalloproteinases (Gerdes et al. 2002). In addition, the serum level of IL-18 has recently been identified as a strong predictor of cardiovascular death in stable and unstable angina (Blankenberg et al. 2002). Furthermore, a large study in healthy, middle-aged European men has shown that total plasma IL-18 concentrations are an independent predictor of coronary events (Blankenberg et al. 2003); importantly, variation within the IL-18 gene has been shown to influence circulating concentrations of IL-18 and clinical outcome in patients with coronary heart disease (Tiret et al. 2005).

## Interleukin- 6

Interleukin-6 (IL-6) is a proinflammatory cytokine that in several mechanisms contributes to the development of atherosclerosis. IL-6 is a powerful inducer of the hepatic acute phase response (Kampoli et al. 2012). Endothelial cells, smooth muscle cells, and macrophages are capable of elaborating IL-6, and its expression has been observed in atherosclerotic lesions in humans, hypercholesterolemic rabbits, and apoE-deficient mice (Rus et al. 1996; Sukovich et al. 1998; Schieffer et al. 2000). Although the endothelium is largely unresponsive to IL-6 (Podor et al. 1989), addition of the soluble IL-6Rα subunit (sIL-6R) enables endothelial cells to mount an inflammatory response to IL-6, by interacting with membrane-bound gp130 (Jones et al. 2001). sIL-6R present in serum and/or elaborated locally by cells in the intima may serve to augment endothelial adhesion and extravasation of leukocytes into the atherosclerotic plaque. In addition, smooth muscle cells are stimulated by IL-6 to express ICAM-1 (Ikeda et al. 2001) and to evolve into foam cells (Klouche et al. 2000). Interestingly, the progression of atherosclerotic lesions to an advanced phenotype appears to be inhibited by IL-6 in apoE-deficient mice, uncovering a potentially biphasic mode of action in atherogenesis (Elhage et al. 2001), which is perhaps partly explained by its observed anti-inflammatory properties (Xing et al. 1998) and its inhibition of macrophage class A scavenger receptor expression (Liao et al. 1999). Elevated concentrations of acute phase reactants, such as C-reactive protein, are found in patients with acute coronary syndromes, and predict future risk in apparently healthy subjects. The acute phase reaction is associated with elevated levels of fibrinogen, a strong risk factor for CHD, with autocrine and paracrine activation of monocytes by IL-6 in the vessel wall contributing to the deposition of fibrinogen (Kampoli et al. 2012). In a large multicenter study, IL-6 gene polymorphisms were found to correlate with the severity of coronary artery disease and the risk of myocardial infarction (Georges et al. 2001). Also, carotid atherosclerosis has been shown to be independently linked with an IL-6 promoter polymorphism (Rundek et al. 2002), as has the risk of coronary artery disease (Humphries et al. 2001).

## Tumor necrosis factor alpha (TNF-a)

Tumor necrosis factor (TNF-α) is a cytokine with a wide range of proinflammatory properties and is a member of a group of cytokines that stimulate the acute phase reaction (Vassali 1992). It is produced chiefly by activated macrophages, although it can be produced by other cell types as well. Increased plasma concentrations of TNF-α have been found in patients with premature coronary artery disease (Jovinge et al. 1998). TNF-α elicits the expression of IL-6, which induces expression of hepatic genes encoding acute phase reactants, as well as the production of other effector molecules in the inflammatory response, such as cellular adhesion molecules for leukocytes (Skooga et al. 2002; Libby et al. 1999b). Adhesion of circulating leukocytes to endothelial cells with ensuing transendothelial migration is considered an important early step in atherogenesis (Cybulsky and Gimbrone Jr. 1991), and increased expression of cellular adhesion molecules may accordingly be one mechanism by which TNF-α is implicated in atherosclerotic disease. The role of TNF-a in premature atherosclerosis was evaluated in a group of patients with SLE (Rho et al. 2008). Concentrations of cytokines (TNF-alpha, IL-1alpha, and VEGF), inflammatory enzymes (MPO and MMP-9), acute-phase reactants (ESR, CRP, and SAA) and adhesion molecules (VCAM, ICAM, and E-selectin) were compared between patients with SLE and controls. In multivariable analyses adjusting for Framingham risk score, cumulative corticosteroid dose, and diabetes, E-selectin (OR 1.90, 95% CI 1.08–3.33), VCAM (OR 1.99, 1.18–3.37), ICAM (OR 2.30, 1.13–4.7), and TNF-alpha (OR 2.36, 1.10–5.06) were significantly associated with the severity of coronary calcium.

## Apolipoproteins

Apolipoproteins are proteins that bind lipids to form lipoproteins. They transport the lipids through the lymphatic and circulatory systems. It is well known that the metabolism of apolipoproteins (apo) is closely associated with the development of atherosclerosis. In lipid transport, apolipoproteins function as structural components of lipoprotein particles, cofactors for enzymes and ligands for cell-surface receptors. In particular, apoA1 is the major protein component of high-density lipoproteins, whereas apoA4 is thought to act primarily in intestinal lipid absorption. Further, apoE is a blood plasma protein that mediates the transport and uptake of cholesterol and lipid by way of its high affinity interaction with different cellular receptors, including the low-density lipoprotein (LDL) receptor. Recent findings with apoA1 and apoE suggest that the tertiary structures of these two members of the human exchangeable apolipoprotein gene family are related (Saito et al. 2004). ApoB is a protein constituent of the atherogenic very low-density lipoproteins (VLDL), intermediate density lipoproteins (IDL), LDL and apoA-I as well as a structural protein component of high density lipoproteins (HDL). As is obvious, measurement of plasma apoB and apoA-I allows an assessment of the total number of atherogenic (LDL, VLDL, and IDL) and antiatherogenic particles (HDL), respectively (Kampoli et al. 2012). Alterations and polymorphisms in Apo-A genes have been associated with the development of premature atherosclerosis (Ikewaki et al. 2004; Hovingh et al. 2004).

One of the most important genes associated with lipid metabolism is the one coding Apo E, a component of chylomicrons and very low density lipoproteins acting as a ligand for LDL and LRP receptors, thus regulating lipid transport and metabolism (Balcerzyk et al. 2007). Particles of Apo E also participate in removal of cholesterol from cells and have multiple pleiotropic effects. Apo E is involved in inflammatory response modulation, regulation of platelet function, apoptosis as well as oxidative stress (Balcerzyk et al. 2007). The association and the role of apolipoprotein E polymorphism in premature coronary artery disease were evaluated in children with family history of premature coronary artery disease (Ciftdoğan et al. 2012). It was found that total cholesterol (Tc) levels are influenced by apoE genotypes in childhood. Also, the frequency of the ε4 allele is greater in children with family history of premature CAD. The ε4 allele may be associated with an increased risk for development of atherosclerosis by elevated levels of Tc in children with family history of CAD. All of the aforementioned indicate that the evaluation of apoE gene polymorhisms may contribute to the assessment of cardiovascular risk in children with a family history of CAD.

## Fibrinogen and fibrin

Fibrinogen is a protein produced by the liver. Normal fibrinogen circulates in the plasma at a concentration of approximately 200 to 400 mg/dl, has numerous functional interactions and plays a pivotal role in the hemostatic balance (Kampoli et al. 2012). There are several mechanisms by which fibrinogen may increase the risk for developing atherosclerotic plaques. Firstly, it binds specifically to activated platelets via glycoprotein IIb/IIIa, contributing to platelet aggregation. Secondly, increased fibrinogen levels promote fibrin formation. Thirdly, it is a major contributor to plasma viscosity. Finally, it is an acute-phase reactant that is increased in inflammatory states (Kampoli et al. 2012). In the European Atherosclerosis Research Study Group (EARS study) (Bara et al. 1994) fibrinogen, factor VIIc and PAI-1 were compared between patients with CAD and controls across European regions. Fibrinogen and factor VIIc were positively correlated with BMI, smoking and contraception. PAI-1 was positively and independently correlated with BMI and waist-to-hip ratio, and negatively with contraception. Factor VIIc and PAI-1 were correlated with cholesterol and triglycerides, and fibrinogen was weakly correlated with LDL-cholesterol. After adjustment for covariates and lipids, fibrinogen level was significantly higher in male cases than in controls (2.38 vs 2.29, $p < 0.01$). No such difference was found in females (2.59 vs 2.57 - NS). There was no significant case/control difference for factor VIIc and PAI-1. These results support the hypothesis that fibrinogen is a transmissible risk factor of coronary artery disease in males. The role of fibrin in premature atherosclerosis was examined in a group of young, post-myocardial infarction patients with angiographic-proven CAD and in healthy volunteers matched for age and sex (Collet et al. 2006). Young CAD patients had a significant increase in plasma concentration of fibrinogen, von Willebrand factor, plasminogen activator inhibitor type 1, and lipoprotein(a) as compared with controls ($P<0.05$). Fibrin of young CAD patients was stiffer ($P=0.002$), made of numerous ($P=0.002$) and shorter fibers ($P=0.04$), and lysed at a slower rate than that of controls ($P=0.03$). Fibrin stiffness was an independent predictor for both premature CAD and hypofibrinolysis.

## Homocysteine (HCY)

Homocysteine is a non-protein amino acid with the formula $HSCH_2CH_2CH(NH_2)CO_2H$. It is a homologue of the amino acid cysteine, differing by an additional methylene ($-CH_2-$) group. It is biosynthesized from methionine by the removal of its terminal $C^\varepsilon$ methyl group. An elevated serum concentration of homocysteine (a state called hyperhomocysteinemia) is a known risk factor for atherosclerosis and is associated with an increased risk of myocardial infarction and death (Kampoli et al. 2012). Homocysteine decreases the availability of endothelial cell-derived NO, impairs endothelial-dependent vasodilatation, induces oxidative stress, and increases the risk of thrombosis (Maron and Loscalzo 2009). It has been suggested that some effects of homocysteine on EC function are mediated by asymmetric dimethylarginine (ADMA), an endogenous inhibitor of NO synthase (Stuhlinger et al. 2003). Homocysteine inhibits the enzyme that metabolizes ADMA and also increases the synthesis of ADMA by activating the endoplasmic reticulum stress pathway that leads to proteolysis of proteins with methylated arginine residues and release of ADMA (Dayal and Lentz 2005). In patients with systemic lupus erythematosus (SLE), ADMA and homocysteine have been proved to be biomarkers for and may be mediators of premature arterial stiffening (Perna et al. 2010). Evidence is accumulating and documents its associations with atherosclerotic disease in both case-control observations and prospective cohort studies, *in vitro* experiments, and *in vivo* experimental models in both animals and human subjects, as well as the successful improvement by homocysteine-lowering of endothelial function as surrogate atherosclerosis endpoints in asymptomatic human and coronary patients (secondary prevention) (Woo et al. 2002). The role of HCY in premature atherosclerosis was pinpointed in a group of patients with family history of premature coronary artery disease (Taraboanta et al. 2008), where it was indicated that noninvasive assessment of carotid artery plaques and intima media thickness and plasma homocysteine measurements may be useful in the prediction of the disease.

## Novel Biomarkers

### *Lipoprotein-associated phospholipase A2 (Lp-PLA$_2$)*

Lipoprotein-associated phospholipase A2 (Lp-PLA$_2$) is a novel biomarker of vascular-specific inflammation that provides information about atherosclerotic plaque inflammation and stability. It is an enzyme hydrolyzing sn-2 side chains of oxidized LDL to yield a free oxidized fatty acid and lysophosphatidylcholine (Kampoli et al. 2012). This activity creates two potentially pro-inflammatory and pro-atherogenic particles from one. It is primarily present on LDL, and small amounts are found in HDL, but it also seems to have a particular predilection for binding to Lp(a), where it is found up to seven times higher content than equimolar amounts of LDL (Kampoli et al. 2012). Lp-PLA$_2$ is produced by hematopoietic cells, primarily from monocytes/macrophages (Asano et al. 1999; Stafforini et al. 1990), and it is located with and highly expressed by macrophages within the necrotic core and the fibrotic cap of advanced rupture-prone plaques (Häkkinen et al. 1999; Kolodgie et al. 2006). The Lp-PLA$_2$ associated with LDL is the major determinant of plasma enzyme levels, and most studies over past years suggest that this enzyme plays a pro-inflammatory role in the artery wall thus promoting vascular inflammation and atherosclerosis (Shi et al. 2007; Lavi et al. 2007). The proinflammatory and proatherogenic role of plasma Lp-PLA$_2$ is primarily attributed to the fact that during the hydrolysis of oxidized phospholipids, this enzyme generates lysophosphatidylcholine and oxidized free fatty acids, both of which exhibit proatherogenic activities (MacPhee et al. 1999). In this regard, a substantial body of peer-reviewed studies in Caucasian populations has supported Lp-PLA$_2$ as a new, independent cardiovascular risk factor (Garza et al. 2007). In contrast to the Lp-PLA$_2$ associated with LDL, several lines of evidence suggest that HDL-associated Lp-PLA$_2$ (HDL-Lp-PLA$_2$), although present at low levels, may contribute to the anti-atherogenic effects of this lipoprotein. In a study conducted in patients with β-thalassemia (exhibiting high plasma Lp-PLA$_2$ levels, attributed to increased enzyme secretion from monocytes/macrophages and to the predominance of sdLDL particles in plasma), it was concluded that plasma Lp-PLA$_2$ is correlated with carotid IMT, suggesting that this enzyme may be implicated in premature carotid atherosclerosis observed in β-thalassemia (Tselepis et al. 2010).

## *Asymmetric dimethylarginine (ADMA)*

As the principal endogenous inhibitor of nitric oxide synthase, ADMA (asymmetric dimethylarginine), regulates rates of nitric oxide (NO) formation. Low circulating concentrations of endogenous methylarginines like ADMA or L-monomethylarginine (L-NMMA) competitively inhibit eNOS within a physiological concentration range and result in a rightward shift of the concentration-response curve of the NOS substrate, L-arginine (Tsikas et al. 2000). Data from experimental studies document that pathophysiologically relevant ADMA blood concentrations, that is between 2 and 10 μmol/l, significantly inhibit NOS in isolated human blood vessels, in cultured macrophages and in endothelial cells, thereby reducing the generation of NO (Faraci et al. 1995; Paravicini et al. 2006). The role of ADMA in premature atherosclerosis was investigated in children (Ayer et al. 2009). It was highlighted that lower HDL-cholesterol, higher levels of ADMA and systolic BP are significantly associated with greater arterial wall thickness in early childhood. A recent study assessed the relationship of ADMA and homocysteine to subclinical vascular disease in patients with SLE which are at high risk of accelerated atherosclerosis and premature arterial stiffening (Perna et al. 2010). ADMA was significantly related to the arterial stiffness index. Independent correlates of arterial stiffening included the ADMA concentration, the presence of diabetes mellitus, older age at the time of diagnosis, longer disease duration, and the absence of anti-Sm or anti-RNP antibodies. Based on these data it was concluded that ADMA is a biomarker for and may be a mediator of premature arterial stiffening in patients with SLE.

## *Myeloperoxidase (MPO)*

Myeloperoxidase (MPO) is one of the major enzymes by which neutrophils, monocytes and macrophages are capable of altering the structural integrity of the vessel wall. MPO has evolved within the last decades from a heme enzyme primarily regarded as a key constituent in innate immunity to an enzyme critically involved in inflammatory vascular disease (Böger 2011; Lau and Baldus 2006). MPO catalyzes the conversion of chloride and hydrogen peroxide to HOCl, resulting in LDL oxidation and conversion into high-uptake forms such as ox-LDL for macrophages, leading to cholesterol deposition and foam cell formation (Steffen et al. 2006; Kraemer et al. 2004; Morrow et al. 2008). Prospective-cohort, case control and cross-sectional studies in a wide range of patients and populations have indicated that MPO, in addition to traditional risk factors, is an important marker of cardiovascular disease closely associated with its prevalence (Schindhelm et al. 2009; Liu et al. 2012). An interesting study revealed that elevated MPO levels in PMNs and whole blood was associated with the presence of coronary artery disease, which supported a potential role of MPO as an inflammatory marker in coronary artery disease and might have implication for atherosclerosis diagnosis and risk assessment (Zhang et al. 2001). Furthermore, MPO, synergistically with other cytokines, serves as an enzymatic source of eicosanoids and bioactive lipids and generates atherogenic forms of both LDL and HDL (Zhang et al. 2002; Wang et al. 2007). Furthermore, MPO accelerates tissue damage of the atherosclerotic artery and affects the transformation from stable coronary artery plaques to unstable lesions (Meuwese et al. 2007). These data clearly confirm a direct impairment of endothelial function by MPO and its interaction in the development of premature atherosclerosis.

## *Cathepsins and Cystatin C*

The lysosomal cysteine cathepsins have been shown to be implicated in the development and progression of atherosclerosis. Cathepsin S is a member of the lysosomal cysteine proteinase family with strong elastolytic and collagenolytic properties, locates in lysosomes and endosomes, but can also function outside lysosomes or endosomes (Gu et al. 2009; Punturieri et al. 2000). In the pathogenesis of atherosclerosis, cathepsin S plays a key role in extracellular matrix (ECM) remodeling by degrading ECM proteins, and behaves as an imbalance factor between proteases and their inhibitors in ECM remodeling (Gu et al. 2009; Sukhova et al. 1998). Cathepsin S is upregulated in atherosclerotic lesions, and may be involved in the destabilization and rupture of atherosclerotic plaques (Rodgers et al. 2006). Cystatin C, on the other hand, is a major endogenous inhibitor of cysteine cathepsins with the highest inhibiting activity to cathepsin L and S (Gu et al. 2009; Liu et al. 2004). Cystatin C is decreased in atherosclerosis site (Dubin 2005; Shi et al. 1999b). Gu et al. examined the association between circulating cathepsin S and cystatin C and coronary plaque morphology (Gu et al. 2009). For this reason, patients with unstable angina (UA) or stable angina (SA) and controls were recruited. Intravascular ultrasound (IVUS) was used to evaluate plaque morphology, whereas plasma cathepsin S and cystatin C were measured. It was concluded that plasma cathepsin S and cystatin C increased significantly in UA patients. In angina patients, higher plasma cathepsin S may suggest the presence of vulnerable plaque, and higher plasma cystatin C may be a clue for larger atherosclerotic coronary plaque. In UA patients with vulnerable plaques, more macrophages and inflammatory cytokines are released into circulation, inducing macrophages to secrete more cathepsin S to accelerate the plaque instability. This positive feedback mechanism of inflammation may exist in angina patients with higher plasma cathepsin S. In addition, cathepsin S is involved in the degradation of modified LDL, resulting in extracellular lipid aggregation in the vessel wall. Cathepsin S also blocks cholesterol efflux from macrophages, contributing to the formation and accumulation of foam cells in atherosclerotic plaques (Lindstedt et al. 2003).

## *Microparticles*

Microparticles (MPs) are small vesicles released from cells upon apoptosis or activation. Microparticles are present in blood of healthy individuals (Baron et al. 2012). Increased MPs have been noted in serum from patients with a number of chronic conditions including atherosclerosis and secondary or accelerated vasculopathies (Chironi et al. 2009). The MP population is heterogeneous and includes several classes of apoptotic bodies, which form during removal of apoptotic cell remnants with characteristic exposure of phosphatidyl serine on the outer membrane leaflet, enabling phagocytosis by neighboring cells (Halicka et al. 2000). However, procoagulant tissue factor (TF) bearing MP, which do not have the characteristics of apoptotic bodies have also been described (Essayagh et al. 2007) suggesting that they may play an active role in disease progression rather than being only a marker of apoptosis or cellular damage. Importantly for patients with cardiovascular disease, MPs have been identified from platelets, monocytes, and ECs, and less frequently from red blood cells and granulocytes. Expression of GpIb (CD42b), platelet endothelium adhesion molecule (PECAM-1; CD31), the integrin $\alpha_{IIb}$ ß3 (GpIIb-IIIa), CD63, CD41a, and CD61 have been used as markers of platelet MP (Piccin et al. 2007). Studies have shown that high, sheer stress alone can induce platelet activation and platelet MP formation (Holme et al. 1997; Nomura et al. 2000; Sakariassen et al. 1998). An interesting study suggested that during a systemic inflammatory response, the interaction of circulating monocytes, platelets and EC could contribute to the production of procoagulant MP that have been described in the circulation of patients with vascular disease (Macey et al. 2010).

## Conclusion

Efforts to identify patients at increased risk for, or with clinically silent but established, atherosclerotic disease will intensify, and serum biomarkers will continue to play a crucial role in the four major domains of screening, diagnosis, prognosis, and management. All biochemical markers should be reliable, well standardized, and accurate, of good sensitivity, specificity and predictive value, be cost effective, and be available and suitable for widespread application. The effectiveness of some biomarkers in the major cardiovascular diseases caused by the development of atherosclerotic plaques, is already well established, but many of them need additional research and testing. The major tasks to accomplish are to define well, the clinical usefulness of various biochemical markers, to standardize the methodological approaches and the reporting results as well as to establish cost-effectiveness of the biomarkers.

## References

Ait-Oufella, H., B.L. Salomon, S. Potteaux, A.K. Robertson, P. Gourdy, J. Zoll, R. Merval, B. Esposito, J.L. Cohen, S. Fisson, R.A. Flavell, G.K. Hansson, D. Klatzmann, A. Tedgui, and Z. Mallat. 2006. Natural regulatory T cells control the development of atherosclerosis in mice. Nat. Med. 12: 178–180.

Aldons, J.L. 2000. Atherosclerosis. Nature. 407: 233–241.

Angeli, V., J. Llodrá, J.X. Rong, K. Satoh, S. Ishii, T. Shimizu, E.A. Fisher, and G.J. Randolph. 2004. Dyslipidemia associated with atherosclerotic disease systemically alters dendritic cell mobilization. Immunity 21: 561–574.

Asano, K., S. Okamoto, K. Fukunaga, T. Shiomi, T. Mori, M. Iwata, Y. Ikeda, and K. Yamaguchi. 1999. Cellular source(s) of platelet-activating-factor acetylhydrolase activity in plasma. Biochem. Biophys. Res. Commun. 261: 511–514.

Atabek, M.E., O. Pirgon, S. Kurtoglu, and H. Imamoglu. 2006. Evidence for an association between type 1 diabetes and premature carotid atherosclerosis in childhood. Pediatr. Cardiol. 27: 428–433.

Ayer, J.G., J.A. Harmer, S. Nakhla, W. Xuan, M.K. Ng, O.T. Raitakari, G.B. Marks, and D.S. Celermajer. 2009. HDL-cholesterol, blood pressure, and asymmetric dimethylarginine are significantly associated with arterial wall thickness in children. Arterioscler. Thromb. Vasc. Biol. 29: 943–949.

Balcerzyk, A., I. Zak, and J. Krauze. 2007. Synergistic effects of apolipoprotein E gene epsilon polymorphism and some conventional risk factors on premature ischaemic heart disease development. Kardiol. Pol. 65: 1058–1065; discussion 1066-7.

Bara, L., V. Nicaud, L. Tiret, F. Cambien, and M.M. Samama. 1994. Expression of a paternal history of premature myocardial infarction on fibrinogen, factor VIIC and PAI-1 in European offspring—the EARS study. European Atherosclerosis Research Study Group. Thromb. Haemost. 71: 434–440.

Baron, M., C.M. Boulanger, B. Staels, A. Tailleux, and M. Simionescu. 2012. Cell-derived microparticles in atherosclerosis: biomarkers and targets for pharmacological modulation? J. Cell. Mol. Med. 16: 1365–1376.

Blake, G.J. and P.M. Ridker. 2003. C-reactive protein and other inflammatory risk markers in acute coronary syndromes. J. Am. Coll. Cardiol. 41(4 Suppl S): 37S–42S.

Blankenberg, S., L. Tiret, C. Bickel, D. Peetz, F. Cambien, J. Meyer, and H.J. Rupprecht. 2002. Interleukin-18 is a strong predictor of cardiovascular death in stable and unstable angina. Circulation 106: 24–30.

Blankenberg, S., G. Luc, P. Ducimetiere, D. Arveiler, J. Ferrieres, P. Amouyel, A. Evans, F. Cambien, and L. Tiret. 2003. Interleukin-18 and the risk of coronary heart disease in European men: the Prospective Epidemiological Study of Myocardial Infarction (PRIME). Circulation 108: 2453–2459.

Bobryshev, Y.V. and R.S. Lord. 1995. A S-100 positive cells in human arterial intima and in atherosclerotic lesions. Cardiovas. Res. 29: 689–696.

Böger, R.H. 2011. Risk factors and vascular endothelium. Dimitris Tousoulis and Christodoulos Stefanadis, Ed.; Nova Science Publishers, V; New York, USA.

Borén, J., K. Olin, I. Lee, A. Chait, T.N. Wight, and T.L. Innerarity. 1998. Identification of the principal proteoglycan-binding site in LDL: A single-point mutation in apo-B100 severely affects proteoglycan interaction without affecting LDL receptor binding. J. Clin. Invest. 101: 2658–2664.

Busse, R. and I. Fleming. 1996. Endothelial Dysfunction in Atherosclerosis. J. Vasc. Res. 33: 181–194.

Calabro, P., J.T. Willerson, and E.T. Yeh. 2003. Inflammatory cytokines stimulated C-reactive protein production by human coronary artery smooth muscle cells. Circulation 108: 1930–1932.

Celermajer, D.S., K.E. Sorensen, D. Georgakopoulos, C. Bull, O. Thomas, J. Robinson, and J.E. Deanfield. 1993. Cigarette smoking is associated with dose-related and potentially reversible impairment of endothelium-dependent dilation in healthy young adults. Circulation 88: 2149–2155.

Chen, J., J. Jin, M. Song, H. Dong, G. Zhao, and L. Huang. 2012. C-reactive protein down-regulates endothelial nitric oxide synthase expression and promotes apoptosis in endothelial progenitor cells through receptor for advanced glycation end-products. Gene 496: 128–135.

Chi, Z. and A.J. Melendez. 2007. Role of cell adhesion molecules and immune-cell migration in the initiation, onset and development of atherosclerosis. Cell. Adh. Migr. 1: 171–175.

Chironi, G.N., C.M. Boulanger, A. Simon, F. Dignat-George, J.M. Freyssinet, and A. Tedgui. 2009. Endothelial microparticles in diseases. Cell Tissue Res. 335: 143–151.

Chiu, B., E. Viira, W. Tucker, and I.W. Fong. 1997. Chlamydia pneumoniae, cytomegalovirus, and herpes simplex virus in atherosclerosis of the carotid artery. Circulation 96: 2144–2148.

Ciftdoğan, D.Y., S. Coskun, C. Ulman, and H. Tıkız. 2012. The association of apolipoprotein E polymorphism and lipid levels in children with a family history of premature coronary artery disease. J. Clin. Lipidol. 6: 81–87.

Collet, J.P., Y. Allali, C. Lesty, M.L. Tanguy, J. Silvain, A. Ankri, B. Blanchet, R. Dumaine, J. Gianetti, L. Payot, J.W. Weisel, and G. Montalescot. 2006. Altered fibrin architecture is associated with hypofibrinolysis and premature coronary atherothrombosis. Arterioscler. Thromb. Vasc. Biol. 26: 2567–2573.

Colling, R.G. 2000. P-selectin or intercellular adhesion molecule (ICAM1) deficiency substantially protects against atherosclerosis in apolipoprotein E-deficient mice. J. Exp. Med. 191: 189–194.

Collins, T. and M.I. Cybulsky. 2001. NF-kappa B: pivotal mediator or innocent bystander in atherogenesis? J. Clin. Investig. 107: 255–264.

Cushman, M., A.M. Arnold, B.M. Psaty, T.A. Manolio, L.H. Kuller, G.L. Burke, J.F. Polak, and R.P. Tracy. 2005. C-reactive protein and the 10-year incedence of coronary heart disease in older men and women: the cardiovascular health study. Circulation 112: 25–31.

Cybulsky, M. and M.A. Gimbrone. 1991. Endothelial expression of a mononuclear leukocyte adhesion molecule during atherosclerosis. Science 251: 788–791.

Danesh, J., J.G. Wheeler, G.M. Hirschfield, S. Eda, G. Eiriksdottir, A. Rumley, G.D. Lowe, M.B. Pepys, and V. Gudnason. 2004. C-reactive protein and other circulating markers of inflammation in the prediction of coronary heart disease. N. Engl. J. Med. 350(14): 1387–1397.

Daugherty, A., N.R. Webb, D.L. Rateri, and V.L. King. 2005. Thematic review series: The immune system and atherogenesis. Cytokine regulation of macrophage functions in atherosclerosis. J. Lipid. Research 46: 1812–1822.

Dayal, S. and S.R. Lentz. 2005. ADMA and hyperhomocysteinemia. Vasc. Med. 10 Suppl 1: S27–33.

Dinarello, C.A. 1997. Interleukin-1. Cytokine Growth Factor Rev. 8: 253–265.

Dinarello, C.A. 1999. IL-18: A T-H1-inducing, proinflammatory cytokine and new member of the IL-1 family. J. Allergy. Clin. Immunol. 103: 11–24.

Dubin, G. 2005. Proteinaceous cysteine protease inhibitors. Cell Mol. Life. Sci. 62: 653–669.

Duval, C., G. Chinetti, F. Trottein, J.C. Fruchart, and B. Staels. 2002. The role of PPARs in atherosclerosis. Trends Mol. Med. 8: 422–430.

Elhage, R., S. Clamens, S. Besnard, Z. Mallat, A. Tedgui, J.F. Arnal, A. Maret, and F. Bayard. 2001. Involvement of interleukin-6 in atherosclerosis but not in the prevention of fatty streak formation by 17 beta-estradiol in apolipoprotein E-deficient mice. Atherosclerosis 156: 315–320.

Essayagh, S., J.M. Xuereb, A.D. Terrisse, L. Tellier-Cirioni, B. Pipy, and P. Sie. 2007. Microparticles from apoptotic monocytes induce transient platelet recruitment and tissue factor expression by cultured human vascular endothelial cells via a redox-sensitive mechanism. Thromb. Haemost. 98: 831–837.

European cardiovascular disease statistics. 2008 edition. http://www.ehnheart.org/component/downloads/downloads/683.html (Accessed October 20, 2011).

Faraci, F.M., J.E.J. Brian, and D.D. Heistad. 1995. Response of cerebral blood vessels to an endogenous inhibitor of nitric oxide synthase. Am. J. Physiol. 269: H1522–H1527.

Fragakis, N., E. Ioannidou, A. Bounda, S. Theodoridou, P. Klonizakis, and V. Garipidou. 2010. Increased levels of proinflammatory cytokines in children with family history of coronary artery disease. Clin. Cardiol. 33(4): E6–10.

Gabrielli, M., L. Santarelli, and A. Gasbarrini. 2002. Role for chronic infections in atherosclerosis? Circulation 106(7): e32.

Galea, J., J. Armstrong, P. Gadsdon, H. Holden, S.E. Francis, and C.M. Holt. 1996. Interleukin-1 beta in coronary arteries of patients with ischemic heart disease. Arterioscler. Thromb. Vasc. Biol. 16: 1000–1006.

Galis, Z.S., M. Muszynski, G.K. Sukhova, E. Simon-Morrissey, and P. Libby. 1995. Enhanced expression of vascular matrix metalloproteinases induced *in vitro* by cytokines and in regions of human atherosclerotic lesions. Ann. N.Y. Acad. Sci. 748501–748507.

Garcia, G.E., Y.Y. Xia, S.Z. Chen, Y.B. Wang, R.D. Ye, J.K. Harrison, K.B. Bacon, H.G. Zerwes, and L.L. Feng. 2000. NF-kappa B-dependent fractalkine induction in rat aortic endothelial cells stimulated by IL-1 beta, TNF-alpha and LPS. J. Leukoc. Biol. 67: 577–584.

Garza, C.A., V.M. Montori, J.P. McConnell, V.K. Somers, I.J. Kullo, and F. Lopez-Jimenez. 2007. Association between lipoprotein-associated phospholipase A2 and cardiovascular disease: a systematic review. Mayo. Clin. Proc. 82: 159–165.

Georges, J.L., V. Loukaci, O. Poirier, A. Evans, G. Luc, D. Arveiler, J.B. Ruidavets, F. Cambien, and L. Tiret. 2001. Interleukin-6 gene polymorphisms and susceptibility to myocardial infarction: the ECTIM study. J. Mol. Med. 79: 300–305.

Gerdes, N., G.K. Sukhova, P. Libby, R.S. Reynolds, J.L. Young, and U. Schonbeck. 2002. Expression of interleukin (IL)-18 and functional IL-18 receptor on human vascular endothelial cells, smooth muscle cells and macrophages: implications for atherogenesis. J. Exp. Med. 195: 245–257.

Goldstein, J.L. and M.S. Brown. 1979. Binding sites on macrophages that mediate uptake and degradation of aceylated low density lipoprotein, producing massive cholesterol deposition. Proc. Natl. Acad. Sci. USA. 76: 333–337.

Goran, K.H. and L. Peter. 2006. The immune response in atherosclerosis: A double-edged sword. Nat. Rev. Immunol. 6: 508–519.

Gu, F.F., S.Z. Lü, Y.D. Chen, Y.J. Zhou, X.T. Song, Z.N. Jin, and H. Liu. 2009. Relationship between plasma cathepsin S and cystatin C levels and coronary plaque morphology of mild to moderate lesions: an *in vivo* study using intravascular ultrasound. Chin. Med. J. (Engl). 122: 2820–2826.

Häkkinen, T., J.S. Luoma, M.O. Hiltunen, C.H. Macphee, K.J. Milliner, L. Patel, S.Q. Rice, D.G. Tew, K. Karkola, and S. Ylä-Herttuala. 1999. Lipoprotein-associated phospholipase A(2), platelet-activating factor acetylhydrolase, is expressed by macrophages in human and rabbit atherosclerotic lesions. Arterioscler. Thromb. Vasc. Biol. 19: 2909–2917.

Halicka, H.D., E. Bedner, and Z. Darzynkiewicz. 2000. Segregation of RNA and separate packaging of DNA and RNA in apoptotic bodies during apoptosis. Exp. Cell. Res. 260: 248–256.

Hegyi, L., S.J. Hardwick, R.C. Siow, and J.N. Skepper. 2001. Macrophage death and the role of apoptosis in human atherosclerosis. J. Hematother. Stem. Cell. Res. 10: 27–42.

Holme, P.A., U. Orvim, M.J. Hamers, N.O. Solum, F.R. Brosstad, R.M. Barstad, and K.S. Sakariassen. 1997. Shear-induced platelet activation and platelet microparticle formation at blood flow conditions as in arteries with a severe stenosis. Arterioscler. Thromb. Vasc. Biol. 17: 646–653.

Hovingh, G.K., A. Brownlie, R.J. Bisoendial, M.P. Dube, J.H. Levels, W. Petersen, R.P. Dullaart, E.S. Stroes, A.H. Zwinderman, E. de Groot, M.R. Hayden, J.A. Kuivenhoven, and J.J. Kastelein. 2004. A novel apoA-I mutation (L178P) leads to endothelial dysfunction, increased arterial wall thickness, and premature coronary artery disease. J. Am. Coll. Cardiol. 44: 1429–1435.

Humphries, S.E., L.A. Luong, M.S. Ogg, E. Hawe, and G.J. Miller. 2001. The interleukin-6-174 G/C promoter polymorphism is associated with risk of coronary heart disease and systolic blood pressure in healthy men. Eur. Heart. J. 22: 2243–2252.

Ikeda, U., T. Ito and K. Shimada. 2001. Interleukin-6 and acute coronary syndrome. Clin. Cardiol. 24: 701–704.

Ikewaki, K., A. Matsunaga, H. Han, H. Watanabe, A. Endo, J. Tohyama, M. Kuno, J. Mogi, K. Sugimoto, N. Tada, J. Sasaki and S. Mochizuki. 2004. A novel two nucleotide deletion in the apolipoprotein A-I gene, apoA-I Shinbashi, associated with high density lipoprotein deficiency, corneal opacities, planar xanthomas, and premature coronary artery disease. Atherosclerosis 172: 39–45.

Itabe, H. 2003. Oxidized low density lipoproteins: What is understood and what remains to be clarified. Biol. Pharm. Bulletin 26: 1–9.

Jabs, W.J., E. Theissing, M. Nitschke, J.F. Bechtel, M. Duchrow, S. Mohamed, B. Jahrbeck, H.H. Sievers, J. Steinhoff, and C. Bartels. 2003.Local generation of C-reactive protein in diseased coronary artery venous bypass grafts and normal vascular tissue. Circulation 108: 1428–1431.

Jacob, H.S., M. Visser, N.S. Key, J.L. Goodman, C.F. Moldow, and G.M. Vercellotti. 1992. Herpes virus infection of endothelium: new insights into atherosclerosis. Trans. Am. Clin. Climatol. Assoc. 103: 95–104.

Jones, S.A., S. Horiuchi, N. Topley, N. Yamamoto, and G.M. Fuller. 2001. The soluble interleukin 6 receptor: mechanisms of production and implications in disease. FASEB. J. 15: 43–58.

Jovinge, S., A. Hamsten, P. Tornvall, A. Proudler, P. Båvenholm, C.G. Ericsson, I. Godsland, U. de Faire, and J. Nilsson. 1998. Evidence for a role of tumor necrosis factor alpha in disturbances of triglyceride and glucose metabolism predisposing to coronary heart disease. Metabolism 47: 113–118.

Kampoli, A.M., D. Tousoulis, K. Marinou, G. Siasos, and C. Stefanadis. 2009a. Vascular effects of diabetes mellitus. Vascular disease prevention 6: 85–90.

Kampoli, A.M., D. Tousoulis, C. Antoniades, G. Siasos, and C. Stefanadis. 2009b. Biomarkers of premature atherosclerosis. Trends. Mol. Med. 15: 323–32.

Kampoli, A.M., D. Tousoulis, N. Papageorgiou, and C. Stefanadis. 2011. Risk factors and vascular endothelium. Nova Science Publishers, New York, USA.

Kampoli, A.M., D. Tousoulis, N. Papageorgiou, Z. Pallatza, G. Vogiatzi, A. Briasoulis, E. Androulakis, C. Toutouzas, P. Stougianos, C. Tentolouris,and C. Stefanadis. 2012. Clinical utility of biomarkers in premature atherosclerosis. Curr. Med. Chem. 19: 2521–2533.

Klouche, M., S. Rose-John, W. Schmiedt, and S. Bhakdi. 2000. Enzymatically degraded, nonoxidized LDL induces human vascular smooth muscle cell activation, foam cell transformation and proliferation. Circulation 101: 1799–1805.

Koba, S. and T. Hirano. 2011. Dyslipidemia and atherosclerosis. Nihon. Rinsho. 69: 138–143.

Kolodgie, F.D., A.P. Burke, K.S. Skorija, E. Ladich, R. Kutys, A.T. Makuria, and R. Virmani. 2006. Lipoprotein-associated phospholipase A2 protein expression in the natural progression of human coronary atherosclerosis. Thromb. Vasc. Biol. 26: 2523–2529.

Koenig, W., H. Löwel, J. Baumert, and C. Meisinger. 2004. C-reactive protein modulates risk prediction based on the Framingham Score: implications for future risk assessment: results from a large cohort study in southern Germany. Circulation 109: 1349–1353.

Kraemer, T., I. Prokosay, R.A. Date, H. Sies, and T. Schewe. 2004. Oxi-dative modification of low-density lipoprotein: lipid per-oxidation by myeloperoxidase in the presence of nitrite. Biol. Chem. 385: 809–818.

Lau, D. and S. Baldus. 2006. Myeloperoxidase and its contributory role in inflammatory vascular disease. Pharmacol. Ther. 111: 16–26.

Lavi, S., J.P. McConnell, C.S. Rihal, A. Prasad, V. Mathew, L.O. Lerman, and A. Lerman. 2007. Local production of lipoprotein-associated phospholipase A2 and lysophosphatidylcholine in the coronary circulation: association with early coronary atherosclerosis and endothelial dysfunction in humans. Circulation 115(21): 2715–2721.

Liao, H.S., A. Matsumoto, H. Itakura, T. Doi, M. Honda, T. Kodama, and Y.J. Geng. 1999. Transcriptional inhibition by interleukin-6 of the class A, macrophage scavenger receptor in macrophages derived from human peripheral monocytes and the THP-1 monocytic cell line. Arterioscler. Thromb. Vasc. Biol. 19: 1872–1880.

Libby, P. 1999a. Changing concepts of atherosclerosis. J. Intern. Med. 247: 349–358.

Libby, P., and P.M. Ridker. 1999b. Novel inflammatory markers of coronary risk: theory versus practice. Circulation 100: 1148–1150.

Lindstedt, L., M. Lee, K. Öörni, D. Bromme, and P.T. Kovanen. 2003. Cathepsins F and S block HDL3-induced cholesterol efflux from macrophage foam cells. Biochem. Biophys. Res. Commun. 312: 1019–1024.

Liu, C., G. Xie, W. Huang, Y. Yang, P. Li, and Z. Tu. 2012. Elevated serum myeloperoxidase activities are significantly associated with the prevalence of ACS and high LDL-C levels in CHD patients. J. Atheroscler. Thromb. 19: 435–443.

Liu, J., G.K. Sukhova, J.S. Sun, W.H. Xu, P. Libby, and G.P. Shi. 2004. Lysosomal cysteine proteases in atherosclerosis. Arterioscler. Thromb. Vasc. Biol. 24: 1359–1366.

Macey, M.G., S.I. Wolf, and C. Lawson. 2010. Microparticle formation after exposure of blood to activated endothelium under flow. Cytometry A. 77: 761–768.

MacPhee, C.H., K.E. Moores, H.F. Boyd, D. Dhanak, R.J. Ife, C.A. Leach, D.S. Leake, K.J. Milliner, R.A. Patterson, K.E. Suckling, D.G. Tew, and D.M. Hickey. 1999. Lipoprotein-associated phospholipase A2, platelet-activating factor acetylhydrolase, generates two bioactive products during the oxidation of low-density lipoprotein: use of a novel inhibitor. Biochem. J. 338 ( Pt 2): 479–487.

Mallat, Z., S. Besnard, M. Duriez, V. Deleuze, F. Emmanuel, M.F. Bureau, F. Soubrier, B. Esposito, H. Duez, C. Fievet, B. Staels, N. Duverger, D. Scherman, and A. Tedgui. 1999. Protective role of interleukin-10 in atherosclerosis. Circ. Res. 85: e17–24.

Maron, B.A. and J. Loscalzo. 2009. The treatment of hyperhomocysteinemia. Annu. Rev. Med. 60: 39–54.

Mayr, M., S. Kiechl, J. Willeit, G. Wick, and Q. Xu. 2000. Infections, immunity, and atherosclerosis: associations of antibodies to Chlamydia pneumoniae, Helicobacter pylori, and cytomegalovirus with immune reactions to heat-shock protein 60 and carotid or femoral atherosclerosis. Circulation 102: 833–839.

McGill, H. Jr. 1988. The pathogenesis of atherosclerosis. Clin. Chem. 34: B33–B39.

Meuwese, M.C., E.S. Stroes, S.L. Hazen, J.N. van Miert, J.A. Kuiv-enhoven, R.G. Schaub, N.J. Wareham, R. Luben, J.J. Kastelein, K.T. Khaw, and S.M. Boekholdt. 2007. Serum myeloper-oxidase levels are associated with the future risk of coro-nary artery disease in apparently healthy individuals: the EPIC-Norfolk Prospective Population Study. J. Am. Coll. Cardiol. 50: 159–165.

Morrow, D.A., M.S. Sabatine, M.L. Brennan, J.A. Lemos, S.A. Murphy, C.T. Ruff, N. Rifai, C.P. Cannon, and S.L. Hazen. 2008. Concurrent evaluation of novel cardiac biomarkers in acute coronary syndrome: myeloperoxidase and soluble CD40 ligand and the risk of recurrent ischaemic events in TACTICS-TIMI 18. European Heart Journal 29: 1096–1102.

Moser, R., B. Schleiffenbaum, P. Groscurth, and J. Fehr. 1989. Interleukin 1 and tumor necrosis factor stimulate human vascular endothelial cells to promote transendothelial neutrophil passage. J. Clin. Investig. 83: 444–455.

Napoli, C., O. Pignalosa, F. de Nigris, and V. Sica. 2005. Childhood infection and endothelial dysfunction: a potential link in atherosclerosis? Circulation 111: 1568–1570.

Noma, K., C. Goto, K. Nishioka, K. Hara, M. Kimura, T. Umemura, D. Jitsuiki, K. Nakagawa, T. Oshima, K. Chayama, M. Yoshizumi, and Y. Higashi. 2005. Smoking, endothelial function, and Rho-kinase in humans. Arterioscler. Thromb. Vasc. Biol. 25: 2630–2635.

Nomura, S., N.N. Tandon, T. Nakamura, and J. Kambayashi. 2000. Morphological differences between GPIb antibody-induced and shear stress-induced platelet aggregates. Haemostasis 30: 174–188.

Pai, J.K., T. Pischon, J. Ma, J.E. Manson, S.E. Hankinson, K. Joshipura, G.C. Curhan, N. Rifai, C.C. Cannuscio, M.J. Stampfer, and E.B. Rimm. 2004. Inflammatory markers and the risk of coronary heart disease in men and women. N. Engl. J. Med. 351: 2599–2610.

Paravicini, T.M., A.A. Miller, G.R. Drummond, and C.G. Sobey. 2006. Flow-induced cerebral vasodilatation *in vivo* involves activation of phosphatidylinositol-3 kinase, NADPH-oxidase, and nitric oxide synthase. J. Cereb. Blood. Flow. Metab. 26: 836–845.

Pearson, T.A., G.A. Mensah, R.W. Alexander, J.L. Anderson, R.O. Cannon 3rd, M. Criqui, Y.Y. Fadl, S.P. Fortmann, Y. Hong, G.L. Myers, N. Rifai, S.C. Smith Jr., K. Taubert, R.P. Tracy, and F. Vinicor. 2003. Markers of inflammation and cardiovascular disease: application to clinical and public health practice: A statement for healthcare professionals from the Centers for Disease Control and Prevention and the American Heart Association. Circulation 107: 499–511.

Perna, M., M.J. Roman, D.R. Alpert, M.K. Crow, M.D. Lockshin, L. Sammaritano, R.B. Devereux, J.P. Cooke, and J.E. Salmon. 2010. Relationship of asymmetric dimethylarginine and homocysteine to vascular aging in systemic lupus erythematosus patients. Arthritis. Rheum. 62: 1718–1722.

Piccin, A., W.G. Murphy, and O.P. Smith. 2007. Circulating microparticles: pathophysiology and clinical implications. Blood. Rev. 21: 157–171.

Podor, T.J., F.R. Jirik, D.J. Loskutoff, D.A. Carson, and M. Lotz. 1989. Human endothelial cells produce IL-6. Lack of responses to exogenous IL-6. Ann. N Y. Acad. Sci. 557374–557385.
Prasad, A., J. Zhu, J.P. Halcox, M.A. Waclawiw, S.E. Epstein, and A.A. Quyyumi. 2002. Predisposition to atherosclerosis by infections: role of endothelial dysfunction. Circulation 106: 184–190.
Pfützner, A. and T. Forst. 2006. High-sensitivity C-reactive protein as cardiovascular risk marker in patients with diabetes mellitus. Diabetes Technol. Ther. 8: 28–36.
Pfützner, A., T. Schöndorf, M. Hanefeld, and T. Forst. 2010. High-sensitivity C-reactive protein predicts cardiovascular risk in diabetic and nondiabetic patients: effects of insulin-sensitizing treatment with pioglitazone. J. Diabetes. Sci. Technol. 4: 706–716.
Punturieri, A., S. Filippov, E. Allen, I. Caras, R. Murray, V. Reddy, and S.J. Weiss. 2000. Regulation of elastinolytic cysteine proteinase activity in normal and cathepsin K-deficient human macrophages. J. Exp. Med. 192: 789–799.
Rho, Y.H., C.P. Chung, A. Oeser, J. Solus, P. Raggi, T. Gebretsadik, A. Shintani and C.M. Stein. 2008. Novel cardiovascular risk factors in premature coronary atherosclerosis associated with systemic lupus erythematosus. J. Rheumatol. 35: 1789–1794.
Ridker, P.M., N. Rifai, L. Rose, J.E. Buring, and N.R. Cook. 2002. Comparison of C-reactive protein and low-density lipoprotein cholesterol levels in the prediction of first cardiovascular events. N. Engl. J. Med. 347: 1557–1565.
Ridker, P.M., J.E. Buring, N.R. Cook, and N. Rifai. 2003. C-reactive protein, the metabolic syndrome, and risk of incident cardiovascular events: an 8-year follow-up of 14,719 initially healthy American women. Circulation 107: 391–397.
Ridker, P.M., P.W. Wilson, and S.M. Grundy. 2004a. Should C-reactive protein be added to metabolic syndrome and to assessment of global cardiovascular risk? Circulation 109: 2818–2825.
Ridker, P.M. and N. Cook. 2004b. Clinical usefulness of very high and very low levels of C-reactive protein across the full range of Framingham Risk Scores. Circulation 109: 1955–1959.
Ridker, P.M. 2004c. High-sensitivity C-reactive protein, inflammation, and cardiovascular risk: from concept to clinical practice to clinical benefit. Am. Heart J. 148(1 Suppl): S19–S26.
Ridker, P.M., S.S. Bassuk, and P.P. Toth. 2005. C-reactive protein and risk of cardiovascular disease: evidence and clinical application. Curr. Atheroscler. Rep. 5: 341–349.
Ridker, P.M. 2007. C-Reactive Protein and the Prediction of Cardiovascular Events Among Those at Intermediate Risk. JACC 49: 2129–2138.
Roger, V.L., A.S. Go, D.M. Lloyd-Jones, R.J. Adams, J.D. Berry, T.D. Brown, M.R. Carnethon, S. Dai, G. de Simone, E.S. Ford, C.S. Fox, H.J. Fullerton, C. Gillespie, K.J. Greenlund, S.M. Hailpern, J.A. Heit, P.M. Ho, V.J. Howard, B.M. Kissela, S.J. Kittner, D.T. Lackland, J.H. Lichtman, L.D. Lisabeth, D.M. Makuc, G.M. Marcus, A. Marelli, D.B. Matchar, M.M. McDermott, J.B. Meigs, C.S. Moy, D. Mozaffarian, M.E. Mussolino, G. Nichol, N.P. Paynter, W.D. Rosamond, P.D. Sorlie, R.S. Stafford, T.N. Turan, M.B. Turner, N.D. Wong, and J. Wylie-Rosett. 2011. Heart Disease and Stroke Statistics-2011 Update. A report from the American Heart Association Statistics Committee and Stroke Statistics Subcommittee. Circulation 123: e1–e192.
Rodgers, K.J., D.J. Watkins, A.L. Miller, P.Y. Chan, S. Karanam, W.H. Brissette, C.J. Long, and C.L. Jackson. 2006. Destabilizing role of cathepsin S in murine atherosclerotic plaques. Arterioscler. Thromb. Vasc. Biol. 26: 851–856.
Rosenfeld, M.E., S. Ylä-Herttuala, B.A. Lipton, V.A. Ord, J.L. Witztum, and D. Steinberg. 1992. Macrophage colony-stimulating factor mRNA and protein in atherosclerotic lesions of rabbits and humans. Am. J. Pathol. 140: 291–300.
Ross, R. 1993. The pathogenesis of atherosclerosis: A perspective for the 1990s. Nature 362: 801–809.
Ross, R. 1999. Atherosclerosis: an inflammatory disease. N. Engl. J. Med. 340: 115–126.

Rundek, T., M.S. Elkind, J. Pittman, B. Boden-Albala, S. Martin, S.E. Humphries, S.H. Juo, and R.L. Sacco. 2002. Carotid intima-media thickness is associated with allelic variants of stromelysin-1, interleukin-6 and hepatic lipase genes: the Northern Manhattan Prospective Cohort Study. Stroke 33: 1420–1423.

Rus, H.G., R. Vlaicu, and F. Niculescu. 1996. Interleukin-6 and interleukin-8 protein and gene expression in human arterial atherosclerotic wall. Atherosclerosis 127: 263–271.

Saito, H., S. Lund-Katz, and M.C. Phillips. 2004. Contributions of domain structure and lipid interaction to the functionality of exchangeable human apolipoproteins. Prog. Lipid Res. 43: 350–380.

Sakariassen, K.S., P.A. Holme, U. Orvim, R.M. Barstad, N.O. Solum, and F.R. 1998. Brosstad. Shear-induced platelet activation and platelet microparticle formation in native human blood. Thromb. Res. 92(6 Suppl 2): S33–S41.

Schieffer, B., E. Schieffer, D. Hilfiker-Kleiner, A. Hilfiker, P.T. Kovanen, M. Kaartinen, J. Nussberger, W. Harringer and H. Drexler. 2000. Expression of angiotensin II and interleukin 6 in human coronary atherosclerotic plaques: potential implications for inflammation and plaque instability. Circulation 101: 1372–1378.

Schindhelm, R.K., L.P. van der Zwan, T. Teerlink, and P.G. Scheffer. 2009. Myeloperoxidase: a useful biomarker for cardiovascu-lar disease risk stratification? Clin. Chem. 55: 1462–1470.

Sica, A., J.M. Wang, F. Colotta,E. Dejana, A. Mantovani, J.J. Oppenheim, C.G. Larsen, C.O. Zachariae, and K. Matsushima. 1990. Monocyte chemotactic and activating factor gene expression induced in endothelial cells by IL-1 and tumor necrosis factor. J. Immunol. 144: 3034–3038.

Shih, P.T. 1999a. Blocking very late antigen-4 integrin decreases leukocyte entry and fatty streak formation in mice fed an atherogenic diet. Circ. Res. 84: 345–351.

Shi, G.P., G.K. Sukhova, A. Grubb, A. Ducharme, L.H. Rhode, R.T. Lee, P.M. Ridker, P. Libby, and H.A. Chapman. 1999b. Cystatin C deficiency in human atherosclerosis and aortic aneurysms. J. Clin. Invest. 104: 1191–1197.

Shi, Y., P. Zhang, L. Zhang, H. Osman, E.R. Mohler 3rd, C. Macphee, A. Zalewski, A. Postle, and R.L. Wilensky. 2007. Role of lipoprotein-associated phospholipase A2 in leukocyte activation and inflammatory responses. Atherosclerosis 19: 54–62.

Skooga, T., W. Dichtlb, S. Boquista, C. Skoglund-Anderssona, F. Karpea, R. Tangd, M.G. Bondd, U. de Fairee,f, J. Nilssonb, P. Erikssona, and A. Hamstena. 2002. Plasma tumor necrosis factor α and early carotid atherosclerosis in healthy middle-aged men. European Heart Journal 23: 376–383.

Springer, T.A. 1990. Adhesion receptors of the immune system. Nature 346: 425–434.

Stafforini, D.M., M.R. Elstad, T.M. McIntyre, G.A. Zimmerman, and S.M. Prescott. 1990. Human macrophages secret platelet-activating factor acetylhydrolase. J. Biol. Chem. 265: 9682–9687.

Stanford, S.J., J.R. Pepper, and J.A. Mitchell. 2000. Cyclooxygenase-2 regulates granulocyte-macrophage colony-stimulating factor, but not interleukin-8, production by human vascular cells—role of cAMP. Arterioscler. Thromb. Vasc. Biol. 20: 677–682.

Stary, H.C., D.H. Blankenhorn, A.B. Chandler, S. Glagov, W. Insull Jr., M. Richardson, M.E. Rosenfeld, S.A. Schaffer, C.J. Schwartz, and W.D. Wagner. 1992. A definition of the intima of human arteries and of its atherosclerosis-prone regions: a report from the Committee on Vascular Lesions of the Council on Arteriosclerosis, American Heart Association. Circulation 85: 391–405.

Steffen, Y., T. Schewe, and H. Sies. 2006. Myeloperoxidase-mediated LDL oxidation and endothelial cell toxicity of oxidized LDL: attenuation by (?)-epicatechin. Free. Radic. Res. 40: 1076–1085.

Stoppa-Vaucher, S., M. Dirlewanger, C.A. Meier, P. de Moerloose, G. Reber, P. Roux-Lombard, C. Combescure, S. Saudan, and V.M. Schwitzgebel. 2012. Inflammatory and prothrombotic states in obese children of European descent. Obesity (Silver Spring) (in press).

Strong, J.P. and H.C. McGill Jr. 1962. The natural history of coronary atherosclerosis. Am. J. Pathol. 40: 37–49.

Stuhlinger, M.C., R.K. Oka, E.E. Graf, I. Schmolzer, B.M. Upson, O. Kapoor, A. Szuba, M.R. Malinow, T.C. Wascher, O. Pachinger, and J.P. Cooke. 2003. Endothelial dysfunction induced by hyperhomocysteinemia: role of asymmetric dimethylarginine. Circulation 108: 933–938.

Sukhova, G.K., G.P. Shi, D.I. Simon, H.A. Chapman, and P. Libby. 1998. Expression of the elastolytic cathepsins S and K in human atheroma and regulation of their production in smooth muscle cells. J. Clin. Invest. 102: 576–583.

Sukovich, D.A., K. Kauser, F.D. Shirley, V. DelVecchio, M. Halks-Miller, and G.M. Rubanyi. 1998. Expression of interleukin-6 in atherosclerotic lesions of male ApoE-knockout mice: inhibition by 17beta-estradiol. Arterioscler. Thromb. Vasc. Biol. 18: 1498–1505.

Suzuki, H., K. Shibano, M. Okane, I. Kono, Y. Matsui, K. Yamane, and H. Kashiwagi. 1989. Interferon-gamma modulates messenger RNA levels of c-sis (PDGF-B chain), PDGF-A chain and IL-1 beta genes in human vascular endothelial cells. Am. J. Pathol. 134: 35–43.

Tanriverdi, H., H. Evrengul, O. Kuru, S. Tanriverdi, D. Seleci, Y. Enli, H.A. Kaftan, and M. Kilic. 2006. Cigarette smoking induced oxidative stress may impair endothelial function and coronary blood flow in angiographically normal coronary arteries. Circ. J. 70: 593–599.

Taraboanta, C., E. Wu, S. Lear, S. DiPalma, J. Hill, G.B. Mancini, and J. Frohlich. 2008. Subclinical atherosclerosis in subjects with family history of premature coronary artery disease. Am. Heart J. 155: 1020–1026.

Tiret, L., T. Godefroy, E. Lubos, V. Nicaud, D.A. Tregouet, S. Barbaux, R. Schnabel, C. Bickel, C. Espinola-Klein, O. Poirier, C. Perret, T. Münzel, H.J. Rupprecht, K. Lackner, F. Cambien, and S. Blankenberg. 2005. Genetic analysis of the interleukin-18 system highlights the role of the interleukin-18 gene in cardiovascular disease. Circulation 112: 643–650.

Tousoulis, D., A.M. Kampoli, N. Papageorgiou, S. Papaoikonomou, C. Antoniades, and C. Stefanadis. 2009. The impact of diabetes mellitus on coronary artery disease: new therapeutic approaches. Curr. Pharm. Des. 15: 2037–2048.

Tselepis, A.D., G. Hahalis, C.C. Tellis, E.C. Papavasiliou, P.T. Mylona, A. Kourakli, and D.C. Alexopoulos. 2010. Plasma levels of lipoprotein-associated phospholipase A(2) are increased in patients with β-thalassemia. J. Lipid. Res. 51: 3331–3341.

Tsikas, D., R.H. Böger, J. Sandmann, S.M. Bode-Böger, and J.C. Frölich. 2000. Hypothesis: Endogenous nitric oxide synthase inhibitors are responsible for the L-arginine paradox. FEBS. Lett. 23810: 1–3.

Varki, A. 1994. Selectin ligands. Proc. Natl. Acad. Sci. USA 91: 7390–7397.

Vassali, P. 1992. The pathophysiology of tumor necrosis factors. Annu. Rev. Immunol. 10: 411–452.

Victor, V.M., M. Rocha, E. Solá, C. Bañuls, K. Garcia-Malpartida, and A. Hernández-Mijares. 2009. Oxidative stress, endothelial dysfunction and atherosclerosis. Curr. Pharm. Des. 15: 2988–3002.

Vidal-Vanaclocha, F., G. Fantuzzi, L. Mendoza, A.M. Fuentes, M.J. Anasagasti, J. Martin, T. Carrascal, P. Walsh, L.L. Reznikov, and S.H. Kim. 2000. IL-18 regulates IL-1 beta-dependent hepatic melanoma metastasis via vascular cell adhesion molecule-1. Proc. Natl. Acad. Sci. USA 97: 734–739.

von der Thüsen, J.H., J. Kuiper, T.J. van Berkel, and E.A. Biessen. 2003. Interleukins in atherosclerosis: molecular pathways and therapeutic potential. Pharmacol. Rev. 55: 133–166.

Wang, X., G.Z. Feuerstein, J.L. Gu, P.G. Lysko, and T.L. Yue. 1995. Interleukin-1 beta induces expression of adhesion molecules in human vascular smooth muscle cells and enhances adhesion of leukocytes to smooth muscle cells. Atherosclerosis 115: 89–98.

Wang, Z., S.J. Nicholls, E.R. Rodriguez, O. Kummu, S. Horkko, J. Barnard, W.F. Reynolds, E.J. Topol, J.A. DiDonato, and S.L. Hazen. 2007. Protein carbamylation links inflammation, smoking, uremia and atherogenesis. Nat. Med. 13: 1176–1184.

Woo, K.S., M. Qiao, P. Chook, P.Y. Poon, A.K. Chan, J.T. Lau, K.P. Fung, and J.L. Woo. 2002. Homocysteine, endothelial dysfunction, and coronary artery disease: emerging strategy for secondary prevention. J. Card. Surg. 17: 432–435.

Xing, Z., J. Gauldie, G. Cox, H. Baumann, M. Jordana, X.F. Lei, and M.K. Achong. 1998. IL-6 is an anti-inflammatory cytokine required for controlling local or systemic acute inflammatory responses. J. Clin. Investig. 101: 311–320.

Yasojima, K., C. Schwab, and E.G. McGeer. 2001. Generation of C-reactive protein and complement components in atherosclerotic plaques. Am. J. Pathol. 158: 1039–1051.

Zhang, R., M.L. Brennan, X. Fu, R.J. Aviles, G.L. Pearce, M.S. Penn, E.J. Topol, D.L. Sprecher, and S.L. Hazen. 2001. Association between myeloperoxidase levels and risk of coronary artery disease. JAMA 286: 2136–2142.

Zhang, R., M.L. Brennan, Z. Shen, J.C. MacPherson, D. Schmitt, C.E. Molenda, and S.L. Hazen. 2002. Myeloperoxidase functions as a major enzymatic catalyst for initiation of lipid peroxida-tion at sites of inflammation. J. Biol. Chem. 277: 46116–46122.

# CHAPTER 3

# Biomarkers in Essential Hypertension

Emmanuel Androulakis, Dimitris Tousoulis,* Nikolaos
Papageorgiou and Christodoulos Stefanadis

## Introduction

Hypertension is a major modifiable risk factor for cardiovascular disease, renal failure, and stroke (Androulakis et al. 2009). Almost 26% of the adult population experiences hypertension, while its prevalence may rise to 29% by the year 2025, according to the available literature (Vedanthan and Fuster 2009; Kearney et al. 2005).

Accordingly, there has been a growing emphasis in preventing hypertension and in understanding the most common mechanisms underlying its development. Increasing evidence has indicated that hypertension-associated vascular disease is an inflammatory process. Thus, chronic inflammation seems to participate in the pathophysiology of hypertension (Li and Chen 2005). It is well-known that the inflammation entails a complex set of interactions which leads to increased expression of adhesion molecules, cytokines, chemokines, matrix metalloproteinases and growth factors. Also, activation of renin-angiotensin-aldosterone system (RAAS) and increase in the local production of angiotensin II (Ang II) have been implicated in this process, while molecular mechanisms, such as reactive oxygen species (ROS) may also play a pivotal role (Touyz 2005; Ruiz Ortega et al. 2006).

Recently, emerging interest has focused on detection of biomarkers, which may increase the ability to predict hypertension. Importantly, it has been suggested that inflammatory markers are not just markers; rather,

1st Department of Cardiology, 'Hippokration' Hospital, University of Athens Medical School, Athens, Greece.
*Corresponding author: drtousoulis@hotmail.com

they could be mediators implicated in the pathogenesis of hypertension. It seems reasonable that inflammatory molecules, such as C-reactive protein (CRP), interleukin-1 (IL-1), intercellular adhesion molecule-1 (ICAM-1) and monocyte chemoattractant protein-1 (MCP-1) could be used as diagnostic objectives in hypertension (Ruiz Ortega et al. 2006). Other biomarkers, including soluble CD40 ligand (sCD40L) levels have been increased and might discriminate hypertensive patients at a high risk of cardiovascular events (Ferroni and Guadagni 2008).

In the present article, we will review recent findings regarding pathophysiological mechanisms in relation to essential hypertension. Furthermore, we will discuss the role of biomarkers in hypertension as screening strategies that can reliably identify individuals most likely to develop hypertension and/or its complications.

## Pathophysiology of Essential Hypertension

Although the pathophysiology of essential hypertension is multi-factorial and still unknown, emerging evidence from basic research studies has shown that vascular inflammation could contribute to the development of hypertension (Fig. 1). During the initial stages, hypertension is believed to be caused by neural or humoral stimuli, leading to changes in pressure-natriuresis in the kidney, autoregulatory response and vasoconstriction, due

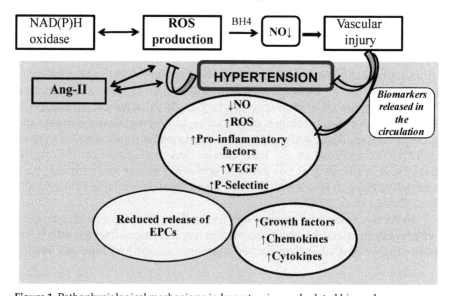

**Figure 1.** Pathophysiological mechanisms in hypertension and related biomarkers.

**Abbreviations**: ROS: reactive oxygen species, VEGF: vascular endothelial cell growth factors, NO: nitric oxide, EPCs: endothelial progenitor cells, BH4: tetrahydrobiopterin.

to vascular hypertrophy (Androulakis et al. 2009; Geza 2004). Recently, a pathway has been proposed, relative to the pathophysiology of essential hypertension, which includes two phases. Episodes of renal vasoconstriction occur in the first phase, induced by hyperactivation of sympathetic nervous system and activation of RAS, while tubulointerstitial inflammation, which is associated with local Ang II formation and generation of ROS, distinguishes the second phase (Johnson et al. 2005). Apparently, these changes favor salt retention and maintain fluid imbalance (Johnson et al. 2008; Johnson et al. 1997). Furthermore, immunopathogenic mechanisms have been demonstrated to play an important role in the pathogenesis of hypertension (Hoch 2009). Interest has also been directed toward oxidative stress and reduced bioavailability of nitric oxide (NO), which leads to vasoconstriction and impaired vascular function (Androulakis et al. 2009). According to an interesting point of view, hypertension is a systemic inflammatory state originating in the perinatal period.

## Renin-angiotensin-aldosterone System

Hypertension may induce proinflammatory and procoagulatory responses. In that state, tissue and systemic inflammatory mediators are increased, including components of the RAAS, endothelial adhesion molecules, chemokines, cytokines and tissue factor (TF) (Das 2006). Independently of the etiology, inflammation is associated with increased vascular permeability, leukocyte extravasation and tissue repair. It has been demonstrated that Ang-II may play a key role in the regulation of inflammatory processes in hypertension (Skultetyova et al. 2007). In particular, it induces increased vascular permeability, cell infiltration and exudation of protein-rich fluid via pressure mediated mechanical injury of the endothelium and local release of second mediators. Other molecules such as prostaglandins and vascular endothelial cell growth factors also stimulated by Ang-II may play significant role (Suzuki et al. 2003; Marchesi et al. 2008). Moreover, Ang-II has been associated with the upregulation of VEGF mRNA expression, primarily through the type 1 Ang II (AT1) receptor. Thus, it seems to participate in the early stages of vascular inflammation beyond hemodynamic effects (Zhao et al. 2004). Cytokine release, adhesion molecule expression, such as ICAM-1, ICAM-2 and vascular cell adhesion molecule-1 (VCAM-1), required for firm adhesion, and the production of plasminogen activator inhibitor-1 (PAI-1) are modulated by Ang-II, thus favoring inflammatory vascular and renal damage (Montecucco et al. 2011; Schieffer et al. 2000; Vaughan et al. 1995). According to experimental data Ang-II infusion has increased leukocyte rolling, adhesion and migration and has upregulated the expression of ICAM-1, VCAM-1 and E-Selectin in endothelial cells (ECs) (Piqueras et al. 2000). Of note, Ang II has been shown to stimulate the production of proinflammatory cytokines and chemokines, such as IL-6, IL-8, osteopontin and MCP-1 (Funakoshi et al. 2001).

It is well-known that chemokines induce migration of specific types of leucocytes (Mateo et al. 2006). Furthermore, Ang-II has stimulated MCP-1 gene expression in renal and vascular cells and the expression of tumor necrosis factor-a (TNF-a) and IL-6 in renal cells and in cardiac fibroblasts (Rose et al. 2008; Ruiz-Ortega 2002). Emerging data have suggested that aldosterone may participate in vascular inflammation via both mineralocorticoid receptor (MR)-dependent and independent pathways. With regard to the pathophysiological mechanisms, it has been shown that aldosterone and MR seem to induce oxidative stress and vascular inflammation (Joffe and Adler 2005). As for the *in vivo* proinflammatory effects of aldosterone, there is a growing interest in the measurement of circulating inflammatory biomarkers (Pickering 2007). It is also worth mentioning that mineralocorticoid receptor antagonism has favorable effects in aortic inflammation, fibrosis, and hypertrophy, as well as oxidative stress and inflammation, according to experimental data (Suzuki et al. 2006).

## Reactive Oxygen Species-transcription Factors

Oxidative stress participates in the pathophysiology of hypertension through several mechanisms and molecular pathways (Table 1). A major source of vascular ROS, such as superoxide anion ($\bullet O2$), hydrogen peroxide ($H_2O_2$), hydroxyl radical (OH-), NO and peroxynitrite (ONOO-), is NADPH, which is produced in ECs, vascular smooth muscle cells (VSMCs), fibroblasts and monocytes/macrophages (Fridovich 1997). Reactive oxygen species participate in every stage of inflammation including vascular permeability, leucocyte recruitment, cell growth and fibrosis, modulating vascular tone and remodeling (Hernandez Schulman et al. 2006). Since NO production by the endothelium maintains the vasculature in a state of vasodilation, it seems reasonable that mice lacking the endothelial nitric oxide synthase (eNOS) gene have mild hypertension (Klahr 2001). Furthermore, Ang-II has been shown to induce ROS production in several cell types through activation of NADH/NADPH oxidase (Touyz 2003). Notably, data

**Table 1.** Oxidative stress-related mechanisms in hypertension.

| |
|---|
| Endothelial dysfunction |
| Decreased bioavailability of nitric oxide |
| Endothelial and vascular smooth muscle cells damage |
| Activation of transcription factors |
| Induction of proinflammatory processes |
| Generation lipid peroxidation products |
| Stimulation of growth signaling events |
| Depletion of tetrahydrobiopterin |

derived from experimental models of hypertension have demonstrated that therapeutic strategies with antioxidants may have a negative effect on blood pressure and Ang II-induced end-organ damage (Harrison 1997; Virdis et al. 2004).

Most proinflammatory responses which are activated by ROS are mediated via several proteins, enzymes and transcription factors, such as nuclear factor-κB (NF-κB), activator protein-1 (AP-1) and hypoxia-inducible factor 1 (HIF-1) (Chiarugi and Cirri 2003). The importance of NF-κB in inflammation has been proposed by experimental models of atherosclerosis and hypertension (Androulakis et al. 2009; Tousoulis et al. 2011a; Briasoulis et al. 2012). In particular, Ang II may activate NF-κB mainly via AT1 and AT2 receptors (Ruiz-Ortega 2000). Accordingly, NF-κB induces Ang-II gene expression, thereby further participating in the subsequent inflammatory responses. Moreover, this factor also has a prominent role in other phases of inflammation, such as in migration and proliferation of VSMCs, both under normal physiological conditions, and following vascular injury (Zahradka et al. 2002).

## Biomarkers in Essential Hypertension

Understanding of mechanisms which are implicated in the development of essential hypertension is vital for establishing biomarkers as potential diagnostic tools and designing prevention or treatment strategies (Table 2).

**Table 2.** Biomarkers associated with essential hypertension.

| Category | Biomarker | State |
|---|---|---|
| Adhesion molecules | ICAM-1<br>VCAM-1 | ↑<br>↑ |
| Cytokines/Chemokines | IL-1-IL-6<br>TNF-a<br>MCP-1 | ↑<br>↑<br>↑ |
| Other pro-inflammatory mediators | Selectins (E-Selectin)<br>Growth factors (VEGF)<br>ET-1 | ↑<br>↑<br>↑ |
| Procoagulant and thrombotic markers | PAI-1<br>Fibrinogen<br>vWF | ↑<br>↑<br>↑ |
| Other biomarkers | EPCs<br>Urinary alpha-1 microglobulin<br>Soluble CD40 ligand<br>Matrix metalloproteinases | ↓<br>↑<br>↑<br>↑/↓ |

**Abbreviations:** ICAM: intercellular cell adhesion molecules, VCAM-1: vascular cell adhesion molecule-1, MCP-1: monocyte chemoattractant protein-1, IL: interleukin, TNF-a: tumor necrosis factor-alpha, VEGF: vascular endothelial cell growth factors, PAI-1: tissue plasminogen activator inhibitor-1, vWF: von Wilebrand factor, EPCs: endothelial progenitor cells.

This approach could be equally attractive in both preclinical and clinical atherosclerosis and accordingly, in cardiovascular disease. According to recent studies, selected biomarkers may be elevated before the onset of hypertension and given the fact that inflammatory mechanisms are implicated in the pathogenesis of hypertension, scientific interest has focused on detection and monitoring of inflammatory markers (Stefanadi et al. 2010; Kampoli et al. 2012a; Tousoulis et al. 2012a).

## C-reactive protein

The most exploited circulating inflammatory marker in plasma is CRP. In addition to its role as an important inflammatory marker, it can exert modulatory effects through its well-demonstrated presence in atherosclerotic plaques (Tousoulis et al. 2012a). C-reactive protein is an acute-phase protein with a homopentameric structure. It is a pentraxin composed of five equal subgroups, and its expression is regulated mainly by IL-6 during the acute phase (Volanakis 2000; Srinivasan et al. 1994; Lefkou et al. 2006). Of the many inflammation markers, CRP has been correlated with higher prevalence of hypertension, independently of other cardiovascular risk factors (Stefanadi et al. 2010). Of note, in a study of a random sample of 300 subjects from the general population, it has been shown that CRP may be an independent risk factor for the development of hypertension. The unadjusted prevalence of hypertension was 58.7% in the highest quartile of CRP, but only 34.7% in the lowest quartile (Bautista et al. 2001). Furthermore, CRP has been positively associated with blood pressure (BP) variability, suggesting that inflammation may be one of the factors that promote increased BP variability. This finding is of major importance, given that elevated BP variability has been associated with target organ damage and adverse cardiovascular events in hypertensive patients (Abramson et al. 2006; Tatasciore et al. 2008). Furthermore, the ATTICA study, a cross-sectional population-based survey conducted during 2001 to 2002 based on a large, random sample of cardiovascular disease-free adults, has revealed an association between prehypertension and inflammatory markers linked to the atherosclerotic process (Chrysohoou et al. 2004).

In the Framingham Offspring Study, serum CRP levels were higher in non-hypertensive offspring of parents with hypertension compared with the offspring without parental hypertension, thus suggesting that inflammation may partly mediate the familial influences on hypertension risk. Regarding prospective studies, it has been demonstrated that elevated baseline CRP value precedes new-onset hypertension at an early stage among an elderly healthy population. The 2-year risk for new-onset hypertension was 52% greater for 1 mg/l increment of CRP. Among 335 individuals, considered normotensive at baseline, with no history of hypertension and no use of antihypertensive treatment, the incidence of hypertension was 9.9% 2 years later, while the 2-year risk for new-onset hypertension was 18% greater for 1 mg/l increment of CRP (Dauphinot et al. 2009). Moreover, CRP can be significantly associated with structural changes of human arteries, such as arterial stiffness (Montecucco et al. 2011), while in a prospective study with a follow-up of four years, an aortic stain, distensibility, and stiffness index were significantly associated with the progression to future hypertension after adjustment to classic risk factors in men and women and in young and old populations (Dernellis and Panaretou 2005).

## *Cytokines*

In addition to CRP, other inflammatory molecules, such as proinflammatory, anti-inflammatory cytokines and leukocyte adhesion molecules have been evaluated as possible determinants of blood pressure levels. However, these markers are still under investigation. Most of the cytokines are glycoproteins with a monomeric molecular mass of 15–25 kDa. Two of the main cytokines participating in atherogenesis are IL-6 and TNF-a; as regards their structure, IL-6 is a four helix bundle, while TNF-a is a trimeric protein (Tousoulis et al. 2011a).

Peeters et al. have shown that circulating levels of IL-1 receptor antagonist (IL-1ra), IL-1 and IL-6 production capacity (*ex vivo* stimulation) were higher in hypertensive patients (Peeters et al. 2001). As for target organ damage, it has been indicated that soluble TNF receptors, IL-6 and IL-1s were significantly correlated with left ventricular mass index in hypertensive patients (Rosello-Lleti et al. 2009). Furthermore, urinary excretion of TNF-a might be an early marker of left ventricular hypertrophy and microalbuminuria. More specifically, regression analysis revealed that urinary TNF-a concentration was independently related to target organ damage in newly diagnosed never-treated hypertensive patients (Navarro-González et al. 2008). Recently, the same investigators, examining the relationship between inflammatory parameters and urinary albumin excretion (UAE) in a total of 65 prehypertensive subjects, have demonstrated that inflammatory parameters are significantly associated with UAE in prehypertensive subjects, suggesting that inflammation may be a pathogenic factor for the early vascular or target organ damage (Navarro-González et al. 2012). According to experimental data, interleukin-8 (IL-8/CXCL8) expression in thoracic aorta tissue and VSMCs were significantly higher in spontaneously hypertensive rats than in controls (Kim et al. 2008). Notably, recent data suggested that cytokines may induce alterations in normal gene expression patterns. To determine whether prenatal exposure to IL-6 influences gene expression of the intrarenal RAS and contributes to renal dysfunction and hypertension in adulthood, Samuelsson et al. exposed female rats to IL-6 and analysed BP in the offspring at 5–20 weeks of age. Specifically, prenatal exposure to IL-6 influenced RAS gene expression and contributed to the development of hypertension, especially in females (Samuelsson et al. 2006).

## Adhesion molecules-chemokines

Cell adhesion molecules (CAMs) are transmembrane glycoproteins which consist of an extracellular component accompanied by a hydrophobic transmembrane component and an intracytoplasmic component. Integrins, the immunoglobulin superfamily and the selectins are the three families of CAMs (Krieglstein and Granger 2001; Price and Loscalzo 1999). On the other hand, chemokines are members of a group of structurally related and secretable, chemotactic cytokines divided into four families (CC, CXC, CX3C, XC) and can be located in different vascular cell types such as endothelial cells, but also inflammatory cells (Braunersreuther et al. 2007).

Arterial hypertension alone, as Madej et al. have demonstrated, may increase the expression of ICAM-1, MCP-1 and contribute to the progression of endothelial injury even without inflammation, lipid and carbohydrate disorders (Madej et al. 2005). Another study has investigated the plasma activity of inflammatory mediators such as granulocyte-macrophage colony-stimulating factor (GM-CSF), C-C chemokines and soluble adhesion molecules, produced by monocyte-endothelial cell adhesive interaction in patients with arterial hypertension and has shown that hypertension is associated with significantly higher levels of GM-CSF, macrophage inflammatory protein-1 alpha (MIP-1a), ICAM-1 and VCAM-1 than those of normotensives (Pickering 2007). Having considered that oxidant stress is implicated in the pathogenesis of atherosclerosis in cardiovascular diseases, Cottone et al. have shown that oxidative stress, as estimated by 8-iso-prostaglandin F2alpha, is correlated with high-sensitivity CRP (hsCRP), TNF-a, ICAM-1 and VCAM-1 (Cottone et al. 2007).

## Pro-thrombotic biomarkers

It is well-established that hypertension is an important risk factor for cardiovascular disease which may be attributed to the development of atherosclerosis. Given that hypertension is usually accompanied by other risk factors, such as hypertriglyceridemia, obesity, and insulin resistance and that myocardial infarction and stroke predominantly occur due to thrombosis, it can be assumed that hypertension may exert its effect mainly via promotion of a prothrombotic state (Stamler et al. 1989; Tousoulis et al. 2011b; Collins et al. 1990). Several studies have focused on fibrinogen, demonstrating a strong association with the presence of cardiovascular disease. Of note, pro-inflammatory cytokines which are produced from the vasculature, adipose tissue and myocardium also have a further regulatory role, inducing synthesis of acute phase proteins, such as fibrinogen, and consequently inflammatory and pro-thrombotic processes (Papageorgiou et al. 2010). Fibrinogen is a glycoprotein that consists of three pairs of polypeptide chains, Aa, Bb, and γ. Beyond coagulation cascade, fibrinogen is known to participate in the formation of atherosclerotic plaque during early stages, given its property to incorporate in the arterial wall and turn into decomposition products (Tousoulis et al. 2011a). It has been recently demonstrated that the concentration of plasma fibrinogen in the hypertension group was significantly higher than those in the control group, though no difference was noted between the hypertension grade 1 and 2 groups (Papageorgiou et al. 2010). Notably, in 229 consecutive, non-diabetic patients with uncomplicated, never-treated essential hypertension, plasma fibrinogen level were independently associated with aortic stiffening, a finding which potentially implies a contribution of this compound to the pathophysiology of cardiovascular disease (Vlachopoulos et al. 2007). Also, PAI-1 is a serine protease inhibitor produced by a variety of cells, including endothelial cells, hepatocytes and vascular smooth muscle cells. Elevated PAI-1 levels, which are regarded as a marker of reduced fibrinolytic potential, have been associated with hypertensive individuals, though it is more likely that hypertension leads to elevated PAI-1 levels rather than the converse (Wall et al. 1995; Lip et al. 2000). Moreover, elevated plasma renin activity, aldosterone, fibrinogen, D-dimer, prothrombin fragments, and PAI-1 levels were associated with clinical evidence of hypertension-related cardiac and renal damage, while fibrinogen and PAI-1 were independent predictors of the presence of organ damage (Sechi et al. 2008).

## *Other biomarkers*

Furthermore, essential hypertension has been associated with other biomarkers of low-grade inflammation. In particular, in a study population consisting of 1,445 nondiabetic patients with newly diagnosed arterial hypertension and no evidence of renal insufficiency serum levels of CRP, serum amyloid alpha, plasma fibrinogen, as well as urinary alpha-1 microglobulin excretion, were estimated. It has been demonstrated that urinary alpha-1 microglobulin may be independently associated with circulating acute phase proteins in patients with newly diagnosed hypertension, whereas it was closely associated with systolic but not diastolic blood pressure. In multivariate analysis, systolic blood pressure was independently associated with alpha-1 microglobulin, CRP, and serum amyloid alpha (Vyssoulis et al. 2007). Furthermore, it has been suggested that biomarkers of endothelial dysfunction and oxidative stress may have diagnostic value in patients with essential hypertension. Specifically, the prostacyclin and oxidized-low density lipoproteine (LDL) cut-off values were associated with a 4.9 higher significant risk of hypertension (Kuklinska et al. 2009). Soluble CD40L has been shown to participate in the pathogenesis of vascular damage and has been regarded as a molecular link between inflammation, thrombosis and angiogenesis. Specifically, the binding of CD40L to CD40 seems to induce inflammatory processes, such as pro-inflammatory cytokines and the expression of adhesion molecules, implying a role in atherosclerosis and its clinical complications (Tousoulis et al. 2010a). Moreover, recent data have indicated that Ang-II could stimulate inflammatory activation via CD40/CD40L ligation in human vascular cells, while sCD40L levels may be elevated in hypertensive patients, thus potentially distinguishing hypertensive patients at a high risk of cardiovascular events (Ferroni et al. 2008). In addition, matrix metalloproteinases (MMPs), a family of zinc metallo-endopeptidases, are responsible for much of the turnover of matrix components. Degradation of extracellular matrix by the MMPs system plays an important role in many physiological and pathological processes. A substantial body of evidence supports the notion that the imbalance between MMPS and tissue inhibitors of metalloproteinases may contribute to the pathogenesis of the atherosclerotic process, from the initial lesion to plaque rupture as well as other cardiovascular diseases including hypertension (Androulakis et al. 2012; Kampoli et al. 2012b). Notably, it has been suggested that serum MMP-1 is decreased in patients with essential hypertension. Also notable is that hypertensives with left ventricular hypertrophy had increased levels of MMP-9 and decreased levels of MMP-2 and MMP-13 (Ahmed et al. 2006), while patients with renal failure exhibited increased MMP-2 and MMP-10 levels compared to uncomplicated hypertensives and normotensives (Friese et al. 2009). It has been recently shown that MMP-9 and tissue inhibitor of metalloproteinase-1 were higher in healthy individuals with positive history of hypertension (Solini et al. 2009).

It is also worth mentioning that the number of circulating endothelial progenitor cells (EPC) has been inversely correlated with cardiovascular risk factors and vascular function. It is well recognized that EPCs play an important role in the maintenance of endothelial integrity and homeostasis. Recently, low circulating numbers and reduced functional capacity of EPCs have been used to predict future cardiovascular events, independently of other cardiovascular risk factors (Bakogiannis et al. 2012). Also, emerging data have linked decreased circulating EPC numbers and adhesive function with the presence of essential hypertension and organ damage, including electrocardiographic left ventricular hyperterophy (Lee et al. 2011).

## Multiple biomarker approach

As we have shown, studies are generally focused on an individual biomarker approach which represents single pathways. There is a growing interest with regard to whether multiple biomarkers are independently associated with hypertension risk. The hypothesis that abnormalities in multiple pathways may precede clinical hypertension could reflect a multiple biomarker approach in the early detection of hypertension and organ damage. This is nicely exemplified by Wang et al., who conducted a prospective study of 1,456 non-hypertensive individuals and measured nine biomarkers: CRP, fibrinogen, PAI-1, aldosterone, renin, B-type natriuretic peptide, and N-terminal proatrial natriuretic peptide, homocysteine and urinary albumin/creatinine ratio. Hypertension was developed in 232 participants over a mean follow-up of three years and the biomarker panel was significantly associated with incident hypertension. More specifically, inflammation, reduced fibrinolytic potential, and low-grade microalbuminuria biomarkers were associated with hypertension risk in non-hypertensive individuals. However, the predictive value of this "multimarker" score seems to be modest, while the overall specificity of the multimarker score was high and the sensitivity was low. Thus, it is apparent that identification of additional, novel biomarkers, which could improve the performance characteristics of this multiple biomarker approach estimating future hypertension risk, requires further investigation (Wang et al. 2007).

## Biomarkers as Therapeutic Targets in Hypertension

### Classic anti-hypertensive agents (Table 3)

As mentioned before, the renin-angiotensin-aldosterone system, beyond its role in BP regulation, vasoconstriction and fluid regulation, participates in inflammation and stimulation of cell growth. Its pharmacological inhibition remains the current 'gold standard' therapy in the management of hypertension (Tousoulis et al. 2012b). Further, inflammatory biomarkers could be potential therapeutic targets. With the aim of reducing cardiovascular risk of hypertensive patients, current anti-hypertensive strategies using angiotensin converting enzyme inhibitors (ACE), angiotensin receptor blockers (ARBs) and calcium antagonists seem to exert additive beneficial effects in hypertensive patients (Savoia and Schiffrin 2006; Androulakis et al. 2011). In particular, according to clinical and experimental data, vascular inflammation could be a potential therapeutic target, and given the potential reciprocal relationship between endothelial dysfunction and hypertension,

**Table 3.** Treatment strategies in hypertension.

| ANTIHYPERTENSIVE AGENTS | - Angiotensin converting enzyme inhibitors<br>-Angiotensin II receptors blockers<br>-$Ca^{+2}$ channel blockers |
|---|---|
| LIPID-LOWERING AGENTS | -Statins |
| PEROXISOME PROLIFERATOR-ACTIVATED RECEPTORS (PPAR) AGONISTS | -Fibrates<br>-Thiazolidinediones<br>-Glitazones |
| OTHER STRATEGIES | -Beta-blockers<br>-5-methyltetrahydrofolate<br>-Folic acid<br>-Physical activity |

these agents which potentially improve vascular function could be proved useful in the management of arterial hypertension (Papageorgiou et al. 2012; Tousoulis et al. 2010b).

ACE inhibitors, besides reducing the formation of Ang II, seem to improve endothelial function via decreased NADPH oxidase expression and diminished •O2 formation by vascular endothelium (Pasini et al. 2007). In addition, emerging evidence has proposed the potential antiinflammatory effects of ACE inhibitors. This is nicely exemplified by enalapril and candesartan which have showed similar effects on circulating adhesion molecules (Rosei et al. 2005a). Specifically, the primary end-point of the study was to evaluate changes of ICAM-1 plasma levels during treatment, while the secondary end-points were: changes in VCAM-1, von Willebrand factor (vWF), fibrinogen and PAI-1 circulating levels and urinary albumin excretion rate. They also exerted comparable effects on coagulation factors, though urinary protein excretion was reduced more by candesartan, in this study. Moreover, eight weeks of antihypertensive treatment with enalapril has been demonstrated to decrease plasma levels of circulating adhesion molecules and MCP-1 in hypertensive patients (Jilma et al. 2002). Another study has demonstrated that 3-month-antihypertensive treatment with either lisinopril or candesartan decreased levels of MMP-9, whereas tissue inhibitor of metalloproteinase-1 levels were increased (Onal et al. 2009). Notably, ramiprilat decreased extracellular matrix molecules, such as collagen deposition, increased elastin and fibrillin-1 deposition and reduced expression of both MMP-2 and MMP-3 *in vitro* (Ahimastos et al. 2005).

According to significant evidence, ARBs suppress cytokine and chemokine production both *in vitro* and *in vivo*. In particular, it has been demonstrated that candesartan can improve endothelial function, fibrinolysis, oxidant stress and can reduce MCP-1 and TNF-a levels (Koh et al. 2003). Later, the same investigators indicated that 16 mg candesartan administration daily during two months to 45 patients with mild to moderate hypertension increased adiponectin levels and improved insulin sensitivity in hypertensive patients (Koh et al. 2003). Specifically, in Japanese hypertensive patients, telmisartan improved vascular inflammation, causing significant reduction in IL-6 and hs-CRP levels (Koh et al. 2006). Moreover, treatment with olmesartan significantly reduced several biomarkers of inflammation, such as of hs-CRP, TNF-a, IL-6 and MCP-1 in 100 hypertensive patients by as early as week 6 of therapy (Fliser et al. 2004). Also, calcium antagonists have been shown to exert additional effects to the lowering BP properties with potential implications for atherogenesis and the reduction of cardiovascular events. Specifically, treatment with both nifedipine and enalapril tended to reduce ICAM-1 and E-selectin levels, given data derived from a randomized, multicenter trial (Rosei et al. 2005b). Of note, in a randomized clinical trial by De Ciuceis et al. patients with mild essential hypertension received barnidipine up to 20 mg or hydrochlorothiazide (HCT) up to 25 mg; barnidipine significantly increased EPCs after three and six months of treatment, whereas no effect was observed with HCT (de Ciuceis et al. 2011).

## *Other pharmacological strategies (Table 3)*

Statins have been shown to possess additional beneficial effects in several cardiovascular disease states beyond cholesterol lowering effects due to their cholesterol-independent or pleiotropic effects (O'Driscoll et al. 1997; Tousoulis et al. 2006; Tousoulis et al. 2005). Moreover, they may also improve the endothelial function by improving oxidative stress and up-regulating the expression and activity of eNOS (Briasoulis et al. 2012). In another study, treatment with either rosuvastatin or metformin led to reductions in IL-6 and, TNF-a, and improved markers of oxidative stress in patients with hypertension and dyslipidaemia (Gómez-García et al. 2007). With regard to the combination of anti-hypertensive treatment with other agents, it has been indicated that 12 weeks administration of valsartan in combination with a high dose of simvastatin was superior in reducing hsCRP; all three therapeutic approaches decreased plasma MCP-1 levels (Rajagopalan et al. 2007).

Anti-inflammatory actions of peroxisome proliferator-activated receptors (PPAR) and agonists (fibrates, thiazolidinediones or glitazones), may have an impact on the inflammatory component of hypertension (Touyz and Schiffrin 2006). They have been shown to exert a broad spectrum of antiatherogenic effects *in vitro*, in animal models of atherosclerosis, and in humans (Hsueh and Bruemmer 2004). Of note, pioglitazone reduced CRP, ICAM-1 and VCAM-1 levels within one month in patients with hypertension who had developed type 2 diabetes mellitus; however CRP levels were decreased after six months of treatment with either pioglitazone or voglibose (Takase et al. 2007). In addition, nebivolol, a third generation selective beta (1)-adrenoceptor antagonist, has shown antioxidant properties, increasing NO also by decreasing its oxidative inactivation. Specifically, it has improved markers of oxidative stress; also, a reduction of ROS and •O2 concentration in endothelial cells exposed to oxidative stress has been observed (Pasini et al. 2005).

A number of studies have focused on the role of several other agents and treatment strategies with respect to their effects on biomarkers and their relation to other parameters in a state of hypertension, though with inconsistent results. For example, folic acid and its circulating metabolite, 5-methyltetrahydrofolate, potentially affect vascular superoxide production both *in vivo* and *ex vivo*, and may improve vascular function in humans, independently of reducing plasma homocysteine (Briasoulis et al. 2012; Antoniades et al. 2009; Antoniades et al. 2006). Furthermore, physical activity, apart from being an effective antihypertensive treatment, may also have beneficial effects on inflammatory status in that state. Of note, it has been demonstrated that physical activity can reduce circulating inflammatory cytokines IL-6 and TNF-a and soluble adhesion molecules (sICAM-1 and sVCAM-1) (Adamopoulos et al. 2002; Adamopoulos et al. 2001), while several other studies have reported significant reductions in CRP and IL-6 levels for hypertensive patients. However, there is a need for larger clinical trials examining the inflammation-lowering effects of exercise alone in untreated hypertension, without the intervention of an anti-hypertensive agent (Edwardsa et al. 2007).

## Conclusion

Recent studies have consistently supported the hypothesis that inflammation plays a key role in the pathophysiology of hypertension along with other mechanisms, such as activation of renin-angiotensin-aldosterone system, the local production of angiotensin II and reactive oxygen species. Accordingly, various biomarkers have been associated with the presence of hypertension. Also, given the fact that selected biomarkers may be elevated before the onset of hypertension, scientific interest has focused on detection and monitoring of these markers. Treatment strategies, which mostly target the endothelium, have been shown to downregulate inflammatory mediators and increase nitric oxide bioavailability with an acceptable cost and safety profile. To date, several agents seem to exert pleiotropic actions, targeting several pathways. However, they have not been extensively studied on an experimental basis, with respect to prognosis and clinical outcome in hypertensive patients.

# References

Abramson, J.L., C. Lewis, N.V. Murrah, G.T. Anderson, and V. Vaccarino. 2006. Relation of C-reactive protein and tumor necrosis factor-alpha to ambulatory blood pressure variability in healthy adults. Am. J. Cardiol. 98: 649–652.

Adamopoulos, S., J. Parissis, C. Kroupis, M. Georgiadis, D. Karatzas, G. Karavolias, K. Koniavitou, A.J. Coats, and D.T. Kremastinos. 2001. Physical training reduces peripheral markers of inflammation in patients with chronic heart failure. Eur. Heart J. 22: 791–797.

Adamopoulos, S., J. Parissis, D. Karatzas, C. Kroupis, M. Georgiadis, G. Karavolias, J. Paraskevaidis, K. Koniavitou, A.J. Coats, and D.T. Kremastinos. 2002. Physical training modulates proinflammatory cytokines and the soluble Fas/soluble Fas ligand system in patients with chronic heart failure. J. Am. Coll. Cardiol. 39: 653–663.

Ahimastos, A.A., A.K. Natoli, A. Lawler, P.A. Blombery, and B.A. Kingwell. 2005. Ramipril reduces large-artery stiffness in peripheral arterial disease and promotes elastogenic remodeling in cell culture. Hypertension 45: 1194–1199.

Ahmed, S.H., L.L. Clark, W.R. Pennington, C.S. Webb, D.D. Bonnema, A.H. Leonardi, C.D. McClure, F.G. Spinale, and M.R. Zile. 2006. Matrix metalloproteinases/Tissue inhibitors of metalloproteinases. Relationship between changes in proteolytic determinants of matrix composition and structural, functional and clinical manifestations of hypertensive heart disease. Circulation 113: 2089–2096.

Androulakis, E., D. Tousoulis, N. Papageorgiou, G. Latsios, G. Siasos, and C. Stefanadis. 2012. The role of matrix metalloproteinases in essential hypertension. Curr. Top. Med. Chem. 12: 1149–1158.

Androulakis, E., D. Tousoulis, N. Papageorgiou, G. Latsios, G. Siasos, C. Tsioufis, A. Giolis, and C. Stefanadis. 2011. Inflammation in hypertension: current therapeutic approaches. Curr. Pharm. Des. 17: 4121–4131.

Androulakis, E.S., D. Tousoulis, N. Papageorgiou, C. Tsioufis, I. Kallikazaros, and C. Stefanadis. 2009. Essential hypertension: is there a role for inflammatory mechanisms? Cardiol. Rev. 17: 216–221.

Antoniades, C., A. Antonopoulos, D. Tousoulis, K. Marinou, and C. Stefanadis. 2009. Homocysteine and coronary atherosclerosis: From folate fortification to the recent clinical trials. Eur. Heart J. 30: 6–15.

Antoniades, C., C. Shirodaria, N. Warrick, S. Cai, J. de Bono, J. Lee, P. Leeson, S. Neubauer, C. Ratnatunga, R. Pillai, H. Refsum, and K.M. Channon. 2006. 5-Methyltetrahydrofolate rapidly improves endothelial function and decreases superoxide production in human vessels: Effects on vascular tetrahydrobiopterin availability and endothelial nitric oxide synthase coupling. Circulation 114: 1193–1201.

Bakogiannis, C., D. Tousoulis, E. Androulakis, A. Briasoulis, N. Papageorgiou, G. Vogiatzi, A.M. Kampoli, M. Charakida, G. Siasos, G. Latsios, C. Antoniades, and C. Stefanadis. 2012. Circulating endothelial progenitor cells as biomarkers for prediction of cardiovascular outcomes. Curr. Med. Chem. 19: 2597–2604.

Bautista, L.E., P. Lopez Jaramillo, L.M. Vera, J.P. Casas, A.P. Otero, and A.I. Guaracao. 2001. Is C-reactive protein an independent risk factor for essential hypertension? J. Hypertens. 19: 857–861.

Braunersreuther, V., F. Mach, and S. Steffens. 2007. The specific role of chemokines in atherosclerosis. Thromb. Haemost. 97: 714–721.

Briasoulis, A., D. Tousoulis, E.S. Androulakis, N. Papageorgiou, G. Latsios, and C. Stefanadis. 2012. Endothelial dysfunction and atherosclerosis: focus on novel therapeutic approaches. Recent. Pat. Cardiovasc. Drug Discov. 7: 21–32.

Chiarugi, P. and P. Cirri. 2003. Redox regulation of protein tyrocine phosphatases during receptor tyrocine kinase signal transduction. Trends Biochem. Sci. 28: 509–514.

Chrysohoou, C., C. Pitsavos, D.B. Panagiotakos, J. Skoumas, and C. Stefanadis. 2004. Association between prehypertension status and inflammatory markers related to atherosclerotic disease. The ATTICA Study. Am. J. Hypertens. 17: 568–573.

Collins, R., R. Peto, S. MacMahon, P. Hebert, N.H. Fiebach, K.A. Eberlein, J. Godwin, N. Qizilbash, J.O. Taylor, and C.H. Hennekens. 1990. Blood pressure, stroke and coronary heart disease, II: short-term reductions in blood pressure: overview of randomized drug trials in their epidemiological context. Lancet 335: 827–838.

Cottone, S., G. Mulè, E. Nardi, A. Vadalà, M.C. Lorito, M. Guarneri, R. Arsena, A. Palermo, and G. Cerasola. 2007. C-reactive protein and intercellular adhesion molecule-1 are stronger predictors of oxidant stress than blood pressure in established hypertension. J. Hypertens. 25: 423–428.

Das, U.N. 2006. Hypertension as a low-grade systemic inflammatory condition that has its origins in the perinatal period. J. Assoc. Physicians India 54: 133–142.

Dauphinot, V., F. Roche, M.P. Kossovsky, A.M. Schott, V. Pichot, J.M. Gaspoz, P. Gosse, and Barthelemy. 2009. C-reactive protein implications in new-onset hypertension in a healthy population initially aged 65 years: the Proof study. J. Hypertens. 27: 736–743.

de Ciuceis, C., A. Pilu, D. Rizzoni, E. Porteri, M.L. Muiesan, M. Salvetti, A. Paini, E. Belotti, F. Zani, G.E. Boari, C.A. Rosei, and E.A. Rosei. 2011. Effect of antihypertensive treatment on circulating endothelial progenitor cells in patients with mild essential hypertension. Blood Press. 20: 77–83.

Dernellis, J. and M. Panaretou. 2005. Aortic stiffness is an independent predictor of progression to hypertension in nonhypertensive subjects. Hypertension 45: 426–431.

Edwardsa, K.M., M.G. Zieglerb, and P.J. Millsa. 2007. The potential anti-inflammatory benefits of improving physical fitness in hypertension. J. Hypertens. 25: 1533–1542.

Ferroni, P. and F. Guadagni. 2008. Soluble CD40L and its role in essential hypertension: diagnostic and therapeutic implications. Cardiovasc. Hematol. Disord. Drug Targets 8: 194–202.

Fliser, D., K. Buchholz, and H. Haller; EUropean Trial on Olmesartan and Pravastatin in Inflammation and Atherosclerosis (EUTOPIA) Investigators. 2004. Antiinflammatory effects of angiotensin II subtype 1 receptor blockade in hypertensive patients with microinflammation. Circulation 110: 1103–1107.

Fratta Pasini, A., U. Garbin, M.C. Nava, C. Stranieri, A. Davoli, T. Sawamura, V. Lo Cascio, and L. Cominacini. 2005. Nebivolol decreases oxidative stress in essential hypertensive patients and increases nitric oxide by reducing its oxidative inactivation. J. Hypertens. 23: 589–596.

Fridovich, I. 1997. Superoxide anion radical, superoxide dismutases, and related matters. J. Biol. Chem. 272: 18515–18517.

Friese, R.S., F. Rao, S. Khandrika, B. Thomas, M.G. Ziegler, G.W. Schmid-Schönbein, and D.T. O'Connor. 2009. Matrix metalloproteinases: Discrete elevations in essential hypertension and hypertensive end-stage renal disease. Clin. Exp. Hypertens. 31: 521–533.

Funakoshi, Y., T. Ichiki, and K. Shimikawa. 2001. Rho-kinase mediates angiotensin-II-monocyte chemoattractant protein-1 expression in rat vascular smooth muscle cells. Hypertension 38: 100–104.

Geza, S. 2004. Pathogenesis of structural vascular changes in hypertension. J. Hypertens. 22: 3–10.

Gómez-García, A., G. Martínez Torres, L.E. Ortega-Pierres, E. Rodríguez-Ayala, and C. Alvarez-Aguilar. 2007. Rosuvastatin and metformin decrease inflammation and oxidative stress in patients with hypertension and dyslipidemia. Rev. Esp. Cardiol. 60: 1242–1249.

Harrison, D.G. 1997. Cellular and molecular mechanisms of endothelial cell dysfunction. J. Clin. Invest. 100: 2153–2157.

Hernandez Schulman, I., M.S. Zhou, and L. Raij. 2006. Interaction between nitric oxide and angiotensin II in the endothelium: role in atherosclerosis and hypertension. J. Hypertens. 24(suppl 1): S45–S50.

Hoch, N.E., T.J. Guzik, W. Chen, T. Deans, S.A. Maalouf, P. Gratze, C. Weyand, and D.G. Harrison. 2009. Regulation of T cell function by endogenously produced angiotensin II. Am. J. Physiol. Regul. Integr. Comp. Physiol. 296: R208–R216.

Hsueh, W.A. and D. Bruemmer. 2004. Peroxisome proliferator-activated receptor gamma: Implications for cardiovascular disease. Hypertension 43: 297–305.

Jilma, B., F.L. Li-Saw-Hee, O.F. Wagner, D.G. Beevers, and G.Y. Lip. 2002. Effects of enalapril and losartan on circulating adhesion molecules and monocytes chemotactic protein-1. Clin. Sci. (Lond.) 103: 131–136.

Joffe, H.V. and G.K. Adler. 2005. Effect of aldosterone and mineralocorticoid receptor blockade on vascular inflammation. Heart. Fail. Rev. 10: 31–37.

Johnson, R.J. and G.F. Schreiner. 1997. Hypothesis: the role of acquired tubulointerstitial disease in the pathogenesis of salt-dependent hypertension. Kidney Int. 52: 1169–1179.

Johnson, R.J., B. Rodriguez-Iturbe, D.H. Kang, D.I. Feig, and J. Herrera-Acosta. 2005. A unifying pathway for essential hypertension. Am. J. Hypertension 18: 431–440.

Johnson, R.J., D.I. Feig, T. Nakagawa, L.G. Sanchez-Lozada, and B. Rodriguez-Iturbe. 2008. Pathogenesis of essential hypertension: historical paradigms and modern insights. J. Hypertens. 26: 381–391.

Kampoli, A.M., D. Tousoulis, N. Papageorgiou, C. Antoniades, E. Androulakis, E. Tsiamis, G. Latsios, and C. Stefanadis. 2012b. Matrix metalloproteinases in acute coronary syndromes: current perspectives. Curr. Top. Med. Chem. 12: 1192–1205.

Kampoli, A.M., D. Tousoulis, N. Papageorgiou, Z. Pallatza, G. Vogiatzi, A. Briasoulis, E. Androulakis, C. Toutouzas, P. Stougianos, C. Tentolouris, and C. Stefanadis. 2012a. Clinical utility of biomarkers in premature atherosclerosis. Curr. Med. Chem. 19: 2521–2533.

Kearney, P.M., M. Whelton, K. Reynolds, P. Muntner, P.K. Whelton, and J. He. 2005. Global burden of hypertension: analysis of worldwide data. Lancet 365: 217–223.

Kim, H.Y., Y.I. Kang, I.H. Song, H.C. Choi, and H.S. Kim. 2008. Upregulation of interleukin-8/ CXCL8 in vascular smooth muscle cells from spontaneously hypertensive rat. Hypertens. Res. 31: 515–523.

Klahr, S. 2001. The role of nitric oxide in hypertension and renal disease progression. Nephrol. Dial. Transplant. 16 Suppl 1: 60–62.

Koh, K.K., J.Y. Ahn, S.H. Han, D.S. Kim, D.K. Jin, H.S. Kim, M.S. Shin, T.H. Ahn, I.S. Choi, and E.K. Shin. 2003. Pleitropic effects of angiotensin II receptor blocker in hypertensive patients. J. Am. Coll. Cardiol. 42: 905–910.

Koh, K.K., M.J. Quon, S.H. Han, W.J. Chung, Y. Lee, and E.K. Shin. 2006. Anti-inflammatory and metabolic effects of candesartan in hypertensive patients. Int. J. Cardiol. 108: 96–100.

Krieglstein, C.F. and D.N. Granger. 2001. Adhesion molecules and their role in vascular disease. Am. J. Hypertens. 14: 44S–54S.

Kuklinska, A.M., B. Mroczko, W.J. Musial, M. Usowicz-Szarynska, R. Sawicki, H. Borowska, M. Knapp, and M. Szmitkowski. 2009. Diagnostic biomarkers of essential arterial hypertension: the value of prostacyclin, nitric oxide, oxidized-LDL, and peroxide measurements. Int. Heart J. 50: 341–351.

Lee, C.W., P.H. Huang, S.S. Huang, H.B. Leu, C.C. Huang, T.C. Wu, J.W. Chen, and S.J. Lin. 2011. Decreased circulating endothelial progenitor cell levels and function in essential hypertensive patients with electrocardiographic left ventricular hypertrophy. Hypertens. Res. 34: 999–1003.

Lefkou, E., N. Fragakis, and G. Varlamis. 2006. The inflammatory process during childhood and the risk of developing atheromatous disease in adult life: the role of C-reactive protein. Hellenic. J. Cardiol. 47: 164–169.

Li, J.J. and J.L. Chen. 2005. Inflammation may be a bridge connecting hypertension and atherosclerosis. Med. Hypotheses 64: 925–929

Lip, G.Y. and A.D. Blann. 2000. Does hypertension confer a prothrombotic state? Virchow's triad revisited. Circulation 101: 218–220.

Madej, A., B. Okopien, J. Kowalski, M. Haberka, and Z.S. Herman. 2005. Plasma concentrations of adhesion molecules and chemokines in patients with essential hypertension. Pharmacol. Rep. 57: 878–881.

Marchesi, C., P. Paradis and E.L. Schiffrin. 2008. Role of the renin-angiotensin system in vascular inflammation. Trends Pharmacol. Sci. 29: 367–374.

Mateo, T., Y.N. Abu Nabah, M. Abu Taha, M. Mata, M. Cerdá-Nicolás, A.E. Proudfoot, R.A. Stahl, A.C. Issekutz, J. Cortijo, E.J. Morcillo, P.J. Jose, and M.J. Sanz. 2006. Angiotensin II-induced mononuclear leukocyte interactions with arteriolar and venular endothelium are mediated by the release of different CC chemokines. J. Immunol. 176: 5577–5586.

Montecucco, F., A. Pende, A. Quercioli, and F. Mach. 2011. Inflammation in the pathophysiology of essential hypertension. J. Nephrol. 24: 23–34.

Navarro-González, J.F., C. Mora, M. Muros, A. Jarque, H. Herrera, and J. García. 2008. Association of tumor necrosis factor-alpha with early target organ damage in newly diagnosed with essential hypertension. J. Hypertens. 26: 2168–2175.

Navarro-González, J.F., C. Mora, M. Muros, J. García, J. Donate and V. Cazaña. 2012. Relationship between inflammation and microalbuminuria in prehypertension. J. Hum. Hypertens. Jan 26. [Epub ahead of print]

O'Driscoll, G., D. Green, and R.R. Taylor. 1997. Simvastatin, an HMG-coenzyme A reductase inhibitor, improves endothelial function within 1 month. Circulation 95: 1126–1131.

Onal, I.K., B. Altun, E.D. Onal, A. Kirkpantur, Oz. S. Gul, and C. Turgan. 2009. Serum levels of MMP-9 and TIMP-1 in primary hypertension and effect of antihypertensive treatment. Eur. J. Intern. Med. 20: 369–372.

Papageorgiou, N., D. Tousoulis, E. Androulakis, A. Giotakis, G. Siasos, G. Latsios, and C. Stefanadis. 2012. Lifestyle factors and endothelial function. Curr. Vasc. Pharmacol. 10: 94–106.

Papageorgiou, N., D. Tousoulis, G. Siasos, and C. Stefanadis. 2010. Is fibrinogen a marker of inflammation in coronary artery disease? Hellenic. J. Cardiol. 51: 1–9.

Pasini, A.F., U. Garbin, M.C. Nava, C. Stranieri, M. Pellegrini, V. Boccioletti, M.L. Luchetta, P. Fabrizzi, V. Lo Cascio, and L. Cominacini. 2007. Effect of sulfhydryl and non-sulfhydryl angiotensin-converting enzyme inhibitors on endothelial function in essential hypertensive patients. Am. J. Hypertens. 20: 443–450.

Peeters, A.C., M.G. Netea, M.C. Janssen, B.J. Kullberg, J.W. Van der Meer, and T. Thien. 2001. Pro-inflammatory cytokines in patients with essential hypertension. Eur. J. Clin. Invest. 31: 31–36.

Pickering, T.G. 2007. Stress, inflammation, and hypertension. J. Clin. Hypertens. (Greenwich) 9: 567–571.

Piqueras, L., P. Kubes, A. Alvarez, E. O'Connor, A.C. Issekutz, J.V. Esplugues, and M.J. Sanz. 2000. Angiotensin II induces leukocyte-endothelial cell interactions *in vivo* via AT(1) and AT(2) receptor-mediated P-selectin upregulation. Circulation 102: 2118–2123.

Price, D.T. and J. Loscalzo. 1999. Cellular adhesion molecules and atherogenesis. Am. J. Med. 107: 85–97.

Rajagopalan, S., F. Zannad, A. Radauceanu, R. Glazer, Y. Jia, M.F. Prescott, M. Kariisa, and B. Pitt. 2007. Effects of valsartan alone vs valsartan/simvastatin combination on ambulatory blood pressure, C-reactive protein, lipoproteins, and monocyte chemoattractant protein-1 in patients with hyperlipidemia and hypertension. Am. J. Cardiol. 100: 222–226.

Rose, P., J. Bond, S. Tighe, M.J. Toth, T.L. Wellman, E.M. Briso de Montiano, M.M. Lewinter, and K.M. Lounsbury. 2008. Genes overexpressed in cerebral arteries following salt-induced hypertensive disease are regulated by angiotensin II, JunB, and CREB. Am. J. Physiol. Heart Circ. Physiol. 294: H1075–H1085.

Rosei, E.A., D. Rizzoni, M.L. Muiesan, I. Sleiman, M. Salvetti, C. Monteduro, E. Porteri; CENTRO (CandEsartaN on aTherosclerotic Risk factors) study investigators. 2005a.

Effects of candesartan cilexetil and enalapril on inflammatory markers of atherosclerosis in hypertensive patients with non-insulin diabetes mellitus. J. Hypertens. 23: 435–444.

Rosei, E.A., P. Morelli, and D. Rizzoni. 2005b. Effects of nifedipine GITS 20 mg or enalapril 20 mg on blood pressure and inflammatory markers in patients with mild-moderate hypertension. Blood Press. 14 Suppl 1: 14–22.

Roselló-Lletí, E., M. Rivera, L. Martínez-Dolz, J.R. González Juanatey, R. Cortés, A. Jordán, P. Morillas, C. Lauwers, J.R. Calabuig, I. Antorrena, B. de Rivas, M. Portolés, and V. Bertomeu. 2009. Inflammatory activation and left ventricular mass in essential hypertension. Am. J. Hypertens. 22: 444–450.

Ruiz-Ortega, M., M. Ruperez, O. Lorenzo, V. Esteban, J. Blanco, S. Mezzano, and J. Egido. 2002. Angiotensin II regulates the synthesis of proinflammatory cytokines and chemokines in the kidney. Kidney Int. Suppl. 82: S12–22.

Ruiz-Ortega, M., O. Lorenzo, M. Rupérez, S. König, B. Wittig, and J. Egido. 2000. Angiotensin II activates nuclear transcription factor kB trough AT1 and AT2 in vascular smooth muscle cells molecular mechanisms. Circ. Res. 86: 1266–1272.

Ruiz-Ortega, M., V. Esteban, M. Rupérez, E. Sánchez-López, J. Rodríguez-Vita, G. Carvajal, and J. Egido. 2006. Renal and vascular hypertension-induced inflammation: role of angiotensin II. Curr. Opin. Nephrol. Hypertens. 15: 159–166.

Samuelsson, A.M., C. Alexanderson, J. Mölne, B. Haraldsson, P. Hansell, and A. Holmäng. 2006. Prenatal exposure to interleukin-6 results in hypertension and alterations in the renin-angiotensin system of the rat. J. Physiol. 575: 855–867.

Savoia, C. and E.L. Schiffrin. 2006. Inflammation in hypertension. Curr. Opin. Nephrol. Hypertens. 15: 152–158.

Schieffer, B., M. Luchtefeld, S. Braun, A. Hilfiker, D. Hilfiker-Kleiner, and H. Drexler. 2000. Role of NAD(P)H oxidase in angiotensin II-induced JAK/STAT signaling and cytokine induction. Circ. Res. 87: 1195–1201.

Sechi, L.A., M. Novello, G. Colussi, A. Di Fabio, A. Chiuch, E. Nadalini, A. Casanova-Borca, A. Uzzau, and C. Catena. 2008. Relationship of plasma renin with a prothrombotic state in hypertension: relevance for organ damage. Am. J. Hypertens. 21: 1347–1353.

Skultetyova, D., S. Filipova, I. Riecansky, and J. Skultety. 2007. The role of angiotensin type 1 receptor in inflammation and endothelial dysfunction. Recent Pat. Cardiovasc. Drug Discov. 2: 23–27.

Solini, A., E. Santini, A. Passaro, S. Madec, and E. Ferrannini. 2009. Family history of hypertension, anthropometric parameters and markers of early atherosclerosis in young healthy individuals. J. Hum. Hypertens. 23: 801–807.

Srinivasan, N., H.E. White, and J. Emsley. 1994. Comparative analyses of pentraxins: implications for promoter assembly and ligand binding. Structure 2: 1017–1027.

Stamler, J., J.D. Neaton, and D.N. Wentworth. 1989. Blood pressure and risk of fatal coronary heart disease. Hypertension 13(suppl I): I-2–I-12.

Stefanadi, E., D. Tousoulis, E.S. Androulakis, N. Papageorgiou, M. Charakida, G. Siasos, C. Tsioufis, and C. Stefanadis. 2010. Inflammatory markers in essential hypertension: potential clinical implications. Curr. Vasc. Pharmacol. 8: 509–516.

Suzuki, J., M. Iwai, M. Mogi, A. Oshita, T. Yoshii, J. Higaki, and M. Horiuchi. 2006. Eplerenone with valsartan effectively reduces atherosclerotic lesion by attenuation of oxidative stress and inflammation. Arterioscler. Thromb. Vasc. Biol. 26: 917–921.

Suzuki, Y., M. Ruiz-Ortega, O. Lorenzo, M. Ruperez, V. Esteban, and J. Egido. 2003. Inflammation and angiotensin II. Int. J. Biochem. Cell. Biol. 35: 881–890.

Takase, H., A. Nakazawa, S. Yamashita, T. Toriyama, K. Sato, R. Ueda, and Y. Dohi. 2007. Pioglitazone produces rapid and persistent reduction of vascular inflammation in patients with hypertension and type 2 diabetes mellitus who are receiving angiotensin II receptor blockers. Metabolism 56: 559–564.

Tatasciore, A., M. Zimarino, G. Renda, M. Zurro, M. Soccio, C. Prontera, M. Emdin, M. Flacco, G. Schillaci, and R. DE Caterina. 2008. Awake blood pressure variability, inflammatory markers and target organ damage in newly diagnosed hypertension. Hypertens. Res. 31: 2137–2146.

Tousoulis, D., A.M. Kampoli, N. Papageorgiou, E. Androulakis, C. Antoniades, K. Toutouzas, and C. Stefanadis. 2011a. Pathophysiology of atherosclerosis: the role of inflammation. Curr. Pharm. Des. 17: 4089–4110.

Tousoulis, D., C. Antoniades, E. Bosinakou, M. Kotsopoulou, C. Pitsavos, C. Vlachopoulos, D. Panagiotakos, and C. Stefanadis. 2005. Effects of atorvastatin on reactive hyperemia and inflammatory process in patients with congestive heart failure. Atherosclerosis 178: 359–363.

Tousoulis, D., C. Antoniades, V. Katsi, E. Bosinakou, M. Kotsopoulou, C. Tsioufis, and C. Stefanadis. 2006. The impact of early administration of low-dose atorvastatin treatment on inflammatory process, in patients with unstable angina and low cholesterol level. Int. J. Cardiol. 109: 48–52.

Tousoulis, D., E. Androulakis, N. Papageorgiou, A. Briasoulis, G. Siasos, C. Antoniades, and C. Stefanadis. 2010a. From atherosclerosis to acute coronary syndromes: the role of soluble CD40 ligand. Trends Cardiovasc. Med. 20: 153–164.

Tousoulis, D., E. Androulakis, N. Papageorgiou, and C. Stefanadis. 2012b. Novel therapeutic strategies in the management of arterial hypertension. Pharmacol. Ther. 135: 168–175.

Tousoulis, D., G. Hatzis, N. Papageorgiou, E. Androulakis, G. Bouras, A. Giolis, C. Bakogiannis, G. Siasos, G. Latsios, C. Antoniades, and C. Stefanadis. 2012a. Assessment of acute coronary syndromes: focus on novel biomarkers. Curr. Med. Chem. 19: 2572–2587.

Tousoulis, D., N. Papageorgiou, E. Androulakis, A. Briasoulis, C. Antoniades, and C. Stefanadis. 2011b. Fibrinogen and cardiovascular disease: genetics and biomarkers. Blood. Rev. 25: 239–245.

Tousoulis, D., N. Papageorgiou, E. Androulakis, K. Paroutoglou, and C. Stefanadis. 2010b. Novel therapeutic strategies targeting vascular endothelium in essential hypertension. Expert Opin. Investig. Drugs 19: 1395–1412.

Touyz, R.M. 2003. Reactive oxygen species in vascular biology: role in arterial hypertension. Expert Rev. Cardiovasc. Ther. 1: 99–106.

Touyz, R.M. 2005. Mollecular and cellular mechanisms in vascular injury in hypertension: role of angiotensin II. Curr. Opin. Nephrol. Hypertens. 14: 125–131.

Touyz, R.M. and E.L. Schiffrin. 2006. Peroxisome proliferator-activated receptors in vascular biology-molecular mechanisms and clinical implications. Vasc. Pharmacol. 45: 19–28.

Vaughan, D.E., S.A. Lazos, and K. Tong. 1995. Angiotensin II regulates the expression of plasminogen activator inhibitor-1 in cultured endothelial cells: a potential link between the renin-angiotensin system and thrombosis. J. Clin. Invest. 95: 995–1001.

Vedanthan, R. and V. Fuster. 2009. Disease Prevention: The moving target of global cardiovascular health. Nat. Rev. Cardiol. 6: 327–328.

Virdis, A., M.F. Neves, F. Amiri, R.M. Touyz, and E.L. Schiffrin. 2004. Role of NAD(P)H oxidase on vascular alterations in angiotensin II infused mice. J. Hypertens. 22: 535–542.

Vlachopoulos, C., P. Pietri, K. Aznaouridis, G. Vyssoulis, C. Vasiliadou, A. Bratsas, D. Tousoulis, P. Xaplanteris, E. Stefanadi, and C. Stefanadis. 2007. Relationship of fibrinogen with arterial stiffness and wave reflections. J. Hypertens. 25: 2110–2116.

Volanakis, J.E. 2001. Human C-reactive protein: expression, structure, and function. Mol. Immunol. 38: 189–197.

Vyssoulis, G.P., D. Tousoulis, C. Antoniades, S. Dimitrakopoulos, A. Zervoudaki, A. and C. Stefanadis. 2007. Alpha-1 microglobulin as a new inflammatory marker in newly diagnosed hypertensive patients. Am. J. Hypertens. 20: 1016–1021.

Wall, U., C. Jern, A. Bergbrant, and S. Jern. 1995. Enhanced levels of tissue-type plasminogen activator in borderline hypertension. Hypertension 26: 796–800.

Wang, T.J., P. Gona, M.G. Larson, D. Levy, E.J. Benjamin, G.H. Tofler, P.F. Jacques, J.B. Meigs, N. Rifai, J. Selhub, S.J. Robins, C. Newton-Cheh, and R.S. Vasan. 2007. Multiple biomarkers and the risk of incident hypertension. Hypertension 49: 432–438.

Zahradka, P., J.P. Werner, S. Buhay, B. Litchie, G. Helwer and S. Thomas. 2002. NF-kappaB activation is essential for angiotensin II-dependent proliferation and migration of vascular smooth muscle cells. J. Mol. Cell. Cardiol. 34: 1609–1621.

Zhao, Q., M. Ishibashi, K. Hiasa, C. Tan, A. Takeshita, and K. Egashira. 2004. Essential role of vascular endothelial growth factor in angiotensin II-induced vascular inflammation and remodeling. Hypertension 44: 264–270.

# CHAPTER 4

# Biomarkers in Hyperlipidemia

Dimitris Tousoulis,* Emmanuel Androulakis,
Nikolaos Papageorgiou, George Latsios, Gerasimos Siasos,
Evangelos Oikonomou and Christodoulos Stefanadis

## Introduction

Hyperlipidemia involves abnormally elevated levels of any or all lipids and/ or lipoproteins in the blood. Hyperlipidemias may basically be classified as familial, caused by specific genetic abnormalities, or acquired resulting from another underlying disorder that leads to alterations in plasma lipid and lipoprotein metabolism. This abnormality may also be classified according to which types of lipids are elevated, that is hypercholesterolemia, hypertriglyceridemia or both in combined hyperlipidemia, while elevated levels of lipoprotein(a) may also be classified as a form of hyperlipidemia (Chait and Brunzell 1990; Stone 1994; Ascaso et al. 2007; Yuan et al. 2007).

The core components of the dyslipidemia—the "lipid triad"—along with several classes of lipids are most likely to initiate atherosclerosis. Cholesterol is recognized in distinct particles containing lipoproteins. Beyond triglycerides and triglycerides-rich lipoproteins, three major classes of lipoproteins are found in the serum: low density lipoproteins (LDL), high density lipoproteins (HDL) and very low density lipoproteins (VLDL) (Kraus 1998; NCEP/Adult Treatment Panel III 2002).

Traditional risk factors such as hyperlipidemia, diabetes mellitus, hypertension, smoking, as well as novel ones represent important factors predisposing to endothelial dysfunction and atherosclerosis. More

1st Department of Cardiology, 'Hippokration' Hospital, University of Athens Medical School, Athens, Greece.
*Corresponding author: drtousoulis@hotmail.com

specifically, oxidative modification of LDL in the arterial wall by reactive oxygen species (ROS) leads to atherosclersosis. Moreover, oxidized LDL (ox-LDL), which is entrapped in the intima and isolated from plasma anti-oxidants, can induce a local inflammatory response. Indeed, there are recent data indicating that lipid disorders are strongly related to inflammatory processes, making inflammation the most important regulator in the development of cardiovascular diseases (Tousoulis et al. 2011; Tousoulis et al. 2006b). Thus, several inflammatory markers such as C reactive protein (CRP) and interleukin-6 (IL-6) acquire prognostic significance (Boekholdt et al. 2006).

This article reviews the current knowledge of the impact of hyperlipidemia in atherogenesis progression as well as potential markers of hyperlipidemia and several established therapeutic strategies, in that state.

## Lipids and Progression of Atherosclerosis

Vascular endothelium has emerged mostly as a paracrine organ responsible for the secretion of several beneficial substances and as a major regulator of vascular tone via manufacturing and balancing vasodilators and vasoconstrictive agents, including nitric oxide (NO), which contributed to the understanding of the pathogenesis of atherosclerosis. Endothelial dysfunction (ED) is considered as a consequence of an imbalance among those factors produced by or acting on endothelial cells (ECs) (Kampoli et al. 2012b; Bonetti et al. 2003; Gonzalez et al. 2003). In turn, impaired endothelial function is considered to play a crucial role in the initiation and progression of atherosclerosis. Traditional risk factors such as hyperlipidemia represent important factors predisposing to ED (Behrendt et al. 2002).

Emerging data have suggested that elevated LDL levels are associated with ED (Osborne et al. 1987; Kugiyama et al. 1990; Creager et al. 1990; Stiko-Rahm et al. 1990). According to experimental data, a high-cholesterol diet for two weeks reduced endothelium-dependent dilation and increased myocardial damage (Jeremy et al. 1996). Ox-LDL seems to exert greater negative effect on endothelial function than native LDL given the vasomotor responses to acetylcholine, serotonin, aggregating platelets, and thrombin (Simon et al. 1990; van der Zwan et al. 2009). On the contrary, HDL has been suggested to hamper the inhibitory effect of LDL on endothelium-mediated vasodilation (Matsuda et al. 1993). Moreover, hypercholesterolemia has been associated with increases in several inflammatory molecules and processes, such as endothelial adhesion, molecule expression (Hackman et al. 1996; Sampietro et al. 1997), platelet aggregation, and adhesion (Lacoste et al. 1995; Nofer et al. 1997). It has been indicated that superoxide production is increased in experimental induced hyperlipidemia (Ohara et al. 1993). Thus, it seems reasonable that hypercholesterolemia or atherosclerosis-associated ED is associated with decreased NO bioavailability even though some data also exist that contradict this assumption (Antoniades et al. 2003; Tousoulis et al. 2012b). More specifically, evidence has shown that antioxidant vitamin administration has no beneficial effect on endothelium function (Gilligan et al. 1994) and cardiovascular events (Jha et al. 1995).

It is now well established that atherosclerosis is a disease of large and medium-sized muscular arteries and is characterized by endothelial dysfunction, vascular inflammation, and the build-up of several components such as lipids, cholesterol, and cellular debris within the intima of the vessel wall. This build-up is considered to contribute to plaque formation, vascular remodeling, acute and chronic luminal obstruction (Tousoulis et al. 2003). Specifically, the primary lesion in atherosclerotic disease has its origins in childhood with the development of fatty streaks in the intima of the artery. Atherosclerotic lesions initiate at regions of low shear stress such as bifurcations of arteries. Also, reduced activity of NO seems to be greater towards high shear stress areas (Le Brocq et al. 2008).

Hyperlipidemia and/or abnormal lipoprotein metabolism are known to be closely related to atherosclerosis, and foam cells (lipid-laden macrophages) are recognized as a characteristic feature of the disease. It has been established that one of the important modifications of LDL which contributed to its recognition by the scavenger LDL receptor on macrophages was oxidative modification (Alexander 1995). In turn, ox-LDL seems to exert various biological effects on the vessel wall, such as stimulation of cytokine and growth factor production and inhibition of endothelial cell vasodilator function (Parhami et al. 1993; Tanner et al. 1991). Importantly, this notion is of crucial importance, as it provides mechanistic links between lipoproteins and cell biology of atherosclerosis; it also provides evidence that abnormalities of the redox state of the vasculature might be an important pathogenic mechanism in atherosclerosis.

## The Role of inflammation and Oxidative Stress in Atherosclerosis

It has become increasingly apparent that leukocytes and endothelial cells secrete several molecules stimulating inflammatory processes, such as migration and proliferation of smooth muscle cells (SMCs). In parallel, ox-LDL that has accumulated in the intima can induce a local inflammatory response (Libby et al. 2002). Moreover, other molecules, such as chemokines, produced in response to modified lipoproteins are involved in the migration of adherent monocytes (Ikeda et al. 2002; Tousoulis et al. 2008), and subsequently, infiltration of leukocytes through the endothelial layer and basement is induced (George 2000). Also, due to accumulation of cholesteryl-esters in the macrophages, they are converted into foam cells, which further stimulate inflammatory responses through the secretion of various growth factors and cytokines (Libby et al. 2002). Novel data suggest that elevated soluble CD40 ligand (sCD40L) levels, which are present in patients exhibiting hypercholesterolemia not only represent a risk factor for cardiovascular disease but also predict future adverse events. It seems that the binding of CD40L to CD40 plays an important role in the progression of atherosclerosis, stimulating inflammatory processes via the release of proinflammatory cytokines and the expression of adhesion molecules (Tousoulis et al. 2010a). Accordingly, cytokines and growth factors participate in the expansion of the lesion and extracellular matrix destabilization through the induction of proliferation and expression of metalloproteinases (MMPs) that degrade elastin and collagen (Kampoli et al. 2012a). The evolution of atherosclerosis ultimately leads to the formation of plaques with fibrous caps. It has been suggested that thinning of the fibrous cap is the consequence of increased collagen breakdown by MMPs stimulated mainly by proinflammatory cytokines; thus, plaques tend to rupture at sites of increased macrophage content leading to acute coronary syndromes as clinical presentations (Martinet and De Meyer 2008; Martinet and De Meyer 2009).

As aforementioned, oxidative stress is implicated in several cardiovascular risk factors including hyperlipidemia, contributing to vasoconstriction, thrombosis, plaque rupture, and vascular remodeling (Pashkow 2011). Superoxide anion radicals are produced in response to oxidative stress, which is stimulated by hypercholesterolemia contributing to the production of foam cells. Moreover, oxidative stress via hydrogen peroxide ($H_2O_2$) increases phosphorylation of tyrosin kinases, which then affects atherogenesis through the production of transcription factors such as nuclear factor-kB (NF-kB) and activator protein 1 (AP-1). These have been suggested to participate in several inflammatory processes, favoring the expression of adhesion molecules, such as vascular cellular adhesion molecules (VCAM-1), intracellular adhesion molecules (ICAM-1), E-selectin and other cytokines (Tousoulis et al. 2007b).

## Biomarkers in Hyperlipidemia

There is a growing interest focusing on the inflammatory response and mechanisms implicated in the pathogenesis and progression of cardiovascular disease. Recent studies have shown that several inflammatory markers may be proved useful in cardiovascular disease, while the possibility of combining circulating biomarkers with other methods such as noninvasive and invasive imaging is clinically attractive because this could contribute to the improved diagnosis, prognosis and understanding of premature atherosclerosis pathogenesis (Stefanadi et al. 2010; Tousoulis et al. 2012a; Naghavi et al. 2003) (Table 1). Concerning hyperlipidemia, recent experimental (Table 2) and clinical data have shed light on its association with inflammatory markers.

**Table 1.** Biomarkers in hyperlipidemia and their role in prognosis.

| Biomarker | Structure | Prognosis |
|---|---|---|
| CRP | Homopentameric structure | CAD, MI, stroke |
| Pro-inflammatory cytokines (IL-6, TNF-a) | Glycoproteins | CVD, MI, cardiovascular death and heart failure |
| Adhesion molecules (ICAM-VCAM) | Transmembrane glycoproteins | Future CAD, IMT |
| Lp-PLA2 | Calcium-independent member of the phospholipase A2 family | CVD, plaque progression |
| Chemokines | Chemotactic cytokines | CAD, coronary calcification, |
| CD40/CD40L | type I transmembrane/ type II membrane protein | CVD |
| Resistin, adiponectin, leptin | Adipocytes-derived cytokines | CV events (adiponectine) |

**Abbreviations:** CVD: cardiovascular disease; CAD: coronary artery disease; MI: myocardial infarction; ACS: acute coronary syndromes; IMT: intima-media thickness; IL: intereukin; CRP: C-reactive protein; Lp-PLA2: Lipoprotein-associated phospholipase A2; CD40L: CD40 ligand

## *C-reactive protein*

CRP is an acute-phase protein with a homopentameric structure the expression of which is regulated mainly by IL-6 during the acute phase (Lefkou et al. 2006; Verma et al. 2004). Of note, CRP may exert its proatherosclerotic effects only when dissociated from a pentameric to a monomeric form (Ridker et al. 2000). Moreover, CRP is a sensitive marker for the detection of subclinical inflammation and is well known for its implication in the pathophysiology of atherosclerosis. In particular, it contributes to the binding of LDL cholesterol from macrophages to form foam cells, while it stimulates the secretion of proinflammatory cytokines and tissue factor (Ridker et al. 2000; Cermak et al. 1993).

Experimental and clinical studies during the last decade have examined the role of inflammatory markers in the pathogenesis of atherosclerosis and have focused on the properties of CRP, its involvement in atherosclerotic process and its interactions with hyperlipidemia. Based on data derived from experimental models, CRP seems to be closely associated with lipid

**Table 2.** Studies investigating the association of biomarkers with hyperlipidemia.

| Biomarkers | Model | Comments |
|---|---|---|
| CRP (Murty et al. 2010) | Hypercholesterolemic rabbits | Cholesterol diet increased TC, TG, LDL-C and CRP while decreased HDL-C |
| CRP (Haghjooyjavanmard et al. 2008) | Hypercholesterolemic rabbits | Cholesterol diet increase TC, LDL-C, vWF and CRP, while CRP was correlated with TC levels. |
| TNF-α (Krauss et al. 1990) | Rats | Administration of TNF-α causes a rapid increase in VLDL and an delayed in LDL levels and mass. |
| TNF-α (Bartolome et al. 2007) | Cultured rat hepatocytes (*ex-vivo*) | TNF-α cause increase in secreted levels of VLDL and in cellular levels of apoB mRNA |
| IL-6, MCP-1, TNF-α (Wang et al. 2010) | LDL receptor null mice | Atherogenic diet increased levels of IL-6, MCP-1, TNF-α accompanied by increased serum non-HDL cholesterol levels |
| ICAM-1, VCAM-1, P-selectin (Scalia et al. 1998) | Rabbits | Cholesterol diet increases TC, ICAM-1, VCAM-1, P-selectin levels in the intestinal microvascular endothelium |
| CD40, CRP, MMP9, vWF (Mathapati et al. 2010) | Rabbits | High fat diet increased TC, TG, LDL-C and CRP serum levels with increased immunolocalization of CD40, MMP9, vWF in developing atherosclerotic plaque. |
| Resistin (Liu et al. 2008) | Mice overexpressing resistin | Resistin levels were accompanied by increased TG and decreased HDL-C and TC levels |

**Abbreviations:** TC: total cholesterol, TG: triglyceride, LDL-C: low density lipoprotein cholesterol, HDL-C: high density lipoprotein cholesterol, VLDL: very low density lipoprotein, apoB: apolipoprotein B, vWF: von Willebrand Factor, MCP-1: monocyte chemoattractant protein-1, ICAM: intercellular adhesion molecule-1, VCAM-1: vascular cell adhesion molecule, MMP9: matrix metalloproteinase 9

levels. For example, an increase in CRP, LDL cholesterol and triglycerides levels has been observed in experimentally induced hypercholesterolemic rabbits (Murty et al. 2010). Similarly, in another model of rabbit hypercholesterolemia, CRP levels were positively correlated with total cholesterol levels (Haghjooyjavanmard et al. 2008). In turn, in diabetic rats treated with chromium niacinate, a decrease in plasma levels of cholesterol and CRP has been observed, thereby highlighting the association between lipids and CRP (Jain et al. 2007).

Regarding clinical studies however, the association between hyperlipidemia and CRP still remains still to be clarified (Siasos et al. 2011). Significant data have supported the finding that the decrease in high sensitive CRP levels with statin administration is not associated with LDL cholesterol (Ray et al. 2005; Ridker et al. 1999; Albert et al. 2001). Moreover, the PROVE IT-TIMI 22 Trial, a cross-sectional study of 2,885 patients, has noted that the use of statin therapy led to a lower CRP level independent of the presence of single or multiple cardiovascular risk factors (Ray et al. 2005). In addition, in a randomized trial of 17,802 apparently healthy men and women with LDL cholesterol levels of less than 130 mg per deciliter, rosuvastatin administration, compared to placebo, decreased levels of high sensitivity CRP (hsCRP) and LDL cholesterol with little correlation between the two variables (Ridker et al. 2008). On the contrary, smaller population studies conducted on subjects with hyperlipidemia have supported an association between CRP and LDL cholesterol levels (Duarte et al. 2009; Alvarez-Sala et al. 2008) or CRP and triglyceride levels (Yudkin et al. 1999; Wang et al. 2011). Importantly, a meta-analysis, which included 23 randomized, placebo-controlled trials reporting changes in LDL and CRP with LDL-lowering interventions, has indicated that the anti-inflammatory effect of LDL-lowering therapies is related to the magnitude of change in LDL. Also, it seems that non-LDL effects of statins on inflammation are of minor magnitude (Kinlay et al. 2007). Indeed, in the light of the new, significant evidence from the JUPITER study, Canadians Cardiovascular Society guidelines have already recommended the initiation of statin therapy in individuals of "intermediate" cardiovascular risk and hsCRP>2mg/l (Genest et al. 2009).

## *Interleukin-6*

It is well established that cytokines are glycoproteins with a monomeric molecular mass of 15–25 kDa (kilo daltons) (Somers et al. 1997); the main cytokines participating in atherogenesis are IL-6 and tumor necrosis factor alpha (TNF-a). Regarding IL-6, it exerts proinflammatory effects participating in several processes, while its role in the pathopsysiology of atherosclerosis is not fully elucidated. Significant data have proposed its role in lipid profile in an experimental model of LDL receptor null mice. Treatment with a dual cyclooxygenase and 5-lipoxygenase inhibitor decreased triglycerides levels and expression of proatherogenic molecules such as IL-6 (Choi et al. 2010). Similarly, an atherogenic diet led to increased levels of non-HDL cholesterol and IL-6 levels, compared to mice fed a non-atherogenic diet (Wang et al. 2010). Recent evidence also suggests that spironolactone may suppress hepatic mRNA expression of IL-6, which is considered responsible for the suppression of serum lipids, in a model with diet-induced diabetes and nonalcoholic fatty liver disease (Wada et al. 2010). According to a population study which included 2,208 men aged 45–64 years, IL-6 may play a role in mediating the associations of circulating inflammatory markers with risk of coronary artery disease (CAD) (Patterson et al. 2010). Furthermore, according to a prospective, randomized study,

IL-6 could be an independent marker of increased mortality in unstable CAD and could discriminate patients who benefit most from a strategy of early invasive management (Lindmark et al. 2001).

## Tumor necrosis factor alpha (TNF-a)

Another cytokine, the TNF-a, which is a trimeric protein is also implicated in atherosclerosis. It is released by several cells participating in the inflammatory process and promotes the expression of cell adhesion molecules and consequently, the migration of leukocytes into and through the vascular wall (Tzoulaki et al. 2005; Szekanecz 2008). With respect to its association with lipids, it has been demonstrated that TNF-α may induce an increase in VLDL apo-B plasma levels (Krauss et al. 1990) and may also induce an increase in hepatocytes, VLDL secreted levels, and cellular levels of apo-B mRNA (Bartolome et al. 2007). Even though the association between lipids metabolism and TNF-α is supported by several experimental studies (Zhang et al. 2010), limited clinical data have been reported in this field.

Of note, most accumulating data have focused on the role of TNF-α as a risk predictor for cardiovascular disease and heart failure (Sun et al. 2007; Kampoli et al. 2009; Oikonomou et al. 2011). According to data derived from the Health, Aging, and Body Composition study, TNF-a may be correlated with coronary heart disease (Cesari et al. 2003). Moreover, plasma TNF-a levels were associated with lipemia, plasma VLDL triglyceride, LDL cholesterol concentrations, and peak LDL particle size, whereas they were not independent determinants of common carotid intima-media thickness (Skoog et al. 2002).

Additionally, a prospective study to determine whether biomarkers can enhance the predictive value of classic atherosclerosis risk factors, has recently demonstrated that TNF-a may have additive value to classic risk factors in cardiovascular risk stratification (Kablak-Ziembicka et al. 2011). It has been observed that circulating levels of TNF-a were elevated in those with previous MI and the association retained after adjustment for classical risk factors, suggesting the hypothesis of a correlation between TNF-α and ongoing heart remodeling (Welsh et al. 2009). However data from other studies contradict the role of TNF-α as cardiovascular risk predictor (Sukhija et al. 2007; Kilic et al. 2006).

## Adhesion molecules

It is well known that intercellular interactions are mediated by transmembrane glycoproteins, the cell adhesion molecules (CAMs). Their basic structure consists of an extracellular component, which is accompanied by a hydrophobic transmembrane component and an intracytoplasmic component. Of note, there are three families of cell adhesion molecules including the integrins, the immunoglobulin superfamily and the selectins (Tousoulis et al. 2011; Price and Loscalzo 1999; Tousoulis et al. 2005b). There are several molecules including vascular cell adhesion molecule-1 (VCAM-1), intercellular adhesion molecule-1 (ICAM-1), as well as selectins that are expressed on the cell surface and participate in the cell's adhesion to other cells and to the extracellular matrix. Interestingly, it is well described that long term hypercholesterolemia increases the expression of P-selectin, VCAM-1, and ICAM-1 from microvascular and aortic endothelial cells (Scalia et al. 1998; Verna et al. 2006). Moreover, given experimental studies, lipid-lowering agents such as statins result in a reduction in cholesterol and triglycerides levels, accompanied by a decrease in the expression of VCAM-1 and ICAM-1 (Nachtigal et al. 2008). In accordance, the effect of hypolipidemic agents has resulted in tissue reduction in the expression of adhesion molecules (Do et al. 2008). Of note, significant reports have indicated the contribution of adhesion molecules in atherosclerotic low-grade inflammation, while VCAM-1 and ICAM-1 levels are considered as predictive markers for coronary events in patients with atherosclerosis (Vaverkova et al. 2008; Blankenberg et al. 2001; de Lemos et al. 2000; Tousoulis et al. 2007a). In addition, adhesion molecules have correlated with lipid levels in human subjects. This is nicely exemplified by data supporting elevated levels of soluble adhesion molecules in young, healthy men with hyper-triglyceridemia compared to a control group (Lundman et al. 2003).

## *Lipoprotein-associated phospholipase A2 (Lp-PLA2)*

Lipoprotein-associated phospholipase A2 (Lp-PLA2), a 45.4-kDa protein, is considered a calcium-independent member of the phospholipase A2 family, which is produced mainly by monocytes, macrophages and T-lymphocytes, and potentially is upregulated in atherosclerotic lesions (Tousoulis et al. 2011). Lp-PLA2, currently recognized as a novel inflammatory biomarker, is also bound predominantly to apo-B containing lipoproteins and has been implicated in the pathogenesis of ED and atherosclerosis (Tousoulis et al. 2012a; Ballantyne et al. 2004). According to experimental data, Lp-PLA2 activity mediates leukocyte activation and inflammatory responses *in vitro*, whereas its inhibition hampers ox-LDL inflammatory responses, thus supporting a proatherogenic role for Lp-PLA2 (Shi et al. 2007). Furthermore, regarding clinical studies, a randomized, double-blind, controlled clinical trial designed to investigate the effects of 3-month treatment with fenofibrate on Lp-PLA2 mass has shown a reduced Lp-PLA2 mass; these changes were associated with fewer, small total LDL particles (Rosenson 2008).

Furthermore, it is worth mentioning that Lp-PLA2 activity is influenced by LDL cholesterol and apo-B100 levels, suggesting a crucial role in lipid peroxidation (Kampoli et al. 2009). Regarding its prognostic role, large clinical trials have demonstrated an increased risk for future cardiovascular events in subjects with high levels of Lp-PLA2 (Siasos et al. 2011).

## sCD40 ligand

CD40 is a type I transmembrane receptor which exists in trimers, property which is crucial for its activation, while CD40L protein is a type II membrane protein, the extracellular region of which belongs to the TNF superfamily. Also, it is stored in platelets' granules, and then rapidly translocates to the platelets surface when they are activated and then cleaved and discarded from the surface as sCD40L. A variety of cells involved in atherosclerosis, though mainly B-cells, express both CD40 and CD40L. As a result of CD40 ligation several inflammatory processes are initiated along with a wide variety of biological functions *in vitro* and *in vivo*, intimately involved in atherogenesis (Tousoulis et al. 2010a; Antoniades et al. 2009). Notably, data from experimental models have shown that interventions to treat hyperlipidemia are followed by reduction in CD40 levels, whereas high cholesterol diet increases CD40 immunopositivity in early atherosclerotic plaque, with analogous increase in total and LDL cholesterol and triglyceride levels (Liu et al. 2009; Mathapati et al. 2010). Moreover, in a mouse model, hypercholesterolemic diet has contributed to leukocyte adhesion in arterioles and arteriolar dysfunction, while in CD40$^-$/CD40L$^-$ mice this diet has had no effect on leukocyte adhesion (Stokes et al. 2009).

## Adipocytes-derived cytokines

Resistin, adiponectin and leptin are cytokines produced by fatty tissue and seem to exert proinflammatory and anti-inflammatory effects (Siasos et al. 2011). Resistin is secreted by adipocytes and has been shown to possess proinflammatory properties by upregulating proinflammatory cytokines, probably via the NF-kB pathway (Bokarewa et al. 2005). Experimental data have shown that over-expression of resistin contributed to increased triglycerides levels and to decreased HDL and total cholesterol concentrations (Liu et al. 2008). In clinical studies, resistin levels have been associated with inflammatory markers (Shetty et al. 2004), while in diabetic patients they were predictive of future cardiovascular events and were correlated with inflammatory markers (Reilly et al. 2005). Moreover, adiponectin, which is exclusively expressed by mature adipocytes, suppresses the secretion of inflammatory molecules, such as TNF-α (Calabro et al. 2004; Fantuzzi et al.

2005). In an experimental model, diet with fibers resulted in a significant reduction in plasma cholesterol and triglyceride levels which was correlated with an increase in plasma adiponectin. Notably, adiponectin was also correlated with HDL and apo-A1 and inversely correlated with triglycerides and apo-B in CAD (Siasos et al. 2011). Leptin is mainly produced by white adipose tissue and mainly controls fat stores and contributes to the regulation of immune cells participating in the inflammatory process (Fantuzzi and Faggioni 2000). Indeed, according to a hamster model, dietary fiber significantly decreased plasma leptin levels which were also correlated with a reduction in plasma cholesterol and triglycerides levels (Hung et al. 2009). Furthermore, omega-6 polyunsaturated fatty acids had increased lipids parameters and increased expression of leptin gene compared to mice fed with omega-3 polyunsaturated fatty acids and normal diet (Magdeldin et al. 2009).

## Hypolipidemic Agents in Relation to Biomarkers

Current data provide encouraging evidence supporting that lifestyle modifications and more specifically diet, physical activity, and behavior therapy may have a significant impact on reducing lipid parameters and biomarkers associated with hyperlipidemia (Papageorgiou et al. 2012). Furthermore, various pharmacological therapies have been studied in respect to their actions in biomarkers, mainly via mechanisms intended to increase NO synthase activity and NO release, and to ameliorate the inflammatory state in atherosclerosis (Briasoulis et al. 2012).

### *Statins*

Statins are inhibitors of the HMG CoA reductase, blocking access of this substrate to the active site of the enzyme (Istvan and Deisenhofer 2001). These agents have been successfully used for lowering the levels of lipids in plasma; however, the beneficial effects of statins are related not only to lipid-lowering but also to direct effects on endothelial function, as well as antithrombotic and anti-inflammatory effects (Tousoulis et al. 2005a; Tousoulis et al. 2006a; Tousoulis et al. 2005b). Indeed, in heart failure patients, atorvastatin treatment decreased serum concentrations of several inflammatory biomarkers, such as IL-6, TNF-α, monocyte chemoattractant protein-1, and sVCAM-1, with an increase in endothelial dependent reactive hyperaemia. Moreover, it has been indicated that statins may negatively affect the activity of AP-1 which has an important role in endothelial inflammatory responses, regulating genes responsible for matrix metalloproteinases, cytokines, chemokines, and adhesion molecules

(Dichtl et al. 2003). The multicenter JUPITER was study designed after the observation that low density lipoprotein and CRP are associated with better outcome than only when LDL is decreased. Rosuvastatin 20 mg has been evaluated in comparison to placebo on the prespecified end points according to on-treatment values of LDL (<70 or ≥ 70 mg/dL) and CRP (<2 mg/L, or <1 mg/L) (Ridker et al. 2009). Moreover, data derived from several other randomized trials have also highlighted the anti-inflammatory and anti-atherogenic properties of statins. This is exemplified by PROVE IT–TIMI 22 study, which included 4,162 patients with acute coronary syndrome, treated with pravastatin (40 mg daily) or atorvastatin (80 mg daily) with a follow-up for 24 months. Notably, this study has indicated that clinical outcomes were better in subjects with LDL cholesterol levels below 70 mg/dl and hsCRP levels below 2 mg/l, thereby suggesting dual mechanisms of the benefit of statin therapy, both via LDL cholesterol lowering and direct anti-inflammatory effect (Ridker et al. 2005).

## *Other pharmacological agents*

Thus, these data have strongly supported the efficacy of intensive statin therapy in secondary prevention and the hypothesis that they exert additive effects in addition to lipid-lowering effects, whereas other agents, including fibrates, omega-3 fatty acids, niacin, ezetimibe, and experimental cholesteryl-ester transfer protein inhibitors have been evaluated for their ability to reduce residual cardiovascular risk (Gotto et al. 2012). Ezetimibe is a cholesterol absorption inhibitor targeting uptake at the jejunal enterocyte brush border, an action which is managed by the selective blockage of the sterol transporter Niemann–Pick C1-like 1 (NPC1L1) protein (Phan et al. 2012). In a study, both treatment with ezetimibe and pravastatin reduced LDL-C and improved endothelial function in hypercholesterolemic patients, suggesting that the observed effect can be explained by the reduction of cholesterol levels (Grigore et al. 2012). Reasonably, beyond lipid lowering effect, ezetimibe may have significant impact on the prevention of atherosclerotic disease, modulating several inflammatory processes. Several studies have supported the notion that ezetimibe may cause a decrease in hsCRP levels as monotherapy or in combination with simvastatin (Sager et al. 2003; Pearson et al. 2007; Davidson et al. 2002). According to a meta-analysis based on currently available data, combining a low-dose statin with ezetimibe provides similar beneficial effects on endothelial function as high-dose statin (Ye et al. 2012). Similarly, combination of ezetimibe with rosuvastatin resulted in a greater reduction in hsCRP levels compared to rosuvastatine alone (Ballantyne et al. 2007). Thus, ezetimibe may have a synergistic effect on hsCRP levels when combined with statins; however, the precise mechanisms of these interactions are still elusive.

There is considerable evidence that n-3 PUFAs (polyunsaturated fatty acids), which are found in fatty fish and in fish oils may decrease the risk of cardiovascular mortality. Notably they have been demonstrated to decrease triglycerides, several anti-inflammatory biomarkers, and total cardiovascular mortality, with a potent effect on sudden death (Calder et al. 2004). According to data summarized from 33 intervention trials investigating the effects of n-3 PUFAs on fasting and/or postprandial endothelial function, they potentially exert a beneficial effect on endothelial function. However, current clinical efficacy does not justify recommendations with respect to a specific dose and the duration of supplementation (Egert and Stehle 2011). Moreover, even though the effects of nicotinic acid on plasma lipoproteins have long been evaluated, the underlying mechanisms of its action are not fully understood. Interestingly, it improves plasma lipids levels as well as proatherogenic lipoprotein(a), while recent human studies have suggested a potential anti-inflammatory role in increasing plasma adiponectin levels which, as aforementioned, has antiatherogenic and anti-inflammatory properties (Digby et al. 2009; Plaisance et al. 2009).

## Conclusion

The current concept of atherosclerosis has evolved beyond the view that these lesions consist of a collection of lipid debris. Also, important evidence advocates a central role for inflammation in all phases of the atherosclerotic process and a potent link between inflammatory status and hyperlipidemia. However, a direct causality relationship between hyperlipidemia and inflammation has not been established yet. Risk assessment of atherosclerosis can be improved through a biochemical marker approach. The effectiveness of some biomarkers in cardiovascular diseases is already well established, but for many of them, additional research and evaluation are needed. Furthermore, several biomarkers are associated with lipid levels, though it remains elusive as to which of them are best associated with the results of hyperlipidemia in the arterial wall and which pharmacological agent(s) can benefit the cardiovascular system and inverse the atherosclerotic process.

## References

Albert, M.A., E. Danielson, N. Rifai, and P.M. Ridker. 2001. Effect of statin therapy on C-reactive protein levels: the pravastatin inflammation/CRP evaluation (PRINCE): a randomized trial and cohort study. JAMA 286: 64–70.

Alexander, R.W. Theodore Cooper Memorial Lecture. 1995. Hypertension and the pathogenesis of atherosclerosis. Oxidative stress and the mediation of arterial inflammatory response: a new perspective. Hypertension 25: 155–161.

Alvarez-Sala, L.A., V. Cachofeiro, L. Masana, C. Suarez, B. Pinilla, N. Plana, F. Trias, M.A. Moreno, G. Gambus, V. Lahera, and X. Pintó. 2008. Effects of fluvastatin extended-release (80 mg) alone and in combination with ezetimibe (10 mg) on low-density lipoprotein cholesterol and inflammatory parameters in patients with primary hypercholesterolemia: a 12-week, multicenter, randomized, open-label, parallel-group study. Clin. Ther. 30: 84–97.

Antoniades, C., D. Tousoulis, C. Tentolouris, P. Toutouzas, and C. Stefanadis. 2003. Oxidative stress, antioxidant vitamins, and atherosclerosis. From basic research to clinical practice. Herz 28: 628–638.

Antoniades, C., C. Bakogiannis, D. Tousoulis, A.S. Antonopoulos, and C. Stefanadis. 2009. The CD40/CD40 ligand system: linking inflammation with atherothrombosis. J. Am. Coll. Cardiol. 54: 669–677.

Ascaso, J., S.P. Gonzalez, P. Santos, A. Hernandez Mijares, A. Mangas Rojas, L. Masana, J. Millan, L.F. Pallardo, J. Pedro-Botet, F. Perez Jimenez, X. Pintó, I. Plaza, J. Rubiés, and M. Zúñiga. 2007. Management of dyslipidemia in the metabolic syndrome: recommendations of the Spanish HDL-Forum. Am. J. Cardiovasc. Drugs. 7: 39–58.

Ballantyne, C.M., R.C. Hoogeveen, H. Bang, J. Coresh, A.R. Folsom, G. Heiss, and A.R. Sharrett. 2004. Lipoprotein-associated phospholipase A2, high-sensitivity C-reactive protein, and risk for incident coronary heart disease in middle-aged men and women in the Atherosclerosis Risk in Communities (ARIC) study. Circulation 109: 837–842.

Ballantyne, C.M., R. Weiss, T. Moccetti, A. Vogt, B. Eber, F. Sosef, and E. Duffield. EXPLORER Study Investigators. 2007. Efficacy and safety of rosuvastatin 40 mg alone or in combination with ezetimibe in patients at high risk of cardiovascular disease (results from the EXPLORER study). Am. J. Cardiol. 99: 673–680.

Bartolome, N., L. Rodriguez, M.J. Martinez, B. Ochoa, and Y. Chico. 2007. Upregulation of apolipoprotein B secretion, but not lipid, by tumor necrosis factor-alpha in rat hepatocyte cultures in the absence of extracellular fatty acids. Ann. N.Y. Acad. Sci. 1096: 55–69.

Behrendt, D. and P. Ganz. 2002. Endothelial function. From vascular biology to clinical applications. Am. J. Cardiol. 90: 40L–48L.

Blankenberg, S., H.J. Rupprecht, C. Bickel, D. Peetz, G. Hafner, L. Tiret, and J. Meyer. 2001. Circulating cell adhesion molecules and death in patients with coronary artery disease. Circulation 104: 1336–1342.

Boekholdt, S.M., C.E. Hack, M.S. Sandhu, R. Luben, S.A. Bingham, N.J. Wareham, R.J. Peters, H.W. Jukema, N.E. Day, J.J. Kastelein, and K.T. Khaw. 2006. C-reactive protein levels and coronary artery disease incidence and mortality in apparently healthy men and women: the EPIC-Norfolk prospective population study 1993–2003. Atherosclerosis 187: 415–422.

Bokarewa, M., I. Nagaev, L. Dahlberg, U. Smith, and A. Tarkowski. 2005. Resistin, an adipokine with potent proinflammatory properties. J. Immunol. 174: 5789–5795.

Bonetti, P.O., L.O. Lerman, and A. Lerman. 2003. Endothelial dysfunction: a marker of atherosclerotic risk. Arterioscler. Thromb. Vasc. Biol. 23: 168–175.

Briasoulis, A., D. Tousoulis, E.S. Androulakis, N. Papageorgiou, G. Latsios, and C. Stefanadis. 2012. Endothelial dysfunction and atherosclerosis: focus on novel therapeutic approaches. Recent Pat. Cardiovasc. Drug Discov. 7: 21–32.

Calabro, P., I. Samudio, J.T. Willerson, and E.T. Yeh. 2004. Resistin promotes smooth muscle cell proliferation through activation of extracellular signal-regulated kinase 1/2 and phosphatidylinositol 3-kinase pathways. Circulation 110: 3335–3340.

Calder, P.C. 2004. n–3 Fatty acids and cardiovascular disease: evidence explained and mechanisms explored. Clin. Sci. (Lond.) 107: 1–11.

Cermak, J., N.S. Key, R.R. Bach, J. Balla, H.S. Jacob, and G.M. Vercellotti. 1993. C-reactive protein induces human peripheral blood monocytes to synthesize tissue factor. Blood 82: 513–520.

Cesari, M., B.W. Penninx, A.B. Newman, S.B. Kritchevsky, B.J. Nicklas, K. Sutton-Tyrrell, S.M. Rubin, J. Ding, E.M. Simonsick, T.B. Harris, and M. Pahor. 2003. Inflammatory markers and onset of cardiovascular events: results from the Health ABC study. Circulation 108: 2317–2322.

Chait, A. and J.D. Brunzell. 1990. Acquired hyperlipidemia (secondary dyslipoproteinemias). Endocrinol. Metab. Clin. North. Am. 19: 259–278.

Choi, J.H., H.J. Jeon, J.G. Park, S.K. Sonn, M.R. Lee, M.N. Lee, H.J. You, G.Y. Kim, J.H. Kim, M.H. Lee, O.S. Kwon, K.H. Nam, H.C. Kim, T.S. Jeong, W.S. Lee, and G.T. Oh. 2010. Anti-atherogenic effect of BHB-TZD having inhibitory activities on cyclooxygenase and 5-lipoxygenase in hyperlipidemic mice. Atherosclerosis 212: 146–152.

Creager, M.A., J.P. Cooke, M.E. Mendelsohn, S.J. Gallagher, S.M. Coleman, J. Loscalzo, and V.J. Dzau. 1990. Impaired vasodilation of forearm resistance vessels in hypercholesterolemic humans. J. Clin. Invest. 86: 228–234.

Davidson, M.H., T. McGarry, R. Bettis, L. Melani, L.J. Lipka, A.P. LeBeaut, R. Suresh, S. Sun, and E.P. Veltri. 2002. Ezetimibe coadministered with simvastatin in patients with primary hypercholesterolemia. J. Am. Coll. Cardiol. 40: 2125–2134.

de Lemos, J.A., C.H. Hennekens, and P.M. Ridker. 2000. Plasma concentration of soluble vascular cell adhesion molecule-1 and subsequent cardiovascular risk. J. Am. Coll. Cardiol. 36: 423–426.

Dichtl, W., J. Dulak, M. Frick, H.F. Alber, S.P. Schwarzacher, M.P. Ares, J. Nilsson, O. Pachinger, and F. Weidinger. 2003. HMG-CoA reductase inhibitors regulate inflammatory transcription factors in human endothelial and vascular smooth muscle cells. Arterioscler. Thromb. Vasc. Biol. 23: 58–63.

Digby, J.E., J.M. Lee, and R.P. Choudhury. 2009. Nicotinic acid and the prevention of coronary artery disease. Curr. Opin. Lipidol. 20: 321–326.

Do, G.M., E.Y. Kwon, H.J. Kim, S.M. Jeon, T.Y. Ha, T. Park, and M.S. Choi. 2008. Long-term effects of resveratrol supplementation on suppression of atherogenic lesion formation and cholesterol synthesis in apo E-deficient mice. Biochem. Biophys. Res. Commun. 374: 55–59.

Duarte, M.M., J.B. Rocha, R.N. Moresco, T. Duarte, I.B. Da Cruz, V.L. Loro, and M.R. Schetinger. 2009. Association between ischemia-modified albumin, lipids and inflammation biomarkers in patients with hypercholesterolemia. Clin. Biochem. 42: 666–671.

Egert, S. and P. Stehle. 2011. Impact of n–3 fatty acids on endothelial function: results from human interventions studies. Curr. Opin. Clin. Nutr. Metab. Care 14: 121–131.

Fantuzzi, G. 2005. Adipose tissue, adipokines, and inflammation. J. Allergy Clin. Immunol. 115: 911–919.

Fantuzzi, G. and R. Faggioni. 2000. Leptin in the regulation of immunity, inflammation, and hematopoiesis. J. Leukoc. Biol. 68: 437–446.

Genest, J., R. McPherson, J. Frohlich, T. Anderson, N. Campbell, A. Carpentier, P. Couture, R. Dufour, G. Fodor, G.A. Francis, S. Grover, M. Gupta, R.A. Hegele, D.C. Lau, L. Leiter, G.F. Lewis, E. Lonn, G.B. Mancini, D. Ng, G.J. Pearson, A. Sniderman, J.A. Stone, and E. Ur. 2009. Canadian Cardiovascular Society/Canadian guidelines for the diagnosis and treatment of dyslipidemia and prevention of cardiovascular disease in the adult—2009 recommendations. Can. J. Cardiol. 25: 567–579.

George, S.J. 2000. Therapeutic potential of matrix metalloproteinase inhibitors in atherosclerosis. Expert. Opin. Investig. Drugs 9: 993–1007.

Gilligan, D.M., M.N. Sack, V. Guetta, P.R. Casino, A.A. Quyyumi, D.J. Rader, J.A. Panza, and R.O. Cannon III. 1994. Effect of antioxidant vitamins on low density lipoprotein oxidation and impaired endothelium-dependent vasodilation in patients with hypercholesterolemia. J. Am. Coll. Cardiol. 24: 1611–1617.

Gonzalez, M.A. and A.P. Selwyn. 2003. Endothelial function, inflammation, and prognosis in cardiovascular disease. Am. J. Med. 115 Suppl. 8A: 99S–106S.

Gotto, A.M. Jr. and J.E. Moon. 2012. Recent clinical studies of the effects of lipid-modifying therapies. Am. J. Cardiol. 110(1 Suppl): 15A–26A.

Grigore, L., S. Raselli, K. Garlaschelli, L. Redaelli, G.D. Norata, A. Pirillo, and A.L. Catapano. 2012. Effect of treatment with pravastatin or ezetimibe on endothelial function in patients with moderate hypercholesterolemia. Eur. J. Clin. Pharmacol. Jul 10.

Hackman, A., Y. Abe, W. Jr. Insull, H. Pownall, L. Smith, K. Dunn, A.M. Jr. Gotto, and C.M. Ballantyne. 1996. Levels of soluble adhesion molecules in patients with dyslipidemia. Circulation 93: 1334–1338.

Haghjooyjavanmardm, S., M. Nematbakhsh, A. Monajemi, and M. Soleimani. 2008. von Willebrand factor, C-reactive protein, nitric oxide, and vascular endothelial growth factor in a dietary reversal model of hypercholesterolemia in rabbit. Biomed. Pap. Med. Fac. Univ. Palacky Olomouc. Czech Repub. 152: 91–95.

Hung, S.C., G. Bartley, S.A. Young, D.R. Albers, D.R. Dielman, W.H. Anderson, and W. Yokoyama. 2009. Dietary fiber improves lipid homeostasis and modulates adipocytokines in hamsters. J. Diabetes 1: 194–206.

Ikeda, U., K. Matsui, Y. Murakami, and K. Shimada. 2002. Monocyte chemoattractant protein-1 and coronary artery disease. Clin. Cardiol. 25: 143–147.

Istvan, E.S. and J. Deisenhofer. Structural mechanism for statin inhibition of HMG-CoA reductase. Science 2001; 292: 1160–1164.

Jain, S.K., J.L. Rains, and J.L. Croad. 2007. Effect of chromium niacinate and chromium picolinate supplementation on lipid peroxidation, TNF-alpha, IL-6, CRP, glycated hemoglobin, triglycerides, and cholesterol levels in blood of streptozotocin-treated diabetic rats. Free Radic. Biol. Med. 43: 1124–1131.

Jeremy, R.W., H. McCarron, and D. Sullivan. 1996. Effects of dietary L-arginine on atherosclerosis and endothelium-dependent vasodilation in the hypercholesterolemic rabbit: Response according to treatment duration, anatomic site, and sex. Circulation 94: 498–506.

Jha, P., M. Flather, E. Lonn, M. Farkouh, and S. Yusuf. 1995. The antioxidant vitamins and cardiovascular disease: A critical review of epidemiologic and clinical trial data. Ann. Intern. Med. 123: 860–872.

Kablak-Ziembicka, A., T. Przewlocki, A. Sokolowski, W. Tracz, and P. Podolec. 2011. Carotid intima-media thickness, hs-CRP and TNF-alpha are independently associated with cardiovascular event risk in patients with atherosclerotic occlusive disease. Atherosclerosis 214: 185–190.

Kampoli, A.M., D. Tousoulis, C. Antoniades, G. Siasos, and C. Stefanadis. 2009. Biomarkers of premature atherosclerosis. Trends Mol. Med. 15: 323–332.

Kampoli, A.M., D. Tousoulis, N. Papageorgiou, C. Antoniades, E. Androulakis, E. Tsiamis, G. Latsios, and C. Stefanadis. 2012a. Matrix metalloproteinases in acute coronary syndromes: current perspectives. Curr. Top. Med. Chem. 12: 1192–1205.

Kampoli, A.M., D. Tousoulis, N. Papageorgiou, Z. Pallatza, G. Vogiatzi, A. Briasoulis, E. Androulakis, C. Toutouzas, P. Stougianos, C. Tentolouris, and C. Stefanadis. 2012b. Clinical utility of biomarkers in premature atherosclerosis. Curr. Med. Chem. 19: 2521–2533.

Kilic, T., D. Ural, E. Ural, Z. Yumuk, A. Agacdiken, T. Sahin, G. Kahraman, G. Kozdag, A. Vural, and B. Komsuoglu. 2006. Relation between proinflammatory to anti-inflammatory cytokine ratios and long-term prognosis in patients with non-ST elevation acute coronary syndrome. Heart 92: 1041–1046.

Kinlay, S. 2007. Low-density lipoprotein-dependent and -independent effects of cholesterol-lowering therapies on C-reactive protein: a meta-analysis. J. Am. Coll. Cardiol. 49: 2003–2009.

Krauss, R.M. Atherogenicity of triglyceride-rich lipoproteins. 1998. Am. J. Cardiol. 81: 13B–7B.

Krauss, R.M., C. Grunfeld, W.T. Doerrler, and K.R. Feingold. 1990. Tumor necrosis factor acutely increases plasma levels of very low density lipoproteins of normal size and composition. Endocrinology 127: 1016–1021.

Kugiyama, K., S.A. Kerns, J.D. Morrisett, R. Roberts, and P.D. Henry. 1990. Impairment of endothelium-dependent relaxation by lysolecithin in modified low-density lipoproteins. Nature 334: 160–162.
Lacoste, L., J.Y. Lam, J. Hung, G. Letchacovski, C.B. Solymoss, and D. Waters. 1995. Hyperlipidemia and coronary disease: Correction of the increased thrombogenic potential with cholesterol reduction. Circulation 92: 3172–3177.
Le Brocq, M., S.J. Leslie, P. Milliken, and I.L. Megson. 2008. Endothelial dysfunction: from molecular mechanisms to measurement, clinical implications, and therapeutic opportunities. Antioxid. Redox Signal. 9: 1631–1674.
Lefkou, E., N. Fragakis, and G. Varlamis. 2006. The inflammatory process during childhood and the risk of developing atheromatous disease in adult life: the role of C-reactive protein. Hellenic J. Cardiol. 47: 164–169.
Libby, P., P.M. Ridker, and A. Maseri. 2002. Inflammation and atherosclerosis. Circulation 105: 1135–1143.
Lindmark, E., E. Diderholm, L. Wallentin, and A. Siegbahn. 2001. Relationship between interleukin 6 and mortality in patients with unstable coronary artery disease: effects of an early invasive or noninvasive strategy. JAMA 286: 2107–2113.
Liu, G., B. Wang, J. Zhang, H. Jiang, and F. Liu. 2009. Total panax notoginsenosides prevent atherosclerosis in apolipoprotein E-knockout mice: Role of downregulation of CD40 and MMP-9 expression. J. Ethnopharmacol. 126: 350–354.
Liu, Y., Q. Wang, Y.B. Pan, Z.J. Gao, Y.F. Liu, and S.H. Chen. 2008. Effects of over-expressing resistin on glucose and lipid metabolism in mice. J. Zhejiang Univ. Sci. B. 9: 44–50.
Lundman, P., M.J. Eriksson, A. Silveira, L.O. Hansson, J. Pernow, C.G. Ericsson, A. Hamsten, and P. Tornvall. 2003. Relation of hypertriglyceridemia to plasma concentrations of biochemical markers of inflammation and endothelial activation (C-reactive protein, interleukin-6, soluble adhesion molecules, von Willebrand factor, and endothelin-1). Am. J. Cardiol. 91: 1128–1131.
Magdeldin, S., Y. Elewa, T. Ikeda, J. Ikei, Y. Zhang, B. Xu, M. Nameta, H. Fujinaka, Y. Yoshida, E. Yaoita, and T. Yamamoto. 2009. Dietary supplementation with arachidonic acid but not eicosapentaenoic or docosahexaenoic acids alter lipids metabolism in C57BL/6J mice. Gen. Physiol. Biophys. 28: 266–275.
Martinet, W. and G.R. De Meyer. 2008. Autophagy in atherosclerosis. Curr. Atheroscler. Rep. 10: 216–223.
Martinet, W. and G.R. De Meyer. 2009. Autophagy in atherosclerosis: a cell survival and death phenomenon with therapeutic potential. Circ. Res. 104: 304–317.
Mathapati, S., S.B. Arumugam, and R.S. Verma. 2010. High cholesterol diet increases MMP9 and CD40 immunopositivity in early atherosclerotic plaque in rabbits. Acta Histochem. 112: 618–623.
Matsuda, Y., K. Hirata, N. Inoue, M. Suematsu, S. Kawashima, H. Akita, and M. Yokoyama. 1993. High density lipoprotein reverses inhibitory effect of oxidized low density lipoprotein on endothelium-dependent arterial relaxation. Circ. Res. 72: 1103–1109.
Murty, D., E. Rajesh, D. Raghava, T.V. Raghavan, and M.K. Surulivel. 2010. Hypolipidemic effect of arborium plus in experimentally induced hypercholestermic rabbits. Yakugaku Zasshi 130: 841–846.
Nachtigal, P., N. Pospisilova, G. Jamborova, K. Pospechova, D. Solichova, C. Andrys, P. Zdansky, S. Micuda, and V. Semecky. 2008. Atorvastatin has hypolipidemic and anti-inflammatory effects in apoE/LDL receptor-double-knockout mice. Life Sci. 82: 708–717.
Naghavi, M., P. Libby, E. Falk, S.W. Casscells, S. Litovsky, J. Rumberger, J.J. Badimon, C. Stefanadis, P. Moreno, G. Pasterkamp, Z. Fayad, P.H. Stone, S. Waxman, P. Raggi, M. Madjid, A. Zarrabi, A. Burke, C. Yuan, P.J. Fitzgerald, D.S. Siscovick, C.L. de Korte, M. Aikawa, K.E. Airaksinen, G. Assmann, C.R. Becker, J.H. Chesebro, A. Farb, Z.S. Galis, C. Jackson, I.K. Jang, W. Koenig, R.A. Lodder, K. March, J. Demirovic, M. Navab, S.G. Priori, M.D. Rekhter, R. Bahr, S.M. Grundy, R. Mehran, A. Colombo, E. Boerwinkle,

C. Ballantyne, W. Jr. Insull, R.S. Schwartz, R. Vogel, P.W. Serruys, G.K. Hansson, D.P. Faxon, S. Kaul, H. Drexler, P. Greenland, J.E. Muller, R. Virmani, P.M. Ridker, D.P. Zipes, P.K. Shah, and J.T. Willerson. 2003. From vulnerable plaque to vulnerable patient: a call for new definitions and risk assessment strategies: Part II. Circulation 108: 1772–1778.

Nofer, J.R., M. Tepel, B. Kehrel, S. Wierwille, M. Walter, U. Seedorf, W. Zidek, and G. Assmann G. 1997. Low-density lipoproteins inhibit the Na–/H + antiport in human platelets: A novel mechanism enhancing platelet activity in hypercholesterolemia. Circulation 95: 1370–1377.

Ohara, Y., T.E. Pederson, and D.G. Harrison. 1993. Hypercholesterolemia increases endothelial superoxide production. J. Clin. Invest. 91: 2546–2551.

Oikonomou, E., D. Tousoulis, G. Siasos, M. Zaromitidou, A.G. Papavassiliou, and C. Stefanadis. 2011. The role of inflammation in heart failure: new therapeutic approaches. Hellenic J. Cardiol. 52: 30–40.

Osborne, J.A., M.J. Siegman, A.W. Sedar, S.U. Mooers, and A.M. Lefer. Lack of endothelium-dependent relaxation in coronary resistance arteries of cholesterol-fed rabbits. 1987. Am. J. Physiol. 256: C591–C597.

Papageorgiou, N., D. Tousoulis, E. Androulakis, A. Giotakis, G. Siasos, G. Latsios, and C. Stefanadis. 2012. Lifestyle factors and endothelial function. Curr. Vasc. Pharmacol. 10: 94–106.

Parhami, F., Z.T. Fang, A.M. Fogelman, A. Andalibi, M.C. Territo, and J.A. Berliner. 1993. Minimally modified low density lipoprotein-induced inflammatory responses in endothelial cells are mediated by cyclic adenosine monophosphate. J. Clin. Invest. 92: 471–478.

Pashkow, F.J. 2011. Oxidative stress and inflammation in heart disease: Do antioxidants have a role in treatment and/or prevention? Int. J. Inflam. 2011: 514–623.

Patterson, C.C., A.E. Smith, J.W. Yarnell, A. Rumley, Y. Ben-Shlomo, and G.D. Lowe. 2010. The associations of interleukin-6 (IL-6) and downstream inflammatory markers with risk of cardiovascular disease: the Caerphilly Study. Atherosclerosis 209: 551–557.

Pearson, T., C. Ballantyne, C. Sisk, A. Shah, E. Veltri, and D. Maccubbin. 2007. Comparison of effects of ezetimibe/simvastatin versus simvastatin versus atorvastatin in reducing C-reactive protein and low-density lipoprotein cholesterol levels. Am. J. Cardiol. 99: 1706–1713.

Phan, B.A., T.D. Dayspring, and P.P. Toth. 2012. Ezetimibe therapy: mechanism of action and clinical update. Vasc. Health Risk Manag. 8: 415–427.

Plaisance, E.P., M. Lukasova, S. Offermanns, Y. Zhang, G. Cao, and R.L. Judd. 2009. Niacin stimulates adiponectin secretion through the GPR109A receptor. Am. J. Physiol. Endocrinol. Metab. 296: E549–558.

Price, D.T. and J. Loscalzo. 1999. Cellular adhesion molecules and atherogenesis. Am. J. Med. 107: 85–97.

Ray, K.K., C.P. Cannon, R. Cairns, D.A. Morrow, N. Rifai, A.J. Kirtane, C.H. McCabe, A.M. Skene, C.M. Gibson, P.M. Ridker, and E. Braunwald. PROVE IT-TIMI 22 Investigators. 2005. Relationship between uncontrolled risk factors and C-reactive protein levels in patients receiving standard or intensive statin therapy for acute coronary syndromes in the PROVE IT-TIMI 22 trial. J. Am. Coll. Cardiol. 46: 1417–1424.

Reilly, M.P., M. Lehrke, M.L. Wolfe, A. Rohatgi, M.A. Lazar, and D.J. Rader. 2005. Resistin is an inflammatory marker of atherosclerosis in humans. Circulation 111: 932–939.

Ridker, P.M., N. Rifai, M.A. Pfeffer, F. Sacks, and E. Braunwald. 1999. Long-term effects of pravastatin on plasma concentration of C-reactive protein. The Cholesterol and Recurrent Events (CARE) Investigators. Circulation 100: 230–235.

Ridker, P.M., C.H. Hennekens, J.E. Buring, and N. Rifai. 2000. C-reactive protein and other markers of inflammation in the prediction of cardiovascular disease in women. N. Engl. J. Med. 342: 836–843.

Ridker, P.M., D.A. Morrow, L.M. Rose, N. Rifai, C.P. Cannon, and E. Braunwald. 2005. Relative efficacy of atorvastatin 80 mg and pravastatin 40 mg in achieving the dual goals of low-density lipoprotein cholesterol < 70 mg/dl and C-reactive protein < 2 mg/l: an analysis of the PROVE-IT TIMI-22 trial. J. Am. Coll. Cardiol. 45: 1644–1648.

Ridker, P.M., E. Danielson, F.A. Fonseca, J. Genest, A.M. Jr. Gotto, J.J. Kastelein, W. Koenig, P. Libby, A.J. Lorenzatti, J.G. MacFadyen, B.G. Nordestgaard, J. Shepherd, J.T. Willerson, and R.J. Glynn. JUPITER Study Group. 2008. Rosuvastatin to prevent vascular events in men and women with elevated C-reactive protein. N. Engl. J. Med. 359: 2195–2207.

Ridker, P.M., E. Danielson, F.A. Fonseca, J. Genest, A.M. Jr. Gotto, J.J. Kastelein, W. Koenig, P. Libby, A.J. Lorenzatti, J.G. Macfadyen, B.G. Nordestgaard, J. Shepherd, J.T. Willerson, and R.J. Glynn. JUPITER Trial Study Group. 2009. Reduction in C-reactive protein and LDL cholesterol and cardiovascular event rates after initiation of rosuvastatin: a prospective study of the JUPITER trial. Lancet 373: 1175–1182.

Rosenson, R.S. 2008. Fenofibrate reduces lipoprotein associated phospholipase A2 mass and oxidative lipids in hypertriglyceridemic subjects with the metabolic syndrome. Am. Heart J. 155: 499 e9–16.

Sager, P.T., L. Melani, L. Lipka, J. Strony, B. Yang, R. Suresh, and E. Veltri. Ezetimibe Study Group. 2003. Effect of coadministration of ezetimibe and simvastatin on high-sensitivity C-reactive protein. Am. J. Cardiol. 92: 1414–1418.

Sampietro, T., M. Tuoni, M. Ferdeghini, A. Ciardi, P. Marraccini, C. Prontera, G. Sassi, M. Taddei, and A. Bionda. 1997. Plasma cholesterol regulates soluble cell adhesion molecule expression in familiar hypercholesterolemia. Circulation 96: 1381–1385.

Scalia, R., J.Z. 3rd Appel, and A.M. Lefer. 1998. Leukocyte-endothelium interaction during the early stages of hypercholesterolemia in the rabbit: role of P-selectin, ICAM-1, and VCAM-1. Arterioscler. Thromb. Vasc. Biol. 18: 1093–1100.

Shetty, G.K., P.A. Economides, E.S. Horton, C.S. Mantzoros, and A. Veves. 2004. Circulating adiponectin and resistin levels in relation to metabolic factors, inflammatory markers, and vascular reactivity in diabetic patients and subjects at risk for diabetes. Diabetes Care 27: 2450–2457.

Shi, Y., P. Zhang, L. Zhang, H. Osman, E.R. Mohler III, C. Macphee, A. Zalewski, A. Postle, and R.L. Wilensky. 2007. Role of lipoprotein-associated phospholipase A2 in leukocyte activation and inflammatory responses. Atherosclerosis 191: 54–62.

Siasos, G., D. Tousoulis, E. Oikonomou, M. Zaromitidou, C. Stefanadis, and A.G. Papavassiliou. 2011. Inflammatory markers in hyperlipidemia: from experimental models to clinical practice. Curr. Pharm. Des. 17: 4132–4146.

Simon, B.C., L.D. Cunningham, and R.A. Cohen. 1990. Oxidized low density lipoproteins cause contraction and inhibit endothelium-dependent relaxation in the pig coronary artery. J. Clin. Invest. 86: 75–79.

Skoog, T., W. Dichtl, S. Boquist, C. Skoglund-Andersson, F. Karpe, R. Tang, M.G. Bond, U. de Faire, J. Nilsson, P. Eriksson, and A. Hamsten. 2002. Plasma tumour necrosis factor-alpha and early carotid atherosclerosis in healthy middle-aged men. Eur. Heart J. 23: 376–383.

Somers, W., M. Stahl, and J.S. Seehra. 1997. A crystal structure of interleukin 6: implications for a novel mode of receptor dimerization and signaling. EMBO J. 16: 989–997.

Stefanadi, E., D. Tousoulis, E.S. Androulakis, N. Papageorgiou, M. Charakida, G. Siasos, C. Tsioufis, and C. Stefanadis. 2010. Inflammatory markers in essential hypertension: potential clinical implications. Curr. Vasc. Pharmacol. 8: 509–516.

Stiko-Rahm, A., B. Wiman, A. Hamsten, and J. Nilsson. 1990. Secretion of plasminogen activator inhibitor-1 from cultured human umbilical vein endothelial cells is induced by very low density lipoprotein. Arteriosclerosis 10: 1067– 1073.

Stokes, K.Y., L. Calahan, C.M. Hamric, J.M. Russell, and D.N. Granger. 2009. CD40/CD40L contributes to hypercholesterolemia-induced microvascular inflammation. Am. J. Physiol. Heart Circ. Physiol. 296: H689–697.

Stone, N.J. Secondary causes of hyperlipidemia. 1994. Med. Clin. North. Am. 78: 117–141.

Sukhija, R., I. Fahdi, L. Garza, L. Fink, M. Scott, W. Aude, R. Pacheco, Z. Bursac, A. Grant, and J.L. Mehta. 2007. Inflammatory markers, angiographic severity of coronary artery disease, and patient outcome. Am. J. Cardiol. 99: 879–884.

Sun, J., G.K. Sukhova, P.J. Wolters, M. Yang, S. Kitamoto, P. Libby, L.A. MacFarlane, J. Mallen-St. Clair, and G.P. Shi. 2007. Mast cells promote atherosclerosis by releasing proinflammatory cytokines. Nat. Med. 13: 719–724.

Szekanecz, Z. 2008. Pro-inflammatory cytokines in atherosclerosis. Isr. Med. Assoc. J. 10: 529–530.

Tanner, F.C., G. Noll, C.M. Boulanger, and T.F. Luscher. 1991. Oxidized low density lipoproteins inhibit relaxations of porcine coronary arteries: role of scavenger receptor and endothelium-derived nitric oxide. Circulation 83: 2012–2020.

Third Report of the National Cholesterol Education Program (NCEP) Expert Panel on Detection, Evaluation, and Treatment of High Blood Cholesterol in Adults (Adult Treatment Panel III) final report. 2002. Circulation 106: 3143–3421.

Tousoulis, D., G. Davies, C. Stefanadis, P. Toutouzas, and J.A. Ambrose. 2003. Inflammatory and thrombotic mechanisms in coronary atherosclerosis. Heart 89: 993–997.

Tousoulis, D., C. Antoniades, E. Bosinakou, M. Kotsopoulou, C. Tsioufis, C. Tentolouris, A. Trikas, C. Pitsavos, and C. Stefanadis. 2005a. Effects of atorvastatin on reactive hyperaemia and the thrombosis-fibrinolysis system in patients with heart failure. Heart 91: 27–31.

Tousoulis, D., M. Charakida and C. Stefanadis. 2005b. Inflammation and endothelial dysfunction as therapeutic targets in patients with heart failure. Int. J. Cardiol. 100: 347–353.

Tousoulis, D., C. Antoniades, V. Katsi, E. Bosinakou, M. Kotsopoulou, C. Tsioufis, and Stefanadis. 2006a. The impact of early administration of low-dose atorvastatin treatment on inflammatory process, in patients with unstable angina and low cholesterol level. Int. J. Cardiol. 109: 48–52.

Tousoulis, D., M. Charakida, and C. Stefanadis. 2006b. Endothelial function and inflammation in coronary artery disease. Heart 92: 441–444.

Tousoulis, D., C. Antoniades, and C. Stefanadis. 2007a. Assessing inflammatory status in cardiovascular diseases. Heart 93: 1001–1007.

Tousoulis, D., C. Antoniades, E. Bosinakou, M. Kotsopoulou, C. Tsoufis, K. Marinou, M. Charakida, E. Stefanadi, M. Vavuranakis, G. Latsios, and C. Stefanadis. 2007b. Differences in inflammatory and thrombotic markers between unstable angina and acute myocardial infarction. Int. J. Cardiol. 115: 203–207.

Tousoulis, D., M. Charakida, and C. Stefanadis. 2008. Endothelial function and inflammation in coronary artery disease. Postgrad. Med. J. 84: 368–371.

Tousoulis, D., E. Androulakis, N. Papageorgiou, A. Briasoulis, G. Siasos, C. Antoniades, and C. Stefanadis. 2010a. From atherosclerosis to acute coronary syndromes: the role of soluble CD40 ligand. Trends Cardiovasc. Med. 20: 153–164.

Tousoulis, D., N. Papageorgiou, E. Androulakis, K. Paroutoglou, and C. Stefanadis. 2010b. Novel therapeutic strategies targeting vascular endothelium in essential hypertension. Expert Opin. Investig. Drugs 19: 1395–1412.

Tousoulis, D., A.M. Kampoli, N. Papageorgiou, E. Androulakis, C. Antoniades, K. Toutouzas, and C. Stefanadis. 2011. Pathophysiology of atherosclerosis: the role of inflammation. Curr. Pharm. Des. 17: 4089–4110.

Tousoulis, D., A.M. Kampoli, C. Tentolouris, N. Papageorgiou, and C. Stefanadis. 2012a. The role of nitric oxide on endothelial function. Curr. Vasc. Pharmacol. 10: 4–18.

Tousoulis, D., G. Hatzis, N. Papageorgiou, E. Androulakis, G. Bouras, A. Giolis, C. Bakogiannis, G. Siasos, G. Latsios, C. Antoniades, and C. Stefanadis. 2012b. Assessment of acute coronary syndromes: focus on novel biomarkers. Curr. Med. Chem. 19: 2572–2587.

Tzoulaki, I., G.D. Murray, A.J. Lee, A. Rumley, G.D. Lowe, and F.G. Fowkes. 2005. C-reactive protein, interleukin-6, and soluble adhesion molecules as predictors of progressive peripheral atherosclerosis in the general population: Edinburgh Artery Study. Circulation 112: 976–983.

van der Zwan, L.P., T. Teerlink, J.M. Dekker, R.M. Henry, C.D. Stehouwer, C. Jakobs, R.J. Heine, and P.G. Scheffer. 2009. Circulating oxidized LDL: determinants and association with brachial flow-mediated dilation. J. Lipid. Res. 50: 342–349.

Vaverkova, H., D. Karasek, D. Novotny, D. Jackuliakova, M. Halenka, J. Lukes, and J. Frohlich. 2008. Positive association of adiponectin with soluble vascular cell adhesion molecule sVCAM-1 levels in patients with vascular disease or dyslipidemia. Atherosclerosis 197: 725–731.

Verma, S., P.E. Szmitko, and E.T. Yeh. 2004. C-reactive protein: structure affects function. Circulation 109: 1914–1917.

Verna, L., C. Ganda, and M.B. Stemerman. 2006. *In vivo* low-density lipoprotein exposure induces intercellular adhesion molecule-1 and vascular cell adhesion molecule-1 correlated with activator protein-1 expression. Arterioscler. Thromb. Vasc. Biol. 26: 1344–1349.

Wada, T., H. Kenmochi, Y. Miyashita, M. Sasaki, M. Ojima, M. Sasahara, D. Koya, H. Tsuneki, and T. Sasaoka. 2010. Spironolactone improves glucose and lipid metabolism by ameliorating hepatic steatosis and inflammation and suppressing enhanced gluconeogenesis induced by high-fat and high-fructose diet. Endocrinology 151: 2040–2049.

Wang, S., D. Wu, N.R. Matthan, S. Lamon-Fava, J.L. Lecker, and A.H. Lichtenstein. 2010. Enhanced aortic macrophage lipid accumulation and inflammatory response in LDL receptor null mice fed an atherogenic diet. Lipids 45: 701–711.

Wang, Y.I., J. Schulze, N. Raymond, T. Tomita, K. Tam, S.I. Simon, and A.G. Passerini. 2011. Endothelial inflammation correlates with subject triglycerides and waist size after a high-fat meal. Am. J. Physiol. Heart Circ. Physiol. 300: H784–791.

Welsh, P., M. Woodward, A. Rumley, and G. Lowe. 2009. Associations of circulating TNFalpha and IL-18 with myocardial infarction and cardiovascular risk markers: the Glasgow Myocardial Infarction Study. Cytokine 47: 143–147.

Ye, Y., X. Zhao, G. Zhai, L. Guo, Z. Tian, and S. Zhang. 2012. Effect of high-dose statin versus low-dose statin plus ezetimibe on endothelial function: a meta-analysis of randomized trials. J. Cardiovasc. Pharmacol. Ther. Jun 18.

Yuan, G., K.Z. Al-Shali, and R.A. Hegele. 2007. Hypertriglyceridemia: its etiology, effects and treatment. CMAJ 176: 1113–1120.

Yudkin, J.S., C.D. Stehouwer, J.J. Emeis, and S.W. Coppack. 1999. C-reactive protein in healthy subjects: associations with obesity, insulin resistance, and endothelial dysfunction: a potential role for cytokines originating from adipose tissue? Arterioscler. Thromb. Vasc. Biol. 19: 972–978.

Zhang, N., Y. Huan, H. Huang, G.M. Song, S.J. Sun, and Z.F. Shen. 2010. Atorvastatin improves insulin sensitivity in mice with obesity induced by monosodium glutamate. Acta Pharmacol. Sin. 31: 35–42.

# CHAPTER 5

# Cardiac Biomarkers in Acute Coronary Syndromes

Konstantinos Toutouzas,* Andreas Synetos, Maria Drakopoulou, Dimitris Tousoulis and Christodoulos Stefanadis

## Introduction

Acute coronary syndromes (ACS) refer to a spectrum of clinical presentations ranging from ST elevation myocardial infarction (STEMI), non-ST elevation myocardial infarction (NSTEMI), and unstable angina. The main cause of ACS is the blocking of a coronary artery, which can be due to the formation of an atherosclerotic plaque or coronary spasm. The major trigger for coronary thrombosis is considered the disruption of the vulnerable atherosclerotic plaque which results in the release of various metalloproteinases (Toutouzas et al. 2012c) from activated inflammatory cells, followed by platelet activation and aggregation. Patients presenting with chest pain should undergo a full workup including physical history, an electrocardiogram (ECG) and biomarkers evaluation, while further investigations should be undertaken if warranted. The standard 12-lead ECG is the single best test to identify patients with acute myocardial infarction (AMI) (Lee et al. 1985); nonetheless, it has low sensitivity for detection of an AMI. The sensitivity for detecting a STEMI is 35–50%, leaving at least half of all AMI patients unidentified (Lee et al. 1985; Selker et al. 1997). Thus, there is a great need for additional strategies and diagnostic modalities that will assist in the prompt detection of an AMI. Newer imaging modalities such as optical

1st Department of Cardiology, 'Hippokration' Hospital, University of Athens Medical School, Athens, Greece.
*Corresponding author: ktoutouz@gmail.com

coherence tomography (OCT) (Toutouzas et al. 2012b; Toutouzas et al. 2011b) and microwave radiometry (MR) (Toutouzas et al. 2012a; Toutouzas et al. 2011a) are under investigation for probable use in identifying vulnerable plaques or other causes of ACS. Cardiac biomarkers are one of the most commonly used investigations to identify patients with a likely ACS and, ideally, a cardiac biomarker should allow early detection of ACS patients that would enable optimal treatment initiation.

The first cardiac biomarker in use was aspartate aminotransferase (AST) which was included in the 1959 World Health Organization (WHO) universal AMI definition (Geneva 1959). Since then, various other markers have been proposed and used and the advancement in technology allowed for detection of more cardiac-specific enzymes. Recently, an updated universal AMI definition was published including the detection of a rise and/or fall in cardiac biomarker values, preferably cardiac troponin (Thygesen et al. 2012) as part of the diagnostic criteria of AMI detection.

The aim of this chapter is to discuss the currently used cardiac biomarkers for ACS detection as well as the newer and emerging biomarkers.

## Currently Used Cardiac Biomarkers

### *Cardiac necrosis biomarkers*

#### *Troponin*

Cardiac troponins (cTn) are structural proteins found solely in the heart, located along the thin filaments of myofibrils that regulate the contraction of cardiac muscle. They are composed of three subunits; cTnC is a calcium binding subunit, cTnT is a tropomyosin binding subunit, and cTnI inhibits the ATPase activity of actomyosin. Their detection in peripheral blood indicates cardiomyocyte damage. It has been shown that cTns are the most diagnostically sensitive and specific biomarkers of myocardial injury (Kavsak et al. 2009; Keller et al. 2009; Reichlin et al. 2009a). Although it has been suggested that cTnI is more sensitive than cTnT (Hamm et al. 1997), both cTnI and cTnT kinetics are almost alike and are detectable in the serum within 4 to 12 hours after the onset of AMI, and depending on the reperfusion status, peak values occur 12 to 48 hours from symptom onset (Boden et al. 2008). Even though cTnI is cleared more quickly from the circulation than cTnT, both isoforms remain elevated in the serum for several days after injury, allowing for diagnostic confirmation, even in patients presenting with delayed symptoms. Of note, due to patent regulations there is solely one cTnT immunoassay system, whereas multiple cTnI immunoassay systems exist (Christenson et al. 2001), so cut-off values used in each laboratory should be clearly demarcated.

Patients with renal failure should be thoroughly evaluated even though increased levels of cTns may not be accompanied by ACS symptoms (Freda et al. 2002), since these patients pose a greater risk of cardiovascular mortality. Furthermore, patients with renal disease should be treated in the same way as patients without renal disease and take advantage of the full armory that we have for treating ACS (cardiac catheterization, IIb/IIIa glycoproteins) (Aviles et al. 2002).

The failure of the current cTn assays to elevate in the initial hours after an AMI and propose a mechanism of the reason behind the cardiomyocyte damage have led to the development of the high-sensitivity cTn assays (hs-cTn) (Twerenbold et al. 2012). The benefit of the hs-cTn was more pronounced in patients presenting early after their symptom onset and had higher diagnostic accuracy compared to standard cTn assays (Keller et al. 2009; Reichlin et al. 2009a).

One of the main disadvantages of using cTn is their long half-life in the circulation, thus not being able to detect a re-infarction post the index event. In addition, cTnT is present in skeletal muscles and is further expressed in clinical entities such as polymoysitis or Duchenne's muscular dystrophy (Bodor et al. 1997), whereas cTnI has been shown to be elevated in more rare clinical entities such as Naxos disease (Lazaros et al. 2009).

## Creatine kinase

Creatine kinase (CK), also known as creatine phosphokinase, is an enzyme that catalyses the conversion of creatine and consumes adenosine triphosphate to create phosphocreatine and adenosine diphosphate. It is considered the next best alternative diagnostic biomarker after cTn (Morrow et al. 2007; Thygesen et al. 2007). The MB fraction of CK is primarily found in the myocardium and it rises 4–6 hours after the onset of AMI and remains elevated for 24 to 48 hours (Loria et al. 2008). A study showed that CK-MB serially measured three hours after an ACS presentation had a sensitivity of 90%–95% of correctly detecting it (Gibler et al. 1990). Thus, successive CK-MB measurements over 6–9 hours have been widely in use and are considered adequate to safely exclude AMI if negative values do occur (Braunwald et al. 2000). Although CK-MB is considered relatively sensitive, it lacks specificity since it can be elevated in various conditions such as skeletal muscle disease, chronic renal failure, acute muscle exertion, and cocaine use (Brogan et al. 1997; Hollander et al. 1998).

## Myoglobin

Myoglobin is an iron- and oxygen-binding protein found in the muscle tissue and its lower molecular weight is the reason why it is released quicker than CK or cTn during an AMI, as quick as one hour post AMI. It reaches twice as normal values within two hours and peaks within four hours of AMI symptom onset compared to CK-MB and achieves its maximal diagnostic sensitivity within five hours of symptom onset (de Winter et al. 2000). Its main advantage is the early rise in blood and studies have suggested that combining myoglobin with cTn or CK-MB would be a useful tool for the early detection of an AMI (Brogan et al. 1994; Ohman et al. 1990). A drawback of myoglobin is its rapid renal clearance that decreases its sensitivity as the infarction progresses.

## Inflammatory biomarkers

### C-reactive protein

Increased local temperature in human coronary atherosclerotic plaques is considered to be an independent predictor of clinical outcome in patients undergoing percutaneous coronary intervention (Stefanadis et al. 2001) and detecting the heat released by activated inflammatory cells of atherosclerotic plaques may predict plaque rupture and thrombosis (Stefanadis et al. 1999). Recent evidence suggests that ACS are the outcome of inflammatory processes (Tousoulis et al. 2012). C-reactive protein (CRP) is an acute phase reactant protein produced mainly in the liver, as well as in atherosclerotic lesions by smooth muscle cells and macrophages (Drakopoulou et al. 2009), and rises in response to inflammation. CRP has been implicated in the progression of atherosclerosis (Tousoulis et al. 2011) and has been shown to predict future cardiovascular events (De Servi et al. 2005). A study showed that patients presenting with unstable angina and a CRP value $\geq 0.3$ mg/dl on admission had a sensitivity of 90% and a specificity of 82% for predicting subsequent cardiac events (cardiac death, AMI, or the urgent need for coronary revascularization) (Liuzzo et al. 1994). Nonetheless, a more recent study in 1,210 patients who suffered ACS failed to show any correlation among ACS patients with elevated CRP and future events (Bogaty et al. 2008). Furthermore, other studies have failed to show the superiority of CRP compared to cTn and its association with the early or late occurrence of AMI (De Servi et al. 2005; Investigators 1999). Evidence suggests that statins decrease CRP levels and provide higher plaque stability (Stefanadis et al. 2007). Recently, high-sensitivity CRP (hs-CRP) assays have been in use (Tousoulis et al. 2008) and are considered to be useful inflammatory markers for clinical practice in the setting of ACS. Even though the data are robust, CRP has not yet found its place in the emergency department (ED).

## CD40-Ligand

CD40-ligand (CD40L) is a member of the TNF superfamily of molecules. The soluble CD40-ligand (sCD40L) is elevated in coronary artery disease and is released from activated platelets during an AMI (Tousoulis et al. 2007). Studies on various genetic polymorphisms have shown an increased level of sCD40L in patients with premature AMI (Antoniades et al. 2006). Furthermore, levels of the sCD40L have also been shown to increase the risk of major adverse cardiovascular events in patients with ACS (Antoniades et al. 2009b). It has been shown that glycoprotein IIb/ IIIa antagonists (abciximab, eptifibatide, tirofiban) are able to inhibit the release of sCD40L and platelet aggregation initially *in vitro* and, afterwards, in clinical studies, affecting a key role in the pathogenesis of ACS (Tousoulis et al. 2009; Tousoulis et al. 2010); nonetheless, it is still unclear whether specific treatments targeting the CD40/CD40L system will prove to be beneficial against atherothrombosis (Antoniades et al. 2009a). In addition, another study showed that patients with ACS and elevated plasma levels of sCD40L had a higher risk of death and recurrent AMI independent of other predictive variables such as cTnI and CRP (Varo et al. 2003).

## Metalloproteinases

Matrix metalloproteinases (MMPs) are proteinases that take part in extracellular matrix remodeling and degradation; alteration in their regulation leads to atherosclerotic plaque development and destabilization, and thus to coronary artery disease and heart failure (Briasoulis et al. 2012; Kampoli et al. 2012; Toutouzas et al. 2012c). A recent study in 266 patients presenting with STEMI and NSTEMI showed that MMP-9 levels, which is regarded as a biomarker of plaque rupture or vulnerability, surged earlier compared to cTnT levels and had a higher diagnostic accuracy in early ACS (≤4h of onset) (Kobayashi et al. 2011).

## Neuroendocrine biomarkers

### B-natriuretic peptide

B-type natriuretic peptide (BNP) is a polypeptide secreted from the heart ventricles in response to ventricular wall stretch (de Lemos et al. 2003). BNP, already established as a marker for heart failure, and its prohormone, NT-proBNP, are elevated in ACS and can be used as a risk stratification tool as well as reflect the degree of left ventricular dysfunction (de Lemos et al. 2001; Jernberg et al. 2002; Morita et al. 1993). Also, patients suffering NSTEMI with preserved systolic function that had high levels of BNP showed coronary artery disease extension (Palazzuoli et al. 2012). However, elevated levels of these peptides can be found in patients with renal failure, primary aldosteronism, congestive heart failure, and thyroid disease (Clerico et al. 1999; Loria et al. 2008). Interestingly, a study in patients suffering ACS showed that impaired renal function was protective against development of systolic dysfunction and that an increase in BNP ameliorated this protective effect (Chrysohoou et al. 2010). Furthermore, BNP was identified as an independent predictor of death in patients suffering AMI with worsening renal function (Lazaros et al. 2012).

## Atrial natriuretic peptide

Atrial natriuretic peptide (ANP) is a powerful vasodilator stored and released by cardiac cells in the atria in response to atrial stretch. Various studies have shown an increase in ANP after AMI; however, they have failed to classify ANP as an important prognostic factor for future events as they have for BNP (Omland et al. 1996; Squire et al. 2005).

A summary of currently used cardiac biomarkers is seen in Table 1.

Table 1. Currently used cardiac biomarkers.

| Biomarker | Available data |
|---|---|
| **Cardiac necrosis** | |
| Troponin | Current gold standard for AMI detection. Highly sensitive and specific for myocardial necrosis. Used for risk stratification post main index event. |
| CK-MB | Second in line marker for AMI detection. No prediction of future events. |
| Myoglobin | Fastest rising biomarker in blood. Lacks specificity. |
| **Inflammatory** | |
| CRP | Good sensitivity in unstable angina patients. |
| CD40L | Has been shown to increase rate of major cardiac events, but more data are needed. |
| MMPs | High diagnostic accuracy if used in combination with cTnT. |
| **Neuroendocrine** | |
| BNP | Predictive of heart failure, but lacks sensitivity and specificity. |
| ANP | Insufficient data. |

# Newer and Emerging Cardiac Biomarkers

## *Growth-differentiation factor-15*

Growth differentiation factor 15 (GDF15) is a protein belonging to the transforming growth factor beta superfamily that has a role in regulating inflammatory and apoptotic pathways in injured tissues. A recent study in 479 patients with acute chest pain showed that GDF 15 was an independent predictor of death or AMI after six months (Eggers et al. 2008); another study pointed out that GDF 15 was the strongest predictor of raised 1-year mortality risk, even though it was not related to the risk of early reinfarction in NSTEMI patients (Wollert et al. 2007).

## *Copeptin*

Copeptin is found at the C-terminal end of vasopressin, the antidiuretic hormone, and due to its long half-life is a good marker of the upregulation of the arginine-vasopressin system in cardiac ischemia, especially in patients presenting early (Morgenthaler et al. 2006; Reichlin et al. 2009b). When copeptin was used together with cTnT, an increased sensitivity of up to 88% was observed in ACS patients who presented <12 h of chest pain onset (Keller et al. 2010).

## *Cystatin C*

Cystatin C is a protein encoded by the CST3 gene and is mainly used as a biomarker of kidney function. Lately, it has been studied for its role in predicting new-onset or deteriorating cardiovascular disease. It has proven its superiority compared to serum creatinine concentration as an independent predictor of heart failure in older adults (Sarnak et al. 2005) and as a mortality predictor in heart failure patients (Shlipak et al. 2005). Finally, results from a study following NSTEMI patients revealed that cystatin C levels had a strong relation with mortality at six months (Jernberg et al. 2004).

## *Micro RNA*

MicroRNAs (miRNAs) are a class of 19–25-nucleotide non-coding RNAs which have been shown to participate in cardiovascular disease pathogenesis including atherosclerosis, coronary artery disease, myocardial infarction, heart failure, and cardiac arrhythmias (Papageorgiou et al. 2012). However, more studies are needed to clarify their role in ACS prognosis.

## *Endothelial progenitor cells*

Endothelial progenitor cells (EPCs) are a population of cells that derive from the bone marrow and circulate in the blood with the ability to differentiate into endothelial cells, the cells that craft the inner layer of the blood vessels. It has been shown that the number of EPCs is inversely correlated with cardiovascular risk factors (Bakogiannis et al. 2012). EPCs positive for CD34 and kinase insert domain could predict the occurrence of cardiovascular events and death from cardiovascular causes (Werner et al. 2005).

A summary of emerging cardiac biomarkers is seen in Table 2.

**Table 2.** Emerging cardiac biomarkers.

| Biomarker | Available data |
| --- | --- |
| GDF 15 | Predictive of AMI and mortality, but not sensitive or specific for AMI. |
| Copeptin | Increased sensitivity when used along with cTnT. Predictor of mortality. |
| Cystatin C | Predictor of heart failure and mortality. Lacks sensitivity and specificity. |
| Micro RNA | Raised in AMI, but insufficient data. |
| EPCs | Used for risk stratification; more data are needed. |

## Are Multiple Cardiac Biomarkers the Future?

Multiple cardiac biomarkers are a strategy that could potentially allow for greater and more thorough risk stratification and provide analytical information in a timely manner for patients suffering from ACS. The emergence of more and different biomarkers focusing on ACS diagnosis will inevitably lead to better comprehension of the pathophysiology behind ACS, which will allow for improved and enhanced biomarker assays. Many studies have already focused on this matter with promising results. In the OPUS-TIMI 16 trial, 450 patients suffering NSTEMI had their cTnI, CRP, and BNP biomarkers assessed (Sabatine et al. 2002). All three biomarkers were independent and prognostic predictors for short- and long-term major cardiac events. Another study evaluating patients with cardiac ischemic signs and using assays for cTnI, CK-MB and myoglobin reported a negative predictive value of 100% in a 90-minute time setting when using all three biomarkers (Ng et al. 2001). A large study involving more than 3000 participants evaluated 10 different biomarkers and although the authors report that multiple biomarkers added only moderately to standard risk factors for risk assessment of individual persons, patients with increased biomarkers had a risk of death four times more than the patients who did not show an increase (Wang et al. 2006).

## Conclusion

ACS pathophysiology is rather complicated involving inflammation, muscle necrosis, and oxidative stress. The availability of cTn and more specific hs-cTn assays has provided clinicians a safe and reliable biomarker for ACS diagnosis. Although the research on diagnostic biomarkers is highly intense, none of the biomarkers available can offer sensitivity, specificity, early detection and cost effectiveness all together. Multiple biomarker assays could potentially be the future, but there is still a lot of ground to be covered.

## References

Antoniades, C., C. Bakogiannis, D. Tousoulis, A.S. Antonopoulos, and C. Stefanadis. 2009a. The CD40/CD40 ligand system: linking inflammation with atherothrombosis. J. Am. Coll. Cardiol. 54: 669–677.

Antoniades, C., D. Tousoulis, C. Vasiliadou, E. Stefanadi, K. Marinou, and C. Stefanadis. 2006. Genetic polymorphisms of platelet glycoprotein Ia and the risk for premature myocardial infarction: effects on the release of sCD40L during the acute phase of premature myocardial infarction. J. Am. Coll. Cardiol. 47: 1959–1966.

Antoniades, C., T. Van-Assche, C. Shirodaria, J. Diesch, A.S. Antonopoulos, J. Lee, C. Cunnington, D. Tousoulis, C. Stefanadis, B. Casadei, D. Taggart, K.M. Channon, and P. Leeson. 2009b. Preoperative sCD40L levels predict risk of atrial fibrillation after off-pump coronary artery bypass graft surgery. Circulation 120: S170–176.

Aviles, R.J., A.T. Askari, B. Lindahl, L. Wallentin, G. Jia, E.M. Ohman, K.W. Mahaffey, L.K. Newby, R.M. Califf, M.L. Simoons, E.J. Topol, P. Berger, and M.S. Lauer. 2002. Troponin T levels in patients with acute coronary syndromes, with or without renal dysfunction. N. Engl. J. Med. 346: 2047–2052.

Bakogiannis, C., D. Tousoulis, E. Androulakis, A. Briasoulis, N. Papageorgiou, G. Vogiatzi, A.M. Kampoli, M. Charakida, G. Siasos, G. Latsios, C. Antoniades, and C. Stefanadis. 2012. Circulating endothelial progenitor cells as biomarkers for prediction of cardiovascular outcomes. Curr. Med. Chem. 19: 2597–2604.

Boden, W.E., P.K. Shah, V. Gupta, and E.M. Ohman. 2008. Contemporary approach to the diagnosis and management of non-ST-segment elevation acute coronary syndromes. Prog. Cardiovasc. Dis. 50: 311–351.

Bodor, G.S., L. Survant, E.M. Voss, S. Smith, D. Porterfield, and F.S. Apple. 1997. Cardiac troponin T composition in normal and regenerating human skeletal muscle. Clin. Chem. 43: 476–484.

Bogaty, P., L. Boyer, S. Simard, F. Dauwe, R. Dupuis, B. Verret, T. Huynh, F. Bertrand, G.R. Dagenais, and J.M. Brophy. 2008. Clinical utility of C-reactive protein measured at admission, hospital discharge, and 1 month later to predict outcome in patients with acute coronary disease. The RISCA (recurrence and inflammation in the acute coronary syndromes) study. J. Am. Coll. Cardiol. 51: 2339–2346.

Braunwald, E., E.M. Antman, J.W. Beasley, R.M. Califf, M.D. Cheitlin, J.S. Hochman, R.H. Jones, D. Kereiakes, J. Kupersmith, T.N. Levin, C.J. Pepine, J.W. Schaeffer, E.E. Smith, 3rd, D.E. Steward, P. Theroux, J.S. Alpert, K.A. Eagle, D.P. Faxon, V. Fuster, T.J. Gardner, G. Gregoratos, R.O. Russell, and S.C. Smith, Jr. 2000. ACC/AHA guidelines for the management of patients with unstable angina and non-ST-segment elevation myocardial infarction. A report of the American College of Cardiology/American Heart Association Task Force on Practice Guidelines (Committee on the Management of Patients With Unstable Angina). J. Am. Coll. Cardiol. 36: 970–1062.

Briasoulis, A., D. Tousoulis, N. Papageorgiou, A.M. Kampoli, E. Androulakis, C. Antoniades, E. Tsiamis, G. Latsios, and C. Stefanadis. 2012. Novel therapeutic approaches targeting matrix metalloproteinases in cardiovascular disease. Curr. Top. Med. Chem. 12: 1214–1221.

Brogan, G.X., Jr., J.E. Hollander, C.F. McCuskey, H.C. Thode, Jr., J. Snow, A. Sama, and J.L. Bock. 1997. Evaluation of a new assay for cardiac troponin I vs creatine kinase-MB for the diagnosis of acute myocardial infarction. Biochemical Markers for Acute Myocardial Ischemia (BAMI) Study Group. Acad. Emerg. Med. 4: 6–12.

Brogan, G.X., Jr., S. Friedman, C. McCuskey, D.S. Cooling, L. Berrutti, H.C. Thode, Jr., and J.L. Bock. 1994. Evaluation of a new rapid quantitative immunoassay for serum myoglobin versus CK-MB for ruling out acute myocardial infarction in the emergency department. Ann. Emerg. Med. 24: 665–671.

Christenson, R.H., S.H. Duh, F.S. Apple, G.S. Bodor, D.M. Bunk, J. Dalluge, M. Panteghini, J.D. Potter, M.J. Welch, A.H. Wu, and S.E. Kahn. 2001. Standardization of cardiac troponin I assays: round Robin of ten candidate reference materials. Clin. Chem. 47: 431–437.

Chrysohoou, C., C. Pitsavos, P. Aggelopoulos, G. Metallinos, E. Tsiamis, D. Panagiotakos, and C. Stefanadis. 2010. Brain natriuretic peptide mediates the effect of creatinine clearance on development of left ventricular systolic dysfunction in patients with acute coronary syndrome. Hellenic J. Cardiol. 51: 413–420.

Clerico, A., G. Iervasi, and G. Mariani. 1999. Pathophysiologic relevance of measuring the plasma levels of cardiac natriuretic peptide hormones in humans. Horm. Metab. Re.s 31: 487–498.

de Lemos, J.A., D.A. Morrow, J.H. Bentley, T. Omland, M.S. Sabatine, C.H. McCabe, C. Hall, C.P. Cannon, and E. Braunwald. 2001. The prognostic value of B-type natriuretic peptide in patients with acute coronary syndromes. N. Engl. J. Med. 345: 1014–1021.

de Lemos, J.A., D.K. McGuire, and M.H. Drazner. 2003. B-type natriuretic peptide in cardiovascular disease. Lancet 362: 316–322.

De Servi, S., M. Mariani, G. Mariani, and A. Mazzone. 2005. C-reactive protein increase in unstable coronary disease cause or effect? J. Am. Coll. Cardiol. 46: 1496–502.

de Winter, R.J., J.G. Lijmer, R.W. Koster, F.J. Hoek, and G.T. Sanders. 2000. Diagnostic accuracy of myoglobin concentration for the early diagnosis of acute myocardial infarction. Ann. Emerg. Med. 35: 113–120.

Drakopoulou, M., K. Toutouzas, E. Stefanadi, E. Tsiamis, D. Tousoulis, and C. Stefanadis. 2009. Association of inflammatory markers with angiographic severity and extent of coronary artery disease. Atherosclerosis 206: 335–339.

Eggers, K.M., T. Kempf, T. Allhoff, B. Lindahl, L. Wallentin, and K.C. Wollert. 2008. Growth-differentiation factor-15 for early risk stratification in patients with acute chest pain. Eur. Heart J. 29: 2327–2335.

Freda, B.J., W.H. Tang, F. Van Lente, W.F. Peacock, and G.S. Francis. 2002. Cardiac troponins in renal insufficiency: review and clinical implications. J. Am. Coll. Cardiol. 40: 2065–2071.

Geneva. 1959. Hypertension and coronary heart disease: classification and criteria for epidemiological studies. Technical report series number 168. Geneva: World Health Organization Expert Committee.

Gibler, W.B., L.M. Lewis, R.E. Erb, P.K. Makens, B.C. Kaplan, R.H. Vaughn, A.V. Biagini, J.D. Blanton, and W.B. Campbell. 1990. Early detection of acute myocardial infarction in patients presenting with chest pain and nondiagnostic ECGs: serial CK-MB sampling in the emergency department. Ann. Emerg. Med. 19: 1359–1366.

Hamm, C.W., B.U. Goldmann, C. Heeschen, G. Kreymann, J. Berger, and T. Meinertz. 1997. Emergency room triage of patients with acute chest pain by means of rapid testing for cardiac troponin T or troponin I. N. Engl. J. Med. 337: 1648–1653.

Hollander, J.E., M.A. Levitt, G.P. Young, E. Briglia, C.V. Wetli, and Y. Gawad. 1998. Effect of recent cocaine use on the specificity of cardiac markers for diagnosis of acute myocardial infarction. Am. Heart J. 135: 245–252.

Investigators. 1999. Long-term low-molecular-mass heparin in unstable coronary-artery disease: FRISC II prospective randomised multicentre study. FRagmin and Fast Revascularisation during InStability in Coronary artery disease. Investigators. Lancet 354: 701–707.

Jernberg, T., B. Lindahl, S. James, A. Larsson, L.O. Hansson, and L. Wallentin. 2004. Cystatin C: a novel predictor of outcome in suspected or confirmed non-ST-elevation acute coronary syndrome. Circulation 110: 2342–2348.

Jernberg, T., M. Stridsberg, P. Venge, and B. Lindahl. 2002. N-terminal pro brain natriuretic peptide on admission for early risk stratification of patients with chest pain and no ST-segment elevation. J. Am. Coll. Cardiol. 40: 437–445.

Kampoli, A.M., D. Tousoulis, N. Papageorgiou, C. Antoniades, E. Androulakis, E. Tsiamis, G. Latsios, and C. Stefanadis. 2012. Matrix metalloproteinases in acute coronary syndromes: current perspectives. Curr. Top. Med. Chem. 12: 1192–1205.

Kavsak, P.A., X. Wang, D.T. Ko, A.R. MacRae, and A.S. Jaffe. 2009. Short- and long-term risk stratification using a next-generation, high-sensitivity research cardiac troponin I (hs-cTnI) assay in an emergency department chest pain population. Clin. Chem. 55: 1809–1815.

Keller, T., S. Tzikas, T. Zeller, E. Czyz, L. Lillpopp, F.M. Ojeda, A. Roth, C. Bickel, S. Baldus, C.R. Sinning, P.S. Wild, E. Lubos, D. Peetz, J. Kunde, O. Hartmann, A. Bergmann, F. Post, K.J. Lackner, S. Genth-Zotz, V. Nicaud, L. Tiret, T.F. Munzel, and S. Blankenberg. 2010. Copeptin improves early diagnosis of acute myocardial infarction. J. Am. Coll. Cardiol. 55: 2096–2106.

Keller, T., T. Zeller, D. Peetz, S. Tzikas, A. Roth, E. Czyz, C. Bickel, S. Baldus, A. Warnholtz, M. Frohlich, C.R. Sinning, M.S. Eleftheriadis, P.S. Wild, R.B. Schnabel, E. Lubos, N. Jachmann, S. Genth-Zotz, F. Post, V. Nicaud, L. Tiret, K.J. Lackner, T.F. Munzel, and S. Blankenberg. 2009. Sensitive troponin I assay in early diagnosis of acute myocardial infarction. N. Engl. J. Med. 361: 868–877.

Kobayashi, N., N. Hata, N. Kume, S. Yokoyama, T. Shinada, K. Tomita, M. Kitamura, A. Shirakabe, T. Inami, M. Yamamoto, Y. Seino, and K. Mizuno. 2011. Matrix metalloproteinase-9 for the earliest stage acute coronary syndrome. Circ. J. 75: 2853–2861.

Lazaros, G., A. Anastasakis, D. Tsiachris, P. Dilaveris, N. Protonotarios, and C. Stefanadis. 2009. Naxos disease presenting with ventricular tachycardia, and troponin elevation. Heart Vessels 24: 63–65.

Lazaros, G., D. Tsiachris, D. Tousoulis, A. Patialiakas, K. Dimitriadis, D. Roussos, E. Vergopoulos, C. Tsioufis, C. Vlachopoulos, and C. Stefanadis. 2012. In-hospital worsening renal function is an independent predictor of one-year mortality in patients with acute myocardial infarction. Int. J. Cardiol. 155: 97–101.

Lee, T.H., E.F. Cook, M. Weisberg, R.K. Sargent, C. Wilson, and L. Goldman. 1985. Acute chest pain in the emergency room. Identification, and examination of low-risk patients. Arch. Intern. Med. 145: 65–69.

Liuzzo, G., L.M. Biasucci, J.R. Gallimore, R.L. Grillo, A.G. Rebuzzi, M.B. Pepys, and A. Maseri. 1994. The prognostic value of C-reactive protein and serum amyloid a protein in severe unstable angina. N. Engl. J. Med. 331: 417–424.

Loria, V., I. Dato, G.L. De Maria, and L.M. Biasucci. 2008. Markers of acute coronary syndrome in emergency room. Minerva Med. 99: 497–517.

Morgenthaler, N.G., J. Struck, C. Alonso, and A. Bergmann. 2006. Assay for the measurement of copeptin, a stable peptide derived from the precursor of vasopressin. Clin. Chem. 52: 112–119.

Morita, E., H. Yasue, M. Yoshimura, H. Ogawa, M. Jougasaki, T. Matsumura, M. Mukoyama, and K. Nakao. 1993. Increased plasma levels of brain natriuretic peptide in patients with acute myocardial infarction. Circulation 88: 82–91.

Morrow, D.A., C.P. Cannon, R.L. Jesse, L.K. Newby, J. Ravkilde, A.B. Storrow, A.H. Wu, R.H. Christenson, F.S. Apple, G. Francis, and W. Tang. 2007. National Academy of Clinical Biochemistry Laboratory Medicine Practice Guidelines: clinical characteristics and utilization of biochemical markers in acute coronary syndromes. Clin. Chem. 53: 552–574.

Ng, S.M., P. Krishnaswamy, R. Morissey, P. Clopton, R. Fitzgerald, and A.S. Maisel. 2001. Ninety-minute accelerated critical pathway for chest pain evaluation. Am. J. Cardiol. 88: 611–617.

Ohman, E.M., C. Casey, J.R. Bengtson, D. Pryor, W. Tormey, and J.H. Horgan. 1990. Early detection of acute myocardial infarction: additional diagnostic information from serum concentrations of myoglobin in patients without ST elevation. Br. Heart J. 63: 335–338.

Omland, T., A. Aakvaag, V.V. Bonarjee, K. Caidahl, R.T. Lie, D.W. Nilsen, J.A. Sundsfjord, and K. Dickstein. 1996. Plasma brain natriuretic peptide as an indicator of left ventricular systolic function and long-term survival after acute myocardial infarction. Comparison with plasma atrial natriuretic peptide and N-terminal proatrial natriuretic peptide. Circulation 93: 1963–1969.

Palazzuoli, A., M. Caputo, M. Fineschi, R. Navarri, A. Calabro, M. Cameli, M.S. Campagna, B. Franci, C. Pierli, R. Nuti, and A. Maisel. 2012. B-type natriuretic peptide as an independent predictor of coronary disease extension in non-ST elevation coronary syndromes with preserved systolic function. Eur. J. Prev. Cardiol. 19: 366–373.

Papageorgiou, N., D. Tousoulis, E. Androulakis, G. Siasos, A. Briasoulis, G. Vogiatzi, A.M. Kampoli, E. Tsiamis, C. Tentolouris, and C. Stefanadis. 2012. The role of microRNAs in cardiovascular disease. Curr. Med. Chem. 19: 2605–2610.

Reichlin, T., W. Hochholzer, C. Stelzig, K. Laule, H. Freidank, N.G. Morgenthaler, A. Bergmann, M. Potocki, M. Noveanu, T. Breidthardt, A. Christ, T. Boldanova, R. Merki, N. Schaub, R. Bingisser, M. Christ, and C. Mueller. 2009b. Incremental value of copeptin for rapid rule out of acute myocardial infarction. J. Am. Coll. Cardiol. 54: 60–68.

Reichlin, T., W. Hochholzer, S. Bassetti, S. Steuer, C. Stelzig, S. Hartwiger, S. Biedert, N. Schaub, C. Buerge, M. Potocki, M. Noveanu, T. Breidthardt, R. Twerenbold, K. Winkler, R. Bingisser, and C. Mueller. 2009a. Early diagnosis of myocardial infarction with sensitive cardiac troponin assays. N. Engl. J. Med. 361: 858–867.

Sabatine, M.S., D.A. Morrow, J.A. de Lemos, C.M. Gibson, S.A. Murphy, N. Rifai, C. McCabe, E.M. Antman, C.P. Cannon, and E. Braunwald. 2002. Multimarker approach to risk stratification in non-ST elevation acute coronary syndromes: simultaneous assessment of troponin I, C-reactive protein, and B-type natriuretic peptide. Circulation 105: 1760–1763.

Sarnak, M.J., R. Katz, C.O. Stehman-Breen, L.F. Fried, N.S. Jenny, B.M. Psaty, A.B. Newman, D. Siscovick, and M.G. Shlipak. 2005. Cystatin C concentration as a risk factor for heart failure in older adults. Ann. Intern. Med. 142: 497–505.

Selker, H.P., R.J. Zalenski, E.M. Antman, T.P. Aufderheide, S.A. Bernard, R.O. Bonow, W.B. Gibler, M.D. Hagen, P. Johnson, J. Lau, R.A. McNutt, J. Ornato, J.S. Schwartz, J.D. Scott, P.A. Tunick, and W.D. Weaver. 1997. An evaluation of technologies for identifying acute cardiac ischemia in the emergency department: a report from a National Heart Attack Alert Program Working Group. Ann. Emerg. Med. 29: 13–87.

Shlipak, M.G., R. Katz, L.F. Fried, N.S. Jenny, C.O. Stehman-Breen, A.B. Newman, D. Siscovick, B.M. Psaty, and M.J. Sarnak. 2005. Cystatin-C and mortality in elderly persons with heart failure. J. Am. Coll. Cardiol. 45: 268–271.

Squire, I.B., S. Orn, L.L. Ng, C. Manhenke, L. Shipley, T. Aarsland, and K. Dickstein. 2005. Plasma natriuretic peptides up to 2 years after acute myocardial infarction and relation to prognosis: an OPTIMAAL substudy. J. Card. Fail. 11: 492–497.

Stefanadis, C., K. Toutouzas, E. Tsiamis, C. Stratos, M. Vavuranakis, I. Kallikazaros, D. Panagiotakos, and P. Toutouzas. 2001. Increased local temperature in human coronary atherosclerotic plaques: an independent predictor of clinical outcome in patients undergoing a percutaneous coronary intervention. J. Am. Coll. Cardiol. 37: 1277–1283.

Stefanadis, C., K. Toutouzas, E. Tsiamis, M. Vavuranakis, C. Tsioufis, E. Stefanadi, and H. Boudoulas. 2007. Relation between local temperature and C-reactive protein levels in patients with coronary artery disease: effects of atorvastatin treatment. Atherosclerosis 192: 396–400.

Stefanadis, C., L. Diamantopoulos, C. Vlachopoulos, E. Tsiamis, J. Dernellis, K. Toutouzas, E. Stefanadi, and P. Toutouzas. 1999. Thermal heterogeneity within human atherosclerotic coronary arteries detected *in vivo*: A new method of detection by application of a special thermography catheter. Circulation 99: 1965–1971.

Thygesen, K., J.S. Alpert, A.S. Jaffe, M.L. Simoons, B.R. Chaitman, and H.D. White. 2012. Third Universal Definition of Myocardial Infarction. Circulation.

Thygesen, K., J.S. Alpert, H.D. White, A.S. Jaffe, F.S. Apple, M. Galvani, H.A. Katus, L.K. Newby, J. Ravkilde, B. Chaitman, P.M. Clemmensen, M. Dellborg, H. Hod, P. Porela, R. Underwood, J.J. Bax, G.A. Beller, R. Bonow, E.E. Van der Wall, J.P. Bassand, W. Wijns, T.B. Ferguson, P.G. Steg, B.F. Uretsky, D.O. Williams, P.W. Armstrong, E.M. Antman, K.A. Fox, C.W. Hamm, E.M. Ohman, M.L. Simoons, P.A. Poole-Wilson, E.P. Gurfinkel, J.L. Lopez-Sendon, P. Pais, S. Mendis, J.R. Zhu, L.C. Wallentin, F. Fernandez-Aviles, K.M. Fox, A.N. Parkhomenko, S.G. Priori, M. Tendera, L.M. Voipio-Pulkki, A. Vahanian, A.J. Camm, R. De Caterina, V. Dean, K. Dickstein, G. Filippatos, C. Funck-Brentano, I. Hellemans, S.D. Kristensen, K. McGregor, U. Sechtem, S. Silber, P. Widimsky, J.L. Zamorano, J. Morais, S. Brener, R. Harrington, D. Morrow, M. Lim, M.A. Martinez-Rios, S. Steinhubl, G.N. Levine, W.B. Gibler, D. Goff, M. Tubaro, D. Dudek, and N. Al-Attar. 2007. Universal definition of myocardial infarction. Circulation 116: 2634–2653.

Tousoulis, D., A. Briasoulis, S.S. Dhamrait, C. Antoniades, and C. Stefanadis. 2009. Effective platelet inhibition by aspirin and clopidogrel: where are we now? Heart 95: 850–858.

Tousoulis, D., A.M. Kampoli, E. Stefanadi, C. Antoniades, G. Siasos, A.G. Papavassiliou, and C. Stefanadis. 2008. New biochemical markers in acute coronary syndromes. Curr. Med. Chem. 15: 1288–1296.

Tousoulis, D., A.M. Kampoli, N. Papageorgiou, E. Androulakis, C. Antoniades, K. Toutouzas, and C. Stefanadis. 2011. Pathophysiology of atherosclerosis: the role of inflammation. Curr. Pharm. Des. 17: 4089–4110.

Tousoulis, D., C. Antoniades, A. Nikolopoulou, K. Koniari, C. Vasiliadou, K. Marinou, N. Koumallos, N. Papageorgiou, E. Stefanadi, G. Siasos, and C. Stefanadis. 2007. Interaction between cytokines and sCD40L in patients with stable and unstable coronary syndromes. Eur. J. Clin. Invest. 37: 623–628.

Tousoulis, D., G. Hatzis, N. Papageorgiou, E. Androulakis, G. Bouras, A. Giolis, C. Bakogiannis, G. Siasos, G. Latsios, C. Antoniades, and C. Stefanadis. 2012. Assessment of acute coronary syndromes: focus on novel biomarkers. Curr. Med. Chem. 19: 2572–2587.

Tousoulis, D., I.P. Paroutoglou, N. Papageorgiou, M. Charakida, and C. Stefanadis. 2010. Recent therapeutic approaches to platelet activation in coronary artery disease. Pharmacol. Ther. 127: 108–120.

Toutouzas, K., A. Karanasos, E. Tsiamis, M. Riga, M. Drakopoulou, A. Synetos, A. Papanikolaou, C. Tsioufis, A. Androulakis, E. Stefanadi, D. Tousoulis, and C. Stefanadis. 2011b. New insights by optical coherence tomography into the differences and similarities of culprit ruptured plaque morphology in non-ST-elevation myocardial infarction and ST-elevation myocardial infarction. Am. Heart J. 161: 1192–1199.

Toutouzas, K., A. Karanasos, M. Riga, E. Tsiamis, A. Synetos, A. Michelongona, A. Papanikolaou, G. Triantafyllou, C. Tsioufis, and C. Stefanadis. 2012b. Optical coherence tomography assessment of the spatial distribution of culprit ruptured plaques and thin-cap fibroatheromas in acute coronary syndrome. EuroIntervention 8: 477–485.

Toutouzas, K., A. Synetos, C. Nikolaou, E. Tsiamis, D. Tousoulis, and C. Stefanadis. 2012c. Matrix metalloproteinases and vulnerable atheromatous plaque. Curr. Top. Med. Chem. 12: 1166–1180.

Toutouzas, K., C. Grassos, M. Drakopoulou, A. Synetos, E. Tsiamis, C. Aggeli, K. Stathogiannis, D. Klettas, N. Kavantzas, G. Agrogiannis, E. Patsouris, C. Klonaris, N. Liasis, D. Tousoulis, E. Siores, and C. Stefanadis. 2012a. First *in vivo* application of microwave radiometry in human carotids: a new noninvasive method for detection of local inflammatory activation. J. Am. Coll. Cardiol. 59: 1645–1653.

Toutouzas, K., H. Grassos, A. Synetos, M. Drakopoulou, E. Tsiamis, C. Moldovan, G. Agrogiannis, E. Patsouris, E. Siores, and C. Stefanadis. 2011a. A new non-invasive method for detection of local inflammation in atherosclerotic plaques: experimental application of microwave radiometry. Atherosclerosis 215: 82–89.

Twerenbold, R., A. Jaffe, T. Reichlin, M. Reiter, and C. Mueller. 2012. High-sensitive troponin T measurements: what do we gain and what are the challenges? Eur. Heart. J. 33: 579–586.

Varo, N., J.A. de Lemos, P. Libby, D.A. Morrow, S.A. Murphy, R. Nuzzo, C.M. Gibson, C.P. Cannon, E. Braunwald, and U. Schonbeck. 2003. Soluble CD40L: risk prediction after acute coronary syndromes. Circulation 108: 1049–1052.

Wang, T.J., P. Gona, M.G. Larson, G.H. Tofler, D. Levy, C. Newton-Cheh, P.F. Jacques, N. Rifai, J. Selhub, S.J. Robins, E.J. Benjamin, R.B. D'Agostino, and R.S. Vasan. 2006. Multiple biomarkers for the prediction of first major cardiovascular events and death. N. Engl. J. Med. 355: 2631–2639.

Werner, N., S. Kosiol, T. Schiegl, P. Ahlers, K. Walenta, A. Link, M. Bohm, and G. Nickenig. 2005. Circulating endothelial progenitor cells and cardiovascular outcomes. N. Engl. J. Med. 353: 999–1007.

Wollert, K.C., T. Kempf, T. Peter, S. Olofsson, S. James, N. Johnston, B. Lindahl, R. Horn-Wichmann, G. Brabant, M.L. Simoons, P.W. Armstrong, R.M. Califf, H. Drexler, and L. Wallentin. 2007. Prognostic value of growth-differentiation factor-15 in patients with non-ST-elevation acute coronary syndrome. Circulation 115: 962–971.

# CHAPTER 6

# Biomarkers in Congestive Heart Failure

Chiara Andreoli and John T. Parissis*

## Introduction

Heart failure (HF) is a knotty clinical syndrome, resulting from any structural or functional cardiac abnormality that impairs the ability of the ventricle to fill with or eject blood. It is characterized by specific symptoms, such as dyspnea and fatigue, and signs, such as those associated with fluid retention.

The prevalence of HF has increased the cause of aging of the population and has started the growth of modern therapeutic innovations. However, in spite of advances in therapy, the mortality rate is still unacceptably high (Ho et al. 1993), requiring improvement in diagnosis, as well as early detection of susceptible subjects who would benefit from preventive measures. Since HF results not only from cardiac overload or injury, but also from a complex interplay among genetic, neurohormonal, inflammatory, and biochemical changes acting on both cardiac myocytes and cardiac interstitium (Cadnapaphornchai et al. 2001; Dzau 1987; Francis et al. 1984; Logeart et al. 2008; Sarraf et al. 2009), an increasing number of enzymes, hormones, biologic substances, and other markers of cardiac stress and malfunction, as well as cardiomyocyte injury, appear to have growing clinical importance (Fig. 1). These are new biomarkers, measured in the blood or urine, in addition to serum levels of hemoglobin, electrolytes, liver enzymes, and creatinine, which are routinely determined as part of clinical process (Braunwald 2008). Some biomarkers may provide important information

Heart Failure Unit, Attikon University Hospital, Athens, Greece.
*Corresponding author: jparissis@yahoo.com

concerning the pathogenesis of HF or the identification of subjects at risk for HF, and appear to be useful in risk stratification, in HF diagnosis or in monitoring therapy; many others may be risk factors themselves, therefore representing potential targets of therapy.

According to Morrow and de Lemos, a biomarker must provide information that is not already available from a detailed clinical assessment; in addition, knowing the measured level should improve medical decision making; finally biomarker measurements should be accurate, available to the clinician at a reasonable cost and with short turnaround times (Morrow and de Lemos 2007).

## Biomarker profile in CHF

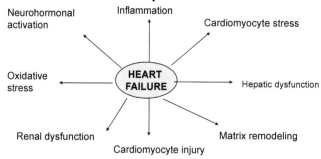

*Modified by E. Braunwald ESC HF Congress 2007*

**Figure 1.** Several biomarkers measured in the blood and in the urine have growing clinical importance playing an important role in genetic, neurohormonal, inflammatory, and biochemical changes visible among the Heart Failure setting.

## Epidemiology of Heart Failure

Precise data of HF prevalence, incidence, and prognosis are lacking (Cowie et al. 1997; Hoes et al. 1998). In addition, pre-symptomatic cardiac dysfunction is now used increasingly as an indicator of forthcoming, if not existing, HF (Bonneux et al. 1994).

A 2010 update from the American Heart Association (AHA) assessed that there were 5.8 million people with HF in the United States in 2006 (Lloyd-Jones et al. 2010). There are an estimated 23 million people with HF worldwide (McMurray et al. 1998). There has been an increase in the prevalence of HF in the population over time, with a boost from 1989 to 1999 of 1/1000 and 0.9/1000 for women and men, respectively (McCullough et al. 2002b). It is now appreciated that HF often occurs with normal left ventricular systolic function, i.e., presumably on the basis of diastolic dysfunction (Elesber and Redfield 2001; Gaasch 1994; Topol et al. 1985). In one review, the estimated prevalence of diastolic dysfunction among patients with HF was 15, 33, and 50 percent at ages less than 50, 50 to 70,

**Table 1.** Biomarkers in Heart Failure.

## Neurohormones

Norepinephrine
Renin
Angiotensin II
Aldosterone
ADH
Endothelin

## Myocyte stress

Brain natriuretic peptide
N-terminal pro-brain natriuretic peptide
Atrial natriuretic peptide
Midregional fragment of proadrenomedullin
Cotransport inhibitory factor
Growth-differentiation factor-15
ST2

## Myocyte injury

Cardiac-specific troponins I and T (4th and 5th generation assays)
Creatine kinase MB fraction
Myosin light-chain kinase I
Heart-type fatty-acid protein

## Inflammation

C-reactive protein
Tumor necrosis factor α
Interleukin 6
Chemokines (MCP-1, macrophage inflammatory protein-1 alpha, RANTES, IL-8, NAP-2, GRO-α)
Cyclooxygenase-2
Fas (APO-1)

## Oxidative stress

Oxidized low-density lipoproteins, myeloperoxidase, urinary biopyrrins, urinary and plasma isoprostanes, plasma malondialdehyde
Xanthine oxidase and uric acid
Allantoin
Biopyrrins
Nitrotyrosine
Hydroxyl radical, myocardial NADPH oxidase and 4-hydroxy-2-nonenal
Phosphodiesterase-5

## Extracellular-matrix remodeling

Matrix metalloproteinases
Tissue inhibitors of metalloproteinases
Collagen propeptides
Propeptide procollagen type I
Plasma procollagen type III

*Table 1. contd....*

*Table 1. contd.*

| Biomarkers of renal injury |
| --- |
| Proximal renal tubular epithelial antigen (HRTE-1) |
| α-Glutathione S-transferase (α-GST) |
| pi-Glutathione S-transferase (pi-GST) |
| γ-Glutamyltranspeptidase (γ-GT) |
| Alanine aminopeptidase (AAP) |
| Lactate dehydrogenase (LDH) |
| N-acetyl-beta-glucosaminidase (NAG) |
| Alkaline phosphatase (ALP) |
| α1-Microglobulin (α1-M) |
| β2-Microglobulin (β2-M) |
| Retinol binding protein (RBP) |
| Adenosine deaminase binding protein (ABP) |
| Urinary cystatin C |
| Neutrophil gelatinase-associated lipocalin (NGAL) |
| Kidney injury molecule-1 (KIM-1) |
| **New biomarkers** |
| Chromogranin |
| Galectin 3 |
| Osteoprotegerin |
| Adiponectin |

and more than 70 years, respectively (Zile and Brutsaert 2002). Also, the prevalence of HF with a preserved ejection fraction (EF) rises with age (Havranek et al. 2002; Masoudi et al. 2003; Zile and Brutsaert 2002), and it is more common among women (Bursi et al. 2006).

Similar to the prevalence, the incidence of HF increases with age (Bleumink et al. 2004). In the Framingham Study, the incidence approximately doubled over each successive decade of life, incrementing more steeply with age in women than in men. The annual incidence in men rose from 2 per 1000 at age 35 to 64 years to 12 per 1000 at age 65 to 94 years (Lloyd-Jones et al. 2002).

Despite reported declines in overall and coronary mortality, national statistics in the United States showed a rise in the death rate cause of HF from 5.8 per 1000 in 1970 to 16.4 per 1000 in 1993 (National Heart Lung and Blood Institute: NIH Bethesda MD 1996). Among individuals with HF, analyses from Scotland, the Framingham Heart Study, and the Mayo Clinic all found a progressive improvement in patient survival after 1980 (Barker et al. 2006; Jhund et al. 2009; Levy et al. 2002; Roger et al. 2004). However, average survival remained poor after hospitalization for a first episode of HF (e.g., in 2002, the median survival in Scotland was 2.3 years in men and 1.7 years in women) (Jhund et al. 2009). According to a meta-analysis of nearly 25,000 patients with HF, the mortality of HF with a preserved systolic function is about half that of HF with an impaired systolic function (Somaratne et al. 2009).

## Difficulties in Diagnosis by Conventional Methods, Physical Exams, x-ray

Though several ways exist to assess cardiac function, there is no diagnostic test for HF; it is a clinical diagnosis, mostly based upon a carefully compiled history and physical examination. In order to assess the presence of HF, at least six scoring methodologies, grounded on symptoms and signs, have been developed. Clinical diagnostic criteria for HF have generally included history, physical examination, and chest radiographs (Mosterd et al. 1997; Roger 2010). A detailed history is the single best discriminator to determine the acuity, etiology, and rate of progression of HF; nonetheless the history alone is insufficient to make a diagnosis of HF. Patients with HF may present with a syndrome of decreased exercise tolerance, fluid retention, or both (Hunt et al. 2009). The physical examination can provide evidence of the presence and extent of cardiac filling pressure elevation, volume overload, ventricular enlargement, pulmonary hypertension, and reduction in cardiac output. Symptoms of HF include those due to excess fluid retention (e.g., dyspnea, orthopnea, edema, pain from hepatic congestion, and abdominal distention from ascites) and those, mostly pronounced with exertion, due to a reduction in cardiac output (e.g., fatigue, weakness); still, many of the symptoms and signs of HF are non-specific. Therefore, various other causes for such symptoms and signs should also be considered.

The chest X-ray is a useful first diagnostic test, particularly in the evaluation of patients who present with dyspnoea, to differentiate HF from primary pulmonary disease (Badgett et al. 1996; Gillespie et al. 1997; Knudsen et al. 2004): cardiomegaly (cardiac-to-thoracic width ratio above 50 percent), as well as cephalization of the pulmonary vessels, Kerley B-lines, and pleural effusions may suggest a diagnosis of HF. In a multicenter study of 880 patients, alveolar edema, interstitial edema, and cephalization all had a specificity of >90 percent for HF, but only cardiomegaly had a sensitivity of >50 percent (Knudsen et al. 2004). A systematic review of the utility of the chest X-ray in diagnosing LV dysfunction showed redistribution and cardiomegaly as the best predictors of increased preload and reduced ejection fraction, respectively (Badgett et al. 1996). Neither finding, however, was sufficient to make a definitive diagnosis of HF.

### Neurohormones

Several reports have shown that the sympathetic nervous system is activated in patients with HF, likely playing a pathogenetic role: in the early 1960s in fact, patients with HF were found to have abnormally elevated levels of plasma norepinephrine at rest and further during exercise (Chidsey et al. 1962); the urinary excretion of norepinephrine was also increased (Chidsey et al. 1965).

The principal neurohumoral systems involved in the response to HF are the sympathetic nervous system, the renin–angiotensin–aldosterone system (RAAS), and antidiuretic hormone (Benedict et al. 1994; Dzau 1987; Francis et al. 1984); other vasoactive substances, such as the vasoconstrictor endothelin and the vasodilatory molecules atrial natriuretic peptide and nitric oxide, may be implicated. These hormonal changes are seen with both systolic and diastolic dysfunction. Although norepinephrine, angiotensin II, aldosterone, endothelin-1, and arginine vasopressin are distinct, all of them are vasoconstrictors, thereby increasing ventricular afterload. Elevated plasma levels of these neurohormones may adversely predict outcomes in patients with HF, even if they can be relatively unstable in plasma and difficult to measure. Furthermore, several of these biomarkers are presumably involved in the causal pathway for HF, since the blockade of the sympathetic nervous system and of the RAAS are cornerstones of current pharmacologic treatment of HF.

## Sympathetic nervous system

The activation of the sympathetic nervous system is one of the first reactions to a decrease in cardiac output, resulting in both increased release and decreased uptake of norepinephrine (NE) at adrenergic nerve endings. In addition, downregulation of peripheral α-2 receptor function, which normally inhibits NE release, may contribute to sympathetic activation in HF (Aggarwal et al. 2001).

Early in HF, catecholamine-induced increase of ventricular contractility and heart rate help maintaining cardiac output, particularly during exercise. However, with progressive worsening of ventricular function, these mechanisms are no longer sufficient (Bhargava et al. 1998). Increased sympathetic activity also participates in the maintenance of blood pressure by augmenting ventricular preload, through systemic and pulmonary vasoconstriction and enhanced venous tone. Glomerular filtration can be relatively well maintained despite a fall in renal blood flow, increasing renal filtration fraction by the efferent arteriole vasoconstriction. In addition, both NE and angiotensin II also contribute to the sodium retention of HF by stimulating proximal tubular sodium reabsorption.

Plasma norepinephrine level has been shown as an independent predictor of mortality correlating directly with the severity of the cardiac dysfunction and inversely with survival (Cohn et al. 1984). In an analysis of the Val-HeFT trial, patients with an initial higher plasma NE concentration (≥572 pg/mL) had a significantly higher mortality rate at two years than those with an initial lower plasma NE concentration (<274 pg/mL) (24.2 versus 13.8 percent) (Anand et al. 2003).

The degree of sympathetic activation can be reduced by proper treatment of HF, as with administration of an ACE inhibitor (Sigurdsson et al. 1994). In the SOLVD trial, patients who had more marked neurohormonal activation had a larger survival benefit with ACE inhibition than patients with less activation (The SOLVD Investigators 1991).

In addition to systemic sympathetic activation, there is an increase in cardiac efferent sympathetic activity in patients with HF, as demonstrated by increased cardiac NE levels in cardiac veins (Kaye et al. 1994; Kaye et al. 1995). Such cardiac NE spillover is decreased by ventricular filling pressures reduction (Azevedo et al. 2000) and with amiodarone (Kaye et al. 1999).

## Role of β-adrenergic receptors

The chronic increase in sympathetic activity also leads to down-regulation and reduction in density of the cardiac β-adrenergic receptors (Nozawa et al. 1998); particularly, in HF, there is a selective reduction in the density of β-1 but not β-2 receptors (Altschuld et al. 1995). As a result, the failing heart is more dependent upon β-2 adrenergic receptors for inotropic support. Myocardial β-2 adrenergic receptors have been demonstrated as being responsible for both beneficial and deleterious effects in HF: the β-2 receptor displays a small decrease in binding affinity for catecholamines and exerts an anti-apoptotic effect (Communal et al. 1999; Noma et al. 2007); on the other hand, stimulation of β-2 receptors may also mediate adverse effects, increasing NE release and consequently heightening the propensity for ventricular fibrillation by afterpotentials (Newton and Parker 1996). In addition, chronically increased stimulation of β-adrenergic receptors may cause molecular and cellular abnormalities, by the re-expression of fetal protein isoforms and the loss of cardiomyocytes due to apoptosis or necrosis (Communal et al. 1998; Nakayama et al. 2007; Vatner et al. 1999). These changes may explain the beneficial effects of β-blockers on both cardiac performance and survival in some patients with chronic HF. Moreover, acute studies suggest that cardiac sympathetic activity is reduced to a greater degree by nonselective than selective β blockers (Newton and Parker 1996).

## Renin–angiotensin system

Decreased stretch of the glomerular afferent arteriole, reduced delivery of chloride to the macula densa, and increased β-1 adrenergic activity, all stimulate renal renin release in HF. Given that angiotensin II can also be synthesized locally at a variety of tissue sites (Mizuno et al. 2001), measurement of the plasma renin activity or angiotensin II concentration may underestimate tissue angiotensin II activity. As an example, the plasma renin activity is often normal in patients with stable, chronic HF, despite persistence of the low output state and renal sodium retention; in opposition, plasma renin levels are usually markedly elevated in patients with recent onset or very symptomatic HF (Dzau et al. 1981).

Angiotensin II has similar actions to NE in HF, increasing sodium reabsorption and inducing systemic and renal vasoconstriction; this hormone can also act directly on myocytes and other cell types in the myocardium promoting pathologic remodeling by myocyte hypertrophy, re-expression of fetal protein isoforms, myocyte apoptosis, and alterations in the interstitial matrix.

## *Aldosterone*

Angiotensin II generates a secondary hyperaldosteronism in HF by stimulation of the adrenal glands and by induction of aldosterone synthase (CYP11B2), leading to a local production of aldosterone in proportion to the severity of HF (Mizuno et al. 2001; Silvestre et al. 1999). In the Randomized Aldactone Evaluation Study (RALES) of patients with severe HF (Zannad et al. 2000), the administration of the aldosterone blocker spironolactone was associated with a reduction of plasma procollagen type III and with a clinical benefit, but only in patients whose baseline levels of the procollagen were above the median. Therefore, blockade of the adverse effects of aldosterone-induced stimulation of cardiac mineralocorticoid receptors, limiting the synthesis of the extracellular matrix, is thought to contribute to the survival benefit associated with the administration of mineralocorticoid receptor antagonists in selected patients with HF.

## *ACE gene polymorphism*

Plasma and tissue concentrations of ACE, and therefore of angiotensin II, are in part determined by the ACE gene. This gene may manifest insertion (I) or deletion (D) polymorphism, and has three genotypes (DD, ID, and II). Plasma and cardiac levels of ACE are 1.5 to 3-fold higher in patients with the DD compared to the II genotype; the values are intermediate in patients with ID genotype (Danser et al. 1995). The DD genotype of the ACE gene has been associated with a number of adverse cardiovascular events: there may be an association between the DD genotype and increased mortality in patients with HF due to idiopathic dilated cardiomyopathy, that may be related to progression of HF rather than to arrhythmic sudden cardiac death (Bedi et al. 2004). This difference may be abolished with β-blocker therapy (McNamara et al. 2001).

## Antidiuretic hormone

Antidiuretic hormone (ADH or vasopressin) is a nonapeptide synthesized in the hypothalamus and stored in the posterior pituitary gland; it has antidiuretic and vasoconstrictor properties, and it is released following the carotid sinus and aortic arch baroreceptors activation by the low cardiac output in HF. Whereas plasma levels of ADH are elevated in patients with acute or chronic HF (Schrier 2006) and are associated with poor clinical outcomes, vasopressin 2 receptor antagonists relieve acute symptoms but do not appear to alter the natural history of severe HF (Konstam et al. 2007). Elevated levels of ADH, by the combination of decreased water excretion and increased water intake via thirst, often lead to a fall in the plasma sodium concentration that parallels the severity of the HF. As a result, the degree of hyponatremia is an important predictor of survival in these patients.

## Endothelin

Endothelin may contribute to the regulation of myocardial function, vascular tone, and peripheral resistance in HF. Endothelin is synthesized as a large pre-pro-hormone (big ET-1) that is cut to pro-ET and then undergoes further modification to yield mature ET. Among the three sequenced isoforms, ET-1 is the predominant isoform synthesized in the human vasculature; ET-1 is a powerful stimulant of vascular smooth-muscle contraction and proliferation and ventricular and vessel fibrosis; it is also a potentiator of other neurohormones (Teerlink 2005). ET-1 plays an important role in the cardiac hypertrophic response to pressure or volume overload (Choukroun et al. 1998). In addition, circulating ET-1 modulates renal function and the activity of the renin-angiotensin and sympathetic nervous systems, helping to maintain volume homeostasis in normal subjects; contra, increased circulating levels of ET-1 in HF may contribute to renal vasoconstriction and sodium retention (Ding et al. 2002). Moreover, ET-1 stimulates the secretion of norepinephrine, angiotensin II, and vasopressin (Wong et al. 2003). Although endothelin exerts a potent inotropic activity in the myocardium, mediated in part by increased sensitivity of the myofilaments to calcium, there is some evidence that endothelin may have a negative inotropic effect in the failing heart (MacCarthy et al. 2000). Finally, ET-1 may be pro-arrhythmic in the setting of HF (Aronson and Burger 2003). Endothelins are rapidly cleared from the circulation by the lung, liver, and kidneys.

There are two major endothelin receptors: ET-A and ET-B. ET-A receptors on vascular smooth muscle cells mediate vasoconstriction and smooth muscle cell proliferation. ET-B receptors are found primarily on vascular endothelial cells and mediate vasodilation via the release of endothelium-derived nitric oxide; they are also located on vascular smooth muscle cells where they mediate vasoconstriction.

The plasma levels of both endothelin-1 and big endothelin-1 are increased in patients with HF and correlate directly with pulmonary artery pressure (Moraes et al. 2000), disease severity, and mortality (Hulsmann et al. 1998). In the Val-HeFT study the prognostic values of plasma neurohormones were compared among 4300 patients: the most powerful predictors of mortality and hospitalization for HF, after BNP, were big endothelin-1, followed by norepinephrine, endothelin-1, plasma renin activity, and aldosterone (Latini et al. 2004). However, trials involving several endothelin-1 receptor antagonists have failed to show any beneficial effects on clinical outcomes, in spite of some evidence of harm including fluid retention and LFT abnormalities.

## Biomarkers of Cardiomyocyte Stress

### *Natriuretic peptides*

The natriuretic peptide system is involved in salt and water handling and pressure regulation; also, it may influence myocardial structure and function.

Brain natriuretic peptide (BNP) is a natriuretic hormone initially recognized in the brain but released primarily from the ventricles; the pro-hormone proBNP is cleaved into a biologically active 32 amino acid BNP as well as biologically inert 76 amino acid N-terminal pro-BNP (NT-proBNP).

Atrial natriuretic peptide (ANP), a 28-amino acid polypeptide, is also released from myocardial cells, mostly the atria and in some cases the ventricles, responding to volume expansion and increased wall stress (Iwanaga et al. 2006). Both ANP and BNP are more released in HF, in response to the high ventricular filling pressures, ventricles dilation, hypertrophy, or increased wall tension (Kinnunen et al. 1993); plasma concentrations of both hormones are greater in patients with asymptomatic and symptomatic left ventricular dysfunction, permitting their use in diagnosis. Both ANP and BNP, through arterial vasodilation, have diuretic, natriuretic, and hypotensive effects; moreover they inhibit the RAAS, endothelin secretion, and systemic and renal sympathetic activity, opposing the physiological abnormalities in HF (Brunner-La Rocca et al. 2001; Elsner et al. 1992; Wilkins et al. 1993). BNP may also protect against collagen accumulation and the pathologic remodeling that contributes to progressive HF (Tamura et al. 2000). Measurement of BNP may be more useful than ANP in evaluation of patients with HF, because the circulating concentration of BNP is less than 20 percent of that of ANP in normal subjects, but can equal or exceed that of ANP in patients with HF.

Assays for BNP and NT-pro-BNP are commercially available, these being the most widely tested biomarkers of HF; such testing is recommended in current guidelines (Heart Failure Society of America 2006; McMurray et al. 2012; Tang et al. 2007). Plasma BNP concentrations have been variously reported in units of pmol/L, ng/L, or pg/mL. The currently available rapid assay reports results in pg/mL; the assay range is 5 to 1300 pg/mL or 1.4 to 376 pmol/L (Tjeerdsma et al. 2002).

Some physiological conditions and disease states must be taken into consideration in the interpretation of natriuretic peptides (Table 2): there is a moderate increase in the level of circulating BNP with increasing age, likely cause of myocardial fibrosis or renal dysfunction (Tang et al. 2007), and with pulmonary hypertension (Sugiura et al. 2005); obese patients tend to have lower plasma BNP and NT-proBNP concentrations than non-obese patients (Das et al. 2005; Horwich et al. 2006; McCord et al. 2004; Tang et al. 2007); natriuretic peptide levels are also elevated in some patients with non-heart failure conditions such as coronary heart disease, valvular heart disease and sepsis.

Several other conditions may influence BNP and NT-proBNP values. In patients with chronic kidney disease, the hypervolemia, left ventricular hypertrophy and hypertension characteristic of renal failure enhance the secretion of BNP, especially the NT-pro-BNP; moreover, since natriuretic peptides are cleared by the kidneys, decreased estimated GFR is associated with increased plasma BNP and even greater elevation in NT-proBNP concentrations (Vickery et al. 2005). As a result, the cut-off values for

Table 2. Conditions that affect plasma levels of natriuretic peptides other than acute or chronic heart failure.

**A. Conditions without heart failure and elevated plasma natriuretic peptides**

- advanced age
- acute coronary syndromes
- acute pulmonary embolism
- acute respiratory distress syndrome
- high output states
- renal failure
- atrial tachy-arrhythmias

**B. Conditions with heart failure and unexpected low plasma natriuretic peptides**

- flash pulmonary oedema
- acute mitral regurgitation
- mitral stenosis
- left cardiac tumors
- constrictive pericarditis
- cardiac tamponade
- obesity

plasma BNP in patients with renal insufficiency are different from those in patients with normal renal function. Measuring plasma NT-proBNP has an added problem since NT-proBNP is cleared by the kidney and its plasma concentration is elevated by renal failure alone. Defined cut-off values for NT-proBNP in this setting do not exist; in a review of 599 dyspnoeic patients with a serum creatinine ≤2.5 mg/dL, cut-off values for patients with an estimated glomerular filtration rate (GFR) of ≥60 mL/min per 1.73 m² were >450 pg/mL in those less than 50 years of age and >900 pg/mL in older patients; the cut-off was instead 1200 pg/mL for patients with a GFR <60 mL/min per 1.73 m² (Anwaruddin et al. 2006).

## *Brain natriuretic peptide*

BNP has an estimated plasma half-life of about 20 minutes (Mair 2008). Since symptoms and physical findings may not be sufficiently sensitive to make an accurate diagnosis of HF, the assay for plasma BNP is a useful test in the evaluation of patients with dyspnoea and suspected HF when the diagnosis is uncertain (Maisel et al. 2002). Patients with HF are more likely to have BNP values above 400 pg/mL, while values below 100 pg/mL have a very high negative predictive value for HF as a cause of dyspnoea; on the other hand, in the range between 100 and 400 pg/mL, other diagnoses, such as pulmonary embolism, LV dysfunction without exacerbation, and cor pulmonale should be considered.

The Breathing Not Properly (BNP) study was conducted among 1,586 patients with acute dyspnoea to appraise the value of rapid bedside measurement of plasma BNP in the differential diagnosis between HF and a pulmonary cause of dyspnoea (Maisel et al. 2002). Plasma BNP levels were higher in patients with clinically diagnosed HF compared to those without HF (mean 675 versus 110 pg/mL); intermediate values were found in asymptomatic patients with baseline LV dysfunction (346 pg/mL). A plasma BNP >100 pg/mL diagnosed HF with a sensitivity, specificity, and predictive accuracy of 90, 76, and 83 percent, respectively; choosing values >125 or >150 pg/mL decreased sensitivity, increased specificity, and did not change overall predictive accuracy. Comparing the predictive accuracy of plasma BNP for HF to other parameters, BNP was equivalent to or better than cardiomegaly on chest X-ray, a history of HF, or rales on physical examination, as well as NHANES and Framingham criteria. Instead, among patients with AF, a cut-off of ≥200 pg/mL had an increased specificity, from 40 percent to 73 percent, with a smaller reduction in sensitivity from 95 percent to 85 percent. Furthermore, comparing plasma BNP to initial clinical judgment, a plasma BNP >100 pg/mL was more sensitive (90 percent versus 49 percent) but less specific (73 percent versus

96 percent) than clinical judgment; BNP in addition to clinical judgment also increased diagnostic accuracy of HF diagnosis from 74 percent to 81 percent (McCullough et al. 2002a).

Regarding diastolic dysfunction, BNP has a similar accuracy to systolic dysfunction, but the values do not differentiate between systolic and diastolic dysfunction (Maisel et al. 2001). However, as with any diagnostic test, individual patient variation is very important: indeed, it has been demonstrated that not all patients with symptomatic HF have high plasma BNP concentrations, and not all asymptomatic patients have low values (Tang et al. 2003). Given that, serial serum BNP concentrations may be more reliable than a single measurement.

Plasma BNP concentration has been demonstrated as a predictor of outcome in both acute and chronic HF. Plasma BNP provides prognostic information in patients with chronic HF (The SOLVD investigators 1991, Berger et al. 2002; de Groote et al. 2004; Koglin et al. 2001; Stanek et al. 2001), and those with asymptomatic or minimally symptomatic LV dysfunction (Maeda et al. 2000; Richards et al. 1999; Tsutamoto et al. 1999); the relative risk of death rises by 35 percent every 100 pg/mL increase in plasma BNP (Doust et al. 2005). Patients with a baseline plasma BNP concentration in the highest quartile ($\geq$238 pg/mL), in an analysis from the Val-HeFT trial, had a significantly greater mortality at two years compared to those with a plasma BNP in the lowest quartile (<41 pg/mL) (32.4 versus 9.7 percent) (Anand et al. 2003). Likewise, persistent elevation of plasma BNP had prognostic significance in patients in optimal medical therapy (Berger et al. 2002; Maeda et al. 2000) and in hospitalized patients prior to discharge (Logeart et al. 2004). In the ADHERE registry also, there was a nearly linear relationship between BNP quartiles and in-hospital mortality (Fonarow et al. 2007). According to both American and European HF guidelines, measurement of natriuretic peptide levels (BNP and NT-proBNP) can be useful in the evaluation and risk stratification of patients in whom the clinical diagnosis of HF is uncertain (Hunt et al. 2009; McMurray et al. 2012). Finally, the STARS-BNP trial (Jourdain et al. 2007) suggested that it may be better to direct therapy by BNP level rather than just medical treatment according to either current guidelines.

Nonetheless, the use of plasma BNP for diagnosis and follow-up of HF has several important limitations: more than one cause of dyspnoea (such as pneumonia and an exacerbation of HF) may exist; in some patients with acute decompensated HF, plasma BNP levels are not diagnostic while some patients with severe chronic HF may have persistently elevated plasma BNP concentrations; elevations in plasma BNP are common in right heart failure and pulmonary hypertension. Therefore, plasma BNP measurements alone are not sufficient to guide therapeutic decision-making in patients with HF, they should instead be used as an addition to clinical assessment.

Natriuretic peptides also appear to be useful in screening asymptomatic subjects at risk of developing HF (Daniels and Maisel 2007; Vickery et al. 2005) and they may also be used to screen for acute or late cardiotoxic effects associated with cancer chemotherapy (Suzuki et al. 1998).

## N-Terminal pro-brain natriuretic peptide

BNP is cleaved from the C-terminal end of its pro-hormone, pro-BNP; thus the N-terminal fragment, N-terminal pro-BNP (NT-proBNP), is also released into the circulation. The plasma half-life of NT-proBNP is longer than that of BNP (approximately 25 to 70 versus 20 minutes) (Mair 2008).

Like BNP, NT-proBNP has been demonstrated as a useful tool in the evaluation of dyspnoeic patients in the acute setting. Among 1,256 patients with and without acute heart failure, differences in plasma NT-proBNP were compared, and for patients <50, 50 to 75, and >75 years of age, the optimal plasma NT-proBNP cut-offs for diagnosing HF were 450 pg/mL, 900 pg/mL, and 1800 pg/mL respectively, with a sensitivity and specificity of 90 percent and 84 percent, respectively. Overall, NT-proBNP levels below 300 pg/mL excluded a diagnosis of HF with a negative predictive value of 98 percent (Januzzi et al. 2006).

Plasma NT-proBNP is a prognostic predictor in patients with acute and chronic HF: in a subgroup of 1,011 patients in the COPERNICUS study, all-cause mortality at one year was significantly higher for patients with a plasma NT-proBNP above the median (1767 pg/mL) than in those with levels below the median (22 versus 7 percent, risk ratio 2.7) (Hartmann et al. 2004). N-terminal pro-BNP may also predict cardiac involvement and prognosis in patients with AL amyloidosis (Palladini et al. 2003) and may be a useful screen for left ventricular dysfunction in children who have received anthracycline chemotherapy (Poutanen et al. 2003).

## Comparison of plasma BNP and N-terminal pro-BNP

Limited data comparing diagnostic and prognostic values of plasma BNP and NT-pro-BNP are available. Whereas in normal subjects, the plasma concentrations of BNP and NT-proBNP are similar (approximately 10 pmol/L), in patients with LV dysfunction, plasma NT-proBNP rises more than BNP. There is no simple conversion factor between BNP and NT-proBNP levels; anyway, an NT-proBNP level >900 pg/mL provides approximately equivalent accuracy as a BNP level of >100 pg/mL for diagnosis of HF. The N-terminal pro-hormone was slightly superior to BNP for predicting death or rehospitalization for HF in two studies which have directly compared BNP and NT-pro-BNP (Masson et al. 2006; Poutanen et al. 2003). The longer half-life of NT-pro-BNP indeed may make it a better predictor of prognosis by a more accurate indexing of ventricular stress. A mortality benefit has been suggested by randomized trials studying the effect on clinical outcomes of BNP or NT-proBNP guided therapy (Berger et al. 2010; Karlstrom et al. 2011; Lainchbury et al. 2009).

Among AHF, since the half-life for BNP is shorter than that for NT-proBNP, serial measurements may be useful in the management of acute decompensated HF. Also, the value of serial natriuretic peptide measurements has not been established in some settings, such as patients with acute HF (first hour or so), with inflow tract obstruction (e.g., mitral stenosis) or constrictive pericarditis (Nohria et al. 2005).

Concerning the hemodynamic response to therapy in patients treated with nesiritide, plasma BNP should not be measured, since nesiritide will be detected as an acute increase in plasma BNP concentration (Heublein et al. 2007; Marcus et al. 1996); instead the NT-proBNP assay does not detect nesiritide (Heublein et al. 2007). Thus, NT-proBNP assay may be helpful in assessing the response to therapy, although there are limited data to support this approach (Miller et al. 2009).

## Atrial natriuretic peptide

Atrial natriuretic peptide (ANP) release is increased in HF, being primarily released from the atria in response to volume expansion, which is sensed as an increase in atrial stretch; ANP levels correlate more closely with atrial volume than atrial pressure (Globits et al. 1998).

The prohormone of atrial natriuretic peptide (pro-ANP) is a polypeptide of 126 amino acids; given that the N-terminal portion of proANP, termed proANP1-98 or NT-proANP, has a much longer half-life than mature ANP, it has been proposed as a more reliable analyte for measurement than ANP. An investigational commercial assay can detect mid-regional pro-atrial natriuretic peptide (MR-proANP), which appeared to be noninferior to BNP for diagnosis of acute HF and to have prognostic value in patients with chronic HF.

The utility of monitoring MR-proANP as compared to BNP was evaluated in the multicenter BACH (Biomarkers in Acute Congestive Heart Failure) trial (Maisel et al. 2010), among 1,641 patients presenting to the emergency department with shortness of breath: MR-proANP (≥120 pmol/l) was noninferior to BNP (≥100 pg/ml) for the diagnosis of AHF (sensitivity 97 percent versus 95.6 percent; specificity 59.9 percent versus 61.9 percent). MR-proANP also provided additional discriminant value to BNP in patients with intermediate BNP (or NT-proBNP levels) or obesity, without renal insufficiency, age >70 yr, or edema (Maisel et al. 2010).

Measurement of MR-proANP also provided prognostic information independent of NT-proBNP in the GISSI-HF trial (Masson et al. 2010). Among 1,237 patients with chronic stable HF, the addition of MR-proANP to models based on clinical risk factors alone or together with NT-proBNP, improved classification for mortality (net reclassification improvement [NRI] = 0.12 and 0.06, respectively); MRproANP levels were also associated with mortality (HR 1.38, 95% CI 0.99–1.93 and HR 1.58, 95% CI 1.13–2.21 in the middle and highest versus lowest tertiles).

## Adrenomedullin

Adrenomedullin was originally isolated from human pheochromocytoma cells; it is a peptide with an amino acid sequence similar to human calcitonin gene-related peptide, a potent vasodilator. In addition to vasodilatory effects on the vasculature, in the heart adrenomedullin enhances myocardial contractility via a cyclic AMP-independent mechanism, suggesting that it may play a role in the compensatory mechanisms in HF (Vickery et al. 2005). Circulating levels of adrenomedullin are indeed elevated in patients with HF, especially in those with evidence of diastolic dysfunction and a restrictive left ventricular filling pattern (Heart Failure Society of America 2006), and correlate with the degree of left ventricular function and elevation in pulmonary artery pressure (Tang et al. 2007). The arterial vasodilation due to adrenomedullin infusion is mediated by a nitric oxide-dependent mechanism, and is attenuated in patients with HF (Maisel et al. 2002); however, even this lesser degree of arterial vasodilation is sufficient to increase the cardiac index and reduce the pulmonary capillary wedge pressure (Januzzi et al. 2005).

An elevated plasma adrenomedullin concentration is an independent predictor of a poor outcome: in the Australia-New Zealand Heart Failure study of 297 patients, patients with supramedian baseline serum adrenomedullin concentrations had, compared to those with inframedian levels, an increased risk of all-cause mortality (risk ratio 3.92), HF mortality (risk ratio 4.87), and hospital admission for HF (risk ratio 2.4) (Richards et al. 1999).

## Cotransport inhibitory factor

Cotransport inhibitory factor (CIF) is a nondigitalis-like natriuretic factor acting like loop diuretic drugs and inhibiting the furosemide-sensitive Na-K-Cl cotransport system. As opposite to ANP, CIF appears to regulate long-term renal sodium excretion. CIF plasma and urinary activities are increased in patients with HF and, although changing in parallel with

ANP and being correlated with the decrease in LVEF, the CIF plasma concentration is twice as great as those of plasma ANP. These changes of CIF may potentially express the triggering of homeostatic, long-term mechanism to counterbalance fluid overload in CHF (Dubois-Rande et al. 1996).

## *Growth-differentiation factor-15*

Growth-differentiation factor-15 (GDF-15) is a member of the transforming growth factor-cytokine superfamily (Bootcov et al. 1997); it is weakly expressed in most tissues under physiological conditions, but its expression may increase significantly in response to inflammation and tissue injury (Kempf et al. 2006). Since increased cardiac expression of GDF-15 has been observed in mouse models of myocardial infarction, pressure overload, and HF (Kempf et al. 2006), it has been hypothesized that GDF-15 shares some of its upstream regulatory stimuli with BNP (LaPointe 2005); but, in contrast to BNP, which is produced almost exclusively in the heart, GDF-15 is not a cardiac-specific factor. Other cardiovascular cell types indeed, such as endothelial cells, vascular smooth muscle cells, and adipocytes, have been found to produce GDF-15 under stressful conditions (Ding et al. 2009).

In addition, circulating levels of GDF-15 are associated with prognosis in patients with cardiovascular disease. In stable and unstable coronary artery disease, GDF-15 was associated with the risk of death and myocardial infarction, independently of clinical variables, renal function, and other biomarkers like C- reactive protein (CRP), N-terminal pro-BNP, and cardiac troponins (Kempf et al. 2007a; Wollert et al. 2007). Referring to the potential role of GDF-15 as a biomarker in HF, baseline levels of GDF-15 were associated with all-cause mortality during one year of follow up in an initial study in 455 HF patients; in that study, GDF-15 also added prognostic information to New York Heart Association (NYHA) class, left ventricular ejection fraction, and N-terminal pro-BNP (Kempf et al. 2007b). GDF-15 was confirmed to add prognostic information to N-terminal pro-BNP in 158 patients with advanced HF receiving resynchronization therapy (Foley et al. 2009). In a subanalysis of the Valsartan Heart Failure Trial (Val-HeFT), GDF-15 levels were associated with several pathological processes connected to the severity and progression of HF, including neurohormonal activation, inflammation, myocyte death, and renal dysfunction; circulating levels of GDF-15 appeared also to be associated with adverse outcomes independently of established clinical and biochemical risk markers (Anand et al. 2010).

Concerning the independent association of GDF-15 with adverse outcome among patients with coronary artery disease (Kempf et al. 2007a; Wollert et al. 2007), it is important to underline that GDF-15 was related to outcome in patients with HF that was or was not attributed to ischemic heart disease; rather, the association was stronger for patients with nonischemic cardiomyopathy. By now, it is still to be investigated how the prognostic information provided by GDF-15 could be used clinically, alone or in combination with other biomarkers, for patient monitoring and deciding treatment strategies.

## *ST2*

ST2 is a novel biomarker of cardiac stress, indicating the presence and severity of adverse cardiac remodeling and tissue fibrosis, which occurs in response to myocardial infarction, acute coronary syndrome, or worsening HF (Rehman et al. 2008; Shah and Januzzi 2010). ST2 is a member of the interleukin 1 receptor family, and consists of both a transmembrane receptor form (the ST2 receptor or ST2L) and a truncated, soluble receptor form (the soluble ST2 or sST2) that can be detected in serum. ST2L, the functionally active transmembrane form, is involved in the development of immunologic tolerance and plays a role in modulating responses of T-helper type 2 cells; the ST2 gene is also markedly upregulated in cardiac myocytes and fibroblasts subjected to mechanical strain (Kohli et al. 2012). ST2 can reduce fibrosis by binding of its ligand Interleukin-33 (IL-33), a cardiac fibroblast product also induced by mechanical strain; the ST2/ IL-33 signaling, similar to BNP, is thought to play an important role in regulating the myocardial response to biomechanical overload in stretched cardiac fibroblasts and cardiomyocytes (Weinberg et al. 2003). Nevertheless, patients with excessively increased soluble ST2 levels may paradoxically experience excessive cardiac fibrosis and worse outcomes in HF, because soluble ST2 binds IL-33 and makes it unavailable to the ST2 receptor for cardioprotective signaling.

Serum levels of ST2 have been reported to increase significantly early after acute myocardial infarction, inversely correlating with ejection fraction (Weinberg et al. 2002). Among patients with chronic HF, serum ST2 levels were associated with adverse outcome (Shimpo et al. 2004; Weinberg et al. 2002), although few detailed data regarding ST2 and acute HF exist. Concentrations of soluble ST2 have appeared as a prognostically meaningful biomarker in HF, paralleling the severity of the disease and strongly predicting mortality: among 346 patients with acutely destabilized HF, ST2 concentrations segregated with more severe HF, and had several independent associations with variables such as left ventricular ejection fraction and renal function (Rehman et al. 2008). In addition, even though these associations were considerably weaker than those reported with natriuretic peptides, several important covariates for natriuretic peptides (such as age, gender, heart rhythm, and body mass index) did not have a significant effect on ST2 concentrations, either in univariable or multivariable linear regression analyses. This relative independence from prevalent comorbidities in patients with HF might represent a potential advantage of ST2 for prognostication over other widely used markers such as the natriuretic peptides. Furthermore, associations between measures of inflammation, such as with temperature, leukocyte count, and CRP, were present with ST2 but not with natriuretic peptides (Rehman et al. 2008).

## Biomarkers of Myocardial Injury

Cardiomyocyte injury, resulting from severe ischemia, but also from inflammation, oxidative stress, and neurohormonal activation, causes the disruption of normal cardiac myocyte membrane integrity, leading to detectable levels of a variety of biologically active cytosolic and structural proteins such as troponin, creatine kinase, myoglobin, heart-type fatty acid binding protein, and lactate dehydrogenase. The cardiac troponins T and I have emerged during the last 20 years as sensitive and specific markers of myocyte injury in acute coronary syndromes; modest elevations of cardiac troponin levels are also found in patients with HF without ischemia (La Vecchia et al. 2000).

### *Troponins*

Cardiac troponin I (cTnI) and T (cTnT) are cardiac regulatory proteins, with cytosolic or early releasable and structural pools, involved in controlling the calcium-mediated interaction of actin and myosin (Adams et al. 1993a). These proteins are products of specific genes and therefore are unique for the heart: studies performed with cTnI have failed to find any cTnI outside of the heart at any stage of neonatal development (Adams et al. 1993b); as opposite, cTnT is expressed to a minor extent in skeletal muscle, and recent data indicate that skeletal muscle can, in some patients, be the source for elevations of cTnT detected in the blood (Jaffe et al. 2011).

Heart failure can generate the release of cardiac troponin via both myocardial strain and myocyte death. Regarding myocardial strain, excessive wall tension due to volume and pressure overload of both the right and left ventricle can result in myofibrillar damage; supporting such a connection between myocardial strain and elevated troponin levels, there is a close correlation between troponin levels and BNP too. Moreover, increased myocardial wall stress may decrease subendocardial perfusion, resulting in troponin elevation and decline in left ventricular systolic function (Wallace et al. 2006). Several factors, such as the activation of the renin-angiotensin system, sympathetic stimulation, inflammatory mediators, as well as integrin stimulation, may contribute to myocyte death.

Furthermore, a link between myocardial wall stretch and programmed cell death, which may also contribute to troponin elevations in this setting, has been demonstrated by *in vitro* experiments (Saenger and Jaffe 2008). Since myocyte loss may play an important role in the progression of cardiac dysfunction, this may explain the bad prognosis of patients with HF and elevated troponin levels.

Cardiac troponin I was detectable ($\geq$0.04 ng per milliliter) in approximately half of 240 patients with advanced, chronic HF without ischemia, and after adjustment for other variables associated with poor prognosis, the presence of cardiac troponin I remained an independent predictor of death (Horwich et al. 2003). In patients with chronic HF, cardiac troponin T levels greater than 0.02 ng per milliliter were found to be associated with a hazard ratio for death bigger than 4 (Hudson et al. 2004). Troponin levels are also reported as a predictor of outcome in hospitalized patients with acute decompensated HF (Peacock et al. 2008). In addition, cardiac troponin T was detectable in 10 percent of patients with chronic HF, but in 92 percent with a new high-sensitivity assay, being associated with an increased risk of death (Latini et al. 2007). As this latter study shows, previously nondetectable levels of cardiac troponin T may provide important additional prognostic information; therefore, with an increased sensitivity of cardiac troponin, the biomarker could be used routinely, along with the natriuretic peptides, to assess the prognosis and response to treatment of patients with HF.

## Creatine kinase

The enzyme creatinine kinase (CK) (called creatinine phosphokinase in the past) exists as isoenzymes, in three combinations of M and B chains dimers: MM, MB, and BB. Whereas CK isoenzyme activity is distributed in many tissues, including skeletal muscle, there is more of the CK-MB fraction in the heart; thus elevations in total serum CK lack specificity for cardiac damage, which improves with measurement of the MB fraction (Roberts et al. 1975).

## Other myocardial proteins

Myosin light chain 1 and heart fatty-acid binding protein are other myocardial proteins, circulating in stable patients with severe HF. Like cardiac troponin T, the presence of these myocardial proteins may be an accurate predictor of death or hospitalization for HF (Sugiura et al. 2005). To determine whether these other biomarkers of myocyte injury add information, future studies should compare their predictive accuracy with that of troponin, measured with a high-sensitivity assay.

## Biomarkers of inflammation

Biomarkers of inflammation have largely been studied in the pathogenesis and progression of many forms of HF (Anker and von Haehling 2004). HF is often characterized by increases in circulating proinflammatory cytokines and their soluble receptor or receptor antagonists, that parallel the myocardial function deterioration (Rauchhaus et al. 2000; Torre-Amione et al. 1996a). Elevated levels of circulating tumor necrosis factor-α (TNF-α) in patients with HF have already been described in 1990 (Levine et al. 1990). According to the cytokine hypothesis of HF, a precipitating event (e.g., ischemic cardiac injury) triggers innate stress responses, including elaboration of proinflammatory cytokines, associated with deleterious effects on left ventricular function, and accelerating the progression of HF (Seta et al. 1996). Proinflammatory cytokines have been shown to cause myocyte apoptosis and necrosis; interleukin-6 (IL-6) induces a hypertrophic response in myocytes (Seta et al. 1996), whereas TNF-α causes left ventricular dilatation through activation of matrix metalloproteinases. Both IL-6 and TNF-α levels may predict the future development of HF in asymptomatic elderly subjects (Lee and Vasan 2005). Moreover, increased levels of proinflammatory cytokines and other inflammatory markers may identify patients at increased risk of developing HF in the future (Murray and Freeman 2003).

## C-reactive protein

C-reactive protein (CRP) is an acute phase protein, mostly produced by hepatocytes under the influence of cytokines such as IL-6 and TNF-α (Castell et al. 1990; Kushner 1982). Even though acute and chronic infections, cigarette smoking, acute coronary syndromes, and active inflammatory states are frequently associated with elevated levels of CRP, these have been found associated with the prevalence of underlying atherosclerosis, risk of recurrent cardiovascular events, and incidence of first cardiovascular events among individuals at risk for atherosclerosis (Pearson et al. 2003; Zacho et al. 2008). Furthermore, several drugs used in the treatment of cardiovascular disease reduce serum CRP; such a property likely contributes to their beneficial effects. However, until now, there is no established role for routine measurement of hsCRP in patients with cardiovascular disease; moreover, it is still under discussion if serum CRP is a nonspecific marker increased as part of the acute phase response to inflammation, or rather a direct participant in the progression of atherosclerosis (Folsom et al. 2001; Nilsson 2005; Scirica and Morrow 2006).

Few data are available regarding the predictive value of serum CRP in patients with HF (Anand et al. 2005; Lamblin et al. 2005). Increased CRP levels were independent predictors of adverse outcomes in patients with acute or chronic HF in a post hoc analysis from the Val-HeFT trial; the median serum CRP was also higher in patients with HF compared to the general population, and the adjusted hazard ratio for mortality was 1.53 in the highest compared to the lowest quartile for serum CRP (Anand et al. 2005). According to a scientific statement from the Center for Disease Control and Prevention and the American Heart Association, a value >3 mg/L is appropriate for predicting outcomes in patients with stable CHD and a threshold >10 mg/L may be more predictive in patients with an acute coronary syndrome (Pearson et al. 2003).

Furthermore, it has been suggested that CRP may play a causal role in vascular disease, by reducing nitric-oxide release, increasing endothelin-1 production, and by inducing expression of endothelial adhesion molecules (Venugopal et al. 2005).

## Tumor necrosis factor-α

Tumor necrosis factor-α (TNF-α), formerly known as cachexin, was at first recognized in 1975 for its ability to lyse tumors (Carswell et al. 1975); it is synthesized initially as a transmembrane precursor protein by activated macrophages and T cells and the cytoplasmic tail is then cleaved to release soluble TNF-α.

Three TNF-α monomers are required to aggregate to form the trimeric biologically active TNF- α, which then binds to one of two types of receptors: TNFR1 or TNFR2, also known as p55 and p75 respectively. Then, TNF-α can exert multiple effects on the immune system, such as stimulation of the release of the inflammatory cytokines interleukin (IL-1, IL-β, IL-6, IL-8, and GM-CSF), upregulation of the expression of endothelial adhesion molecules (ICAM-1, VCAM-1, E-selectin) and chemokines (MCP-1, MIP-2, RANTES and MIP-1α), and coordination of the migration of leukocytes to targeted organs (Roach et al. 2002).

Plasma TNF-α have been often found elevated in patients with HF, also correlating with NYHA functional class (Torre-Amione et al. 1996a). In the setting of HF, the heart itself may be in part the source of TNF-α, given that the nonfailing heart does not express TNF-α, while elevated levels are present in HF patients (Torre-Amione et al. 1996b). Pressure and volume overload (e.g., due to aortic stenosis or mitral regurgitation) can increase plasma TNF-α and other cytokine concentrations (Kapadia et al. 2000).

Increased TNF-α production by peripheral blood mononuclear cells is also predictive of risk (Vasan et al. 2003), and this could be explained by the toxic effects that TNF-α may have on the myocardium (Bryant et al. 1998). Elevated levels of TNF-α and other proinflammatory cytokines may have prognostic importance in patients with HF: higher plasma levels at baseline were associated with increased mortality in a study of 1200 patients with advanced HF enrolled in the VEST trial, with a relationship influenced by age, sex, and the cause of HF (Deswal et al. 2001).

Even though the deleterious actions of TNF-α may be inhibited by antioxidants, since they are in part mediated by the generation of reactive oxygen intermediates (Nakamura et al. 1998), randomized trials (RENAISSANCE trial and the RECOVER trial) of anti-TNF-α therapy in patients with HF did not show any improvement in morbidity or mortality (Anker and Coats 2002).

## Interleukin-6

Interleukin-6 (IL-6) levels are increased in patients with HF and may contribute to disease progression: higher plasma levels of IL-6 at baseline were significantly associated with enhanced mortality among 1200 patients with advanced HF (Deswal et al. 2001). The increase in plasma IL-6 correlates with the severity of the HF, to the contra it can be reduced by optimal medical therapy. Elevated IL-6 levels may also contribute to the decline in cardiac function: in fact, persistently high IL-6 levels have been shown to be associated with increased mortality (Wollert and Drexler 2001). As with TFN-α, it has been suggested that IL-6 can predict the development of HF: in a subanalysis from the Framingham Heart Study among 732 elderly subjects without history of myocardial infarction or HF, a 68 percent increase in HF risk was observed for each tertile increment in plasma IL-6 (Vasan et al. 2003). Finally, plasma IL-6 correlated with CRP levels, potentially displaying the central role of IL-6 in the acute phase response.

## Chemokines

Chemokines (chemotactic cytokines), a family of low molecular weight proteins, are powerful chemoattractants of monocytes and are involved in the formation of reactive oxygen species and cytokines (Baggiolini et al. 1994). All chemokines, including MCP-1, macrophage inflammatory protein-1-α, and RANTES, have been found elevated in patients with HF compared with healthy control subjects, inversely correlating with LVEF (Aukrust et al. 1998); chemokines levels were not related to the cause of HF, but were highest in patients with NYHA class IV functional status.

CXC chemokines, identified by the first two cysteines typically separated by one amino acid, are specific for neutrophils, and include IL-8 (formerly referred to as neutrophil activating peptide-1), neutrophil activating peptide-2 (NAP-2), and GRO-α. Many different cell types, such as endothelial cells, platelets, neutrophils, T lymphocytes, and monocytes, are able to produce CXC chemokines. Concentrations of all three CXC cytokines were significantly increased in 47 patients with HF compared to 20 healthy controls, directly correlating with the severity of HF evaluated both clinically and hemodynamically (Damas et al. 2000).

Treatment with intravenous immunoglobulin has been shown to reduce the level of chemokines and chemokine gene expression, suggesting a potential beneficial effect of this therapy in HF (Damas et al. 2001).

## *Cyclooxygenase-2*

Cyclooxygenase-2 (COX-2), an enzyme catalyzing the conversion of arachidonic acid to prostaglandins and thromboxane A2, is induced in many cells in response to cytokines such as TNF-α, and ischemia.

According to a study among end-stage HF patients, COX-2 may play an important role in myocardial inflammation and injury: in this study, elevated levels of COX-2 were found in the myocytes and inflammatory cells within the area of fibrotic scar of patients with HF, but not in areas of normal myocardium; moreover, the myocardium of normal controls did not express COX-2 (Wong et al. 1998).

## *Fas (AOP-1)*

Fas (also referred to as APO-1), a member of the TNF-α receptor family, is expressed on a variety of cells, including myocytes. Fas, through activation by the Fas ligand, mediates apoptosis being involved in the development and progression of HF. It has already been reported that serum levels of a soluble form of Fas are elevated in patients with HF, being associated with severe disease (Okuyama et al. 1997). Moreover, postinfarction ventricular remodeling can be reduced by the inhibition of soluble Fas, with an improvement in survival (Li et al. 2004). Hence, pharmacologic efforts to reduce Fas levels are still under investigation, but may be of benefit in the treatment or prevention of HF: indeed, in patients with ischemic or dilated cardiomyopathy, plasma levels of Fas and CRP were reduced by the administration of nonspecific immunomodulating agent, like pentoxifylline (Sliwa et al. 2004) or intravenous immunoglobulin (Gullestad and Aukrust 2005), leading to an improvement in left ventricular function.

## Biomarkers of Oxidative Stress

Further to normal aerobic metabolism, reactive oxygen species (ROS), such as superoxide anion (O2-) and hydrogen peroxide, are produced (Griendling and FitzGerald 2003). While oxygen free radicals and peroxides production is physiologically balanced by an effective antioxidant system of molecules capable of "scavenging" ROS, in pathological states there is an imbalance between ROS and endogenous antioxidant defense mechanisms. This deleterious process, referred to as "oxidative stress," may damage endothelial function and proteins by apoptosis and necrosis (Kanani et al. 1999), and take part in the pathogenesis and progression of HF (Ungvari et al. 2005); it is also associated with both arrhythmias and endothelial dysfunction through reduction of nitric oxide synthase activity as well as inactivation of nitric oxide (Grieve and Shah 2003). Inflammation and immune activation, as well as stimulation of the RAAS and the sympathetic nervous system, and catecholamine and peroxynitrite level increases, all may increase oxidative stress (Zimmet and Hare 2006). Thus, free radical scavengers (e.g., antioxidants) may reverse these toxic effects (Lopez et al. 1997; Pinsky et al. 1995).

Since it is well known that risk factors for cardiovascular disease such as hypertension, diabetes, and obesity are associated with increased oxidative stress (Griendling and Alexander 1997), it has been suggested that oxidative stress is increased in patients with chronic HF and may predict outcome. Several potential mechanisms for increased oxidative stress in chronic HF may exist, such as decreased myocardial antioxidant activity, free radical generation during prostaglandin biosynthesis, autooxidation of catecholamines, repetitive ischemia and reperfusion, activation of proinflammatory cytokines, increased activity of xanthine oxidase, thiamine deficiency, post-translational modification in the ubiquitin proteasome system, and reduced availability of L-arginine. Elevated plasma levels of malondialdehyde-like activity (MDA), a marker of lipid peroxidation, correlated inversely with the LVEF and directly with both the chronicity and severity of HF symptoms and reduction in functional capacity, in patients with ischemic and nonischemic cardiomyopathy (McMurray et al. 1993). However, plasma levels of MDA did not correlate with survival in patients with nonischemic cardiomyopathy in the PRAISE-2 trial (Wijeysundera et al. 2003).

## *Oxidized low-density lipoproteins, myeloperoxidase, urinary biopyrrins, urinary and plasma isoprostanes, plasma malondialdehyde*

Since it is difficult to measure ROS directly in humans, indirect markers of oxidative stress have been sought, such as plasma-oxidized low-density lipoproteins, malondialdehyde and myeloperoxidase (MPO, a leukocyte-derived enzyme indexing leukocyte activation), urinary levels of biopyrrins (oxidative metabolites of bilirubin) (Hokamaki et al. 2004) and isoprostane levels in plasma and urine (Polidori et al. 2004). Levels of plasma MPO (Tang et al. 2006) and isoprostane excretion have been shown to correlate with the severity of HF and to be independent predictors of death from HF; in addition urinary excretion of 8-isoprostane correlated with the plasma levels of matrix metalloproteinases, potentially accelerating adverse ventricular remodeling and increasing the severity of HF (Polidori et al. 2004).

MPO can produce a cascade of ROS, which may lead to lipid peroxidation, scavenging of nitric oxide, and nitric oxide synthase inhibition. Patients with chronic systolic HF have shown elevated levels of MPO (Lu et al. 2010), associated with an increased likelihood of more advanced HF, and potentially predictors of a higher rate of adverse clinical outcomes (Halliwell 1989).

## Xanthine oxidase and uric acid

Xanthine oxidase, catalyzing the production of two oxidants, hypoxanthine and xanthine, is thought to play a pathologic role in HF (Berry and Hare 2004). Given that levels of uric acid, associated with increased xanthine oxidase activity, correlate with impaired hemodynamics (Kittleson et al. 2007) and independently predict an adverse prognosis in HF (Anker et al. 2003), uric acid may be a simple, useful, even though unspecific, marker of oxidative stress. In addition, further oxidation of uric acid produces allantoin, that has been shown to be 75 percent higher in hyperuricemic HF patients compared to healthy controls (Doehner et al. 2002).

## Biopyrrins

These are oxidative metabolites of bilirubin. In a study of 94 patients with decompensated HF, the urinary biopyrrin-to-creatinine ratio has been shown to correlate with symptoms and with BNP levels, and also to decrease with adequate treatment (Hokamaki et al. 2004).

## Nitrotyrosine

Nitrotyrosine immunoreactivity is another intracellular marker of oxidative stress. In patients with acute decompensated HF, nitrotyrosine immunoreactivity measured *ex vivo* in venous endothelial cells, was increased in association with abnormal flow-mediated dilation, and decreased by restoring of a compensated state (Colombo et al. 2005).

## *Hydroxyl radical, myocardial NADPH oxidase and 4-hydroxy-2-nonenal*

Hydroxyl radical (OH⁻) production can be examined determining the 2,3-dihydorxybenzoic acid/salicylic acid (DHBA/SA) ratio, after intravenous lysine acetylsalicylate administration. In a study of patients with ST segment elevation myocardial infarction, OH⁻ production peaked at 24 hours and remained elevated in the subgroup of patients who developed HF (Valgimigli et al. 2004). Levels of O2⁻ in failing human hearts have been demonstrated by electroparamagnetic resonance (EPR) with spin trapping (Sam et al. 2005).

Myocardial NADPH oxidase activity has been found elevated in end-stage failing hearts explanted at the time of cardiac transplantation, compared to non-failing control hearts (Heymes et al. 2003).

Myocardial expression of 4-hydroxy-2-nonenal (HNE)-modified protein, which is a major lipid peroxidation product, is increased in biopsy samples from patients with hypertrophic cardiomyopathy compared to controls, inversely correlating with LVEF (Nakamura et al. 2005).

## *Phosphodiesterase-5*

Myocardial phosphodiesterase (PDE)-5 protein expression is 4.5 fold increased in end-stage HF, correlating with markers of oxidative stress such as 3-nitrotryosine and 4-hydoxynonenal. Sildenafil, a PDE-5 inhibitor, has been hypothesized to improve contractile function through reduction of myocardial oxidative stress (Lu et al. 2010).

## Biomarkers of Extracellular Matrix Remodeling

The extracellular matrix provides a "skeleton" for myocytes by determining their size and shape; remodeling of the ventricles plays an important role in the progression of HF (Pfeffer and Braunwald 1990). In physiological conditions there is a balance between matrix metalloproteinases (proteolytic enzymes capable to degrade fibrillar collagen) and tissue inhibitors of metalloproteinases; contra, ventricular dilatation and remodeling are characterized by an imbalance, with dominance of matrix metalloproteinases over tissue inhibitors of metalloproteinases; moreover, an abnormal increase in collagen synthesis may also impair ventricular function through the resultant excessive fibrosis. The propeptide procollagen type I, a serum biomarker of collagen biosynthesis, had a positive correlation with the fractional volume of fibrous tissue determined from cardiac biopsies in patients with essential hypertension (Querejeta et al. 2000). Furthermore, levels of plasma procollagen type III were independent predictors of adverse outcomes among patients with HF (Cicoira et al. 2004). Markers of these processes may become important targets of therapy: by now, at least 15 matrix metalloproteinases and several forms of procollagen and of tissue inhibitors of metalloproteinases have been identified; which of these are the most informative and appropriate for routine measurement should be further investigated (King et al. 2003).

## Biomarkers of Renal Injury

Heart performance and kidney function are closely linked, both in health and in disease, including a variety of conditions, either acute or chronic, in which the primary failing organ can be either the heart or the kidney. This interaction between cardiac and renal failure, in which combined cardiac and renal dysfunction amplifies the progression of failure of the individual organs and has an extremely bad prognosis, has been denominated the cardiorenal syndrome (CRS) (Ronco et al. 2008). Renal insufficiency is also common in patients with HF, with both acute kidney injury (AKI) and worsening renal function being associated with poor prognosis (Hoste et al. 2006; Schiffrin et al. 2007).

Thus, in AKI and chronic renal disease, several promising biomarkers, both in plasma and in urine, have become useful for assessment and prognostication of HF patients, including cystatin C, neutrophil gelatinase-associated lipocalin (NGAL), kidney injury molecule-1, interleukin-18, asymmetric dimethylarginine, and liver-type fatty acid-binding protein (Nickolas et al. 2008a; Trof et al. 2006).

### Urinary tubular enzymes

Urinary tubular enzymes consist of proximal renal tubular epithelial antigen (HRTE-1), α-Glutathione S-transferase (α-GST), pi-Glutathione S-transferase (pi-GST), γ-Glutamyltranspeptidase (γ-GT), Alanine aminopeptidase (AAP), Lactate dehydrogenase (LDH), N-acetyl-beta-glucosaminidase (NAG), and Alkaline phosphatase (ALP). Most of these are released from proximal tubular epithelial cells within 12 hours and four days earlier than a detectable rise in serum creatinine (Herget-Rosenthal et al. 2004; Trof et al. 2006).

### Urinary low molecular weight proteins

These are: α1-Microglobulin (α1-M), β2-Microglobulin (β2-M), retinol binding protein (RBP), adenosine deaminase binding protein (ABP), and urinary cystatin C. They are produced at different sites, filtered at the glomerulus, and reabsorbed at the proximal tubule without secretion.

## Cystatin C

Cystatin C, a member of the family of cystein proteinase inhibitors, is a novel endogenous marker of renal function that may be more sensitive for detecting mild to moderate decrements in glomerular filtration rate (GFR). It is freely filtered across the glomerular membrane and completely catabolized in the proximal tubules. Most studies have suggested that it is not affected by age, sex, or muscle mass and has superior diagnostic accuracy compared with serum creatinine for early renal impairment (Dharnidharka et al. 2002). Serum concentrations of cystatin C may be used to detect renal dysfunction in critically ill patients with AKI 24 to 48 hr earlier than with creatinine measurements.

This marker seems to offer more complete prognostic information than other markers of renal function: in acute HF, the prognostic value of the serum cystatin C measurement has been compared with other markers of renal function and NT-proBNP, revealing that the mortality rate increased significantly with each tertile of cystatin C, and also that risk stratification improved further by combining the tertiles of NT-proBNP and cystatin C; moreover, in patients with normal plasma creatinine levels, elevated cystatin C was associated with significantly higher mortality at 12 months: 40.4% vs 12.6% (Lassus et al. 2007). Cystatin C has also been proven to be a risk factor for HF and cardiovascular disease mortality in the general population (Taglieri et al. 2009).

Although increased concentrations of cystatin C appear to be indicative of preclinical kidney disease being associated with adverse outcomes (Taglieri et al. 2009), further studies are required to ascertain whether atherogenic mechanisms other than renal dysfunction like inflammation, potentially accounting for its predictive value for future cardiovascular risk, may also exist.

## Neutrophil gelatinase-associated lipocalin

Neutrophil gelatinase-associated lipocalin (NGAL) is markedly upregulated and largely expressed in the kidney after renal ischemia (Herget-Rosenthal 2005). In this setting, NGAL may exert a protective function by reducing apoptosis, increasing the normal proliferation of kidney tubule cells, and also by enhancing the delivery of iron (Herget-Rosenthal 2005). Among 635 consecutive patients assessed in the emergency department, mean urinary NGAL levels were significantly higher in patients with AKI compared to those with normal kidney function, chronic kidney disease or prerenal azotemia (Nickolas et al. 2008b).

### *Kidney injury molecule-1*

Kidney injury molecule-1 (KIM-1), a type 1 transmembrane glycoprotein, is markedly upregulated in the proximal tubule in patients with AKI (Vaidya et al. 2006). KIM-1 has been found very useful in differentiating ATN from other forms of AKI and chronic kidney disease: after adjustment for age, gender, time between the initial insult, and sampling of the urine, a one-unit increase in normalized KIM-1 was associated with a greater than 12-fold risk for the presence of ATN (Han et al. 2002). Among 20 children who underwent cardiopulmonary bypass surgery, urinary KIM-1 was increased 6 to 12 hours following cardiopulmonary bypass, and remained elevated up to 48 hours in patients who sustained a greater than 50 percent increase in serum creatinine within the first 48 hours, but not in children who had normal renal function; the increase in KIM-1 was also paralleled by that of NGAL (Han et al. 2008).

## New Biomarkers

Biomarkers other than those already discussed are under investigation. These include chromogranin A, a polypeptide hormone produced by the myocardium, with potent negative inotropic properties, that has been found increased in patients with HF (Pieroni et al. 2007). A second is galectin-3, a protein produced by activated macrophages, that seemed to predict adverse outcomes in patients with HF (van Kimmenade et al. 2006). A third is osteoprotegerin, a member of the TNF receptor superfamily, that has been implicated in the development of left ventricular dysfunction (Omland et al. 2007) and in predicting survival in patients with HF after myocardial infarction (Ueland et al. 2004).

Biomarkers well known in other pathological states may also be helpful in diagnosing HF. Levels of adiponectin, a 244-amino-acid peptide, inversely related to body-mass index, are elevated in patients with advanced HF (McEntegart et al. 2007) (especially those with cardiac cachexia), being a predictor of death in patients with HF (Kistorp et al. 2005).

## Conclusion

Biomarkers may be a valuable addition to the classification of HF. In this chapter, various biomarkers have been discussed but none satisfies all the criteria of an ideal biomarker. Natriuretic peptides have scored over others in prognostic, diagnostic, and therapeutic values and have indeed found a place in the guidelines for the diagnosis of congestive heart failure. However, these too need to be carefully interpreted in the presence of clinical settings and various confounding factors as discussed above. Since the combined use of biomarkers reaches a better risk prediction than that obtained with any single biomarker, a multimarker approach may be better for risk stratification. Finally, other underway biomarkers may open up a new paradigm in the diagnosis and management of HF, succeeding the currently available biomarkers.

# References

Adams, J.E., III, D.R. Abendschein, and A.S. Jaffe. 1993a. Biochemical markers of myocardial injury. Is MB creatine kinase the choice for the 1990s? Circulation 88: 750–763.

Adams, J.E., III, G.S. Bodor, V.G. Davila-Roman, J.A. Delmez, F.S. Apple, J.H. Ladenson, and A.S. Jaffe. 1993b. Cardiac troponin I. A marker with high specificity for cardiac injury. Circulation 88: 101–106.

Aggarwal, A., M.D. Esler, F. Socratous, and D.M. Kaye. 2001. Evidence for functional presynaptic alpha-2 adrenoceptors and their down-regulation in human heart failure. J. Am. Coll. Cardiol. 37: 1246–1251.

Altschuld, R.A., R.C. Starling, R.L. Hamlin, G.E. Billman, J. Hensley, L. Castillo, R.H. Fertel, C.M. Hohl, P.M. Robitaille, L.R. Jones et al. 1995. Response of failing canine and human heart cells to beta 2-adrenergic stimulation. Circulation 92: 1612–1618.

Anand, I.S., L.D. Fisher, Y.T. Chiang, R. Latini, S. Masson, A.P. Maggioni, R.D. Glazer, G. Tognoni, and J.N. Cohn. 2003. Changes in brain natriuretic peptide and norepinephrine over time and mortality and morbidity in the Valsartan Heart Failure Trial (Val-HeFT). Circulation 107: 1278–1283.

Anand, I.S., R. Latini, V.G. Florea, M.A. Kuskowski, T. Rector, S. Masson, S. Signorini, P. Mocarelli, A. Hester, R. Glazer, and J.N. Cohn. 2005. C-reactive protein in heart failure: prognostic value and the effect of valsartan. Circulation 112: 1428–1434.

Anand, I.S., T. Kempf, T.S. Rector, H. Tapken, T. Allhoff, F. Jantzen, M. Kuskowski, J.N. Cohn, H. Drexler, and K.C. Wollert. 2010. Serial measurement of growth-differentiation factor-15 in heart failure: relation to disease severity and prognosis in the Valsartan Heart Failure Trial. Circulation 122: 1387–1395.

Anker, S.D. and A.J. Coats. 2002. How to RECOVER from RENAISSANCE? The significance of the results of RECOVER, RENAISSANCE, RENEWAL and ATTACH. Int. J. Cardiol. 86: 123–130.

Anker, S.D., W. Doehner, M. Rauchhaus, R. Sharma, D. Francis, C. Knosalla, C.H. Davos, M. Cicoira, W. Shamim, M. Kemp, R. Segal, K.J. Osterziel, F. Leyva, R. Hetzer, P. Ponikowski, and A.J. Coats. 2003. Uric acid and survival in chronic heart failure: validation and application in metabolic, functional, and hemodynamic staging. Circulation 107: 1991–1997.

Anker, S.D. and S. von Haehling. 2004. Inflammatory mediators in chronic heart failure: an overview. Heart 90: 464–470.

Anwaruddin, S., D.M. Lloyd-Jones, A. Baggish, A. Chen, D. Krauser, R. Tung, C. Chae, and J.L. Januzzi, Jr. 2006. Renal function, congestive heart failure, and amino-terminal pro-brain natriuretic peptide measurement: results from the ProBNP Investigation of Dyspnea in the Emergency Department (PRIDE) Study. J. Am. Coll. Cardiol. 47: 91–97.

Aronson, D. and A.J. Burger. 2003. Neurohumoral activation and ventricular arrhythmias in patients with decompensated congestive heart failure: role of endothelin. Pacing Clin. Electrophysiol. 26: 703–710.

Aukrust, P., T. Ueland, F. Muller, A.K. Andreassen, I. Nordoy, H. Aas, J. Kjekshus, S. Simonsen, S.S. Froland, and L. Gullestad. 1998. Elevated circulating levels of C-C chemokines in patients with congestive heart failure. Circulation 97: 1136–1143.

Azevedo, E.R., G.E. Newton, J.S. Floras, and J.D. Parker. 2000. Reducing cardiac filling pressure lowers norepinephrine spillover in patients with chronic heart failure. Circulation 101: 2053–2059.

Badgett, R.G., C.D. Mulrow, P.M. Otto, and G. Ramirez. 1996. How well can the chest radiograph diagnose left ventricular dysfunction? J. Gen. Intern. Med. 11: 625–634.

Baggiolini, M., B. Dewald, and B. Moser. 1994. Interleukin-8 and related chemotactic cytokines--CXC and CC chemokines. Adv. Immunol. 55: 97–179.

Barker, W.H., J.P. Mullooly, and W. Getchell. 2006. Changing incidence and survival for heart failure in a well-defined older population, 1970–1974 and 1990–1994. Circulation 113: 799–805.

Bedi, M.S., L.A. Postava, S. Murali, G.A. Macgowan, M. Mathier, D.M. McNamara, and B. London. 2004. Interaction of implantable defibrillator therapy with angiotensin-converting enzyme deletion/insertion polymorphism. J. Cardiovasc. Electrophysiol. 15: 1162–1166.

Benedict, C.R., D.E. Johnstone, D.H. Weiner, M.G. Bourassa, V. Bittner, R. Kay, P. Kirlin, B. Greenberg, R.M. Kohn, J.M. Nicklas et al. 1994. Relation of neurohumoral activation to clinical variables and degree of ventricular dysfunction: a report from the Registry of Studies of Left Ventricular Dysfunction. SOLVD Investigators. J. Am. Coll. Cardiol. 23: 1410–1420.

Berger, R., M. Huelsman, K. Strecker, A. Bojic, P. Moser, B. Stanek, and R. Pacher. 2002. B-type natriuretic peptide predicts sudden death in patients with chronic heart failure. Circulation 105: 2392–2397.

Berger, R., D. Moertl, S. Peter, R. Ahmadi, M. Huelsmann, S. Yamuti, B. Wagner, and R. Pacher. 2010. N-terminal pro-B-type natriuretic peptide-guided, intensive patient management in addition to multidisciplinary care in chronic heart failure a 3-arm, prospective, randomized pilot study. J. Am. Coll. Cardiol. 55: 645–653.

Berry, C.E. and J.M. Hare. 2004. Xanthine oxidoreductase and cardiovascular disease: molecular mechanisms and pathophysiological implications. J. Physiol. 555: 589–606.

Bhargava, V., R. Shabetai, R.A. Mathiasen, N. Dalton, J.J. Hunter, and J. Ross, Jr. 1998. Loss of adrenergic control of the force-frequency relation in heart failure secondary to idiopathic or ischemic cardiomyopathy. Am. J. Cardiol. 81: 1130–1137.

Bleumink, G.S., A.M. Knetsch, M.C. Sturkenboom, S.M. Straus, A. Hofman, J.W. Deckers, J.C. Witteman, and B.H. Stricker. 2004. Quantifying the heart failure epidemic: prevalence, incidence rate, lifetime risk and prognosis of heart failure The Rotterdam Study. Eur. Heart J. 25: 1614–1619.

Bonneux, L., J.J. Barendregt, K. Meeter, G.J. Bonsel, and P.J. van der Maas. 1994. Estimating clinical morbidity due to ischemic heart disease and congestive heart failure: the future rise of heart failure. Am. J. Public Health 84: 20–28.

Bootcov, M.R., A.R. Bauskin, S.M. Valenzuela, A.G. Moore, M. Bansal, X.Y. He, H.P. Zhang, M. Donnellan, S. Mahler, K. Pryor, B.J. Walsh, R.C. Nicholson, W.D. Fairlie, S.B. Por, J.M. Robbins, and S.N. Breit. 1997. MIC-1, a novel macrophage inhibitory cytokine, is a divergent member of the TGF-beta superfamily. Proc. Natl. Acad. Sci. USA 94: 11514–11519.

Braunwald, E. 2008. Biomarkers in heart failure. N. Engl. J. Med. 358: 2148–2159.

Brunner-La Rocca, H.P., D.M. Kaye, R.L. Woods, J. Hastings, and M.D. Esler. 2001. Effects of intravenous brain natriuretic peptide on regional sympathetic activity in patients with chronic heart failure as compared with healthy control subjects. J. Am. Coll. Cardiol. 37: 1221–1227.

Bryant, D., L. Becker, J. Richardson, J. Shelton, F. Franco, R. Peshock, M. Thompson, and B. Giroir. 1998. Cardiac failure in transgenic mice with myocardial expression of tumor necrosis factor-alpha. Circulation 97: 1375–1381.

Bursi, F., S.A. Weston, M.M. Redfield, S.J. Jacobsen, S. Pakhomov, V.T. Nkomo, R.A. Meverden, and V.L. Roger. 2006. Systolic and diastolic heart failure in the community. JAMA 296: 2209–2216.

Cadnapaphornchai, M.A., A.K. Gurevich, H.D. Weinberger, and R.W. Schrier. 2001. Pathophysiology of sodium and water retention in heart failure. Cardiology 96: 122–131.

Carswell, E.A., L.J. Old, R.L. Kassel, S. Green, N. Fiore, and B. Williamson. 1975. An endotoxin-induced serum factor that causes necrosis of tumors. Proc. Natl. Acad. Sci. USA 72: 3666–3670.

Castell, J.V., M.J. Gomez-Lechon, M. David, R. Fabra, R. Trullenque, and P.C. Heinrich. 1990. Acute-phase response of human hepatocytes: regulation of acute-phase protein synthesis by interleukin-6. Hepatology 12: 1179–1186.

Chidsey, C.A., D.C. Harrison, and E. Braunwald. 1962. Augmentation of the plasma norepinephrine response to exercise in patients with congestive heart failure. N. Engl. J. Med. 267: 650–654.

Chidsey, C.A., E. Braunwald, and A.G. Morrow. 1965. Catecholamine Excretion and Cardiac Stores of Norepinephrine in Congestive Heart Failure. Am. J. Med. 39: 442–451.

Choukroun, G., R. Hajjar, J.M. Kyriakis, J.V. Bonventre, A. Rosenzweig, and T. Force. 1998. Role of the stress-activated protein kinases in endothelin-induced cardiomyocyte hypertrophy. J. Clin. Invest. 102: 1311–1320.

Cicoira, M., A. Rossi, S. Bonapace, L. Zanolla, G. Golia, L. Franceschini, B. Caruso, P.N. Marino, and P. Zardini. 2004. Independent and additional prognostic value of aminoterminal propeptide of type III procollagen circulating levels in patients with chronic heart failure. J. Card. Fail. 10: 403–411.

Cohn, J.N., T.B. Levine, M.T. Olivari, V. Garberg, D. Lura, G.S. Francis, A.B. Simon, and T. Rector. 1984. Plasma norepinephrine as a guide to prognosis in patients with chronic congestive heart failure. N. Engl. J. Med. 311: 819–823.

Colombo, P.C., J.E. Banchs, S. Celaj, A. Talreja, J. Lachmann, S. Malla, N.B. DuBois, A.W. Ashton, F. Latif, U.P. Jorde, J.A. Ware, and T.H. LeJemtel. 2005. Endothelial cell activation in patients with decompensated heart failure. Circulation 111: 58–62.

Communal, C., K. Singh, D.R. Pimentel, and W.S. Colucci. 1998. Norepinephrine stimulates apoptosis in adult rat ventricular myocytes by activation of the beta-adrenergic pathway. Circulation 98: 1329–1334.

Communal, C., K. Singh, D.B. Sawyer, and W.S. Colucci. 1999. Opposing effects of beta(1)- and beta(2)-adrenergic receptors on cardiac myocyte apoptosis : role of a pertussis toxin-sensitive G protein. Circulation 100: 2210–2212.

Cowie, M.R., A. Mosterd, D.A. Wood, J.W. Deckers, P.A. Poole-Wilson, G.C. Sutton, and D.E. Grobbee. 1997. The epidemiology of heart failure. Eur. Heart J. 18: 208–225.

Damas, J.K., L. Gullestad, T. Ueland, N.O. Solum, S. Simonsen, S.S. Froland, and P. Aukrust. 2000. CXC-chemokines, a new group of cytokines in congestive heart failure—possible role of platelets and monocytes. Cardiovasc. Res. 45: 428–436.

Damas, J.K., L. Gullestad, H. Aass, S. Simonsen, J.G. Fjeld, L. Wikeby, T. Ueland, H.G. Eiken, S.S. Froland, and P. Aukrust. 2001. Enhanced gene expression of chemokines and their corresponding receptors in mononuclear blood cells in chronic heart failure—modulatory effect of intravenous immunoglobulin. J. Am. Coll. Cardiol. 38: 187–193.

Daniels, L.B. and A.S. Maisel. 2007. Natriuretic peptides. J. Am. Coll. Cardiol. 50: 2357–68.

Danser, A.H., M.A. Schalekamp, W.A. Bax, A.M. van den Brink, P.R. Saxena, G.A. Riegger, and H. Schunkert. 1995. Angiotensin-converting enzyme in the human heart. Effect of the deletion/insertion polymorphism. Circulation 92: 1387–1388.

Das, S.R., M.H. Drazner, D.L. Dries, G.L. Vega, H.G. Stanek, S.M. Abdullah, R.M. Canham, A.K. Chung, D. Leonard, F.H. Wians, Jr., and J.A. de Lemos. 2005. Impact of body mass and body composition on circulating levels of natriuretic peptides: results from the Dallas Heart Study. Circulation 112: 2163–2168.

de Groote, P., J. Dagorn, B. Soudan, N. Lamblin, E. McFadden, and C. Bauters. 2004. B-type natriuretic peptide and peak exercise oxygen consumption provide independent information for risk stratification in patients with stable congestive heart failure. J. Am. Coll. Cardiol. 43: 1584–1589.

Deswal, A., N.J. Petersen, A.M. Feldman, J.B. Young, B.G. White, and D.L. Mann. 2001. Cytokines and cytokine receptors in advanced heart failure: an analysis of the cytokine database from the Vesnarinone trial (VEST). Circulation 103: 2055–2059.

Dharnidharka, V.R., C. Kwon, and G. Stevens. 2002. Serum cystatin C is superior to serum creatinine as a marker of kidney function: a meta-analysis. Am. J. Kidney Dis. 40: 221–226.

Ding, Q., T. Mracek, P. Gonzalez-Muniesa, K. Kos, J. Wilding, P. Trayhurn, and C. Bing. 2009. Identification of macrophage inhibitory cytokine-1 in adipose tissue and its secretion as an adipokine by human adipocytes. Endocrinology 150: 1688–1696.

Ding, S.S., C. Qiu, P. Hess, J.F. Xi, J.P. Clozel, and M. Clozel. 2002. Chronic endothelin receptor blockade prevents renal vasoconstriction and sodium retention in rats with chronic heart failure. Cardiovasc. Res. 53: 963–970.

Doehner, W., N. Schoene, M. Rauchhaus, F. Leyva-Leon, D.V. Pavitt, D.A. Reaveley, G. Schuler, A.J. Coats, S.D. Anker, and R. Hambrecht. 2002. Effects of xanthine oxidase inhibition with allopurinol on endothelial function and peripheral blood flow in hyperuricemic patients with chronic heart failure: results from 2 placebo-controlled studies. Circulation 105: 2619–2624.

Doust, J.A., E. Pietrzak, A. Dobson, and P. Glasziou. 2005. How well does B-type natriuretic peptide predict death and cardiac events in patients with heart failure: systematic review. BMJ 330: 625.

Dubois-Rande, J.L., O. Montagne, M. Alvarez-Guerra, C. Nazaret, B. Crozatier, P. Gueret, A. Castaigne, and R.P. Garay. 1996. Endogenous sodium-potassium-chloride cotransport inhibitor in congestive heart failure. J. Am. Coll. Cardiol. 28: 1464–1470.

Dzau, V.J., W.S. Colucci, N.K. Hollenberg, and G.H. Williams. 1981. Relation of the renin-angiotensin-aldosterone system to clinical state in congestive heart failure. Circulation 63: 645–651.

Dzau, V.J. 1987. Renal and circulatory mechanisms in congestive heart failure. Kidney Int. 31: 1402–1415.

Elesber, A.A. and M.M. Redfield. 2001. Approach to patients with heart failure and normal ejection fraction. Mayo Clin. Proc. 76: 1047–1052.

Elsner, D., A. Muntze, E.P. Kromer, and G.A. Riegger. 1992. Effectiveness of endopeptidase inhibition (candoxatril) in congestive heart failure. Am. J. Cardiol. 70: 494–498.

Foley, P.W., B. Stegemann, K. Ng, S. Ramachandran, A. Proudler, M.P. Frenneaux, L.L. Ng, and F. Leyva. 2009. Growth differentiation factor-15 predicts mortality and morbidity after cardiac resynchronization therapy. Eur. Heart J. 30: 2749–2757.

Folsom, A.R., J.S. Pankow, R.P. Tracy, D.K. Arnett, J.M. Peacock, Y. Hong, L. Djousse, and J.H. Eckfeldt. 2001. Association of C-reactive protein with markers of prevalent atherosclerotic disease. Am. J. Cardiol. 88: 112–117.

Fonarow, G.C., W.F. Peacock, C.O. Phillips, M.M. Givertz, and M. Lopatin. 2007. Admission B-type natriuretic peptide levels and in-hospital mortality in acute decompensated heart failure. J. Am. Coll. Cardiol. 49: 1943–1950.

Francis, G.S., S.R. Goldsmith, T.B. Levine, M.T. Olivari, and J.N. Cohn. 1984. The neurohumoral axis in congestive heart failure. Ann. Intern. Med. 101: 370–377.

Gaasch, W.H. 1994. Diagnosis and treatment of heart failure based on left ventricular systolic or diastolic dysfunction. JAMA 271: 1276–1280.

Gillespie, N.D., G. McNeill, T. Pringle, S. Ogston, A.D. Struthers, and S.D. Pringle. 1997. Cross sectional study of contribution of clinical assessment and simple cardiac investigations to diagnosis of left ventricular systolic dysfunction in patients admitted with acute dyspnoea. BMJ 314: 936–940.

Globits, S., H. Frank, B. Pacher, M. Huelsmann, E. Ogris, and R. Pacher. 1998. Atrial natriuretic peptide release is more dependent on atrial filling volume than on filling pressure in chronic congestive heart failure. Am. Heart J. 135: 592–597.

Griendling, K.K. and R.W. Alexander. 1997. Oxidative stress and cardiovascular disease. Circulation 96: 3264–3265.

Griendling, K.K. and G.A. FitzGerald. 2003. Oxidative stress and cardiovascular injury: Part I: basic mechanisms and *in vivo* monitoring of ROS. Circulation 108: 1912–1916.

Grieve, D.J., and A.M. Shah. 2003. Oxidative stress in heart failure. More than just damage. Eur. Heart J. 24: 2161–2163.

Gullestad, L. and P. Aukrust. 2005. Review of trials in chronic heart failure showing broad-spectrum anti-inflammatory approaches. Am. J. Cardiol. 95: 17C–23C; discussion 38C–40C.

Halliwell B, G.J., ed. Free Radicals in Biology and Medicine. 1989, Oxford University Press: Oxford.

Han, W.K., V. Bailly, R. Abichandani, R. Thadhani, and J.V. Bonventre. 2002. Kidney Injury Molecule-1 (KIM-1): a novel biomarker for human renal proximal tubule injury. Kidney Int. 62: 237–244.

Han, W.K., S.S. Waikar, A. Johnson, R.A. Betensky, C.L. Dent, P. Devarajan, and J.V. Bonventre. 2008. Urinary biomarkers in the early diagnosis of acute kidney injury. Kidney Int. 73: 863–869.

Hartmann, F., M. Packer, A.J. Coats, M.B. Fowler, H. Krum, P. Mohacsi, J.L. Rouleau, M. Tendera, A. Castaigne, S.D. Anker, I. Amann-Zalan, S. Hoersch, and H.A. Katus. 2004. Prognostic impact of plasma N-terminal pro-brain natriuretic peptide in severe chronic congestive heart failure: a substudy of the Carvedilol Prospective Randomized Cumulative Survival (COPERNICUS) trial. Circulation 110: 1780–1786.

Havranek, E.P., F.A. Masoudi, K.A. Westfall, P. Wolfe, D.L. Ordin, and H.M. Krumholz. 2002. Spectrum of heart failure in older patients: results from the National Heart Failure project. Am. Heart J. 143: 412–417.

Heart Failure Society of America 2006. HFSA 2006 Comprehensive Heart Failure Practice Guideline. J. Card. Fail. 12: e1–2.

Herget-Rosenthal, S., D. Poppen, H. Husing, G. Marggraf, F. Pietruck, H.G. Jakob, T. Philipp, and A. Kribben. 2004. Prognostic value of tubular proteinuria and enzymuria in nonoliguric acute tubular necrosis. Clin. Chem. 50: 552–558.

Herget-Rosenthal, S. 2005. One step forward in the early detection of acute renal failure. Lancet 365: 1205–1206.

Heublein, D.M., B.K. Huntley, G. Boerrigter, A. Cataliotti, S.M. Sandberg, M.M. Redfield, and J.C. Burnett, Jr. 2007. Immunoreactivity and guanosine 3′,5′-cyclic monophosphate activating actions of various molecular forms of human B-type natriuretic peptide. Hypertension 49: 1114–1119.

Heymes, C., J.K. Bendall, P. Ratajczak, A.C. Cave, J.L. Samuel, G. Hasenfuss, and A.M. Shah. 2003. Increased myocardial NADPH oxidase activity in human heart failure. J. Am. Coll. Cardiol. 41: 2164–2171.

Ho, K.K., J.L. Pinsky, W.B. Kannel, and D. Levy. 1993. The epidemiology of heart failure: the Framingham Study. J. Am. Coll. Cardiol. 22: 6A–13A.

Hoes, A.W., A. Mosterd, and D.E. Grobbee. 1998. An epidemic of heart failure? Recent evidence from Europe. Eur. Heart J. 19 Suppl L: L2–9.

Hokamaki, J., H. Kawano, M. Yoshimura, H. Soejima, S. Miyamoto, I. Kajiwara, S. Kojima, T. Sakamoto, S. Sugiyama, N. Hirai, H. Shimomura, Y. Nagayoshi, K. Tsujita, I. Shioji, S. Sasaki, and H. Ogawa. 2004. Urinary biopyrrins levels are elevated in relation to severity of heart failure. J. Am. Coll. Cardiol. 43: 1880–1885.

Horwich, T.B., J. Patel, W.R. MacLellan, and G.C. Fonarow. 2003. Cardiac troponin I is associated with impaired hemodynamics, progressive left ventricular dysfunction, and increased mortality rates in advanced heart failure. Circulation 108: 833–838.

Horwich, T.B., M.A. Hamilton, and G.C. Fonarow. 2006. B-type natriuretic peptide levels in obese patients with advanced heart failure. J. Am. Coll. Cardiol. 47: 85–90.

Hoste, E.A., G. Clermont, A. Kersten, R. Venkataraman, D.C. Angus, D. De Bacquer, and J.A. Kellum. 2006. RIFLE criteria for acute kidney injury are associated with hospital mortality in critically ill patients: a cohort analysis. Crit. Care 10: R73.

Hudson, M.P., C.M. O'Connor, W.A. Gattis, G. Tasissa, V. Hasselblad, C.M. Holleman, L.H. Gaulden, F. Sedor, and E.M. Ohman. 2004. Implications of elevated cardiac troponin T in ambulatory patients with heart failure: a prospective analysis. Am. Heart J. 147: 546–552.

Hulsmann, M., B. Stanek, B. Frey, B. Sturm, D. Putz, T. Kos, R. Berger, W. Woloszczuk, G. Maurer, and R. Pacher. 1998. Value of cardiopulmonary exercise testing and big endothelin plasma levels to predict short-term prognosis of patients with chronic heart failure. J. Am. Coll. Cardiol. 32: 1695–1700.

Hunt, S.A., W.T. Abraham, M.H. Chin, A.M. Feldman, G.S. Francis, T.G. Ganiats, M. Jessup, M.A. Konstam, D.M. Mancini, K. Michl, J.A. Oates, P.S. Rahko, M.A. Silver, L.W. Stevenson, and C.W. Yancy. 2009. Focused update incorporated into the ACC/AHA 2005 Guidelines for the Diagnosis and Management of Heart Failure in Adults: a report of the American College of Cardiology Foundation/American Heart Association Task Force on Practice Guidelines: developed in collaboration with the International Society for Heart and Lung Transplantation. Circulation 119: e391–479.

Iwanaga, Y., I. Nishi, S. Furuichi, T. Noguchi, K. Sase, Y. Kihara, Y. Goto, and H. Nonogi. 2006. B-type natriuretic peptide strongly reflects diastolic wall stress in patients with chronic heart failure: comparison between systolic and diastolic heart failure. J. Am. Coll. Cardiol. 47: 742–748.

Jaffe, A.S., V.C. Vasile, M. Milone, A.K. Saenger, K.N. Olson, and F.S. Apple. 2011. Diseased skeletal muscle: a noncardiac source of increased circulating concentrations of cardiac troponin T. J. Am. Coll. Cardiol. 58: 1819–1824.

Januzzi, J.L., R. van Kimmenade, J. Lainchbury, A. Bayes-Genis, J. Ordonez-Llanos, M. Santalo-Bel, Y.M. Pinto, and M. Richards. 2006. NT-proBNP testing for diagnosis and short-term prognosis in acute destabilized heart failure: an international pooled analysis of 1256 patients: the International Collaborative of NT-proBNP Study. Eur. Heart J. 27: 330–337.

Januzzi, J.L., Jr., C.A. Camargo, S. Anwaruddin, A.L. Baggish, A.A. Chen, D.G. Krauser, R. Tung, R. Cameron, J.T. Nagurney, C.U. Chae, D.M. Lloyd-Jones, D.F. Brown, S. Foran-Melanson, P.M. Sluss, E. Lee-Lewandrowski, and K.B. Lewandrowski. 2005. The N-terminal Pro-BNP investigation of dyspnea in the emergency department (PRIDE) study. Am. J. Cardiol. 95: 948–954.

Jhund, P.S., K. Macintyre, C.R. Simpson, J.D. Lewsey, S. Stewart, A. Redpath, J.W. Chalmers, S. Capewell, and J.J. McMurray. 2009. Long-term trends in first hospitalization for heart failure and subsequent survival between 1986 and 2003: a population study of 5.1 million people. Circulation 119: 515–523.

Jourdain, P., G. Jondeau, F. Funck, P. Gueffet, A. Le Helloco, E. Donal, J.F. Aupetit, M.C. Aumont, M. Galinier, J.C. Eicher, A. Cohen-Solal, and Y. Juilliere. 2007. Plasma brain natriuretic peptide-guided therapy to improve outcome in heart failure: the STARS-BNP Multicenter Study. J. Am. Coll. Cardiol. 49: 1733–1739.

Kanani, P.M., C.A. Sinkey, R.L. Browning, M. Allaman, H.R. Knapp, and W.G. Haynes. 1999. Role of oxidant stress in endothelial dysfunction produced by experimental hyperhomocyst(e)inemia in humans. Circulation 100: 1161–1168.

Kapadia, S.R., K. Yakoob, S. Nader, J.D. Thomas, D.L. Mann, and B.P. Griffin. 2000. Elevated circulating levels of serum tumor necrosis factor-alpha in patients with hemodynamically significant pressure and volume overload. J. Am. Coll. Cardiol. 36: 208–212.

Karlstrom, P., U. Alehagen, K. Boman, and U. Dahlstrom. 2011. Brain natriuretic peptide-guided treatment does not improve morbidity and mortality in extensively treated patients with chronic heart failure: responders to treatment have a significantly better outcome. Eur. J. Heart Fail. 13: 1096–1103.

Kaye, D.M., G.W. Lambert, J. Lefkovits, M. Morris, G. Jennings, and M.D. Esler. 1994. Neurochemical evidence of cardiac sympathetic activation and increased central nervous system norepinephrine turnover in severe congestive heart failure. J. Am. Coll. Cardiol. 23: 570–578.

Kaye, D.M., J. Lefkovits, G.L. Jennings, P. Bergin, A. Broughton, and M.D. Esler. 1995. Adverse consequences of high sympathetic nervous activity in the failing human heart. J. Am. Coll. Cardiol. 26: 1257–1263.

Kaye, D.M., A.M. Dart, G.L. Jennings, and M.D. Esler. 1999. Antiadrenergic effect of chronic amiodarone therapy in human heart failure. J. Am. Coll. Cardiol. 33: 1553–1559.

Kempf, T., M. Eden, J. Strelau, M. Naguib, C. Willenbockel, J. Tongers, J. Heineke, D. Kotlarz, J. Xu, J.D. Molkentin, H.W. Niessen, H. Drexler, and K.C. Wollert. 2006. The transforming growth factor-beta superfamily member growth-differentiation factor-15 protects the heart from ischemia/reperfusion injury. Circ. Res. 98: 351–360.

Kempf, T., E. Bjorklund, S. Olofsson, B. Lindahl, T. Allhoff, T. Peter, J. Tongers, K.C. Wollert, and L. Wallentin. 2007a. Growth-differentiation factor-15 improves risk stratification in ST-segment elevation myocardial infarction. Eur. Heart J. 28: 2858–2865.

Kempf, T., S. von Haehling, T. Peter, T. Allhoff, M. Cicoira, W. Doehner, P. Ponikowski, G.S. Filippatos, P. Rozentryt, H. Drexler, S.D. Anker, and K.C. Wollert. 2007b. Prognostic utility of growth differentiation factor-15 in patients with chronic heart failure. J. Am. Coll. Cardiol. 50: 1054–1060.

King, M.K., M.L. Coker, A. Goldberg, J.H. McElmurray, 3rd, H.R. Gunasinghe, R. Mukherjee, M.R. Zile, T.P. O'Neill, and F.G. Spinale. 2003. Selective matrix metalloproteinase inhibition with developing heart failure: effects on left ventricular function and structure. Circ. Res. 92: 177–185.

Kinnunen, P., O. Vuolteenaho, and H. Ruskoaho. 1993. Mechanisms of atrial and brain natriuretic peptide release from rat ventricular myocardium: effect of stretching. Endocrinology 132: 1961–1970.

Kistorp, C., J. Faber, S. Galatius, F. Gustafsson, J. Frystyk, A. Flyvbjerg, and P. Hildebrandt. 2005. Plasma adiponectin, body mass index, and mortality in patients with chronic heart failure. Circulation 112: 1756–1762.

Kittleson, M.M., M.E. St. John, V. Bead, H.C. Champion, E.K. Kasper, S.D. Russell, I.S. Wittstein, and J.M. Hare. 2007. Increased levels of uric acid predict haemodynamic compromise in patients with heart failure independently of B-type natriuretic peptide levels. Heart 93: 365–367.

Knudsen, C.W., T. Omland, P. Clopton, A. Westheim, W.T. Abraham, A.B. Storrow, J. McCord, R.M. Nowak, M.C. Aumont, P. Duc, J.E. Hollander, A.H. Wu, P.A. McCullough, and A.S. Maisel. 2004. Diagnostic value of B-Type natriuretic peptide and chest radiographic findings in patients with acute dyspnea. Am. J. Med. 116: 363–368.

Koglin, J., S. Pehlivanli, M. Schwaiblmair, M. Vogeser, P. Cremer, and W. vonScheidt. 2001. Role of brain natriuretic peptide in risk stratification of patients with congestive heart failure. J. Am. Coll. Cardiol. 38: 1934–1941.

Kohli, P., M.P. Bonaca, R. Kakkar, A.Y. Kudinova, B.M. Scirica, M.S. Sabatine, S.A. Murphy, E. Braunwald, R.T. Lee, and D.A. Morrow. 2012. Role of ST2 in non-ST-elevation acute coronary syndrome in the MERLIN-TIMI 36 trial. Clin. Chem. 58: 257–266.

Konstam, M.A., M. Gheorghiade, J.C. Burnett, Jr., L. Grinfeld, A.P. Maggioni, K. Swedberg, J.E. Udelson, F. Zannad, T. Cook, J. Ouyang, C. Zimmer, and C. Orlandi. 2007. Effects of oral tolvaptan in patients hospitalized for worsening heart failure: the EVEREST Outcome Trial. JAMA 297: 1319–1331.

Kushner, I. 1982. The phenomenon of the acute phase response. Ann. N.Y. Acad. Sci. 389: 39–48.

La Vecchia, L., G. Mezzena, L. Zanolla, M. Paccanaro, L. Varotto, C. Bonanno, and R. Ometto. 2000. Cardiac troponin I as diagnostic and prognostic marker in severe heart failure. J. Heart Lung Transplant 19: 644–652.

Lainchbury, J.G., R.W. Troughton, K.M. Strangman, C.M. Frampton, A. Pilbrow, T.G. Yandle, A.K. Hamid, M.G. Nicholls, and A.M. Richards. 2009. N-terminal pro-B-type natriuretic peptide-guided treatment for chronic heart failure: results from the BATTLESCARRED (NT-proBNP-Assisted Treatment To Lessen Serial Cardiac Readmissions and Death) trial. J. Am. Coll. Cardiol. 55: 53–60.

Lamblin, N., F. Mouquet, B. Hennache, J. Dagorn, S. Susen, C. Bauters, and P. de Groote. 2005. High-sensitivity C-reactive protein: potential adjunct for risk stratification in patients with stable congestive heart failure. Eur. Heart J. 26: 2245–2250.

LaPointe, M.C. 2005. Molecular regulation of the brain natriuretic peptide gene. Peptides 26: 944–956.

Lassus, J., V.P. Harjola, R. Sund, K. Siirila-Waris, J. Melin, K. Peuhkurinen, K. Pulkki, and M.S. Nieminen. 2007. Prognostic value of cystatin C in acute heart failure in relation to other markers of renal function and NT-proBNP. Eur. Heart J. 28: 1841–1847.

Latini, R., S. Masson, I. Anand, M. Salio, A. Hester, D. Judd, S. Barlera, A.P. Maggioni, G. Tognoni, and J.N. Cohn. 2004. The comparative prognostic value of plasma neurohormones at baseline in patients with heart failure enrolled in Val-HeFT. Eur. Heart J. 25: 292–299.

Latini, R., S. Masson, I.S. Anand, E. Missov, M. Carlson, T. Vago, L. Angelici, S. Barlera, G. Parrinello, A.P. Maggioni, G. Tognoni, and J.N. Cohn. 2007. Prognostic value of very low plasma concentrations of troponin T in patients with stable chronic heart failure. Circulation 116: 1242–1249.

Lee, D.S. and R.S. Vasan. 2005. Novel markers for heart failure diagnosis and prognosis. Curr. Opin. Cardiol. 20: 201–210.

Levine, B., J. Kalman, L. Mayer, H.M. Fillit, and M. Packer. 1990. Elevated circulating levels of tumor necrosis factor in severe chronic heart failure. N. Engl. J. Med. 323: 236–241.

Levy, D., S. Kenchaiah, M.G. Larson, E.J. Benjamin, M.J. Kupka, K.K. Ho, J.M. Murabito, and R.S. Vasan. 2002. Long-term trends in the incidence of and survival with heart failure. N. Engl. J. Med. 347: 1397–1402.

Li, Y., G. Takemura, K. Kosai, T. Takahashi, H. Okada, S. Miyata, K. Yuge, S. Nagano, M. Esaki, N.C. Khai, K. Goto, A. Mikami, R. Maruyama, S. Minatoguchi, T. Fujiwara, and H. Fujiwara. 2004. Critical roles for the Fas/Fas ligand system in postinfarction ventricular remodeling and heart failure. Circ. Res. 95: 627–636.

Lloyd-Jones, D., R.J. Adams, T.M. Brown, M. Carnethon, S. Dai, G. De Simone, T.B. Ferguson, E. Ford, K. Furie, C. Gillespie, A. Go, K. Greenlund, N. Haase, S. Hailpern, P.M. Ho, V. Howard, B. Kissela, S. Kittner, D. Lackland, L. Lisabeth, A. Marelli, M.M. McDermott, J. Meigs, D. Mozaffarian, M. Mussolino, G. Nichol, V.L. Roger, W. Rosamond, R. Sacco, P. Sorlie, T. Thom, S. Wasserthiel-Smoller, N.D. Wong, and J. Wylie-Rosett. 2010. Heart disease and stroke statistics—2010 update: a report from the American Heart Association. Circulation 121: e46–e215.

Lloyd-Jones, D.M., M.G. Larson, E.P. Leip, A. Beiser, R.B. D'Agostino, W.B. Kannel, J.M. Murabito, R.S. Vasan, E.J. Benjamin, and D. Levy. 2002. Lifetime risk for developing congestive heart failure: the Framingham Heart Study. Circulation 106: 3068–3072.

Logeart, D., G. Thabut, P. Jourdain, C. Chavelas, P. Beyne, F. Beauvais, E. Bouvier, and A.C. Solal. 2004. Predischarge B-type natriuretic peptide assay for identifying patients at high risk of re-admission after decompensated heart failure. J. Am. Coll. Cardiol. 43: 635–641.

Logeart, D., J.Y. Tabet, L. Hittinger, G. Thabut, P. Jourdain, P. Maison, J.M. Tartiere, and A.C. Solal. 2008. Transient worsening of renal function during hospitalization for acute heart failure alters outcome. Int. J. Cardiol. 127: 228–232.

Lopez, B.L., G.L. Liu, T.A. Christopher, and X.L. Ma. 1997. Peroxynitrite, the product of nitric oxide and superoxide, causes myocardial injury in the isolated perfused rat heart. Coron. Artery Dis. 8: 149–153.

Lu, Z., X. Xu, X. Hu, S. Lee, J.H. Traverse, G. Zhu, J. Fassett, Y. Tao, P. Zhang, C. dos Remedios, M. Pritzker, J.L. Hall, D.J. Garry, and Y. Chen. 2010. Oxidative stress regulates left ventricular PDE5 expression in the failing heart. Circulation 121: 1474–1483.

MacCarthy, P.A., R. Grocott-Mason, B.D. Prendergast, and A.M. Shah. 2000. Contrasting inotropic effects of endogenous endothelin in the normal and failing human heart: studies with an intracoronary ET(A) receptor antagonist. Circulation 101: 142–147.

Maeda, K., T. Tsutamoto, A. Wada, N. Mabuchi, M. Hayashi, T. Tsutsui, M. Ohnishi, M. Sawaki, M. Fujii, T. Matsumoto, and M. Kinoshita. 2000. High levels of plasma brain natriuretic peptide and interleukin-6 after optimized treatment for heart failure are independent risk factors for morbidity and mortality in patients with congestive heart failure. J. Am. Coll. Cardiol. 36: 1587–1593.

Mair, J. 2008. Biochemistry of B-type natriuretic peptide—where are we now? Clin. Chem. Lab Med 46: 1507–1514.

Maisel, A., C. Mueller, R. Nowak, W.F. Peacock, J.W. Landsberg, P. Ponikowski, M. Mockel, C. Hogan, A.H. Wu, M. Richards, P. Clopton, G.S. Filippatos, S. Di Somma, I. Anand, L. Ng, L.B. Daniels, S.X. Neath, R. Christenson, M. Potocki, J. McCord, G. Terracciano, D. Kremastinos, O. Hartmann, S. von Haehling, A. Bergmann, N.G. Morgenthaler and S.D. Anker. 2010. Mid-region pro-hormone markers for diagnosis and prognosis in acute dyspnea: results from the BACH (Biomarkers in Acute Heart Failure) trial. J. Am. Coll. Cardiol. 55: 2062–2076.

Maisel, A.S., J. Koon, P. Krishnaswamy, R. Kazenegra, P. Clopton, N. Gardetto, R. Morrisey, A. Garcia, A. Chiu, and A. De Maria. 2001. Utility of B-natriuretic peptide as a rapid, point-of-care test for screening patients undergoing echocardiography to determine left ventricular dysfunction. Am. Heart J. 141: 367–374.

Maisel, A.S., P. Krishnaswamy, R.M. Nowak, J. McCord, J.E. Hollander, P. Duc, T. Omland, A.B. Storrow, W.T. Abraham, A.H. Wu, P. Clopton, P.G. Steg, A. Westheim, C.W. Knudsen, A. Perez, R. Kazanegra, H.C. Herrmann, and P.A. McCullough. 2002. Rapid measurement of B-type natriuretic peptide in the emergency diagnosis of heart failure. N. Engl. J. Med. 347: 161–167.

Marcus, L.S., D. Hart, M. Packer, M. Yushak, N. Medina, R.S. Danziger, D.F. Heitjan, and S.D. Katz. 1996. Hemodynamic and renal excretory effects of human brain natriuretic peptide infusion in patients with congestive heart failure. A double-blind, placebo-controlled, randomized crossover trial. Circulation 94: 3184–3189.

Masoudi, F.A., E.P. Havranek, G. Smith, R.H. Fish, J.F. Steiner, D.L. Ordin, and H.M. Krumholz. 2003. Gender, age, and heart failure with preserved left ventricular systolic function. J. Am. Coll. Cardiol. 41: 217–223.

Masson, S., R. Latini, I.S. Anand, T. Vago, L. Angelici, S. Barlera, E.D. Missov, A. Clerico, G. Tognoni, and J.N. Cohn. 2006. Direct comparison of B-type natriuretic peptide (BNP) and amino-terminal proBNP in a large population of patients with chronic and symptomatic heart failure: the Valsartan Heart Failure (Val-HeFT) data. Clin. Chem. 52: 1528–1538.

Masson, S., R. Latini, E. Carbonieri, L. Moretti, M.G. Rossi, S. Ciricugno, V. Milani, R. Marchioli, J. Struck, A. Bergmann, A.P. Maggioni, G. Tognoni, and L. Tavazzi. 2010. The predictive value of stable precursor fragments of vasoactive peptides in patients with chronic heart failure: data from the GISSI-heart failure (GISSI-HF) trial. Eur. J. Heart Fail. 12: 338–347.

McCord, J., B.J. Mundy, M.P. Hudson, A.S. Maisel, J.E. Hollander, W.T. Abraham, P.G. Steg, T. Omland, C.W. Knudsen, K.R. Sandberg, and P.A. McCullough. 2004. Relationship between obesity and B-type natriuretic peptide levels. Arch. Intern. Med. 164: 2247–2252.

McCullough, P.A., R.M. Nowak, J. McCord, J.E. Hollander, H.C. Herrmann, P.G. Steg, P. Duc, A. Westheim, T. Omland, C.W. Knudsen, A.B. Storrow, W.T. Abraham, S. Lamba,

A.H. Wu, A. Perez, P. Clopton, P. Krishnaswamy, R. Kazanegra and A.S. Maisel. 2002a. B-type natriuretic peptide and clinical judgment in emergency diagnosis of heart failure: analysis from Breathing Not Properly (BNP) Multinational Study. Circulation 106: 416–422.

McCullough, P.A., E.F. Philbin, J.A. Spertus, S. Kaatz, K.R. Sandberg, and W.D. Weaver. 2002b. Confirmation of a heart failure epidemic: findings from the Resource Utilization Among Congestive Heart Failure (REACH) study. J. Am. Coll. Cardiol. 39: 60–69.

McEntegart, M.B., B. Awede, M.C. Petrie, N. Sattar, F.G. Dunn, N.G. MacFarlane, and J.J. McMurray. 2007. Increase in serum adiponectin concentration in patients with heart failure and cachexia: relationship with leptin, other cytokines, and B-type natriuretic peptide. Eur. Heart J. 28: 829–835.

McMurray, J., M. Chopra, I. Abdullah, W.E. Smith, and H.J. Dargie. 1993. Evidence of oxidative stress in chronic heart failure in humans. Eur. Heart J. 14: 1493–1498.

McMurray, J.J., M.C. Petrie, D.R. Murdoch, and A.P. Davie. 1998. Clinical epidemiology of heart failure: public and private health burden. Eur. Heart J. 19 Suppl P: P9–16.

McMurray, J.J., S. Adamopoulos, S.D. Anker, A. Auricchio, M. Bohm, K. Dickstein, V. Falk, G. Filippatos, C. Fonseca, M.A. Gomez-Sanchez, T. Jaarsma, L. Kober, G.Y. Lip, A.P. Maggioni, A. Parkhomenko, B.M. Pieske, B.A. Popescu, P.K. Ronnevik, F.H. Rutten, J. Schwitter, P. Seferovic, J. Stepinska, P.T. Trindade, A.A. Voors, F. Zannad, A. Zeiher, J.J. Bax, H. Baumgartner, C. Ceconi, V. Dean, C. Deaton, R. Fagard, C. Funck-Brentano, D. Hasdai, A. Hoes, P. Kirchhof, J. Knuuti, P. Kolh, T. McDonagh, C. Moulin, Z. Reiner, U. Sechtem, P.A. Sirnes, M. Tendera, A. Torbicki, A. Vahanian, S. Windecker, L.A. Bonet, P. Avraamides, H.A. Ben Lamin, M. Brignole, A. Coca, P. Cowburn, H. Dargie, P. Elliott, F.A. Flachskampf, G.F. Guida, S. Hardman, B. Iung, B. Merkely, C. Mueller, J.N. Nanas, O.W. Nielsen, S. Orn, J.T. Parissis, and P. Ponikowski. 2012. ESC guidelines for the diagnosis and treatment of acute and chronic heart failure 2012: The Task Force for the Diagnosis and Treatment of Acute and Chronic Heart Failure 2012 of the European Society of Cardiology. Developed in collaboration with the Heart Failure Association (HFA) of the ESC. Eur. J. Heart Fail. 14: 803–869.

McNamara, D.M., R. Holubkov, K. Janosko, A. Palmer, J.J. Wang, G.A. MacGowan, S. Murali, W.D. Rosenblum, B. London and A.M. Feldman. 2001. Pharmacogenetic interactions between beta-blocker therapy and the angiotensin-converting enzyme deletion polymorphism in patients with congestive heart failure. Circulation 103: 1644–1648.

Miller, W.L., J.C. Burnett, Jr., K.A. Hartman, D.O. Hodge, I. Giuliani, F. Minard, C. Larue, and A.S. Jaffe. 2009. Role for precursor Pro-B type natriuretic peptide in assessing response to therapy and prognosis in patients with decompensated heart failure treated with nesiritide. Clin. Chim. Acta. 406: 119–123.

Mizuno, Y., M. Yoshimura, H. Yasue, T. Sakamoto, H. Ogawa, K. Kugiyama, E. Harada, M. Nakayama, S. Nakamura, T. Ito, Y. Shimasaki, Y. Saito, and K. Nakao. 2001. Aldosterone production is activated in failing ventricle in humans. Circulation 103: 72–77.

Moraes, D.L., W.S. Colucci, and M.M. Givertz. 2000. Secondary pulmonary hypertension in chronic heart failure: the role of the endothelium in pathophysiology and management. Circulation 102: 1718–1723.

Morrow, D.A. and J.A. de Lemos. 2007. Benchmarks for the assessment of novel cardiovascular biomarkers. Circulation 115: 949–952.

Mosterd, A., J.W. Deckers, A.W. Hoes, A. Nederpel, A. Smeets, D.T. Linker, and D.E. Grobbee. 1997. Classification of heart failure in population based research: an assessment of six heart failure scores. Eur. J. Epidemiol. 13: 491–502.

Murray, D.R. and G.L. Freeman. 2003. Proinflammatory cytokines: predictors of a failing heart? Circulation 107: 1460–1462.

Nakamura, K., K. Fushimi, H. Kouchi, K. Mihara, M. Miyazaki, T. Ohe, and M. Namba. 1998. Inhibitory effects of antioxidants on neonatal rat cardiac myocyte hypertrophy induced by tumor necrosis factor-alpha and angiotensin II. Circulation 98: 794–799.

Nakamura, K., K.F. Kusano, H. Matsubara, Y. Nakamura, A. Miura, N. Nishii, K. Banba, S. Nagase, K. Miyaji, H. Morita, H. Saito, T. Emori, and T. Ohe. 2005. Relationship between oxidative stress and systolic dysfunction in patients with hypertrophic cardiomyopathy. J. Card. Fail. 11: 117–123.

Nakayama, H., X. Chen, C.P. Baines, R. Klevitsky, X. Zhang, H. Zhang, N. Jaleel, B.H. Chua, T.E. Hewett, J. Robbins, S.R. Houser, and J.D. Molkentin. 2007. Ca2+- and mitochondrial-dependent cardiomyocyte necrosis as a primary mediator of heart failure. J. Clin. Invest. 117: 2431–2444.

National Heart Lung and Blood Institute: NIH Bethesda MD. 1996. Morbidity and Mortality Chartbook on Cardiocascular, Lung and blood Diseases.

Newton, G.E. and J.D. Parker. 1996. Acute effects of beta 1-selective and nonselective beta-adrenergic receptor blockade on cardiac sympathetic activity in congestive heart failure. Circulation 94: 353–358.

Nickolas, T.L., J. Barasch, and P. Devarajan. 2008a. Biomarkers in acute and chronic kidney disease. Curr Opin Nephrol. Hypertens. 17: 127–132.

Nickolas, T.L., M.J. O'Rourke, J. Yang, M.E. Sise, P.A. Canetta, N. Barasch, C. Buchen, F. Khan, K. Mori, J. Giglio, P. Devarajan, and J. Barasch. 2008b. Sensitivity and specificity of a single emergency department measurement of urinary neutrophil gelatinase-associated lipocalin for diagnosing acute kidney injury. Ann. Intern. Med. 148: 810–819.

Nilsson, J. 2005. CRP—marker or maker of cardiovascular disease? Arterioscler. Thromb. Vasc. Biol. 25: 1527–1528.

Nohria, A., L.M. Mielniczuk, and L.W. Stevenson. 2005. Evaluation and monitoring of patients with acute heart failure syndromes. Am. J. Cardiol. 96: 32G–40G.

Noma, T., A. Lemaire, S.V. Naga Prasad, L. Barki-Harrington, D.G. Tilley, J. Chen, P. Le Corvoisier, J.D. Violin, H. Wei, R.J. Lefkowitz, and H.A. Rockman. 2007. Beta-arrestin-mediated beta1-adrenergic receptor transactivation of the EGFR confers cardioprotection. J. Clin. Invest. 117: 2445–2458.

Nozawa, T., A. Igawa, N. Yoshida, M. Maeda, M. Inoue, Y. Yamamura, H. Asanoi, and H. Inoue. 1998. Dual-tracer assessment of coupling between cardiac sympathetic neuronal function and downregulation of beta-receptors during development of hypertensive heart failure of rats. Circulation 97: 2359–2367.

Okuyama, M., S. Yamaguchi, N. Nozaki, M. Yamaoka, M. Shirakabe, and H. Tomoike. 1997. Serum levels of soluble form of Fas molecule in patients with congestive heart failure. Am. J. Cardiol. 79: 1698–1701.

Omland, T., M.H. Drazner, T. Ueland, M. Abedin, S.A. Murphy, P. Aukrust, and J.A. de Lemos. 2007. Plasma osteoprotegerin levels in the general population: relation to indices of left ventricular structure and function. Hypertension 49: 1392–1398.

Palladini, G., C. Campana, C. Klersy, A. Balduini, G. Vadacca, V. Perfetti, S. Perlini, L. Obici, E. Ascari, G.M. d'Eril, R. Moratti, and G. Merlini. 2003. Serum N-terminal pro-brain natriuretic peptide is a sensitive marker of myocardial dysfunction in AL amyloidosis. Circulation 107: 2440–2445.

Peacock, W.F.t., T. De Marco, G.C. Fonarow, D. Diercks, J. Wynne, F.S. Apple, and A.H. Wu. 2008. Cardiac troponin and outcome in acute heart failure. N. Engl. J. Med. 358: 2117–2126.

Pearson, T.A., G.A. Mensah, R.W. Alexander, J.L. Anderson, R.O. Cannon, 3rd, M. Criqui, Y.Y. Fadl, S.P. Fortmann, Y. Hong, G.L. Myers, N. Rifai, S.C. Smith, Jr., K. Taubert, R.P. Tracy, and F. Vinicor. 2003. Markers of inflammation and cardiovascular disease: application to clinical and public health practice: A statement for healthcare professionals from the Centers for Disease Control and Prevention and the American Heart Association. Circulation 107: 499–511.

Pfeffer, M.A. and E. Braunwald. 1990. Ventricular remodeling after myocardial infarction. Experimental observations and clinical implications. Circulation 81: 1161–1172.

Pieroni, M., A. Corti, B. Tota, F. Curnis, T. Angelone, B. Colombo, M.C. Cerra, F. Bellocci, F. Crea, and A. Maseri. 2007. Myocardial production of chromogranin A in human heart: a new regulatory peptide of cardiac function. Eur. Heart J. 28: 1117–1127.

Pinsky, D.J., B. Cai, X. Yang, C. Rodriguez, R.R. Sciacca, and P.J. Cannon. 1995. The lethal effects of cytokine-induced nitric oxide on cardiac myocytes are blocked by nitric oxide synthase antagonism or transforming growth factor beta. J. Clin. Invest. 95: 677–685.

Polidori, M.C., D. Pratico, K. Savino, J. Rokach, W. Stahl, and P. Mecocci. 2004. Increased F2 isoprostane plasma levels in patients with congestive heart failure are correlated with antioxidant status and disease severity. J. Card. Fail. 10: 334–338.

Poutanen, T., T. Tikanoja, P. Riikonen, A. Silvast, and M. Perkkio. 2003. Long-term prospective follow-up study of cardiac function after cardiotoxic therapy for malignancy in children. J. Clin. Oncol. 21: 2349–2356.

Querejeta, R., N. Varo, B. Lopez, M. Larman, E. Artinano, J.C. Etayo, J.L. Martinez Ubago, M. Gutierrez-Stampa, J.I. Emparanza, M.J. Gil, I. Monreal, J.P. Mindan, and J. Diez. 2000. Serum carboxy-terminal propeptide of procollagen type I is a marker of myocardial fibrosis in hypertensive heart disease. Circulation 101: 1729–1735.

Rauchhaus, M., W. Doehner, D.P. Francis, C. Davos, M. Kemp, C. Liebenthal, J. Niebauer, J. Hooper, H.D. Volk, A.J. Coats, and S.D. Anker. 2000. Plasma cytokine parameters and mortality in patients with chronic heart failure. Circulation 102: 3060–3067.

Rehman, S.U., T. Mueller, and J.L. Januzzi, Jr. 2008. Characteristics of the novel interleukin family biomarker ST2 in patients with acute heart failure. J. Am. Coll. Cardiol. 52: 1458–1465.

Richards, A.M., R. Doughty, M.G. Nicholls, S. Macmahon, H. Ikram, N. Sharpe, E.A. Espiner, C. Frampton, and T.G. Yandle. 1999. Neurohumoral prediction of benefit from carvedilol in ischemic left ventricular dysfunction. Australia-New Zealand Heart Failure Group. Circulation 99: 786–792.

Roach, D.R., A.G. Bean, C. Demangel, M.P. France, H. Briscoe, and W.J. Britton. 2002. TNF regulates chemokine induction essential for cell recruitment, granuloma formation, and clearance of mycobacterial infection. J. Immunol. 168: 4620–4627.

Roberts, R., K.S. Gowda, P.A. Ludbrook, and B.E. Sobel. 1975. Specificity of elevated serum MB creatine phosphokinase activity in the diagnosis of acute myocardial infarction. Am. J. Cardiol. 36: 433–437.

Roger, V.L., S.A. Weston, M.M. Redfield, J.P. Hellermann-Homan, J. Killian, B.P. Yawn, and S.J. Jacobsen. 2004. Trends in heart failure incidence and survival in a community-based population. JAMA 292: 344–350.

Roger, V.L. 2010. The heart failure epidemic. Int. J. Environ. Res. Public Health 7: 1807–1830.

Ronco, C., M. Haapio, A.A. House, N. Anavekar, and R. Bellomo. 2008. Cardiorenal syndrome. J. Am. Coll. Cardiol. 52: 1527–1539.

Saenger, A.K. and A.S. Jaffe. 2008. Requiem for a heavyweight: the demise of creatine kinase-MB. Circulation 118: 2200–2206.

Sam, F., D.L. Kerstetter, D.R. Pimental, S. Mulukutla, A. Tabaee, M.R. Bristow, W.S. Colucci, and D.B. Sawyer. 2005. Increased reactive oxygen species production and functional alterations in antioxidant enzymes in human failing myocardium. J. Card. Fail. 11: 473–480.

Sarraf, M., A. Masoumi, and R.W. Schrier. 2009. Cardiorenal syndrome in acute decompensated heart failure. Clin. J. Am. Soc. Nephrol. 4: 2013–2026.

Schiffrin, E.L., M.L. Lipman, and J.F. Mann. 2007. Chronic kidney disease: effects on the cardiovascular system. Circulation 116: 85–97.

Schrier, R.W. 2006. Water and sodium retention in edematous disorders: role of vasopressin and aldosterone. Am. J. Med. 119: S47–53.

Scirica, B.M. and D.A. Morrow. 2006. Is C-reactive protein an innocent bystander or proatherogenic culprit? The verdict is still out. Circulation 113: 2128–2134; discussion 2151.

Seta, Y., K. Shan, B. Bozkurt, H. Oral, and D.L. Mann. 1996. Basic mechanisms in heart failure: the cytokine hypothesis. J. Card. Fail. 2: 243–249.

Shah, R.V. and J.L. Januzzi, Jr. 2010. ST2: a novel remodeling biomarker in acute and chronic heart failure. Curr. Heart Fail. Rep. 7: 9–14.

Shimpo, M., D.A. Morrow, E.O. Weinberg, M.S. Sabatine, S.A. Murphy, E.M. Antman, and R.T. Lee. 2004. Serum levels of the interleukin-1 receptor family member ST2 predict mortality and clinical outcome in acute myocardial infarction. Circulation 109: 2186–2190.

Sigurdsson, A., O. Amtorp, T. Gundersen, B. Nilsson, J. Remes, and K. Swedberg. 1994. Neurohormonal activation in patients with mild or moderately severe congestive heart failure and effects of ramipril. The Ramipril Trial Study Group. Br Heart J 72: 422–427.

Silvestre, J.S., C. Heymes, A. Oubenaissa, V. Robert, B. Aupetit-Faisant, A. Carayon, B. Swynghedauw, and C. Delcayre. 1999. Activation of cardiac aldosterone production in rat myocardial infarction: effect of angiotensin II receptor blockade and role in cardiac fibrosis. Circulation 99: 2694–2701.

Sliwa, K., A. Woodiwiss, V.N. Kone, G. Candy, D. Badenhorst, G. Norton, C. Zambakides, F. Peters, and R. Essop. 2004. Therapy of ischemic cardiomyopathy with the immunomodulating agent pentoxifylline: results of a randomized study. Circulation 109: 750–755.

Somaratne, J.B., C. Berry, J.J. McMurray, K.K. Poppe, R.N. Doughty and G.A. Whalley. 2009. The prognostic significance of heart failure with preserved left ventricular ejection fraction: a literature-based meta-analysis. Eur. J. Heart Fail. 11: 855–862.

Stanek, B., B. Frey, M. Hulsmann, R. Berger, B. Sturm, J. Strametz-Juranek, J. Bergler-Klein, P. Moser, A. Bojic, E. Hartter, and R. Pacher. 2001. Prognostic evaluation of neurohumoral plasma levels before and during beta-blocker therapy in advanced left ventricular dysfunction. J. Am. Coll. Cardiol. 38: 436–442.

Sugiura, T., H. Takase, T. Toriyama, T. Goto, R. Ueda and Y. Dohi. 2005. Circulating levels of myocardial proteins predict future deterioration of congestive heart failure. J. Card. Fail. 11: 504–509.

Suzuki, T., D. Hayashi, T. Yamazaki, T. Mizuno, Y. Kanda, I. Komuro, M. Kurabayashi, K. Yamaoki, K. Mitani, H. Hirai, R. Nagai, and Y. Yazaki. 1998. Elevated B-type natriuretic peptide levels after anthracycline administration. Am. Heart J. 136: 362–363.

Taglieri, N., W. Koenig, and J.C. Kaski. 2009. Cystatin C and cardiovascular risk. Clin. Chem. 55: 1932–1943.

Tamura, N., Y. Ogawa, H. Chusho, K. Nakamura, K. Nakao, M. Suda, M. Kasahara, R. Hashimoto, G. Katsuura, M. Mukoyama, H. Itoh, Y. Saito, I. Tanaka, H. Otani, and M. Katsuki. 2000. Cardiac fibrosis in mice lacking brain natriuretic peptide. Proc. Natl. Acad. Sci. U S A 97: 4239–4244.

Tang, W.H., J.P. Girod, M.J. Lee, R.C. Starling, J.B. Young, F. Van Lente, and G.S. Francis. 2003. Plasma B-type natriuretic peptide levels in ambulatory patients with established chronic symptomatic systolic heart failure. Circulation 108: 2964–2966.

Tang, W.H., M.L. Brennan, K. Philip, W. Tong, S. Mann, F. Van Lente, and S.L. Hazen. 2006. Plasma myeloperoxidase levels in patients with chronic heart failure. Am. J. Cardiol. 98: 796–799.

Tang, W.H., G.S. Francis, D.A. Morrow, L.K. Newby, C.P. Cannon, R.L. Jesse, A.B. Storrow, R.H. Christenson, F.S. Apple, J. Ravkilde, and A.H. Wu. 2007. National Academy of Clinical Biochemistry Laboratory Medicine practice guidelines: Clinical utilization of cardiac biomarker testing in heart failure. Circulation 116: e99–109.

Teerlink, J.R. 2005. Endothelins: pathophysiology and treatment implications in chronic heart failure. Curr. Heart Fail. Rep. 2: 191–197.

The SOLVD Investigators. 1991. Effect of enalapril on survival in patients with reduced left ventricular ejection fractions and congestive heart failure. N. Engl. J. Med. 325: 293–302.

Tjeerdsma, G., R.A. de Boer, F. Boomsma, M.P. van den Berg, Y.M. Pinto, and D.J. van Veldhuisen. 2002. Rapid bedside measurement of brain natriuretic peptide in patients with chronic heart failure. Int. J. Cardiol. 86: 143–9; discussion 149–152.

Topol, E.J., T.A. Traill, and N.J. Fortuin. 1985. Hypertensive hypertrophic cardiomyopathy of the elderly. N. Engl. J. Med. 312: 277–283.

Torre-Amione, G., S. Kapadia, C. Benedict, H. Oral, J.B. Young, and D.L. Mann. 1996a. Proinflammatory cytokine levels in patients with depressed left ventricular ejection fraction: a report from the Studies of Left Ventricular Dysfunction (SOLVD). J. Am. Coll. Cardiol. 27: 1201–1206.

Torre-Amione, G., S. Kapadia, J. Lee, J.B. Durand, R.D. Bies, J.B. Young, and D.L. Mann. 1996b. Tumor necrosis factor-alpha and tumor necrosis factor receptors in the failing human heart. Circulation 93: 704–711.

Trof, R.J., F. Di Maggio, J. Leemreis, and A.B. Groeneveld. 2006. Biomarkers of acute renal injury and renal failure. Shock 26: 245–253.

Tsutamoto, T., A. Wada, K. Maeda, T. Hisanaga, N. Mabuchi, M. Hayashi, M. Ohnishi, M. Sawaki, M. Fujii, H. Horie, Y. Sugimoto, and M. Kinoshita. 1999. Plasma brain natriuretic peptide level as a biochemical marker of morbidity and mortality in patients with asymptomatic or minimally symptomatic left ventricular dysfunction. Comparison with plasma angiotensin II and endothelin-1. Eur. Heart J. 20: 1799–1807.

Ueland, T., R. Jemtland, K. Godang, J. Kjekshus, A. Hognestad, T. Omland, I.B. Squire, L. Gullestad, J. Bollerslev, K. Dickstein, and P. Aukrust. 2004. Prognostic value of osteoprotegerin in heart failure after acute myocardial infarction. J. Am. Coll. Cardiol. 44: 1970–1976.

Ungvari, Z., S.A. Gupte, F.A. Recchia, S. Batkai, and P. Pacher. 2005. Role of oxidative-nitrosative stress and downstream pathways in various forms of cardiomyopathy and heart failure. Curr. Vasc. Pharmacol. 3: 221–229.

Vaidya, V.S., V. Ramirez, T. Ichimura, N.A. Bobadilla, and J.V. Bonventre. 2006. Urinary kidney injury molecule-1: a sensitive quantitative biomarker for early detection of kidney tubular injury. Am J. Physiol. Renal Physiol 290: F517–529.

Valgimigli, M., E. Merli, P. Malagutti, O. Soukhomovskaia, G. Cicchitelli, A. Antelli, D. Canistro, G. Francolini, G. Macri, F. Mastrorilli, M. Paolini, and R. Ferrari. 2004. Hydroxyl radical generation, levels of tumor necrosis factor-alpha, and progression to heart failure after acute myocardial infarction. J. Am. Coll. Cardiol. 43: 2000–2008.

van Kimmenade, R.R., J.L. Januzzi, Jr., P.T. Ellinor, U.C. Sharma, J.A. Bakker, A.F. Low, A. Martinez, H.J. Crijns, C.A. MacRae, P.P. Menheere, and Y.M. Pinto. 2006. Utility of amino-terminal pro-brain natriuretic peptide, galectin-3, and apelin for the evaluation of patients with acute heart failure. J. Am. Coll. Cardiol. 48: 1217–1224.

Vasan, R.S., L.M. Sullivan, R. Roubenoff, C.A. Dinarello, T. Harris, E.J. Benjamin, D.B. Sawyer, D. Levy, P.W. Wilson, and R.B. D'Agostino. 2003. Inflammatory markers and risk of heart failure in elderly subjects without prior myocardial infarction: the Framingham Heart Study. Circulation 107: 1486–1491.

Vatner, D.E., K. Asai, M. Iwase, Y. Ishikawa, R.P. Shannon, C.J. Homcy, and S.F. Vatner. 1999. Beta-adrenergic receptor-G protein-adenylyl cyclase signal transduction in the failing heart. Am. J. Cardiol. 83: 80H–85H.

Venugopal, S.K., S. Devaraj, and I. Jialal. 2005. Effect of C-reactive protein on vascular cells: evidence for a proinflammatory, proatherogenic role. Curr. Opin. Nephrol. Hypertens. 14: 33–37.

Vickery, S., C.P. Price, R.I. John, N.A. Abbas, M.C. Webb, M.E. Kempson and E.J. Lamb. 2005. B-type natriuretic peptide (BNP) and amino-terminal proBNP in patients with CKD: relationship to renal function and left ventricular hypertrophy. Am. J. Kidney Dis. 46: 610–620.

Wallace, T.W., S.M. Abdullah, M.H. Drazner, S.R. Das, A. Khera, D.K. McGuire, F. Wians, M.S. Sabatine, D.A. Morrow, and J.A. de Lemos. 2006. Prevalence and determinants of troponin T elevation in the general population. Circulation 113: 1958–1965.

Weinberg, E.O., M. Shimpo, G.W. De Keulenaer, C. MacGillivray, S. Tominaga, S.D. Solomon, J.L. Rouleau, and R.T. Lee. 2002. Expression and regulation of ST2, an interleukin-1 receptor family member, in cardiomyocytes, and myocardial infarction. Circulation 106: 2961–2966.

Weinberg, E.O., M. Shimpo, S. Hurwitz, S. Tominaga, J.L. Rouleau, and R.T. Lee. 2003. Identification of serum soluble ST2 receptor as a novel heart failure biomarker. Circulation 107: 721–726.

Wijeysundera, H.C., M.S. Hansen, E. Stanton, A.S. Cropp, C. Hall, N.S. Dhalla, J. Ghali, and J.L. Rouleau. 2003. Neurohormones and oxidative stress in nonischemic cardiomyopathy: relationship to survival and the effect of treatment with amlodipine. Am. Heart J. 146: 291–297.

Wilkins, M.R., R.J. Unwin, and A.J. Kenny. 1993. Endopeptidase-24.11 and its inhibitors: potential therapeutic agents for edematous disorders and hypertension. Kidney Int. 43: 273–285.

Wollert, K.C. and H. Drexler. 2001. The role of interleukin-6 in the failing heart. Heart Fail. Rev. 6: 95–103.

Wollert, K.C., T. Kempf, B. Lagerqvist, B. Lindahl, S. Olofsson, T. Allhoff, T. Peter, A. Siegbahn, P. Venge, H. Drexler, and L. Wallentin. 2007. Growth differentiation factor 15 for risk stratification and selection of an invasive treatment strategy in non ST-elevation acute coronary syndrome. Circulation 116: 1540–1548.

Wong, N.L., M. Sonntag, and J.K. Tsui. 2003. Attenuation of renal vasopressin V2 receptor upregulation by bosentan, an ETA/ETB receptor antagonist. Metabolism 52: 1141–1146.

Wong, S.C., M. Fukuchi, P. Melnyk, I. Rodger, and A. Giaid. 1998. Induction of cyclooxygenase-2 and activation of nuclear factor-kappaB in myocardium of patients with congestive heart failure. Circulation 98: 100–103.

Zacho, J., A. Tybjaerg-Hansen, J.S. Jensen, P. Grande, H. Sillesen, and B.G. Nordestgaard. 2008. Genetically elevated C-reactive protein and ischemic vascular disease. N. Engl. J. Med. 359: 1897–1908.

Zannad, F., F. Alla, B. Dousset, A. Perez, and B. Pitt. 2000. Limitation of excessive extracellular matrix turnover may contribute to survival benefit of spironolactone therapy in patients with congestive heart failure: insights from the randomized aldactone evaluation study (RALES). Rales Investigators. Circulation 102: 2700–2706.

Zile, M.R. and D.L. Brutsaert. 2002. New concepts in diastolic dysfunction and diastolic heart failure: Part I: diagnosis, prognosis, and measurements of diastolic function. Circulation 105: 1387–1393.

Zimmet, J.M. and J.M. Hare. 2006. Nitroso-redox interactions in the cardiovascular system. Circulation 114: 1531–1544.

## CHAPTER 7

# Biomarkers of Pulmonary Hypertension and Pulmonary Embolism

### Helias Stamatopoulos and Stella Brili*

## Introduction

Pulmonary hypertension (PH) is a hemodynamic and pathophysiological condition defined as an increase in mean pulmonary arterial pressure (PAP) ≥25 mmHg at rest as assessed by right heart catheterization (Galie et al. 2009). PH can be found in multiple clinical conditions classified into five clinical groups with specific characteristics (Table 1) (McLaughlin et al. 2009).

PAH is a syndrome resulting from pathological increases in pulmonary vascular resistance (PVR) leading ultimately to right heart failure (Voelkel et al. 2006). The predominant cause of increased PVR is vascular remodeling including intimal hyperplasia, medial hypertrophy, adventitial proliferation, thrombosis *in situ*, varying degrees of inflammation and plexiform arteriopathy (McLaughlin et al. 2009). As presenting symptoms are nonspecific, diagnosis is often delayed by, on average, two yrs (Peacock 2003); median survival in untreated disease is 2.8 yrs with an estimated 5-yr survival of 34 percent (D'Alonzo et al. 1991).

1st Department of Cardiology, 'Hippokration' Hospital, University of Athens Medical School, Athens, Greece.
*Corresponding author: stlbrili@gmail.com

**Table 1.** Revised WHO Classification of PH.

| |
|---|
| **1. Pulmonary arterial hypertension (PAH)**<br>  1.1. Idiopathic<br>  1.2. Familial<br>  1.3. Associated with:<br>    1.3.1. Connective tissue disorder<br>    1.3.2. Congenital systemic-to-pulmonary shunts<br>    1.3.3. Portal hypertension<br>    1.3.4. HIV infection<br>    1.3.5. Drugs and toxins<br>    1.3.6. Other (thyroid disorders, glycogen storage disease, Gaucher's disease, hereditary hemorrhagic telangiectasia, hemoglobinopathies, chronic myeloproliferative disorders, splenectomy)<br>  1.4. Associated with significant venous or capillary involvement<br>    1.4.1. Pulmonary veno-occlusive disease<br>    1.4.2. Pulmonary capillary hemangiomatosis<br>  1.5. Persistent pulmonary hypertension of the newborn |
| **2. Pulmonary hypertension with left heart disease**<br>  2.1. Left-sided atrial or ventricular heart disease<br>  2.2. Left-sided valvular heart disease |
| **3. Pulmonary hypertension associated with lung diseases and/or hypoxemia**<br>  3.1. Chronic obstructive pulmonary disease<br>  3.2. Interstitial lung disease<br>  3.3. Sleep disordered breathing<br>  3.4. Alveolar hypoventilation disorders<br>  3.5. Chronic exposure to high altitude<br>  3.6. Developmental abnormalities |
| **4. Pulmonary hypertension due to chronic thrombotic and/or embolic disease**<br>  4.1. Thromboembolic obstruction of proximal pulmonary arteries<br>  4.2. Thromboembolic obstruction of distal pulmonary arteries<br>  4.3. Nonthrombotic pulmonary embolism (tumor, parasites, foreign material) |
| **5. Miscellaneous**<br>  Sarcoidosis, histiocytosis X, lymphangiomatosis, compression of pulmonary vessels (adenopathy, tumor, fibrosing mediastinitis) |

In order to diagnose the specific disease state leading to PH, routine biochemistry, haematology, and thyroid function tests are required in all patients, as well as a number of other essential blood tests (Galie et al. 2009). Serological testing is important to detect underlying connective tissue disease, HIV infection and hepatitis. Systemic sclerosis (scleroderma) is the most important connective tissue disease to exclude because this condition has a high prevalence of PAH. Anti-centromere antibodies are typically positive (in 60–80 percent of patients) in limited scleroderma, while other anti-nuclear antibodies include dsDNA, anti-Ro, U3-RNP, B23, Th/To, and U1-RNP. In the diffuse variety of scleroderma, antitopoisomerase I antibodies, originally called anti-Scl-70, are positive in 40 percent of patients, while U3-RNP are associated with the occurrence of PAH. Other laboratory tests might include antinuclear and anti-dsDNA (systemic lupus erythematosus), rheumatoid factor (rheumatoid arthritis), anti- Ro and anti-La (Sjogren's syndrome), anti-Jo-1 (dermatomyositis/polymyositis) and anti-U1 RNP (mixed connective tissue disease). It should be noted that up to 40 percent of patients with idiopathic PAH have elevated anti-nuclear antibodies, usually in low titre (1:80) (Rich et al. 1986). HIV testing should be considered in all patients, especially those with a compatible history or risk factors. Up to 2 percent of individuals with liver disease will manifest PAH and therefore liver function tests and hepatitis serology should be examined if clinical abnormalities are noted. Thrombophilia screening including anti-phospholipid antibodies, lupus anticoagulant and anti-cardiolipin antibodies should be performed in chronic thromboembolic PH. HIV testing is mandatory. Thyroid disease is commonly seen in PAH and should always be considered, especially if abrupt changes in the clinical course occur.

Recently there has been increasing interest in the use of biomarkers as a non-invasive tool for delineating the severity and prognosis of disease, for monitoring the clinical course of patients and their response to therapy, as well as for screening purposes. These include markers of heart failure, cardiac myocyte damage, endothelial and/or platelet dysfunction and oxidative stress (Warwick et al. 2008).

## Natriuretic Peptides

Atrial natriuretic peptide (ANP) and brain (B-type) natriuretic peptide (BNP) are peptide hormones that are released from cardiac myocytes in response to cardiac pressure and volume overload (Yap et al. 2004). ANP is released from storage granules in atrial tissue, while BNP is secreted mainly from ventricular myocardium. ANP and BNP cause natriuresis and vasodilation and counteract the water-retaining and vasoconstricting effects of the adrenergic, renin-angiotensin-aldosterone and AVP systems. Interest in the clinical application of natriuretic peptides in monitoring RV failure due to chronic PH has focused on BNP.

The final step of BNP synthesis consists of a high molecular weight precursor, proBNP cleaved into biologically inactive N-terminal segment (NT-proBNP) and the proper low molecular weight BNP. NT-proBNP has a longer half-life and a better stability both in circulating blood and after sampling. RV failure is the main cause of death in PAH, and BNP/NT-proBNP levels reflect the severity of RV dysfunction. BNP is elevated in PH of various classes: idiopathic (IPAH) (Nagaya et al. 2000).

PAH associated with connective tissue disease (Wilkins et al. 2005), congenital systemic-to-pulmonary shunts (Nagaya et al. 1998b), PH associated with chronic obstructive pulmonary disease (Bozkanat et al. 2005), interstitial lung disease (Leuchte et al. 2004), chronic thromboembolic disease (Nagaya et al. 2002; Nagaya et al. 2003) and acute pulmonary embolus (Kucher et al. 2003b; ten Wolde et al. 2003). BNP levels correlate with hemodynamic parameters (Leuchte et al. 2004; Nagaya et al. 2002; Nagaya et al. 2000; Nagaya et al. 2003), exercise capacity and NYHA functional class (Andreassen et al. 2006). BNP is of prognostic significance in IPAH, in chronic lung disease (Leuchte et al. 2006) and in pulmonary embolism (Kucher et al. 2003b; ten Wolde et al. 2003). Nagaya et al. (Nagaya et al. 2000) showed that in idiopathic PAH the baseline median value of BNP (150 pg/mL) distinguished patients with a good or bad prognosis. In 49 out of 60 patients, BNP measurement was repeated after three months of targeted therapy and again the supramedian level (>180 pg/mL) was related to worse long-term outcome. Plasma BNP significantly decreased in survivors but increased in non-survivors despite treatment. Similarly, plasma levels of NT-proBNP were found to be higher in patients PH of different classes compared with controls, correlated with hemodynamic parameters and were related to a poor prognosis (Andreassen et al. 2006; Fijalkowska et al. 2006). Serum NT-proBNP below 1400 pg/mL has been suggested for identification of patients with good prognosis. Larger outcome trials are still required to verify the suggested cut-off levels for NT-proBNP. NT-proBNP has also been evaluated as a tool to stratify disease severity.

The role of serial measurement of NT-proBNP during acute vasoreactivity testing of IPAH patients has been evaluated in a small study (Souza et al. 2005). An increase in NT-proBNP after a 60-min inhalation of nitric oxide distinguished responders from nonresponders with a 50 percent specificity and 100 percent sensitivity (positive predictive value 38 percent, negative predictive value 100 percent). Plasma ANP is significantly increased in PH of varying aetiologies and, like BNP, is correlated with hemodynamic variables (Adnot et al. 1987; Wiedemann et al. 2001).

Natriuretic peptides have also been used to monitor response to therapy. BNP levels parallel the course of pulmonary hemodynamics and exercise capacity in patients with PAH on vasodilator therapy (Leuchte et al. 2005; Nagaya et al. 1998a), and can predict response to therapy with epoprostenol (Park et al. 2004). BNP levels mirror hemodynamic improvements in patients with chronic thromboembolic pulmonary hypertension undergoing pre-operative prostacyclin therapy prior to endarterectomy (Nagaya et al. 2003); a persistently elevated BNP following endarterectomy is indicative of persistent PH in the post-operative period (Nagaya et al. 2002). ANP levels also decrease after inhalation of iloprost (Wiedemann et al. 2001).

There are a number of confounding factors in the interpretation of natriuretic peptide levels, including left heart disease, sex, age, weight, and renal dysfunction. Since most studies exclude patients with left heart disease and renal dysfunction, there may be problems with extrapolating these results to a less selected population. A recent study assessed the effect of renal impairment on the diagnostic accuracy of NT-proBNP as a parameter of hemodynamic status (Leuchte et al. 2007). It found that, while the diagnostic accuracy of NT-proBNP as a noninvasive follow-up parameter of the hemodynamic variables seems to be inferior to that of BNP in patients with a creatinine clearance of <60 mL/min, it was superior to BNP as a survival parameter, because it integrated hemodynamic impairment with renal insufficiency, which was itself a sign of increased mortality. It should also be noted that, since elevated levels of natriuretic peptides signify high ventricular wall stress, they should be considered "late" markers of disease. A normal BNP/NT-proBNP level cannot, therefore, be used to exclude the presence of PH.

## Troponin

Cardiac troponins are regulatory proteins of the thin actin filaments of cardiac muscle. Disruption of the cardiac myocyte membrane causes their release and they can then be detected by highly sensitive assays in the peripheral blood. Elevated plasma levels of cardiac troponin T and troponin I are established specific markers of myocardial damage and are prognostic indicators in acute coronary syndromes and acute pulmonary embolism. Torbicki et al. reported that patients with severe PH of varying classes with detectable troponin T had higher heart rates, lower mixed venous oxygen saturation, higher NT-proBNP and shorter 6MWT (Torbicki et al. 2003). They had significantly higher mortality at six, 12 and 24 months. Troponin became undetectable with successful treatment and returned as disease progressed, although this was shown only in a small group of patients. As with the natriuretic peptides, elevated troponin levels represent more advanced disease, since they are indicative of myocardial ischemia. They cannot be expected, therefore, to be a sensitive marker of early disease. Interpretation of troponin T levels in PH may be confounded by concurrent left heart disease and renal impairment.

## Endothelin-1

ET-1 is a potent vasoconstrictor peptide and mitogenic cytokine released by endothelial cells throughout the circulation, as well as a variety of other cell types such as cardiac myocytes. ET-1 is the predominant isoform of endothelin in the cardiovascular system, generated through the cleavage of pre-pro ET-1 to big ET-1 and then to ET-1. ET levels are elevated in the plasma of patients with PH (Cacoub et al. 1997; Rubens et al. 2001) and there is increased expression of ET-1 protein and mRNA in the endothelial cells of affected vessels in patients with PH (Giaid et al. 1993). Rubens et al. found significantly elevated plasma ET-1 and big ET-1 levels in 16 patients with IPAH; plasma ET-1 and big ET-1 showed a significant positive correlation with pulmonary vascular resistance and mean pulmonary artery pressure, and a significant negative correlation with cardiac output, cardiac index and the 6MWT and, hence, disease severity (Rubens et al. 2001). A number of potential confounders must be considered when interpreting ET-1 levels: African ethnicity, male sex and older age are associated with a raised plasma ET-1; while angiotensin-converting enzyme inhibitors, statins, b-blockers and vasodilators reduce the level of ET-1 in plasma.

## Nitric Oxide

Nitric oxide is produced by the enzyme nitric oxide synthase (NOS) in endothelial cells by the conversion of L-arginine to L-citruline and nitric oxide (NO). It diffuses into adjacent vascular smooth muscle cells and binds to soluble guanylate cyclase, stimulating the production of cyclic guanosine monophosphate (cGMP), resulting in muscle relaxation. NO is measurable in exhaled air; it is now accepted as reflecting airway inflammation and represents pulmonary NO production. It comes primarily from airway epithelial cells with a component from the pulmonary vasculature (Warwick et al. 2008). Exhaled NO (eNO) has been shown to be lower in patients with IPAH (Girgis et al. 2005; Kaneko et al. 1998; Ozkan et al. 2001), although there are conflicting studies that show either no difference in eNO compared with controls (Forrest et al. 1999; Riley et al. 1997) or an increased level (Archer et al. 1998). These divergent results could be explained by methodological differences. When measured in the bronchoalveolar lavage fluid of patients with IPAH, biochemical reaction products of NO (nitrate, nitrite and S-nitrosothiol proteins) are significantly lower than in control patients and correlate inversely with pulmonary artery pressures and duration of PAH (Kaneko et al. 1998). eNO levels increase after initiation of intravenous (Ozkan et al. 2001) and inhaled (Forrest et al. 1999) prostacyclin therapy, and also after established treatment with bosentan, a nonselective ET receptor antagonist (Girgis et al. 2005). In interpreting eNO results, the physician must take into account a number of potential confounding factors including age, sex, atopy, infection, and some drugs, such as L-arginine.

## Asymmetric Dimethylarginine

There has recently been increasing interest in asymmetric dimethylarginine (ADMA) as a marker and potential mediator of endothelial dysfunction in pulmonary vascular disease. ADMA is derived from the catabolism of proteins containing methylated arginine residues and acts as a potent competitive inhibitor of NOS. ADMA is excreted by the kidneys or is metabolized by the enzyme dimethylarginine dimethylaminohydrolase, whose activity is inhibited by homocystein. As a biomarker, ADMA has been evaluated in several different classes of PH. In IPAH, plasma levels are significantly higher than in controls and correlate positively with right atrial pressure and negatively with mixed venous oxygen saturation, stroke volume, cardiac index, and survival (Kielstein et al. 2005). In patients with congenital heart disease and PH, plasma ADMA is higher when compared with patients with congenital heart disease but normal pulmonary pressures, and with controls (Gorenflo et al. 2001). In chronic thromboembolic PH, plasma ADMA was significantly elevated in patients compared with controls and correlated with a number of hemodynamic parameters and survival and fell to levels of healthy controls after pulmonary endarterectomy (Skoro-Sajer et al. 2007). By receiver operating curve analysis, ADMA predicted death with a sensitivity of 81.1 percent and a specificity of 79.3 percent at a cut-off value of 0.64 mmol/L.

## cGMP

cGMP is an intracellular second messenger of NO, bradykinin and the natriuretic peptides. It is produced by the activation of the enzyme guanylate cyclase and is an indirect marker of natriuretic peptide or NO production. Urinary and plasma cGMP levels are higher in PH compared with patients and correlate with hemodynamic parameters and, hence, may provide an indicator of disease severity (Bogdan et al. 1998; Ghofrani et al. 2002). A marked increase in cGMP levels was observed in response to NO inhalation, but the magnitude of this response did not discriminate between responders and nonresponders. Hence, cGMP measurement could not be substituted for right heart catheterisation when testing for NO responsiveness. Plasma cGMP levels are reported to decrease after inhalation of iloprost (Wiedemann et al. 2001).

## D-dimer

*In situ* thrombosis is a prominent finding in patients with PAH. D-dimer, a degradation product of crosslinked fibrin, may represent microvascular thrombosis. In patients with IPAH, D-dimer has been shown to be elevated compared with controls (Shitrit et al. 2002a; Shitrit et al. 2002b) and correlates with disease severity, as measured by NYHA class, resting oxygen saturation and pulmonary artery pressure. D-dimer also correlated inversely with survival at one year. However, these results remain to be validated in a larger cohort of patients.

## Serotonin

Serotonin is a pulmonary vasoconstrictor and vascular smooth muscle mitogen. It is released from pulmonary neuroendocrine cells and neuroepithelial bodies distributed throughout the airways. Elevated levels have been demonstrated in patients with IPAH with a concomitant decrease in the serotonin content of platelets, probably due to abnormal handling of serotonin by platelets, with a normal whole blood serotonin level (Herve et al. 1995). In another study, (serotonin levels were elevated in patients with IPAH compared with controls, correlated positively with total pulmonary resistance (Kereveur et al. 2000). However, they were not found to be a predictive marker of PAH severity, and their evolution was independent of clinical and hemodynamic status. Despite a therapeutic benefit, treatment with the potent anti-aggregating agent epoprostenol did not prevent further increases in plasma serotonin.

## Plasma von Willebrand Factor

von Willebrand factor (vWF) is a multimeric glycoprotein that is stored in the Weibel-Palade bodies of vascular endothelium and, to a lesser extent, in granules of platelets. Released after endothelial perturbation, functional vWF multimers bind to exposed collagen in areas of vascular injury and initiate platelet plug formation. The ristocetin cofactor activity (RCA)/vWF ratio reflects the relative platelet-binding activity of vWF. vWF also contributes to thrombogenesis by stabilizing Factor VIII and preserving its function. Elevated plasma vWF and its antigen (vWF: Ag) have been used as markers of endothelial cell injury in a variety of conditions. Levels of vWF: Ag is elevated in PH and baseline vWF: Ag correlates with the risk of death in the subsequent year (Friedman et al. 1997; Lopes and Maeda 1998). In another study, increased vWF levels at baseline and follow-up were associated with worse survival in patients with PAH; however, RCA/vWF ratios were not. vWF was also found to be elevated in severe PAH, and paralleled improvements in hemodynamics on prostacyclin therapy, although this was only in 10 patients (Veyradier et al. 2000).

## Uric Acid

Uric acid is the final oxidation product of purine metabolism. Serum uric acid is a marker of impaired oxidative metabolism of ischemic peripheral tissue. Elevated uric acid levels have long been known to be a poor prognostic sign in acute illness. Several studies have demonstrated that elevated uric acid levels in PH correlate with the severity of disease and mortality and decrease after successful treatment with prostacyclin (Bendayan et al. 2003;

Hoeper et al. 1999; Nagaya et al. 1999). Uric acid levels are dependent on age and sex, and are affected by renal impairment and diuretic therapy; hence, their interpretation may be difficult in some patients.

## BNP/NT-proBNP

Regarding screening, the incidence of PAH in the general population is so low that this obviates the need for a general screening programme. However, there are certain high-risk groups in which the likelihood of PAH is greater and in these individuals there is a need for screening. Current recommendations advise Doppler echocardiography for patients in recognized high-risk groups (McLaughlin et al. 2009). Small studies have evaluated the role of biomarkers in screening programs. NT-proBNP has been evaluated as a marker of early PAH in patients with systemic sclerosis and it has been found that both high levels of NT-pro-BNP and a low DLCO (corrected for alveolar volume) independently predicted the development of PAH in univariate and multivariate analysis (Allanore et al. 2003; Williams et al. 2006). NT-proBNP has also been used to screen for PH in patients with sickle cell disease and pulmonary fibrosis (Leuchte et al. 2004).

In conclusion, several circulating biomarkers convey prognostic information in patients with PAH, but their value in everyday clinical practice is still not established. For a biomarker to become accepted in clinical use, its utility should be consistently demonstrated in a large, prospective study; that is certainly not the case regarding biomarkers in the context of PH. Furthermore, cut-off values suggested by small studies may not apply to the wider PH population, particularly if the multiple confounding factors affecting biomarkers are taken into consideration.

According to current guidelines (Galie et al. 2009), BNP/NT-proBNP plasma levels should be recommended for initial risk stratification and re-evaluated at 3–6 month intervals (the interval should be adjusted to individual patients needs). BNP/NT-proBNP plasma levels may also be considered for monitoring the effects of treatment at 3–4 months after initiation or changes in therapy in view of their prognostic implications, as well as in cases of clinical worsening. Low and stable or decreasing BNP/NT-proBNP may be a useful marker of successful disease control in PAH. Certainly, initial and serial evaluation of patients with PH involves looking at a panel of data derived from clinical evaluation, exercise tests, biochemical markers, and echocardiographic and hemodynamic assessments avoiding relying on a single parameter, so that the clearest possible picture of the patient's status can be obtained.

## Pulmonary Embolism

In diagnostic evaluation of patients with suspected pulmonary embolism (PE), plasma D-dimer, a degradation product of crosslinked fibrin, plays a major role. This blood-screening test relies on the principle that most patients with PE have ongoing endogenous fibrinolysis that is not effective enough to prevent PE but does break down some of the fibrin clots to D-dimers. Hence, elevated plasma levels of D-dimers are sensitive for the presence of PE and the negative predictive value (NPV) is high. On the other hand, although D-dimer is very specific for fibrin, the specificity of fibrin for venous thromboembolism is poor because fibrin is produced in a wide variety of conditions, such as cancer, inflammation, infection, necrosis, dissection of the aorta and postoperatively for at least one week leading to a low positive predictive value (PPV) (Torbicki et al. 2008). Therefore, D-dimer is particularly suited for outpatients with suspected PE but no coexisting acute systemic illness and is less useful for hospitalized patients. The specificity of D-dimer in suspected PE decreases steadily with age and may reach ≤10 percent in patients above 80 years. D-dimer is also more frequently elevated during pregnancy. Therefore, the number of patients with suspected PE in whom D-dimer must be measured to exclude one PE (also referred to as the number needed to test) varies between three in the emergency department and 10 or above in the specific situations listed above. Deciding whether measuring D-dimer is worthwhile in a given situation remains a matter of clinical judgement.

There are a number of available assays with different characteristics (Di Nisio et al. 2007; Stein et al. 2004). The highly sensitive assays include quantitative enzyme-linked immunoabsorbent assay (ELISA) and ELISA-derived assays, which have a sensitivity of >95 percent and a specificity around 40 percent. They can therefore be used to exclude PE in patients with either a low or a moderate probability of PE as estimated implicitly by the clinician or with the use of a clinical prediction rule (Torbicki et al. 2008). In the emergency department, a negative ELISA D-dimer test can exclude PE without further testing in approximately 30 percent of patients. Outcome studies using the Vidas D-dimer assay showed that the 3-month thromboembolic risk in patients was below 1 percent in patients left untreated on the basis of a negative test result. Quantitative latex-derived assays and a whole-blood agglutination assay have lower sensitivity, in the range of 85–90 percent, and are often referred to as moderately sensitive assays (Di Nisio et al. 2007; Stein et al. 2004). The most extensively studied outcome studies to date are the Tinaquant and the SimpliRED assays, which yield a 3-month thromboembolic risk of <1% in patients with a low clinical probability who are left untreated. However, their safety for ruling out PE has not been established in the moderate clinical probability category when using a three-level probability scheme. When using the dichotomous Wells rule, which classifies patients as 'PE unlikely' and 'PE likely', moderately sensitive assays are safe for the exclusion of PE in patients categorized as PE unlikely, i.e., those with a score of ≤4 points.

D-dimer should not be measured in patients with a high clinical probability as a normal result does not safely exclude PE even when using a highly sensitive assay. Such patients should be further evaluated with an imaging technique such as multidetector CT angiography. Similarly, in high-risk PE, as indicated by the presence of shock or hypotension, emergency CT or bedside echocardiography (depending on availability and clinical circumstances) is recommended for diagnostic purposes without measurement of D-dimer.

Besides their usefulness in PE diagnosis, a prognostic role has been suggested by a study reporting that D-dimer levels below 1500 mg/L have a 99 percent NPV in predicting all-cause 3-month mortality (Aujesky et al. 2006).

An additional role of D-dimer in the management of patients with PE is related to the estimation of risk of recurrence of venous thromboembolism. Palareti et al. reported that a negative D-dimer test one month after discontinuation of treatment with a vitamin K antagonist due to a first unprovoked proximal deep-vein thrombosis or pulmonary embolism seems to be a protective factor for venous thromboembolism recurrence (relative risk 0.4) (Palareti et al. 2006).

Laboratory tests are also used for the identification of patients with acquired thrombophilia, such as antiphospholipid syndrome and hyperhomocysteinemia, and hereditary thrombophilia, including factor V Leiden resulting in activated protein C resistance, prothrombin gene mutation 20120 and deficiency of antithrombin III, protein C or protein S. These tests are more useful in patients with a positive family history of venous thromboembolism, recurrent episodes, or absence of specific predisposing factors, such as surgery, trauma, medical illness, estrogen therapy, and pregnancy. Among carriers of molecular thrombophilia, patients with lupus anticoagulant, those with confirmed deficit of protein C or protein S, and patients homozygous for factor V Leiden or homozygous for PTG20210A may be candidates for indefinite anticoagulant treatment after a first unprovoked event of venous thromboembolism. No evidence of a clinical benefit of extended anticoagulant treatment is currently available for heterozygous carriers of factor V Leiden or the prothrombin. The clinician should not rely solely on the results of blood tests to diagnose molecular thrombophilia as these tests may be misleading. Consumption coagulopathy can depress the levels of antithrombin III, protein C or protein S. Both oral contraceptives and pregnancy can depress the levels of protein S. Heparin can decrease the levels of antithrombin III and warfarin causes a mild decrease of protein C or S levels. Therefore, a careful family history remains the most rapid and cost-effective method of identifying patients with hereditary thrombophilia.

Several variables included in routine clinical and laboratory evaluation have prognostic significance in PE. Many of them, such as elevated creatinine levels, are related to the pre-existing condition and the comorbidities of the individual patient rather than to the severity of the index PE episode.

Biochemical markers of right ventricular dysfunction (RVD) and myocardial injury are also useful for the estimation of PE-related risk of early death (in-hospital or 30-day mortality). Risk stratification in PE is crucial for the choice of the optimal diagnostic strategy and initial management. Immediate bedside clinical assessment for the presence or absence of clinical markers, i.e., hypotension or shock, identifies high-risk patients, who require thromblolysis or embolectomy. Non-high-risk PE can be further stratified into intermediate- and low-risk according to the presence of markers of RVD or myocardial injury. The former include, besides imaging and hemodynamic data, the elevation of BNP and pro-BNP, while myocardial injury is detected with the elevation of troponin I or T.

BNP is released due to increased myocardial stretch associated with RVD. There is growing evidence that in acute PE, levels of BNP or N-terminal proBNP (NT-proBNP) reflect the severity of RVD and hemodynamic compromise (Kruger et al. 2003; Pruszczyk et al. 2003b). Recent reports suggest that BNP or NT-proBNP as markers of RVD provide prognostic information additional to that derived from echocardiography (Kostrubiec et al. 2005; Kucher et al. 2003b; Pieralli et al. 2006).

Kucher et al., who measured BNP concentration in 73 PE patients, showed that most patients who died or required therapy intensification during hospitalization had elevated serum BNP (Kucher et al. 2003b). However, the cutoff value to diagnose 95 percent patients with mild disease course was not < 100 pg/mL (which is required to exclude left ventricular failure) but was < 50 pg/mL. In studies by Pieralli et al., a BNP level of < 85 pg/mL was shown to exclude echocardiographical findings of RVO in 61 normotensive patients with first APE incident, with a high degree of accuracy (Pieralli et al. 2006). However, all patients with BNP over 527 pg/mL revealed RVO features in echocardiography; all acute complications and deaths occurred in this patient group. Kucher et al., who analyzed pro-BNP in PE patients, demonstrated that low levels, i.e., < 500 ng/mL, are prognostic of an uncomplicated in-hospital course (Kucher et al. 2003a).

Troponin increase in PE is mild, may occur from 6–12 hours after the acute embolic incident and is transient, usually lasting no longer than 2–3 days. Elevated plasma troponin levels have been repeatedly reported as associated with worse prognosis in patients with PE (Becattini et al. 2007). In an early study, the prevalence of a positive troponin T test, defined as

>0.1 ng/mL, was reported in 0–35 percent and 50 percent of patients with non-massive, submassive, and clinically massive PE, respectively (Giannitsis et al. 2000). Positive troponin T was related to an in-hospital mortality of 44 percent, compared with 3 percent for negative troponin T (odds ratio) (OR, 15, 2; 95% CI, 1,2–190, 4). In another study, levels of troponins I and T correlated both with in-hospital mortality and a complicated clinical course (Konstantinides et al. 2002). Increased in-hospital mortality has also been reported in normotensive patients with PE with the use of cutoff values for troponin T as low as 0.01 ng/mL (OR, 21,0; 95% CI, 1,2–389,0) (Pruszczyk et al. 2003a). Most trials reported PPV of elevated troponin for PE-related early mortality in the range of 12 percent–44 percent, with very high NPV (99 percent–100 percent), irrespective of various methods and cutoff values applied. A recent meta-analysis confirmed that elevated troponin levels were associated with increased mortality in the subgroup of hemodynamically stable patients (OR, 5.9; 95% CI, 2.7–12.9) (Becattini et al. 2007).

Recently, heart type fatty acid binding protein (H-FABP), an early marker of myocardial injury, was reported to be superior to troponin for risk stratification of PE on admission. Fatty acid binding proteins—small proteins located in the cytoplasm—facilitate transport of fatty acids and other lipids within the cell. H-FABP is a sensitive marker of myocyte damage and, unlike troponin, is released by both ischemia and necrosis. It is also released more rapidly than troponin, although the advent of hs-cTn will overcome the advantage of H-FABP in this regard. H-FABP is released from the damaged cell within 1–3 hours, returning to normal by 12–24 hours. It is now available as a point of care test for the diagnosis of AMI in many countries and also has good data to support its use in prognostication from MI. H-FABP >6 ng/mL had a PPV and NPV for early PE-related mortality of 23 percent–37 percent and 96 percent–100 percent, respectively (Kaczynska et al. 2006; Puls et al. 2007).

A combination of troponin and NT-proBNP was found to stratify normotensive patients with PE more accurately. PE-related 40-day mortality in the group with high levels of both cardiac troponin T and NT-proBNP exceeded 30 percent. Patients with an isolated elevation of NT-proBNP had an intermediate mortality rate (3.7 percent), while low levels of both biomarkers indicated a good short-term prognosis (Kostrubiec et al. 2005).

The currently available data do not allow the proposal of specific cutoff levels of markers that could be used for therapeutic decision-making in patients with non-high-risk PE.

In conclusion, current guidelines state that initial risk stratification of suspected or confirmed PE should be based on the presence of shock and hypotension to distinguish between patients with high and non-high-risk (class of recommendation I). In non-high-risk PE patients, further stratification to an intermediate- or low-risk PE subgroup should be based on the presence of imaging or biochemical markers of RVD and myocardial injury (class of recommendation IIa).

Regarding treatment strategy, routine use of thrombolysis in non-high-risk patients is not recommended, but may be considered in selected patients with intermediate-risk PE and after thorough consideration of conditions increasing the risk of bleeding. Thrombolytic therapy should be not used in patients with low-risk PE. Therefore, the determination of biomarkers may contribute to the selection of hemodynamically stable patients for fibrinolysis.

# References

Adnot, S., P.E. Chabrier, P. Andrivet, I. Viossat, J. Piquet, C. Brun-Buisson, Y. Gutkowska, and P. Braquet. 1987. Atrial natriuretic peptide concentrations and pulmonary hemodynamics in patients with pulmonary artery hypertension. Am. Rev. Respir. Dis. 136: 951–956.

Allanore, Y., D. Borderie, C. Meune, L. Cabanes, S. Weber, O.G. Ekindjian, and A. Kahan. 2003. N-terminal pro-brain natriuretic peptide as a diagnostic marker of early pulmonary artery hypertension in patients with systemic sclerosis and effects of calcium-channel blockers. Arthritis Rheum. 48: 3503–3508.

Andreassen, A.K., R. Wergeland, S. Simonsen, O. Geiran, C. Guevara, and T. Ueland. 2006. N-terminal pro-B-type natriuretic peptide as an indicator of disease severity in a heterogeneous group of patients with chronic precapillary pulmonary hypertension. Am. J. Cardiol. 98: 525–529.

Archer, S.L., K. Djaballah, M. Humbert, K.E. Weir, M. Fartoukh, J. Dall'ava-Santucci, J.C. Mercier, G. Simonneau, and A.T. Dinh-Xuan. 1998. Nitric oxide deficiency in fenfluramine- and dexfenfluramine-induced pulmonary hypertension. Am. J. Respir. Crit. Care Med. 158: 1061–1067.

Aujesky, D., P.M. Roy, M. Guy, J. Cornuz, O. Sanchez, and A. Perrier. 2006. Prognostic value of D-dimer in patients with pulmonary embolism. Thromb. Haemost. 96: 478–482.

Becattini, C., M.C. Vedovati, and G. Agnelli. 2007. Prognostic value of troponins in acute pulmonary embolism: a meta-analysis. Circulation 116: 427–433.

Bendayan, D., D. Shitrit, M. Ygla, M. Huerta, G. Fink, and M.R. Kramer. 2003. Hyperuricemia as a prognostic factor in pulmonary arterial hypertension. Respir. Med. 97: 130–133.

Bogdan, M., M. Humbert, J. Francoual, C. Claise, P. Duroux, G. Simonneau, and A. Lindenbaum. 1998. Urinary cGMP concentrations in severe primary pulmonary hypertension. Thorax 53: 1059–1062.

Bozkanat, E., E. Tozkoparan, O. Baysan, O. Deniz, F. Ciftci, and M. Yokusoglu. 2005. The significance of elevated brain natriuretic peptide levels in chronic obstructive pulmonary disease. J. Int. Med. Res. 33: 537–544.

Cacoub, P., R. Dorent, P. Nataf, A. Carayon, M. Riquet, E. Noe, J.C. Piette, P. Godeau, and I. Gandjbakhch. 1997. Endothelin-1 in the lungs of patients with pulmonary hypertension. Cardiovasc. Res. 33: 196–200.

D'Alonzo, G.E., R.J. Barst, S.M. Ayres, E.H. Bergofsky, B.H. Brundage, K.M. Detre, A.P. Fishman, R.M. Goldring, B.M. Groves, J.T. Kernis et al. 1991. Survival in patients with primary pulmonary hypertension. Results from a national prospective registry. Ann. Intern. Med. 115: 343–349.

Di Nisio, M., A. Squizzato, A.W. Rutjes, H.R. Buller, A.H. Zwinderman, and P.M. Bossuyt. 2007. Diagnostic accuracy of D-dimer test for exclusion of venous thromboembolism: a systematic review. J. Thromb. Haemost. 5: 296–304.

Fijalkowska, A., M. Kurzyna, A. Torbicki, G. Szewczyk, M. Florczyk, P. Pruszczyk, and M. Szturmowicz. 2006. Serum N-terminal brain natriuretic peptide as a prognostic parameter in patients with pulmonary hypertension. Chest 129: 1313–1321.

Forrest, I.A., T. Small, and P.A. Corris. 1999. Effect of nebulized epoprostenol (prostacyclin) on exhaled nitric oxide in patients with pulmonary hypertension due to congenital heart disease and in normal controls. Clin. Sci. (Lond.) 97: 99–102.

Friedman, R., J.G. Mears, and R.J. Barst. 1997. Continuous infusion of prostacyclin normalizes plasma markers of endothelial cell injury and platelet aggregation in primary pulmonary hypertension. Circulation 96: 2782–2784.

Galie, N., M.M. Hoeper, M. Humbert, A. Torbicki, J.L. Vachiery, J.A. Barbera, M. Beghetti, P. Corris, S. Gaine, J.S. Gibbs, M.A. Gomez-Sanchez, G. Jondeau, W. Klepetko, C. Opitz, A. Peacock, L. Rubin, M. Zellweger, and G. Simonneau. 2009. Guidelines for the diagnosis and treatment of pulmonary hypertension: the Task Force for the Diagnosis and Treatment of Pulmonary Hypertension of the European Society of Cardiology (ESC) and the European Respiratory Society (ERS), endorsed by the International Society of Heart and Lung Transplantation (ISHLT). Eur. Heart J. 30: 2493–2537.

Ghofrani, H.A., R. Wiedemann, F. Rose, N. Weissmann, R.T. Schermuly, K. Quanz, F. Grimminger, W. Seeger, and H. Olschewski. 2002. Lung cGMP release subsequent to NO inhalation in pulmonary hypertension: responders versus nonresponders. Eur. Respir. J. 19: 664–671.

Giaid, A., M. Yanagisawa, D. Langleben, R.P. Michel, R. Levy, H. Shennib, S. Kimura, T. Masaki, W.P. Duguid, and D.J. Stewart. 1993. Expression of endothelin-1 in the lungs of patients with pulmonary hypertension. N. Engl. J. Med. 328: 1732–1739.

Giannitsis, E., M. Muller-Bardorff, V. Kurowski, B. Weidtmann, U. Wiegand, M. Kampmann, and H.A. Katus. 2000. Independent prognostic value of cardiac troponin T in patients with confirmed pulmonary embolism. Circulation 102: 211–217.

Girgis, R.E., H.C. Champion, G.B. Diette, R.A. Johns, S. Permutt, and J.T. Sylvester. 2005. Decreased exhaled nitric oxide in pulmonary arterial hypertension: response to bosentan therapy. Am. J. Respir. Crit. Care Med. 172: 352–357.

Gorenflo, M., C. Zheng, E. Werle, W. Fiehn, and H.E. Ulmer. 2001. Plasma levels of asymmetrical dimethyl-L-arginine in patients with congenital heart disease and pulmonary hypertension. J Cardiovasc. Pharmacol. 37: 489–492.

Herve, P., J.M. Launay, M.L. Scrobohaci, F. Brenot, G. Simonneau, P. Petitpretz, P. Poubeau, J. Cerrina, P. Duroux, and L. Drouet. 1995. Increased plasma serotonin in primary pulmonary hypertension. Am. J. Med. 99: 249–254.

Hoeper, M.M., J.M. Hohlfeld, and H. Fabel. 1999. Hyperuricaemia in patients with right or left heart failure. Eur. Respir. J. 13: 682–685.

Kaczynska, A., M.M. Pelsers, A. Bochowicz, M. Kostrubiec, J.F. Glatz, and P. Pruszczyk. 2006. Plasma heart-type fatty acid binding protein is superior to troponin and myoglobin for rapid risk stratification in acute pulmonary embolism. Clin. Chim. Acta. 371: 117–123.

Kaneko, F.T., A.C. Arroliga, R.A. Dweik, S.A. Comhair, D. Laskowski, R. Oppedisano, M.J. Thomassen, and S.C. Erzurum. 1998. Biochemical reaction products of nitric oxide as quantitative markers of primary pulmonary hypertension. Am. J. Respir. Crit. Care Med. 158: 917–923.

Kereveur, A., J. Callebert, M. Humbert, P. Herve, G. Simonneau, J.M. Launay, and L. Drouet. 2000. High plasma serotonin levels in primary pulmonary hypertension. Effect of long-term epoprostenol (prostacyclin) therapy. Arterioscler. Thromb. Vasc. Biol. 20: 2233–2239.

Kielstein, J.T., S.M. Bode-Boger, G. Hesse, J. Martens-Lobenhoffer, A. Takacs, D. Fliser, and M.M. Hoeper. 2005. Asymmetrical dimethylarginine in idiopathic pulmonary arterial hypertension. Arterioscler. Thromb. Vasc. Biol. 25: 1414–1418.

Konstantinides, S., A. Geibel, M. Olschewski, W. Kasper, N. Hruska, S. Jackle, and L. Binder. 2002. Importance of cardiac troponins I and T in risk stratification of patients with acute pulmonary embolism. Circulation 106: 1263–1268.

Kostrubiec, M., P. Pruszczyk, A. Bochowicz, R. Pacho, M. Szulc, A. Kaczynska, G. Styczynski, A. Kuch-Wocial, P. Abramczyk, Z. Bartoszewicz, H. Berent, and K. Kuczynska. 2005. Biomarker-based risk assessment model in acute pulmonary embolism. Eur. Heart J. 26: 2166–2172.

Kruger, S., M.W. Merx, and J. Graf. 2003. Utility of brain natriuretic peptide to predict right ventricular dysfunction and clinical outcome in patients with acute pulmonary embolism. Circulation 108: e94; author reply e94–95.

Kucher, N., G. Printzen, T. Doernhoefer, S. Windecker, B. Meier, and O.M. Hess. 2003a. Low pro-brain natriuretic peptide levels predict benign clinical outcome in acute pulmonary embolism. Circulation 107: 1576–1578.

Kucher, N., G. Printzen, and S.Z. Goldhaber. 2003b. Prognostic role of brain natriuretic peptide in acute pulmonary embolism. Circulation 107: 2545–2547.

Leuchte, H.H., C. Neurohr, R. Baumgartner, M. Holzapfel, W. Giehrl, M. Vogeser, and J. Behr. 2004. Brain natriuretic peptide and exercise capacity in lung fibrosis and pulmonary hypertension. Am. J. Respir. Crit. Care Med. 170: 360–365.

Leuchte, H.H., M. Holzapfel, R.A. Baumgartner, C. Neurohr, M. Vogeser, and J. Behr. 2005. Characterization of brain natriuretic peptide in long-term follow-up of pulmonary arterial hypertension. Chest 128: 2368–2374.

Leuchte, H.H., R.A. Baumgartner, M.E. Nounou, M. Vogeser, C. Neurohr, M. Trautnitz, and J. Behr. 2006. Brain natriuretic peptide is a prognostic parameter in chronic lung disease. Am. J. Respir. Crit. Care Med. 173: 744–750.

Leuchte, H.H., M. El Nounou, J.C. Tuerpe, B. Hartmann, R.A. Baumgartner, M. Vogeser, O. Muehling, and J. Behr. 2007. N-terminal pro-brain natriuretic peptide and renal insufficiency as predictors of mortality in pulmonary hypertension. Chest 131: 402–409.

Lopes, A.A. and N.Y. Maeda. 1998. Circulating von Willebrand factor antigen as a predictor of short-term prognosis in pulmonary hypertension. Chest 114: 1276–1282.

McLaughlin, V.V., S.L. Archer, D.B. Badesch, R.J. Barst, H.W. Farber, J.R. Lindner, M.A. Mathier, M.D. McGoon, M.H. Park, R.S. Rosenson, L.J. Rubin, V.F. Tapson, J. Varga, R.A. Harrington, J.L. Anderson, E.R. Bates, C.R. Bridges, M.J. Eisenberg, V.A. Ferrari, C.L. Grines, M.A. Hlatky, A.K. Jacobs, S. Kaul, R.C. Lichtenberg, D.J. Moliterno, D. Mukherjee, G.M. Pohost, R.S. Schofield, S.J. Shubrooks, J.H. Stein, C.M. Tracy, H.H. Weitz, and D.J. Wesley. 2009. ACCF/AHA 2009 expert consensus document on pulmonary hypertension: a report of the American College of Cardiology Foundation Task Force on Expert Consensus Documents and the American Heart Association: developed in collaboration with the American College of Chest Physicians, American Thoracic Society, Inc., and the Pulmonary Hypertension Association. Circulation 119: 2250–2294.

Nagaya, N., T. Nishikimi, Y. Okano, M. Uematsu, T. Satoh, S. Kyotani, S. Kuribayashi, S. Hamada, M. Kakishita, N. Nakanishi, M. Takamiya, T. Kunieda, H. Matsuo, and K. Kangawa. 1998a. Plasma brain natriuretic peptide levels increase in proportion to the extent of right ventricular dysfunction in pulmonary hypertension. J. Am. Coll. Cardiol. 31: 202–208.

Nagaya, N., T. Nishikimi, M. Uematsu, S. Kyotani, T. Satoh, N. Nakanishi, H. Matsuo, and K. Kangawa. 1998b. Secretion patterns of brain natriuretic peptide and atrial natriuretic peptide in patients with or without pulmonary hypertension complicating atrial septal defect. Am. Heart J. 136: 297–301.

Nagaya, N., M. Uematsu, T. Satoh, S. Kyotani, F. Sakamaki, N. Nakanishi, M. Yamagishi, T. Kunieda, and K. Miyatake. 1999. Serum uric acid levels correlate with the severity and the mortality of primary pulmonary hypertension. Am. J. Respir. Crit. Care Med. 160: 487–492.

Nagaya, N., T. Nishikimi, M. Uematsu, T. Satoh, S. Kyotani, F. Sakamaki, M. Kakishita, K. Fukushima, Y. Okano, N. Nakanishi, K. Miyatake, and K. Kangawa. 2000. Plasma brain natriuretic peptide as a prognostic indicator in patients with primary pulmonary hypertension. Circulation 102: 865–870.

Nagaya, N., M. Ando, H. Oya, Y. Ohkita, S. Kyotani, F. Sakamaki, and N. Nakanishi. 2002. Plasma brain natriuretic peptide as a noninvasive marker for efficacy of pulmonary thromboendarterectomy. Ann. Thorac. Surg. 74: 180–184; discussion 184.

Nagaya, N., N. Sasaki, M. Ando, H. Ogino, F. Sakamaki, S. Kyotani, and N. Nakanishi. 2003. Prostacyclin therapy before pulmonary thromboendarterectomy in patients with chronic thromboembolic pulmonary hypertension. Chest 123: 338–343.

Ozkan, M., R.A. Dweik, D. Laskowski, A.C. Arroliga, and S.C. Erzurum. 2001. High levels of nitric oxide in individuals with pulmonary hypertension receiving epoprostenol therapy. Lung 179: 233–243.

Palareti, G., B. Cosmi, C. Legnani, A. Tosetto, C. Brusi, A. Iorio, V. Pengo, A. Ghirarduzzi, C. Pattacini, S. Testa, A.W. Lensing, and A. Tripodi. 2006. D-dimer testing to determine the duration of anticoagulation therapy. N. Engl. J. Med. 355: 1780–1789.

Park, M.H., R.L. Scott, P.A. Uber, H.O. Ventura, and M.R. Mehra. 2004. Usefulness of B-type natriuretic peptide as a predictor of treatment outcome in pulmonary arterial hypertension. Congest. Heart Fail. 10: 221–225.

Peacock, A.J. 2003. Treatment of pulmonary hypertension. BMJ 326: 835–836.

Pieralli, F., I. Olivotto, S. Vanni, A. Conti, A. Camaiti, G. Targioni, S. Grifoni, and G. Berni. 2006. Usefulness of bedside testing for brain natriuretic peptide to identify right ventricular dysfunction and outcome in normotensive patients with acute pulmonary embolism. Am. J. Cardiol. 97: 1386–1390.

Pruszczyk, P., A. Bochowicz, A. Torbicki, M. Szulc, M. Kurzyna, A. Fijalkowska, and A. Kuch-Wocial. 2003a. Cardiac troponin T monitoring identifies high-risk group of normotensive patients with acute pulmonary embolism. Chest 123: 1947–1952.

Pruszczyk, P., M. Kostrubiec, A. Bochowicz, G. Styczynski, M. Szulc, M. Kurzyna, A. Fijalkowska, A. Kuch-Wocial, I. Chlewicka, and A. Torbicki. 2003b. N-terminal pro-brain natriuretic peptide in patients with acute pulmonary embolism. Eur. Respir. J. 22: 649–653.

Puls, M., C. Dellas, M. Lankeit, M. Olschewski, L. Binder, A. Geibel, C. Reiner, K. Schafer, G. Hasenfuss, and S. Konstantinides. 2007. Heart-type fatty acid-binding protein permits early risk stratification of pulmonary embolism. Eur. Heart J. 28: 224–229.

Rich, S., K. Kieras, K. Hart, B.M. Groves, J.D. Stobo, and B.H. Brundage. 1986. Antinuclear antibodies in primary pulmonary hypertension. J Am. Coll. Cardiol. 8: 1307–1311.

Riley, M.S., J. Porszasz, J. Miranda, M.P. Engelen, B. Brundage, and K. Wasserman. 1997. Exhaled nitric oxide during exercise in primary pulmonary hypertension and pulmonary fibrosis. Chest 111: 44–50.

Rubens, C., R. Ewert, M. Halank, R. Wensel, H.D. Orzechowski, H.P. Schultheiss, and G. Hoeffken. 2001. Big endothelin-1 and endothelin-1 plasma levels are correlated with the severity of primary pulmonary hypertension. Chest 120: 1562–1569.

Shitrit, D., D. Bendayan, A. Bar-Gil-Shitrit, M. Huerta, B. Rudensky, G. Fink, and M.R. Kramer. 2002a. Significance of a plasma D-dimer test in patients with primary pulmonary hypertension. Chest 122: 1674–1678.

Shitrit, D., D. Bendayan, B. Rudensky, G. Izbicki, M. Huerta, G. Fink, and M.R. Kramer. 2002b. Elevation of ELISA D-dimer levels in patients with primary pulmonary hypertension. Respiration 69: 327–329.

Skoro-Sajer, N., F. Mittermayer, A. Panzenboeck, D. Bonderman, R. Sadushi, R. Hitsch, J. Jakowitsch, W. Klepetko, M.P. Kneussl, M. Wolzt, and I.M. Lang. 2007. Asymmetric dimethylarginine is increased in chronic thromboembolic pulmonary hypertension. Am. J. Respir. Crit. Care Med. 176: 1154–1160.

Souza, R., H.B. Bogossian, M. Humbert, C. Jardim, R. Rabelo, M.B. Amato, and C.R. Carvalho. 2005. N-terminal-pro-brain natriuretic peptide as a hemodynamic marker in idiopathic pulmonary arterial hypertension. Eur. Respir. J. 25: 509–513.

Stein, P.D., R.D. Hull, K.C. Patel, R.E. Olson, W.A. Ghali, R. Brant, R.K. Biel, V. Bharadia, and N.K. Kalra. 2004. D-dimer for the exclusion of acute venous thrombosis and pulmonary embolism: a systematic review. Ann. Intern. Med. 140: 589–602.

ten Wolde, M., Tulevski, II, J.W. Mulder, M. Sohne, F. Boomsma, B.J. Mulder, and H.R. Buller. 2003. Brain natriuretic peptide as a predictor of adverse outcome in patients with pulmonary embolism. Circulation 107: 2082–2084.

Torbicki, A., M. Kurzyna, P. Kuca, A. Fijalkowska, J. Sikora, M. Florczyk, P. Pruszczyk, J. Burakowski, and L. Wawrzynska. 2003. Detectable serum cardiac troponin T as a marker of poor prognosis among patients with chronic precapillary pulmonary hypertension. Circulation 108: 844–848.

Torbicki, A., A. Perrier, S. Konstantinides, G. Agnelli, N. Galie, P. Pruszczyk, F. Bengel, A.J. Brady, D. Ferreira, U. Janssens, W. Klepetko, E. Mayer, M. Remy-Jardin, and J.P. Bassand. 2008. Guidelines on the diagnosis and management of acute pulmonary embolism: the Task Force for the Diagnosis and Management of Acute Pulmonary Embolism of the European Society of Cardiology (ESC). Eur. Heart J. 29: 2276–2315.

Veyradier, A., T. Nishikubo, M. Humbert, M. Wolf, O. Sitbon, G. Simonneau, J.P. Girma, and D. Meyer. 2000. Improvement of von Willebrand factor proteolysis after prostacyclin infusion in severe pulmonary arterial hypertension. Circulation 102: 2460–2462.

Voelkel, N.F., R.A. Quaife, L.A. Leinwand, R.J. Barst, M.D. McGoon, D.R. Meldrum, J. Dupuis, C.S. Long, L.J. Rubin, F.W. Smart, Y.J. Suzuki, M. Gladwin, E.M. Denholm, and D.B. Gail. 2006. Right ventricular function and failure: report of a National Heart, Lung, and Blood Institute working group on cellular and molecular mechanisms of right heart failure. Circulation 114: 1883–1891.

Warwick, G., P.S. Thomas, and D.H. Yates. 2008. Biomarkers in pulmonary hypertension. Eur. Respir. J. 32: 503–512.

Wiedemann, R., H.A. Ghofrani, N. Weissmann, R. Schermuly, K. Quanz, F. Grimminger, W. Seeger, and H. Olschewski. 2001. Atrial natriuretic peptide in severe primary and nonprimary pulmonary hypertension: response to iloprost inhalation. J. Am. Coll. Cardiol. 38: 1130–1136.

Wilkins, M.R., G.A. Paul, J.W. Strange, N. Tunariu, W. Gin-Sing, W.A. Banya, M.A. Westwood, A. Stefanidis, L.L. Ng, D.J. Pennell, R.H. Mohiaddin, P. Nihoyannopoulos, and J.S. Gibbs. 2005. Sildenafil versus Endothelin Receptor Antagonist for Pulmonary Hypertension (SERAPH) study. Am. J. Respir. Crit. Care Med. 171: 1292–1297.

Williams, M.H., C.E. Handler, R. Akram, C.J. Smith, C. Das, J. Smee, D. Nair, C.P. Denton, C.M. Black, and J.G. Coghlan. 2006. Role of N-terminal brain natriuretic peptide (N-TproBNP) in scleroderma-associated pulmonary arterial hypertension. Eur. Heart J. 27: 1485–1494.

Yap, L.B., D. Mukerjee, P.M. Timms, H. Ashrafian, and J.G. Coghlan. 2004. Natriuretic peptides, respiratory disease, and the right heart. Chest 126: 1330–1336.

# CHAPTER 8

# Coronary Artery Bypass Grafting and Biomarkers

Constantina Aggeli,* Ioannis Felekos, Emanuel Poulidakis, Iosif Koutagiar, Erifili Venieri and Christodoulos Stefanadis

## Introduction

Coronary artery bypass grafting (CABG) is regarded as the mainstay of treatment in selected patients with coronary artery disease including those with left-main, three-vessel disease, or diabetics. Most of the retrospective studies have proven that revascularization with CABG not only ameliorates symptoms, but also contributes to the improvement of survival.

As a major intervention, there are adverse events in the post-operative period that surgeons and cardiologists are often confronted with. Myocardial injury, acute kidney injury, and neurological deficiencies are of major concern, since they affect mortality and morbidity rates. At the cellular level, neuro-hormonal, metabolic, and inflammatory events affect patient outcomes in a complex systemic fashion.

In this chapter, we aim to describe the interrelation between biomarkers and the postoperative events.

## Screening for Myocardial Injury

One of the major complications of this therapeutic approach is the occurrence of post-surgery myocardial infarction. According to the Coronary Artery Trial its incidence varies between 2 percent and 8 percent (Birdi et

1st Department of Cardiology, 'Hippokration' Hospital, University of Athens Medical School, Athens, Greece.
*Corresponding author: caggeli@otenet.gr

al. 1997). Other studies, whose values are based on ECG assessment and CK-MB measurements, report an incident rate of 2 percent–26 percent. Perioperative myocardial infarction is now recognized as a common cause of postoperative morbidity and mortality. Therefore the preservation of myocardial function has been central to the practice of cardiac surgery.

The improvement of cardio-protective surgery is of paramount importance and can only be demonstrated by the accurate documentation of myocardial injury. This could be achieved by means of commonly used tools such as echocardiography, ECG or even more invasive methods such as ventriculography. However, there are drawbacks that limit their use for this purpose. Although echocardiography is a widely-available, bedside, cost-effective technique for the quantification of global and regional abnormalities of myocardial function it lacks the ability to detect subtle degrees of injury. Ventriculography on the other hand is an invasive method, while ECG detects ischemia at a more advanced stage (Muehlschlegel et al. 2009).

Therefore, biomarkers could represent a simple, easily accessible, non-invasive, clinical, and research tool. Biomarker evaluation is based on postoperative release from tissue other than the heart, or due to postoperative infarct. Several theories have been postulated with regards to the mechanisms of this complication. During CABG, numerous additional factors can lead to peri-procedural necrosis. These include direct myocardial trauma from sewing needles or manipulation of the heart, coronary dissection, global or regional ischemia related to inadequate cardiac protection, microvascular events related to reperfusion, myocardial damage induced by oxygen free radical generation, or failure to reperfuse areas of the myocardium that are not subtended by graftable vessels (Lim et al. 2011b).

Two of the most extensively studied biomarkers in the context of perioperative myocardial infarction are CK-MB and cardiac troponins. Identification of troponin in the circulation may reflect release from cytoplasmic stores or damage of the myocardial contractile apparatus and myofibrillar destruction. Although these two phenomena may occur concurrently, cytosolic leakage may be a relatively "benign" early phenomenon and may not be associated with adverse outcomes. In contrast myocyte necrosis (either early or late) may immediately worsen cardiac function or herald adverse ventricular remodeling and ultimately heart failure (Peivandi et al. 2004).

Cardiac biomarkers have been used as surrogates of adverse events even when measured preoperatively. In a small-sample research consisting of only 448 patients, Carrier et al. studied the pre-operative clinical value of troponin T (Carrier et al. 2000). The authors concluded that the specific biomarker could identify a subgroup of patients with increased risk of postoperative complications including myocardial infarction, heart failure and prolonged hospitalizations. It was implied that the study population suffered from unrecognized myocardial injury prior to surgery. In addition, Paparella et al. found that preoperative levels of troponin I in patients who had suffered from recent acute myocardial infarction were highly associated with adverse events during a 6-month follow-up (Paparella et al. 2010). Perioperative myocardial damage was more pronounced in patients with cTnI exceeding 0.15 ng/ml. All of these results seem to be in accordance with the findings of Amin et al., who postulated the theory that early cTnI measurement after admission can predict adverse outcomes after CABG (Amin et al. 2009). This association extends to long-term adverse events after CABG.

Recently, Lim et al. conducted a research combining troponin I assessment and CMR with gadolinium enhancement (Lim et al. 2011a). The study protocol involved the measurement of cardiac troponin I and CK-MB one hour after the surgical procedure (early measurement as opposed to standard clinical practice guidelines). According to their results, troponin I exhibited excellent correlation with CMR for the identification of myocardial injury. This finding could have significant clinical implication for the early diagnosis of postoperative infarction and consequently for prompt intervention for myocardial salvage. Additionally, troponin I illustrated superiority to CK-MB for the early detection of cardiac injury. Geene et al. (van Geene et al. 2010) verified those findings and suggested that postoperative cTnI level, measured within the first hour after cardiac surgery, can identify a subgroup of patients with increased risk for hospital mortality.

Those findings were also verified by Selvanayagam et al. in a similar protocol where CMR and biomarkers were utilized (Selvanayagam et al. 2005). Moreover, the investigators found that there was no significant difference in the amount of injury whether on-pump or off-pump procedure was undertaken. However, the latter statement is contradicted by Kathresan et al. who illustrated that the off-pump technique results in smaller release of troponin and thus offers better cardio-protection (Kathiresan et al. 2004). Again troponin was a better predictor of adverse events as compared to CK-MB. Troponins T and I were further compared in another study (Carrier et al. 2000). It was shown that troponin I peaks earlier and declines earlier than troponin T.

Another important issue that must be addressed is which cTnI troponometric is best able to determine midterm survival after CABG. This was recently looked into by Ranasinghe et al. who found that serial troponin I data should be collected until 72 hours postoperatively to calculate cumulative area under the curve at 72 hours (CAUC72), as this troponometric best predicts midterm mortality (Ranasinghe et al. 2011).

Furthermore, in the presence of other known predictors of postoperative survival (CrCl and logistic EuroSCOREs, which contain important comorbidities including age, gender, peripheral vascular disease, left ventricular function, recent myocardial infarction, and urgency of surgery), measurement of cTnI up to 72 hours provides the best independent predictor of midterm survival after CABG.

A meta-analysis performed by Petaja et al. provided useful insights into the clinical implementation of troponins and CK-MB (Petaja et al. 2009). The authors concluded that CTNI is a better predictor of cardiovascular events due to better specificity for myocardial injury, than CK-MB. However, they highlighted the fact that there are diversities in the various research protocols since they consist of different populations; also, there is a lack of assay standardization and biomarkers exhibit different kinetics. Furthermore, off-pump surgery may require different cut-off values than on-pump operation for the more accurate assessment of postsurgical injury.

In 2007 the ESC issued the universal definition of myocardial infarction (Thygesen et al. 2007). According to the consensus statement, myocardial infarction associated with CABG is defined as type V in the clinical classification of different types of myocardial infarction. The Task Force suggests, by arbitrary convention, that biomarker values more than five times the 99th percentile of the normal reference range during the first 72 h following CABG, when associated with the appearance of new pathological Q-waves or new LBBB, or angiographically documented new graft or native coronary artery occlusion, or imaging evidence of new loss of viable myocardium, should be considered as diagnostic of a CABG-related myocardial infarction.

## Novel biomarkers of myocardial injury

BNP secretion after cardiac surgery is associated with acute myocardial infarction even in the absence of left ventricular dysfunction. However, postoperative NT-proBNP levels may increase significantly also after major noncardiac surgery, whereas cardiac troponin-T concentrations are within normal limits. Although the predominant stimulus controlling the synthesis and release of BNP from cardiac myocytes is wall stretch, a variety of endocrine factors, such as norepinephrine and angiotensin II have been shown to affect BNP release. Like congestive heart failure, the surgical trauma results in elevated circulating levels of norepinephrine and angiotensin II, which cannot only increase the hemodynamic stress on the ventricle but also directly stimulate BNP release from cardiac myocytes. This point can be supported by the study of Velissaris et al. which showed that, despite the avoidance of Cardio Pulmonary Bypass (CPB), Off-Pump Coronary Artery Bypass (OPCAB) surgery triggers a systematic stress hormone response that is comparable to conventional surgical revascularization (Velissaris et al. 2004). Additionally, a study by Crescenzi et al. showed that no significant differences were found in the postoperative levels of NT-proBNP between on-pump CABG and OPCAB surgery (Crescenzi et al. 2009).

Recent studies have found that pre- and postoperative BNP concentrations have interesting prognostic value. The high preoperative values are associated with the need for postoperative inotropic support, a more prolonged intensive care unit (ICU) stay and a higher hospital mortality after coronary artery bypass graft (CABG) surgery, whereas postoperatively elevated BNP levels were associated with prolonged hospital stay, increased 1-year mortality (Fox et al. 2011; Hutfless et al. 2004; Provenchere et al. 2006) and worse longer-term physical function (Nozohoor et al. 2011). Hutfless et al. demonstrated that preoperative BNP levels >385 pg/ml predict the postoperative complications and one-year mortality after heart surgery and that elevated BNP levels at day 1 were associated with prolonged hospital stay and increased long-term mortality within one year (Hutfless et al. 2004). A limitation of this study was the relatively small, all-male population sample. Moreover, Provenchère et al. (Provenchere et al. 2006) observed that elevated BNP concentration at day 1 was a strong predictor of postoperative cardiac dysfunction and that BNP at day 1 was also significantly associated with 1-year mortality. However, in their study, no discrimination was made between the types of surgery and no multivariate analysis was performed. A larger study was conducted by Nozohoor et al. who showed that an increase in BNP level during the first day after cardiac surgery is an independent predictor of late mortality and postoperative complications (Nozohoor et al. 2011). For patients undergoing isolated CABG surgery, a BNP level of 172 pg/mL on the first postoperative day had a sensitivity of 80 percent and a specificity of 79 percent in predicting 1-year mortality (95 percent CI, 0.75–0.93). Moreover postoperative complications, especially prolonged requirement for inotropic support, were significantly more frequent in patients presenting with a high postoperative BNP on admittance to the ICU (Nozohoor et al. 2011).

Apart from the BNP the metabolite NT-pro BNP has also been studied in the context of CABG because of its longer plasma half-life. Considering the longer half-life of NT-proBNP (60–120 min) compared to that of BNP (20 min), NT-proBNP may be superior in the perioperative setting allowing a single postoperative sample to be used, although impaired renal function may reduce its specificity. Crescenzi et al. showed for the first time that postoperative NT-proBNP levels are associated with in-hospital mortality and prolonged ICU stay after CABG surgery (Crescenzi et al. 2009). The study, though, was not powered to find correlations between NT-proBNP and single postoperative complications and multivariate analysis was not performed as well. In addition, Rothenburger et al. suggested that NT-proBNP could be a useful marker for recovery after a high-risk CABG procedure, with significant correlation to clinical parameters (Rothenburger et al. 2004).

In the future one might speculate that the combination of a necrosis marker (troponin I) and a volume and pressure marker (BNP or NT-proBNP) might be useful for further risk stratification of patients before and in the early postoperative state (Hutfless et al. 2004). The real clinical breakthrough however would be to show that the elevated postoperative natriuretic peptides, if identified early, can prompt clinical interventions to reverse the adverse outcomes and the elevated mortality and to improve the poor physical activity, all associated with high natriuretic peptide levels.

HeartFatty-Acid–Binding Protein (h-FABP) is released during myocardial ischemia even in the absence of irreversible myocardial necrosis, and is a sensitive marker of MI in the ambulatory setting. The rapid time course of hFABP suggests that it may be uniquely capable of rapidly identifying postoperative patients with acute occlusion of a graft or native vessel (Rade and Hogue 2010). h-FABP is an early diagnostic parameter reflecting perioperative myocardial ischemia and injury in cardiac surgery. Quantitative h-FABP monitoring could predict the severity of myocardial ischemia and injury early during cardiac surgery. By comparative analysis, hFABP was superior to cTnI and other biomarkers when added to multivariate models predicting 5-year mortality and hospital length of stay (Cihan et al. 2004).

Myoglobin is a low-molecular weight cytoplasmic heme protein found in cardiac and skeletal muscle. It is released rapidly after myocyte cell membrane disruption and therefore, is of value in the timing of cell injury to the perioperative period. Myoglobin should be measured in the coronary sinus to improve specificity when skeletal muscle injury is present. Simultaneous measurement of carbonic anhydrase III and myoglobin may also improve specificity. In addition, determination of pericardial/serum MG ratio may be a useful tool for the early diagnosis of the perioperative MI after CABG (Sinha et al. 2003).

Ischemia Modified Albumin (IMA) is a new marker used to detect myocardial ischemia and it shows an early change. The most important characteristic that differentiates IMA from other cardiac ischemia markers is that it increases in the early phase particularly. It elevates in just minutes, peaks within 2 to 4 hours and returns to normal in 6 to 12 hours (Muehlschlegel et al. 2010). IMA may be used as not only an indicator of myocardial ischemia-reperfusion injury, but also as a useful indicator of the cardioprotective effect of N-Acetylo-cysteine in CABG (Karahan et al. 2010). The most important factor that restricts IMA is that IMA may elevate in other pathologies as well, which brings about a requirement for subgroup studies.

Glycogen 6-phosphorylase (G6P) is the key enzyme for glycogenolysis and exists as three isoenzymes: BB(brain), MM (muscle), and LL (liver). The isoenzyme G6-BB is also found in the myocardium, where it is the predominant phenotype, whereas skeletal muscle contains only G6P-MM. Glycogenolysis is significantly increased in the ischemic myocardium and large amounts of G6P-BB are released into the circulation under these circumstances (Rabitzsch et al. 1993). Preliminary data indicate that G6P and G6P-BB catalytic concentrations are sensitive markers of perioperative myocardial injury in patients undergoing coronary artery bypass grafting. G6P-BB mass concentrations are even more sensitive for myocardial injury than G6P-BB catalytic activity.

## Biochemical Markers Associated with Reperfusion Injury

The glycocalyx covering the endothelium is shed during ischemia and reperfusion. The shedding is accompanied by increased levels of the glycocalyx component syndecan-1 in the circulation. Plasma levels of syndecan-1 increased significantly during CABG, with or without the use of CPB (Svennevig et al. 2008).

Plasma levels of asymmetric dimethylarginine (ADMA), symmetric dimethylarginine (SDMA), L-arginine, and L-arginine/ADMA ratio, which are reliable and feasible markers of an early ischemia-reperfusion injury, increase significantly and remain elevated until the first postoperative day due to extensive ischemia-reperfusion injury caused by CPB (Cziraki et al. 2011). The postoperative release of lactate, glutathione peroxidase (GPX), and superoxide dismutase (SOD) in patients undergoing the CABG (on-pump) technique is significantly higher compared to those subjected to OPCABG or CPB with a tissue stabilizing device (SUP.CPB) (Chandrasena et al. 2009). The G-SH levels are decreased and Catalase activity is increased significantly in both on-pump and off-pump CABG patients (Akila et al. 2007). Urinary isoprostane iPF2alpha-III is a new marker reflecting oxidative stress; it has emerged as the most reliable marker of oxidative stress status *in vivo*. In one randomized study in low-risk coronary patients, OPCAB revealed less perioperative oxidative stress, as reflected by lack of excretion of iPF2alpha-III in urine, by lack of increase of plasma free malondialdehyde, and by lower decreases in plasma total antioxidant status (Cavalca et al. 2006). Smoking patients have weaker anti-oxidation capability during CABG, therefore they have higher incidence of low cardiac function, as indicated by the change in plasma free 15-F2t-isoprostane fn(15-F(2t)-IsoP) concentration (Yao et al. 2005).

## Prediction of Cardiac Arrhythmias after CABG

A common complication following CABG operation is the occurrence of arrhythmic events. Atrial tachy-arrhythmias early in the recovery period after cardiothoracic surgery are common; they develop in 11 percent to 40 percent of patients after coronary-artery bypass grafting and in over 50 percent of patients after valvular surgery. Postoperative atrial fibrillation and atrial flutter (POAF) are the most common complications of cardiac surgery that require intervention or prolonged intensive care unit and total hospital stay (Mitchell 2011).

On the other hand, cardiac troponins have been investigated as predictors of AF in various clinical circumstances. The GISSI- AF study has shown that circulating biomarkers including ctNI are related to recurrence of AF in patients in sinus rhythm with a history of recent AF. Moreover, cTnI elevation predicts new-onset AF on 24-hour Holter measurement in patients with acute ischemic stroke or TIA and may indicate a poorer prognosis and a higher risk of stroke, MI, and death at three months (Leal et al. 2012).

Atrial fibrillation could be attributed to inflammatory process occurring after surgery and according to some investigators, this event could be predicted peri-operatively by implementing a cut-off value of 0.901 ng/ml. However, this is contradicted by other authors who claim that AF is not correlated to CTNi levels (Knayzer et al. 2007).

## Acute Kidney Injury and Cardiac Surgery

Assessment of kidney function prior to cardiac surgery is critical, as preoperative chronic renal dysfunction is an independent predictor of mortality in cardiac surgery (Chonchol et al. 2007). In coronary artery bypass grafting (CABG), the previous presence of renal dysfunction, evidenced by elevated preoperative serum creatinine (SCr), is considered an independent risk factor for operative and postoperative mortality and morbidity (Anderson et al. 1999; Asimakopoulos et al. 2005; Cooper et al. 2006; Devbhandari et al. 2006; Durmaz et al. 1999; Hillege et al. 2006; Lok et al. 2004; Roques et al. 1999; Witczak et al. 2005; Zakeri et al. 2005) as well as for longer hospital stay (Santopinto et al. 2003). Apart from higher in-hospital mortality, chronic kidney disease has also been associated with worse long-term outcomes, even three years after bypass grafting (Chonchol et al. 2007).

However, SCr alone is not sufficiently precise to identify renal dysfunction in large population samples (Duncan et al. 2001; Santopinto et al. 2003; Wijeysundera et al. 2006), because serum levels of this marker are determined by age, race, muscle mass and dietary intake, other than creatinine filtration. Up to one quarter of the patients who undergo isolated coronary artery bypass surgery with "normal" serum creatinine (<1.5 mg/dL) show significant impairment of renal function, defined as creatinine clearance <60 mL/min (Miceli et al. 2011; Volkmann et al. 2011). Estimated creatinine clearance (eCrCl) has been described as a more sensitive screening test in predicting renal reserve and a better predictor of mortality and morbidity than SCr in cardiac surgery (Holzmann et al. 2005; Noyez et al. 2006; Wang et al. 2003; Wijeysundera et al. 2006) and is recommended by the National Kidney Foundation and the American Society of Nephrology for assessing preoperative renal function, removing the interpretation of SCr values alone (Haase et al. 2011).

Besides the preoperative assessment of kidney function, an even more important application of renal biomarkers in the perioperative period, is the diagnosis of acute kidney injury (AKI), after CABG, as this complication remains a significant cause of perioperative morbidity and mortality (Bansal 2012), extends the patient's need for hospitalization and increases the associated costs (Jarvela et al. 2011). More specifically, hospital mortality of patients with AKI was 6.8 percent compared to 0.5 percent in the group of patients with normal renal function. Mortality of patients with acute renal failure (ARF) requiring dialysis was 79.7 percent, while among controls, it was 4.8 percent (Bahar et al. 2005). Post-surgical AKI not only leads to higher short-term mortality, but it also leads to increased long-term mortality (Hobson et al. 2009; Lafrance and Miller. 2010; Loef et al. 2005; Lok et al. 2004) and this rise in mortality is unrelated to the recovery of renal function after surgery. Moreover, patients who develop AKI, which requires dialysis, often remain dialysis dependent (Leacche et al. 2004). On the other hand, even subclinical increases in serum creatinine that do not meet acute renal injury criteria, are independently associated with 30-day all-cause mortality in patients with normal renal function or preoperative renal insufficiency undergoing coronary artery bypass grafting (Ishani et al. 2009; Lassnigg et al. 2004; Tolpin et al. 2012).

The incidence of AKI after cardiac surgery is different across studies, depending on the adopted definitions, ranging from 1 percent to 30 percent of the patients, whereas the frequency of AKI requiring dialysis is generally lower, occurring in 1 percent to 5 percent (Arnaoutakis et al. 2007; Aronson et al. 2007; Bove et al. 2004; Brown et al. 2007; Brown et al. 2010; Chertow et al. 2005; Chertow et al. 1997; Coca et al. 2009; Hobson et al. 2009; Hoste et al. 2008; Karkouti et al. 2009; Mangano et al. 1998; Mehta et al. 2006 Ostermann et al. 2000; Palomba et al. 2007; Rosner and Okusa 2006; Thakar et al. 2005; Tuttle et al. 2003; Wijeysundera et al. 2007). The incidence of AKI is influenced by the type of cardiac operation (Chertow et al. 2005; Chertow et al. 1997; Karkouti et al. 2009; Tuttle et al. 2003; Wijeysundera et al. 2007), with patients undergoing (CABG) presenting the lowest incidence (2 percent to 5 percent), while patients undergoing valvular or combined procedures showing a higher rate (as high as 30 percent) (Leacche et al. 2004; Seabra et al. 2010). AKI is believed to be mainly related to the adverse effects of cardiopulmonary bypass (CPB), such as hypoperfusion and activation of both innate and adaptive immune responses that can extend renal injury (Benedetto et al. 2010) (Fig. 1). Several definitions of AKI have been proposed, and the adopted measurements include absolute creatinine value, absolute or percentage changes in serum creatinine (sCr) or estimated glomerular filtration (eGFR) values, and reduction in urine output (Arnaoutakis et al. 2007; Aronson et al. 2007; Brown et al. 2007; Brown et al.; Chertow et al. 2005; Chertow et al. 1997; Coca et al. 2009; Hobson et al. 2009; Karkouti et al. 2009; Mangano et al. 1998; Mehta et al. 2006; Palomba et al. 2007; Rosner and Okusa 2006; Thakar et al. 2005; Wijeysundera et al. 2007). With the introduction, however, of RIFLE and AKIN criteria for diagnosis of AKI, there has been uniformity in definition of AKI (Table 1) (Bellomo et al. 2004; Mehta et al. 2007). These criteria are useful diagnostic and monitoring tools for AKI, and have also been validated in post cardiac surgery patients (Bagshaw et al. 2008).

Serum creatinine is a useful prognostic marker and is used by recent criteria to define AKI (Lassnigg et al. 2008; Lassnigg et al. 2004), but its clinical usefulness for timely diagnosis and assessment of the severity of AKI is limited, as SCr only rises 24 hr–48 hr after the renal injurious event. This relates to the fact that the GFR must decline to approximately half the normal level before the SCr concentration begins to rise above the upper normal limit (Haase et al. 2011).

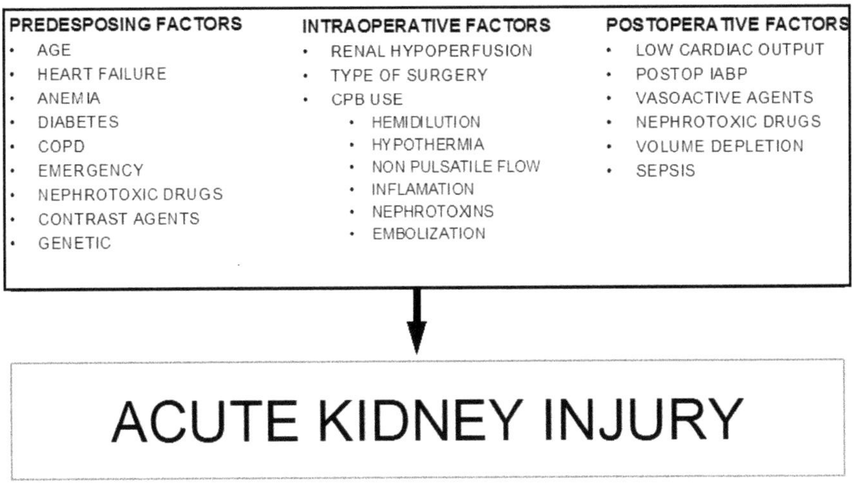

| PREDESPOSING FACTORS | INTRAOPERATIVE FACTORS | POSTOPERATIVE FACTORS |
|---|---|---|
| • AGE | • RENAL HYPOPERFUSION | • LOW CARDIAC OUTPUT |
| • HEART FAILURE | • TYPE OF SURGERY | • POSTOP IABP |
| • ANEMIA | • CPB USE | • VASOACTIVE AGENTS |
| • DIABETES | • HEMIDILUTION | • NEPHROTOXIC DRUGS |
| • COPD | • HYPOTHERMIA | • VOLUME DEPLETION |
| • EMERGENCY | • NON PULSATILE FLOW | • SEPSIS |
| • NEPHROTOXIC DRUGS | • INFLAMATION | |
| • CONTRAST AGENTS | • NEPHROTOXINS | |
| • GENETIC | • EMBOLIZATION | |

# ACUTE KIDNEY INJURY

**Figure 1.** Pathogenesis of acute kidney injury.
Adopted from reference (Bansal 2012).
**Abbreviations:** COPD: chronic obstructive pulmonary disease, CPB: cardiopulmonary bypass, IABP: intraaortic balloon pump, POSTOP: postoperative.

**Table 1.** RIFLE and AKIN criteria for the diagnosis of AKI.

| Risk, Injury, Failure, Loss and End Stage Kidney Disease (RIFLE) Criteria | |
|---|---|
| Stage | GFR Criteria |
| Risk | Increased SCr x1.5 or GFR decrease > 25% |
| Injury | Increased SCr x 2 or GFR decrease > 50% |
| Failure | Increased SCr x 3 or GFR decrease > 75% or SCr ≥ 4 mg/dl with an acute rise ≥ 0.5 mg/dl |
| Loss | Persistent Acute Renal Failure = complete loss of kidney function > 4 weeks |
| ESKD | End Stage Kidney Disease (> 3 months) |
| **Acute Kidney Injury Network (AKIN) Classification** | |
| Stage | Serum Creatinine Criteria |
| 1 | Increase in SCr ≥ 0.3 mg/dl or Increased SCr x1.5 from baseline |
| 2 | Increased SCr x 2 from baseline |
| 3 | Increased SCr x 3 from baseline or SCr ≥ 4 mg/dl with an acute rise ≥ 0.5 mg/dl |

Adopted from references Thomas et al. (Thomas et al. 2011) and Haase-Fielitz et al. (Haase-Fielitz et al. 2009a)

In any case, the clinician should keep in mind that, in hospitalized patients, the interpretation of GFR is difficult, because of potential confounders due to the non-steady state of serum-creatinine metabolism and tubular excretion additional to filtration, co-morbid conditions that cause malnutrition and the frequent use of medications that can interfere with the measurement of SCr. Further, in AKI, none of the eGFR equations should be used due to lack of validation in this non-steady state (Haase et al. 2011a). Caution is advised in the interpretation of serum markers, as the considerable hemodilution induced by cardiac surgery with the use of cardiopulmonary bypass (CPB) may lead to underestimation of the severity of AKI as it increases the time required to identify a 50 percent relative increase in SCr (Jarvela et al. 2011; Macedo et al. 2010).

## Emerging biomarkers

Markers of reduced renal function, such as serum creatinine or urine output, increase between 24 hours and 72 hours after the injurious event to the kidney (Haase et al. 2011b) and do not indicate either the nature or the site of the kidney injury (Ricci et al. 2011). Thus, the early diagnosis of AKI has been problematic and this leads to the inability to apply adequate preventive measures and probably delayed treatment (Fig. 2). In response to this problem, several biomarkers have been recently investigated as possible tools for the early detection of AKI (Table 2).

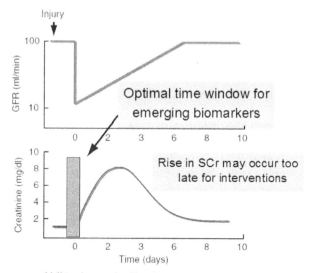

**Figure 2.** Time course of fall in glomerular filtration rate as compared to rise in serum creatinine and rise in a hypothetical biomarker (bar) in relation to an injurious insult.
Adopted from reference Rosner 2009 (Rosner 2009).

**Table 2.** Emerging Biomarkers for the early diagnosis of AKI.

| Biomarker | Sample Source | After Cardiopulmonary Bypass | Commercial Assay |
|---|---|---|---|
| Cystatin C | Plasma | 12 h post-CPB | Nephelometry |
| NGAL | Plasma | 2 h post-CPB | ELISA |
| NGAL | Urine | 2 h post-CPB | ELISA |
| NAG | Urine | 6 h post-CPB | Colorimetric assay |
| L-FABP | Urine | 4 h post-CPB | ELISA |
| KIM-1 | Urine | 12–24 h post-CPB | ELISA |
| IL-18 | Urine | 4–6 h post-CPB | ELISA |

Modified from references Thomas et al. (Thomas et al. 2011), Han et al. (Han et al. 2008) and Katagiri et al. (Katagiri et al. 2012)

**Abbreviations:** CPB: cardiopulmonary bypass, ELISA: enzyme-linked immunosorbent assay.

Cystatin C is a cysteine protease inhibitor that is produced and released at a constant rate by all nucleated cells and is also freely filtered by the glomerulus (Lane et al. 2009; Thomas et al. 2011). For these reasons, serum cystatin C has been proposed as a superior marker than SCr, although it appears to be influenced by factors other than renal function alone, and, despite initial expectations, it does seem to be associated with old age and larger body size, in healthy individuals at least (Knight et al. 2004). It has been proven in a variety of clinical settings that serum cystatin C-based equation is a reliable marker of GFR, with a very high diagnostic accuracy and ability to predict patients with CKD, while an increasing number of studies are focusing on the use of cystatin C in the evaluation of cardiac surgery patients. Cystatin C has, in this patient group, a stronger correlation to iohexol clearance than creatinine, taking into account that clearance of iohexol, a non-ionic contrast agent, is an established and accurate method for GFR assessment (Bronden et al. 2011). Cystatin C and cystatin C-based estimation of GFR may be useful and more sensitive than creatinine and the creatinine-based estimation of GFR in detecting mild acute renal insufficiency in diabetic patients after coronary artery bypass grafting surgery (Jarvela et al. 2011). In pediatric population undergoing CPB, serum levels were significantly increased in AKI patients at 12 hours after CPB and remained elevated at 24 hours, and cystatine C proved to be an earlier and more accurate AKI marker compared to conventional SCr, correlating also with severity and duration of renal dysfunction (Knight et al. 2004). However, urinary cystatin C is also a significant predictor of AKI and long term renal replacement therapy in adult cardiac surgery patients. Moreover, urinary cystatine C might be an even better marker for early detection of AKI than serum concentration, as urinary levels are elevated within the first six hours of surgery, whereas the serum levels are not predictive of AKI within this time window (Koyner et al. 2008). Finally, cystatin C is not specific for ischemic AKI and increase in serum occurs much later than other emerging markers (Bansal 2012).

Neutrophil gelatinase-associated lipocalin (NGAL) is one of the most intensively investigated novel renal biomarkers with predominantly promising data from clinical studies comprising more than 3500 cardiac surgery or critically ill patients (Haase et al. 2010). It measures tubular stress and is involved in the ischemic renal injury and repair process (Haase et al. 2009b; Schmidt-Ott et al. 2006); NGAL increases dramatically in response to tubular injury and precedes rises in SCr by more than 24 hours (Schmidt-Ott et al. 2006). Several studies directly compared NGAL with SCr for the early diagnosis of AKI, finding NGAL to be superior in various clinical settings (Haase-Fielitz et al. 2009b; Makris et al. 2009; Martensson et al. 2010; Shapiro et al. 2010).

In critically ill patients, those who are NGAL-positive are 16 times more likely to undergo dialysis, three times more likely to die during hospitalization and spend on average three extra days in intensive care and eight extra days in the hospital (Haase et al. 2011a). Furthermore, in patients undergoing cardiopulmonary bypass serum and/or urine, NGAL has the ability to predict the development of postsurgical AKI before elevations in SCr (Mishra et al. 2005; Wagener et al. 2006). Its diagnostic utility has been verified in children undergoing CPB, where both urine and plasma NGAL were strong independent predictors of AKI with area under the curve (AUC) of 0.998 and 0.91 for 2-hour urine and plasma NGAL measurements, respectively. Within 2–6 hours of bypass, there is a >10-fold increase in NGAL levels in children with AKI, whereas increase in SCr is first detected between one and three days after surgery (Mishra et al. 2005). In adult patients who undergo CPB, NGAL levels at 1, 3, and 18 hours after cardiac surgery, are significantly higher in patients who develop clinically significant AKI (Wagener et al. 2006); also, the use of urinary NGAL after cardiac surgery appears to be cost-effective in the early diagnosis of AKI (Shaw et al. 2011). However, in adult cardiac surgery, the predictive value for AKI of NGAL appears to have wide variability. There are reports which have failed to confirm the diagnostic utility of NGAL (Wagener et al. 2008), especially in patients with pre-operative GFR < 60 mL/min (McIlroy et al. 2010). This indicates that NGAL may be affected by pre-existing renal disease while it also influenced by infections (Mariscalco et al. 2011). The better performance in children compared with adults most likely reflects the less frequent comorbidities in pediatric population (Haase et al. 2011b). In addition, the choice of definition of AKI might, at least in part, account for such variability in adult population, as the predictive value of plasma NGAL is higher for more severe AKI and increases with increasing RIFLE classes or AKIN stages (Haase-Fielitz et al. 2009a). Given the somewhat variable predictive performance of NGAL in the clinical setting and several known confounders, a meta-analysis was performed based on single-centre studies, a uniform though creatinine-based AKI definition, and a standardized timing of NGAL measurement in relation to the occurrence of renal insult. In 2,538 patients, the sensitivity and specificity of NGAL—measured several days prior to current AKI diagnosis—range on average between 75 percent and 95 percent (Haase et al. 2009b). NGAL's predictive value for AKI was an area under the receiver operating curve (AUC) of 0.80 and, for subsequent initiation of renal replacement therapy, an AUC of 0.78. The effort to combine NGAL and cystatin C did not add any predictive value for AKI after adult cardiac surgery, although it did verify that each marker separately is an independent predictor of duration and severity of AKI and duration of intensive care stay, with an AUC of 0.77 and 0.76 for NGAL and cystatin C respectively (Haase et al. 2009a).

N-acetyl-β-D-glucosaminidase (NAG), a proximal tubule lysosomal enzyme, has been extensively studied in both the adult and pediatric population and has proven to be a sensitive, persistent, and robust indicator of AKI. Increased NAG levels have been reported in a variety of clinical settings, as well as following cardiopulmonary bypass procedures (Ascione et al. 1999). The receiver-operator characteristic area under the curve (AUC) for the diagnosis of AKI, evaluated in a heterogeneous population using urinary NAG, can reach the impressive 0.97 mark. In patients developing AKI after cardiac bypass surgery, the rise in urinary NAG levels occurs six hours after surgery and approximately 18 hours before the rise in SCr (Han et al. 2008). The area under curve for the diagnosis of AKI with urinary NAG is estimated to be 0.75 (Katagiri et al. 2012).

Urinary NAG activity has been found to be inhibited by endogenous urea (Bondiou et al. 1985) as well as by a number of nephrotoxicants and heavy metals (Vaidya and Bonventre 2006), thus representing possible confounding factors. In addition, given the various conditions that have been associated with increased NAG excretion, nonspecificity for AKI may limit its use as a biomarker and further research is required to determine its role as diagnostic marker (Rosner 2009).

Fatty Acid Binding Proteins (FABP) are small cytoplasmic proteins abundantly expressed in tissues with an active fatty-acid binding metabolism. They carry endogenous antioxidant activity by promoting free-fatty-acid metabolism and by binding long-chain fatty-acid-oxidation products (Hofstra et al. 2008). Two types of FABP have been identified in the human kidney, liver-type FABP (L-FABP) in the proximal tubule and heart-type FABP (H-FABP) in the distal tubule (Ek-Von Mentzer et al. 2001; Maatman et al. 1992). Increased levels of cystosolic free fatty acids with attendant increased FABP expression may be seen in response to a variety of pathophysiologic tubular stresses (Pelsers et al. 2005), such as ischemic and nephrotoxin-induced kidney injury (Negishi et al. 2009). Urinary L-FABP increase within four hours after cardiac surgery in children undergoing CPB and can predict the subsequent development of AKI with an AUC of 0.81 (Portilla et al. 2008). In a heterogeneous group of hospitalized adult patients, the diagnostic performance of urinary L-FABP for AKI, assessed by the area under the receiver operating characteristic curve, was 0.93 (Ferguson et al. 2010). Urinary L-FABP has been recently evaluated for adult post-cardiac surgery AKI. The use of L-FABP alone yielded an AUC-ROC of 0.72 at four hours after surgery, whereas the combination of L-FABP with NAG can detect AKI with higher accuracy than either biomarker measurement alone (AUC-ROC 0.81). Combining these two markers presents a reasonable strategy to improve the diagnostic performance of biomarkers, which takes advantage of the different sensitivity and specificity of each marker for AKI detection (Katagiri et al. 2012). On the other hand, a drawback of urine L-FABP appears to be the delayed rise in comparison to other biomarkers such as NGAL (Moore et al. 2010).

Kidney injury molecule-1 (KIM-1) is an immunoglobulin super family transmembrane protein normally present at low levels in proximal renal tubular cells. It significantly increases following acute ischemic or nephrotoxic insult (Han et al. 2002). Thus, structure and expression data suggest that KIM-1 is an epithelial adhesion molecule upregulated in dedifferentiating and regenerating tubule epithelial cells following injury and may play a role in the restoration of morphological integrity of the tubule (Ichimura et al. 1998). Several studies have established a potential clinical role for KIM-1 as a predictive biomarker for AKI after cardiac surgery (Han et al. 2008; Liangos et al. 2009). Urinary levels of KIM-1 that are markedly elevated in children 12 hours after surgery are better indicators of AKI than NAG levels at all time points (Han et al. 2008). In adult patients undergoing CPB, KIM-1 levels increase significantly at both 2 and 24 hours post-operatively in patients with AKI (Liangos 2006), and the marker has demonstrated superiority over other emerging biomarkers, at several time points after cardiac surgery, with the achieved AUC ranging from 0.65 to 0.78 (Han et al. 2009; Liangos et al. 2009). However, kidney injury molecule-1 seems to be specific for ischemic and nephrotoxic AKI and may not be useful for other causes of renal injury.

Interleukin-18 (IL-18) is a proinflammatory cytokine that belongs to the IL-1 super family. It is activated within proximal tubule cells and then excreted in the urine after ischemic injury (Parikh and Devarajan 2008). In children undergoing CPB, urine IL-18 is detectable at 4–6 hours, peaks at 12 hours and remains elevated for >48 hours, and may be a marker of AKI severity (Parikh et al. 2006). In contrast, in adult cardiac surgical patients undergoing cardiopulmonary bypass, urinary IL-18 was not predictive of AKI either immediately postoperative (AUC–ROC 0.55) or 24 hours after surgery (AUC–ROC 0.53), but was significantly correlated with duration of cardiopulmonary bypass, thus representing a nonspecific marker of cardiopulmonary bypass-associated systemic inflammation (Haase et al. 2008).

## Serologic Neurological Markers of Brain Injury and Cognitive Function after Cardiopulmonary Bypass

Brain injury following cardiopulmonary bypass (CPB) remains a common and serious complication that is often misdiagnosed. The American College of Cardiology/American Heart Association guidelines for CABG surgery divide postoperative neurologic deficits into two categories. Type 1 deficits include major focal neurologic events, stupor and coma. Type 2 deficits describe more global cognitive deficits such as deterioration in intellectual function, memory and confusion without evidence of focal injury. While physical examination and neuroimaging modalities have proven valuable for the detection and treatment of acute focal brain damage postoperatively, mild and diffuse injuries such as neurocognitive decline (NCD) seen in Type 2 deficits would benefit from improvements in the early diagnosis and identification of these patients. Identifying biochemical surrogate markers for brain dysfunction would greatly assist in the diagnosis and timely treatment of patients with this complication. Recent studies have focused on proteins that are expressed predominantly in cerebral cells and released into the cerebrospinal fluid and blood following an acute insult to the brain such as CPB (Kalman et al. 2006). After brain injury, several substances, including S-100β, NSE, brain-specific creatine phosphokinase, and glutamate, have been shown to be released from brain tissue. The metalloproteinase and ubiquitin C terminal hydroxylase-L1 (UCH-L1) are the most recently researched markers; however, their usefulness is still unclear (Cata et al. 2011). Unfortunately, owing to one or more limitations in sensitivity and specificity, none of these has emerged as a widely used diagnostic or prognostic clinical tool or a validated surrogate endpoint measure for irreversible brain damage. Consequently, the need remains for new, highly sensitive biomarkers which may be combined with neurologic and neuroradiologic methods to improve the diagnosis, prognosis and experimental therapeutic evaluation of acute brain injuries.

Neuron specific enolase (NSE), one of the five isozymes of the glycolytic enzyme, enolase, is a serum marker of brain injury. NSE release has also been found to occur from hemolyzed erythrocytes. Serum NSE is a sensitive marker in the early diagnosis of brain injuries after CPB. Detection of both S100B and NSE is the most specific marker. Furthermore, postoperative serum concentrations and kinetics of NSE have a high predictive value with respect to the early neuropsychological and neuropsychiatric outcome after cardiac surgery (Herrmann et al. 2000). CABG with CPB causes a significantly greater increase in NSE serum levels than off-pump surgery and correlates with CPB duration (Bonacchi et al. 2006).

One of the most studied potential protein markers of brain injury is S-100β, an astroglial protein that leaks from damaged cells and across the blood-brain barrier. S-100β has been studied in the cardiac surgical setting with inconclusive results. Following early enthusiasm about the ability of S-100β to predict severity of neuronal damage in cardiac surgical patients, it has been recently discredited due to its lack of specificity. Notwithstanding this finding, S-100β could potentially facilitate assessment of experimental neuroprotectant treatment strategies and comparison between different surgical techniques (Herrmann et al. 2000).

Tau is a microtubule-associated protein that is released following axonal injury. Recent studies show that concentration of tau in the cerebro-spinal fluid markedly increases six months postoperatively, and is potentially a better marker of brain injury compared with S-100β. Moreover, it appears to be more specific than the previous two biomarkers since it is only found in the central nervous system (Bonacchi et al. 2006). Unfortunately, tau protein is limited so far by our ability to reliably detect it quantitatively in serum due to our lack of understanding of its pharmacokinetics and other technical issues.

## Metabolic–red Cells Biomarkers

Low-density-lipoprotein (LDL) is associated with symptomatic CABG disease at one year after surgery (Harris et al. 2004). Apart from this, it has been demonstrated that positive family history for coronary atherosclerosis in the presence of high serum triglyceride level significantly predicts the need for future CABG reoperation (Mennander et al. 2005). Also, increased preoperative triglyceride levels predict increased late mortality and cardiac event risk in diabetic post-CABG patients, more strongly in women than in men (Sprecher et al. 2000). High-density-lipoprotein(HDL-C) is an important predictor of survival in post-CABG patients so the measurement of HDL-C provides a compelling strategy for the identification of high-risk subsets of patients who undergo CABG (Foody et al. 2000). In patients with chronic ischemic heart disease (IHD), high level of Lp(a) can serve as independent predictor of unfavorable events including death, nonfatal MI during 10 years after CABG and of developing saphenous vein (SV) graft stenosis after CABG (Harris et al. 2004).

Paraoxonase-1 (PON-1) is an HDL-bound enzyme which has anti-atherogenic properties and protects LDL cholesterol from oxidative modification. Reduced PON-1 activity may lead to acceleration of SV graft occlusion (Cagirci et al. 2009).

In a recent study, it has been identified that metabolic profiles composed of short-chain dicarboxylacylcarnitines, ketone-related metabolites, and short-chain acylcarnitines predict adverse outcomes after CABG, independent of standard clinical predictors of risk (Shah et al. 2012). Metabolic changes at the interstitial concentrations of the pyruvate and lactate after coronary bypass surgery occur with some differences related to CPB use (Cossu et al. 2012). These changes have no impact on the postoperative clinical outcome (Pojar et al. 2008).

In diabetic patients undergoing CABG surgery, aggressive glycemic control does not result in any significant improvement in clinical outcomes that can be achieved with moderate control (Bhamidipati et al. 2011). Glucose levels greater than 8.8 mmol/L on postoperative day 1 and having three or more hypoglycaemic episodes in the postoperative period are predictive of mortality and morbidity among patients undergoing cardiovascular surgery (Frioud et al. 2010). A target blood glucose range of 121 to 180 mg/dL is recommended for patients after CABG as advocated by the Society of Thoracic Surgeons (Desai et al. 2012).

Fasting plasma glucose (FPG) alone does not seem sufficient for diagnosing approximately half of the patients with dysglycemia. The use of FPG and HbA1c measurements in combination may be a useful strategy to preoperatively identify coronary patients with unknown diabetes (Tekumit et al. 2010). Poor preoperative glycemic control, as measured by an elevated HbA1c, is associated with reduced long-term survival after CABG. However, one study refers that preoperative HbA1c levels might not predict long-term outcomes for diabetic patients undergoing off-pump coronary artery bypass (OPCAB) (Tsuruta et al. 2011). Finally, preoperative HbA1c independently predicts the occurrence of atrial fibrillation (AF) after isolated OPCAB (Kinoshita et al. 2012).

Because of their extremely limited coronary reserve, patients undergoing CABG surgery represent a population potentially most sensitive to the impact of low Hb levels. Specifically, preoperative hemoglobin level less than 10–11 g/dL was an independent predictor for postoperative renal and central nervous system outcome, and the association with increased cardiac adverse events was caused by concomitant risk factors prevalent in anemic patients (Kulier et al. 2007). Importantly, the extent of pre-existing comorbidities substantially affects perioperative anemia tolerance. Therefore, preoperative risk assessment and subsequent therapeutic strategies, such as blood transfusion, should take into account both the individual level of preoperative Hb and the extent of concomitant risk factors (Westenbrink et al. 2011). Postoperative anemia is common, frequently persists for months after CABG surgery and is associated with an impaired outcome. In patients with anemia, ACE inhibitors slowed recovery from postoperative anemia and increased the incidence of cardiovascular events after CABG (Wenzel et al. 2011).

## Hormones Alterations and CABG

Appropriate to the decrease in Hb concentration and platelet count, clear alterations of serum EPO and thrombopoietin levels could postoperatively be observed. EPO levels showed an inverse correlation to hemoglobin concentrations, whereas a disturbed thrombopoietin feedback mechanism resulted in the phenomenon of reactive thrombocytosis (Pearce et al. 2010).

CABG is associated with thyroid hormone changes consistent with euthyroid sick syndrome. It has been argued that some aspects inherent to CPB may contribute to the changes of the thyroid hormone profile, such as hemodilution, hypothermia, and nonpulsatile flow. Furthermore, it has been shown that changes in thyroid hormone levels are likely to be detected in the immediate postoperative period, since at this stage, there is an active state of catabolism associated with high rates of oxygen consumption (Cerillo et al. 2003). Some authors have postulated that the decrease in thyroid hormones in the post-CPB favours the occurrence of global myocardial dysfunction and arrhythmias, especially AF (Gungor et al. 2011).

Resistin is a newly identified adipocyte-secreted hormone belonging to a cysteine-rich protein family. Patients with an elevated postoperative resistin level may have high risk for AF after CABG (Dashwood et al. 2011). Perivascular tissue (PVT) surrounding many blood vessels, including those used as bypass conduits, is the source of adipocyte-derived relaxing factors, one of which is leptin. PVF-derived leptin may also play an important role both at harvesting and in the improved long-term performance of no-touch SVs in patients undergoing CABG (Eyileten et al. 2007). Decreased plasma adiponectin, which functions as an anti-inflammatory and anti-atherogenic factor, is a risk factor for coronary disease. CABG ameliorates the decreased plasma adiponectin level in atherosclerotic patients and as a consequence decreases the cardiac risk factors in atherosclerotic patients (Novella et al. 2007).

Recent studies have suggested that endogenous vasopressin (AVP) acts as a spasmogen during coronary artery bypass grafting (CABG). In addition, AVP, among other factors, produces vasospasm in internal mammary artery (IMA) grafts (Maggio et al. 2012).

After CABG testosterone concentration frequently declines to less than 200ng/dl, a situation suggestive of overt hypogonadism. Since men with low testosterone levels have a high probability of developing mobility limitations, the perioperative use of testosterone treatment has been proposed in older men undergoing cardiac revascularization surgery (Yin et al. 2005).

High serum cortisol level or the disordered cortisol is associated with increased risk of postoperative delirium and neuropsychological deficits (Yin et al. 2005). Apart from this, intravenous hydrocortisone reduces the incidence of AF after cardiac surgery (Yin et al. 2005).

Elevated pre-operative homocysteine level is independently associated with increased morbidity and mortality, particularly in patients undergoing CABG. Specific post-operative antithrombotic strategies may be advisable in hyperhomocysteinanemic patients (Halonen et al. 2007). Moreover, elevated levels of plasma homocysteine are related to atherosclerotic lesions of SV grafts after CABG as well as coronary atherosclerosis (Halonen et al. 2007).

## Coagulation Biomarkers

CABG with CPB is associated with intense activation of hemostatic mechanisms. Both extensive contacts between blood and nonendothelial surfaces of the bypass circuit and the release and reinfusion of tissue factor lead to increased thrombin generation during the CPB procedure. This results in fibrin formation, fibrinolysis, and platelet activation, despite full heparinization. As a part of the systemic inflammatory response, hemostatic activation may lead to generation of microthrombi and may contribute to CPB-related organ dysfunction. Another, more common, complication due to hemostatic activation after CPB is abnormal postoperative bleeding, because abnormal activation of the hemostatic system can lead to consumption of coagulation factors, increased fibrinolysis, and destruction of platelets (Despotis and Joist 1999).

Clinical studies have shown that thrombin formation during cardiac surgery, especially during myocardial reperfusion, is involved with myocardial damage and impaired hemodynamic recovery. Thrombin generation during reperfusion after CABG is associated with pulmonary vascular resistance and postoperative myocardial damage (Raivio et al. 2006). It has been argued that preoperative thrombin-antithrombin (TAT) complexes values might be good markers of early graft occlusion (Rifon et al. 1994).

Levels of preoperative activated protein C and activated protein C measured after heparin neutralization have been associated with unfavourable hemodynamic recovery postoperatively. The balance of activated protein C with thrombin is associated dynamically with postoperative hemodynamic recovery (Raivio et al. 2007). CABG high preoperative prothrombin fragment F1+2 and low preoperative protein S levels have been associated with a less favourable hemodynamic profile postoperatively but not with postoperative cardiac biomarker levels. Baseline activation of coagulation and the balance between pro-coagulant and anti-coagulant factors preoperatively might have implications for postoperative hemodynamic recovery after CABG (Raivio et al. 2011). In contrast to standard coagulation testing, platelet function predicts both bleeding and thrombosis after OPCAB. Titration of perioperative platelet function according to these tests may minimize thrombosis without increasing bleeding (Poston et al. 2005).

Postoperative increase in plasminogen activator inhibitor-1 (PAI-1) levels has been associated with an increased risk of graft occlusion after coronary artery bypass surgery (CABG) (Rifon et al. 1997). Increase in PAI-1 antigen values in patients who undergo off-pump CABG is significantly higher than in those who undergo conventional CABG with cardiopulmonary bypass (Ozkara et al. 2007). Increased PAI activity might have a predictive value for early thrombosis in patients undergoing CABG (Sjoland et al. 2007). However, PAI-1, when evaluated prior to the operative procedure, is not strongly associated with increased mortality in the long-term after CABG, when other comorbidity factors are simultaneously considered (Vallely et al. 2009). Finally, tissue PAI-1 has been used in order to assess the efficacy of different therapeutics interventions during and after CABG.

## Biomarkers of Endothelial Dysfunction

The use of extracorporeal circulation causes a significant increase in the vonWillebrand (vWF) level which suggests a marked endothelial injury caused by CABG procedure (Kaireviciute et al. 2011). The predictive role of vWf regarding the postoperative AF is controversial (Kaireviciute et al. 2010).

The number of circulating endothelial cells (CEC) is another marker of endothelial activation and damage during CABG (Schmid et al. 2006). CEC have been used in order to compare different CABG techniques.

Furthermore, endothelial Nitric oxide synthase (eNOS) has been used as an indicator of endothelial integrity and damage during different techniques of preparation and processing of vein or artery grafts (Nowicki et al. 2011).

Endothelins are a family of potent vasoconstrictor peptides with cell-growth promoting and mitogenic properties. These peptides are produced not only by endothelial cells but also by smooth muscle cells and macrophages. Plasma endothelin levels are raised in patients with significant atherosclerosis and show a positive relation to the extent and severity of coronary artery disease (Dashwood et al. 2009).

ET-1 has been used in order to compare off-pump versus on-pump CABG. ET-1 does not have value in predicting postoperative mortality and complication (Lu et al. 2007). However, it has been related with vein graft occlusion and the therapeutic potential of endothelin receptor antagonists (ERA) in improving graft performance has been discussed (Dashwood 2009). In addition, ET-1 might be an important mediator of ischemia-reperfusion injury in patients with diabetes. Consequently, the use of ERA might be a novel strategy for improving the resistance of the diabetic heart to cardioplegic arrest and reperfusion (Verma et al. 2002).

## Inflammatory Biomarkers

The presence of cardiopulmonary bypass (CPB) support during the traditional coronary artery bypass surgery (on-pump coronary artery bypass grafting, OnPCABG), has been recognized as the main origin of a complex systemic inflammatory response, which drastically contributes to several adverse postoperative outcomes such as renal, pulmonary, and neurological complications, bleeding and even multiple organ dysfunction. Other factors associated with the inflammatory response are surgical trauma, reperfusion injury and preexisting comorbidities. The inflammatory response associated with CPB plays a central role in leukocyte recruitment and the production of cytokines and chemokines. Cytokines, being produced and secreted by the cells in the immune system, are small soluble polypeptides and proteins that act as paracrine messengers. Chemokines, a cytokine subfamily, are chemotactic and participate in recognizing, recruiting, removing, and repairing tissue undergoing inflammation. Proinflammatory cytokines (including tumor necrosis factors (TNF)-α, interleukin (IL)-1, IL-6, IL-8, and IL-12) facilitate a wide range of inflammatory processes. Anti-inflammatory cytokines (including IL-4, IL-6, IL-10, IL-11, and IL-13) inhibit inflammatory processes by reducing the production of the proinflammatory cytokines or counteracting their effects. Cytokines may also influence the production of other cytokines and furthermore induce expression of cytokine receptors as well as enzymes like inducible nitric oxide synthase and cyclooxygenase-2. These enzymes in turn contribute to the inflammatory response (Castellheim et al. 2008).

After the recent development of effective devices for target vessel exposure and stabilization, beating heart techniques, such as off-pump coronary artery bypass grafting (OffPCABG), have gained widespread dissemination as an alternative technique to conventional OnPCABG. Several supporters of the OffPCABG procedure claim that by avoiding CPB the probability and severity of the systemic inflammatory response syndrome reduces markedly (Serrano et al. 2010).

Serrano et al. (Serrano et al. 2010) studied the inflammatory response between the two aforementioned types of bypass surgery. The inflammatory biomarkers selected for evaluation were leukocyte count, CRP, IL-6, IL-8, sP-selectin, and sICAM-1. Their results suggested an overall greater inflammatory response during OnPCABG without however a difference in survival. The only marker that had significant positive correlations with events (e.g., need for dobutamine use and longer ICU stay) was the increased 24-hour postoperative CRP levels (N3.0 mg/L) and occurred just for the OnPCABG patients (Serrano et al. 2010).

The profile of cytokines and chemokines in low-risk patients undergoing OffPCABG and OnPCABG was also investigated by Castelheim et al. (Castellheim et al. 2008). Their study showed that the differences in the cytokine inflammatory response between patients operated with the on-pump or off-pump techniques are small and probably have a limited impact on outcome in low-risk patients. Of the 25 biomarkers analyzed, 11 were not detected while 14 increased significantly in both groups. Only three mediators, eotaxin (a small cytokine produced by a variety of cell types, acts as a potent eosinophil chemoattractant), macrophage inflammatory protein (MIP)-1β, and interleukin (IL)-12 were significantly different between the two groups, increasing more in the on-pump than in the off-pump group ($p < 0.001$, $p < 0.01$, and $p < 0.05$, respectively). There was a marked, comparable increase in the concentrations of the cytokines IL-6, IL-10, IL-15, and IL-1Ra as well as the chemokines inducible protein (IP)-10, monokine induced by interferon gamma (MIG), monocyte chemoattractant protein 1 (MCP-1), and regulated on activation, normal T cell expressed and secreted (RANTES) in both groups (p < 0.001 for all). There was only a modest, but still statistically significant, increase in IL-8, tumor necrosis factors-α, and IL-2R, without any intergroup differences. When corrected for hemodilution, the production of the anti-inflammatory biomarkers IL-1Ra and IL-10 were significantly higher in the on-pump group (p < 0.001 for both) (Castellheim et al. 2008).

The same researchers in a previous study in high-risk patients had showed that the two types of surgery differ strikingly with respect to complement activation. The C3 activation products and sC5b-9 increased markedly in the on-pump, but not in the off-pump group. This difference was not seen for neutrophil and platelet activation. The role of the activation of the complement system induced by CPB in high-risk patients and its correlation with the long-term clinical outcome, however, remains to be investigated.

One of the features of acute inflammation after cardiac surgery with CPB is endothelial injury. The endothelium may become activated in response to inflammatory insults such as cytokines and physical insults such as shear stress and hypoxia. Cell adhesion molecules (CAMs) mediate the contact and subsequent subendothelial penetration of leucocytes with the endothelium. These interactions are critical in maintaining vascular homeostasis and mediating inflammatory injury. If they are dysregulated they can contribute to adverse inflammatory tissue injury and pathologic thrombosis. Endothelial cell adhesion molecules (CAMs) are expressed and released when the endothelium is activated. Vallely et al. (Vallely et al. 2010) compared plasma CAMs (E-selectin, ICAM-1 and VCAM-1) and HUVEC (Human Umbilical Vein Endothelial Cell) expression of the same CAMs when exposed to plasma taken before, during and after OffPCABG or OnPCABG surgery. They found that plasma E-selectin was unchanged, soluble ICAM-1 and VCAM-1 were elevated post-operatively but did not discriminate between OffPCABG and OnPCABG and that in HUVEC exposed to peri-operative plasma, VCAM-1 expression is decreased after CABG and increased after OffPCABG. These different inflammatory responses with respect to endothelial activation may have clinical implications. Harvesting the vein graft is known to activate the graft endothelium. Cell adhesion molecules have been shown to play a role in graft occlusion after coronary artery surgery, with their suppression shown to be protective. The activation of HUVEC by peri-operative OffPCABG plasma with the suppression of HUVEC by OnPCABG plasma suggests that the OffPCABG patients may be at higher risk of adverse vein graft events. Patients having long saphenous vein grafts performed off-pump may require more aggressive antiplatelet or statin therapy. Alternatively, it may be better to adopt a total-arterial revascularisation approach when performing coronary artery surgery off-pump (Vallely et al. 2010).

In conclusion, both types of CABG surgery are associated with an inflammatory response, which with regard to the inflammatory markers measured, is greater during OnPCABG. However this difference has not been shown to have significant impact on clinical outcome. In addition, the different inflammatory response with respect to endothelial activation may contribute to adverse prothrombotic events after OffPCABG surgery.

Studies in cardiac surgical patients suggest that increased preoperative CRP levels are associated with increased short- and long-term morbidity and mortality. Kangasniemi et al. (Kangasniemi et al. 2006) showed significantly increased 12-year mortality in patients with preoperative CRP levels more than 10mg/l and undergoing isolated on-pump CABG surgery. Moreover, Perry et al. (Perry et al. 2010) not only observed an association between preoperative CRP levels more than 10 mg/l and increased postoperative mortality, but also demonstrated that preoperative CRP levels as low as 3mg/l are associated with decreased survival and extended hospital length of stay (HLOS) in lower-acuity cardiac surgical patients free of ongoing myocardial ischemia or infarction at the time of nonemergent surgical revascularization. In the study, it was shown that increased preoperative CRP levels predict long-term (7 year), all-cause mortality and extended HLOS in a cohort of lower-acuity CABG-only surgical patients. These findings may allow for more objective risk stratification of patients who present for nonemergent, surgical coronary revascularization. The results do not support or refute the important question of whether or not CRP is a proinflammatory mediator of postoperative mortality or merely a marker. Confirmation of a direct causal relationship between CRP and cardiovascular outcomes will require further investigation (Perry et al. 2010).

Apart from the preoperative value of CRP, the CRP genes contribute in the modulation of susceptibility to cognitive decline after CABG, as it was shown in a study which sought to investigate the hypothesis that candidate gene polymorphisms in biologic pathways regulating inflammation are associated with postoperative cognitive deficit (POCD). It was found that minor alleles of the CRP 1059G/C SNP (Single-Nucleotide Polymorphisms) (odds ratio [OR] 0.37, 95 percent confidence interval [CI] 0.16 to 0.78; p =0.013) were associated with a reduction in cognitive deficit in European Americans (n=443). The absolute risk reduction in the observed incidence of POCD was 20.6 percent for carriers of the CRP 1059C allele. These results could have potential implications for identifying populations at risk, who might benefit from targeted perioperative anti-inflammatory strategies (Mathew et al. 2007).

Another aspect of CRP regarding CABG is its association with postoperative atrial fibrillation (AF). Postoperative AF occurs in approximately 25 percent to 40 percent of patients undergoing CABG and is associated with significantly increased risk of short-term mortality and perioperative stroke. Inflammation and oxidative stress play a key role in the initiation and prolongation of atrial fibrillation. Kaireviciute et al. (Kaireviciute et al. 2010) showed that increased peripheral (p=0.018) and intracardiac levels (from the right atrial appendage (p=0.029) and the left atrium (p=0.026)) of hs-CRP were associated with the presence of AF after CABG. Furhthermore Girerd et al. (Girerd et al. 2009) found that male patients younger that 65 years with increased waist circumference (WC>102cm) combined with elevated C-reactive protein levels (CRP≥1.5 mg/L) were at a high risk of developing post-op AF (OR : 2.32, P : 0.02). It remains to be determined whether it is possible to acutely modify the risk of post-op AF of abdominally obese patients planned for CABG surgery (Kaireviciute et al. 2010).

Increased IL-6 levels following surgery are also associated with poorer outcome. Sanders et al. (Sanders et al. 2009) sought to determine whether levels of IL-6 and functional common variants of the IL6 gene at baseline and at 6 hours and 24 hours post CABG are associated with post-operative outcome. They found that IL-6 levels pre- and at 24 hours following CABG were greater (>148pg/ml at 24 hours) in post-operative complications (POC). Pre- IL-6 levels independently predicted POC (OR 1.4, 95 percent CI 1.1–1.7, p = 0.008) outcomes. IL6 gene variants didn't predict outcome. Post-operative complications included post-operative bleeding (>500ml and/or returned to theatre), prolonged respiratory or circulatory support (ventilation time >12 h, inotropic support >1day, IABP), prolonged ICU stay (>2 days), renal failure requiring hemofiltration, infections (wound, respiratory and urinary) documented as such within medical notes and requiring antibiotic therapy, pyrexia (>38.0°C), neurological deficit and ventricular dysrhythmia. Atrial fibrillation (AF) is very common following cardiac surgery (25 percent to 40 percent of patients) and thus, was not characterized as a post-operative complication. However, results remain the same if it had been included (Ishida et al. 2006).

Pre-operative therapeutic anti-inflammatory strategies may be more valuable given the lack of association of 6-hour post-operative inflammatory response with post-operative complications. As these results suggest, if preoperative assessment indicates a higher risk for post-surgery complication, greater emphasis on pre-, intra- and post-operative anti-inflammatory treatment to reduce the risk and favourably influence post-operative outcome would be indicated. Pre-operative assessment of IL-6 levels could provide additional, and modifiable, risk assessment information to aid post-operative patient outcome (Mathew et al. 2007).

With respect to AF it has been shown that increased levels of IL-6 immediately after off-pump CABG independently predict postoperative

AF (odds ratio 7.63; 95 percent CI, 1.06-54.9; P=0.04)10 and that raised intracardiac levels of IL-6 in samples from the right atrial appendage, left atrium and left atrial appendage taken before CABG were also associated with postoperative AF (p=0.007) (Kaireviciute et al. 2010).

In CABG patients, Dacey et al. (Dacey et al. 2003) showed that patients with a preoperative WBC count of ≥12 x109/L had an adjusted mortality rate threefold higher than those with a WBC less than 6x109/L (4.8% v 1.7 percent). Moreover, an elevated preoperative WBC count was significantly associated with increasing rates of perioperative strokes and the need for intra-aortic balloon pump counterpulsation, suggestive of an interaction among WBC counts, endothelial dysfunction, and hypercoagulability (Dacey et al. 2003).

To better understand the relationship between humoral and cellular markers of inflammation and postoperative atrial fibrillation, Fontes et al. studied sixty adult patients >60 years of age presenting for elective coronary artery bypass surgery with cardiopulmonary bypass. Serial measurements of WBC and hsCRP were performed pre- and postoperatively. Preoperative WBC values were higher for patients who developed AF and according to stepwise logistic regression, it was the sole independent predictor of postoperative AF (odds ratio =6.7; 95 percent confidence interval, 1.6–29.0; p = 0.01). A twofold higher preoperative WBC was associated with a nearly sevenfold higher risk of developing AF, and WBC >7 x 109/L was associated with a nearly fourfold higher risk of AF (odds ratio = 3.8, p = 0.03). In conclusion, preoperative but not postoperative leukocytosis was associated with AF after conventional CABG surgery independent of CRP. This finding requires confirmation in a larger study. The preoperative WBC count is routinely obtained, adds no further costs, and is easy to combine with other preoperative risk indices of AF (Fontes et al. 2009).

Pentraxin 3 (PTX3), a marker of vascular inflammation that belongs to the same family as C-reactive protein, was recently found to be a predictor of short-term functional recovery and 1-year major adverse cardiovascular event in patients undergoing rehabilitation after cardiac surgery, regardless of clinical and instrumental parameters. Ferratini et al. evaluated several circulating markers of cardiac stress, inflammation, and endothelial function to investigate their ability to predict short-term functional recovery and long-term clinical outcome in heart surgery patients undergoing inpatient rehabilitation. Apart from PTX3, the other biomarkers measured were brain natriuretic peptide, high-sensitivity cardiac troponin-T, high-sensitivity C-reactive protein, creatine kinase and myoglobin. PTX3 showed the closest association with a six minute walk distance (P =.01) and was the only predictor of major adverse cardiovascular event, also in the subgroup of CABG patients (OR [95 percent CI] = 1.14 [1.03-1.27]; P =.015) (Ferratini et al. 2012).

New technologies to generate high-dimensional data provide unprecedented opportunities for unbiased identification of biomarkers that can be used to optimize pre-operative planning, with the goal of avoiding costly post-operative complications and prolonged hospitalization. To identify such markers, Hägg et al. (Hagg et al. 2011) studied the global gene expression profiles of three organs central to the metabolic and inflammatory homeostasis isolated from coronary artery disease (CAD) patients during coronary artery bypass grafting (CABG) surgery. A total of 198 whole-genome expression profiles of liver, skeletal muscle and visceral fat from 66 CAD patients of the Stockholm Atherosclerosis Gene Expression (STAGE) cohort were analyzed. Of ~50,000 mRNAs measured in each patient, the mRNA levels of the anti-inflammatory gene, dual-specificity phosphatase-1 (DUSP1) correlated independently with post-operative stay, discriminating patients with normal ($\leq$8 days) from those with prolonged (>8 days) hospitalization ($p<0.004$). To validate DUSP1 as a marker of risk for post-operative complications, the researchers prospectively analyzed 181 patients undergoing CABG for DUSP1 protein levels in pre-operative blood samples. The pre-operative plasma levels of DUSP1 clearly discriminated patients with normal from those with prolonged hospitalization ($p=2\times10-13$; odds ratio = 5.1, $p<0.0001$; receiver operating characteristic area under the curve = 0.80). Taken together, these results indicate that blood levels of the anti-inflammatory protein DUSP1 can be used as a biomarker for post-operative complications leading to prolonged hospitalization after CABG, and therefore merit further testing in longitudinal studies of patients eligible for CABG (Girerd et al. 2009).

DUSP1 is a highly conserved gene that encodes a protein expressed in the nucleus. As its name implies, DUSP1 has a dual specificity for tyrosine and threonine and inactivates mitogen-activated protein kinases (MAPKs). MAPKs are part of a phosphorylation signaling pathway that induces a cascade of genes that trigger innate immune responses. Regulation of these genes at the transcriptional and post-transcriptional level by DUSP1 affects cell proliferation and differentiation as well as apoptosis. The MAPK pathway has a negative feedback loop that upregulates DUSP1 in response to pro-inflammatory stimuli, such as oxidative stress and cytokine secretion.

## Other Biomarkers

There has been constant research activity on novel biomarkers and their implementation for the screening of patients who have undergone CABG, as illustrated in Table 3. There is a significant association between increased serum uric acid levels, serum gamma-glutamyltransferase GGT levels, and SVG disease in patients undergoing CABG (Tavil et al. 2008; Ulus et al. 2011).

**Table 3.** Biomarkers and their use or association with CABG.

| Biomarkers | Use-association with CABG |
|---|---|
| HeartFatty-Acid–Binding Proteins | Perioperative myocardial ischemia and injury |
| Myoglobin | Perioperative myocardial injury |
| Ischemia-modified albumin | Myocardial ischemia-reperfusion injury, indicator of the cardioprotective effect of N-Acetylo-cysteine in CABG |
| Glycogen Phosphorylase | Perioperative myocardial injury |
| NSE | Brain injuries after CABG, high predictive value with respect to the early neuropsychological and neuropsychiatric outcome after cardiac surgery |
| S-100b | Brain injury during CABG,assessment of experimental neuroprotectant treatment strategies and comparison between different surgical techniques |
| Tau protein | Brain injury during CABG |
| Lipids (LDL, triglycerides, HDL, Lp(a)) | Predictors of unfavorable events after CABG |
| Paraoxonase-1 | SV graft occlusion |
| Metabolic profiles (carnitines, ketonerelated metabolites) | Predictors of adverse outcomes after CABG |
| Glucose | Predictive of mortality and morbidity among patients undergoing CABG |
| Hemoglobin A1c | Long-term survival after CABG, occurrence of atrial fibrillation (AF) after isolated OPCAB |
| Thyroids hormones | Decrease in thyroid hormones in the post-CPB favours the occurrence of global myocardial dysfunction and arrhythmias, especially AF |
| Adipokines (Resistin, leptin, adiponectin) | High risk for AF after CABG, cardiac risk factors after CABG |
| Testosterone | High probability of developing mobility limitations after CABG |
| Cortisol | Risk of postoperative delirium and neuropsychological deficits after CABG |
| Von Willebrand factor | Endothelial injury during CABG, predictive role of vWf regarding the postoperative AF is controversial |
| Circulating endothelial cells | Endothelial activation and damage during CABG, comparison of different CABG techniques |
| Endothelial nitric oxide synthase | Endothelial integrity and damage during different techniques of preparation and processing of vein or artery grafts |
| Endothelin | Comparison of different CABG techniques, vein graft occlusion, therapeutic potential of endothelin receptor antagonists (ERA) |
| Haemoglobin | Predictor of postoperative renal and central nervous system outcome, preoperative risk assessment and subsequent therapeutic strategies |
| Homocysteine | Morbidity and mortality after CABG, atherosclerotic lesions of SV grafts after CABG |
| VEGF | Polymorphism might be a prognostic factor of an adverse postoperative course, pathogenesis of post-CABG PEs |
| SFlt-1 | Occurrence of post-CABG complication |

*Table 3. contd....*

*Table 3. contd.*

| Biomarkers | Use-association with CABG |
|---|---|
| Syndecan-1 | Reperfusion injury during CABG |
| Asymmetric dimethylarginine (ADMA), symmetric dimethylarginine (SDMA), L-arginine, and L-arginine/ADMA ratio | Early ischemia-reperfusion injury caused by CABG |
| Glutathione peroxidase (GPX), superoxide dismutase (SOD) | Ischemia-reperfusion injury caused by CABG |
| Glutathione, catalase activity | Ischemia-reperfusion injury caused by CABG |
| Urinary isoprostane iPF2alpha-III, free malondialdehyde | Perioperative oxidative stress |
| Plasma free 15-F2t-isoprostane fn | Assessment of cardiac function after CABG |
| Metalloproteinases, tissue inhibitors of metalloproteinases (MMP-2, MMP-9, TIMP-2) | State of the left ventricle after CABG, graft patency, disorders of brain Mg homeostasis and postoperative neuropsychological disorders, postoperative atrial fibrillation |
| FAS LG, FAS promoter and TP53 | Polymorphisms indicate an increased risk of coronary artery disease (CAD) after CABG |
| Bcl-2, p53, Bax and CPP32 | Bypass grafts occlusion |
| Heat shock proteins (HSP-70, HSP60, anti-HSP65 antibody) | Incidence of postoperative AF, structural markers of 'myocardial stunning' after CABG, comparison of different CABG techniques |
| Thrombin | Myocardial damage and impaired hemodynamic recovery, early graft occlusion |
| Protein C | Hemodynamic recovery postoperatively |
| Protein S, prothrombin fragment 1.2 | Hemodynamic profile postoperatively |
| Platelets function | Prediction of both bleeding and thrombosis after CABG |
| Tissue plasminogen activator-plasminogen activator inhibitor-1 (PAI-1) | Risk of graft occlusion after CABG, comparison of different CABG techniques, predictive value for early thrombosis after CABG, assessment of the efficacy of different therapeutics interventions during and after CABG |
| Uric acid | Saphenous vein graft (SVG) disease |
| Gamma-glutamyltransferase | Saphenous vein graft (SVG) disease |
| Tumor marker CA 125 | Unknown clinical significance |
| Osteopontin | Vascular injury |
| ST2 | Immune suppression during CABG |
| sCD163 scavenger molecule | Immune suppression during CABG |
| Serum magnesium levels | Perioperative and postoperative myocardial infarction in patients undergoing CABG |
| Osteoprotegerin | Perioperative and postoperative myocardial infarction in patients undergoing CABG |
| Myocardial G protein-coupled receptor kinase-2 (GRK2) activity | Need for inotropic support in patients with severe ventricular dysfunction during and after CABG |
| Clara-cell 16 | Monitor in alveolar capillary damage during CABG |

Biomarkers which are influenced by CABG are tumor marker CA 125, total proteins and osteopontin (OPN), a multifunctional protein associated with vascular injury (Battaloglu et al. 2002; Gabriel et al. 2011; Sbarouni et al. 2012). Furthermore, CABG induces a massive long-lasting secretion of ST2 and sCD163 scavenger molecule for hemoglobin, proteins related to immune suppression (Kolackova et al. 2009; Szerafin et al. 2009). Low serum magnesium levels and osteoprotegerin (OPG), an important regulator in the genesis of arteriosclerosis and bone formation, might be related with perioperative and postoperative myocardial infarction in patients undergoing CABG (Booth et al. 2003; Hermann-Arnhof et al. 2006). Increased myocardial G protein-coupled receptor kinase-2 (GRK2) activity, a serine-threonine kinase which phosphorylates and desensitizes agonist-occupied beta-AReceptors, is related with greater need for inotropic support in patients with severe ventricular dysfunction during and after CABG (Bulcao et al. 2008). One study finally refers to the value of the lung-specific biomarkers Clara-cell 16 (CC16) as marker in monitor in alveolar capillary damage during CABG (van Boven et al. 2005).

Matrix metalloproteinases (MMPs) are zinc-dependent endopeptidases; they are capable of degrading all kinds of extracellular matrix proteins, but also can process a number of bioactive molecules. MMPs are thought to play a major role in cell behaviors such as cell proliferation, migration (adhesion/dispersion), differentiation, angiogenesis, apoptosis, and host defence. Kinetics of MMPs concentrations in serum after cardiac surgery appears to depend on many factors. Plasma MMP-2 concentrations reflect the state of the left ventricle, and changes in plasma MMP-2 and TIMP-2 concentrations during CPB may play an important role in LV remodeling after cardiac surgery (Watanabe et al. 2002). MMPs and tissue inhibitors of metalloproteinases (TIMPs) regulate vascular structure and may have important influence on graft patency. MMP-2 levels and activity are significantly more abundant in the SV compared to the IMA. These differences may contribute to the early pathologic remodeling of the SV vs IMA conduit following CABG surgery (Anstadt et al. 2004). The plasma MMP-9 concentration has been correlated with disorders of brain Mg homeostasis and with postoperative neuropsychological disorders PNPDs (Dabrowski et al. 2010). Intracardiac levels of MMP-9 in the left atrial appendage (LAA) sample have been also associated with AF (Kaireviciute et al. 2010).

Progression of coronary artery disease (CAD) after CABG is frequent and may lead to recurrent symptoms. Various data indicate that apoptosis is the main event occurring during development and progression of atherosclerotic plaque. Plaque vascular smooth muscle cells (VSMCs) are more sensitive than regular VSMCs to TP53-mediated apoptosis. Patients presenting with polymorphisms of FAS LG, FAS promoter and TP53 have an increased risk of CAD progression, as they have a higher rate of re-interventions (Beiras-Fernandez et al. 2011). Moreover, observations of the expression of both anti-apoptotic (Bcl-2) and pro-apoptotic proteins (p53, Bax and CPP32) indicate that in atherosclerotic areas of SV bypass grafts, cell death occurs mainly by necrosis, while in nonatherosclerotic areas, cell death occurs by both necrosis and apoptosis (Wang et al. 2001).

Heat shock proteins (HSPs) are a family of chaperone proteins which assist in preservation of cellular integrity by maintaining proteins in their correctly folded state. High circulating HSP-70 levels are connected with higher incidence of postoperative AF after open heart surgery (Pizon et al. 2006). Intracellular, but not serum, HSP70 level is negatively correlated with postoperative AF. This suggests a cardioprotective and an antiarrhythmic role for intracellular HSP70 (Mandal et al. 2005). Heat shock protein expression is not, however, correlated with the onset of AF and the resistance to administered medications (St. Rammos et al. 2002). Serum HSP60 and anti-HSP65 antibody levels may also be markers for subsequent development of AF (Mandal et al. 2004; Oc et al. 2007). Increased HSP-70i expression, as well as a deranged actin cross-striation pattern, might be structural markers to determine 'myocardial stunning' after CABG and cold crystalloid cardioplegia. Serum levels of HSP are increased in patients undergoing on-pump CABG operation as compared with off-pump CABG technique (Szerafin et al. 2008).

Myocardial ischemia secondary to CPB-CABG and OP-CABG consists of a potent stimulator of vascular endothelial growth factor production, which may have implications for graft endothelialization and cardiovascular hemodynamics post-operatively (Burton et al. 2000). The vascular endothelial growth factor (VEGF) gene +405 G>C polymorphism might be a prognostic factor of an adverse postoperative course in patients undergoing CABG (Pastuszczak et al. 2009). In post-CABG pleural effusions (PE), preferential local production of VEGF in the pleural cavity is most commonly observed a month or later after surgery. A recent study suggests that VEGF may be involved in the pathogenesis of post-CABG Pes (Kalomenidis et al. 2007).

SFlt-1, the soluble form of VEGF, is released during CABG with extracorporeal circulation (ECC). It has been suggested that sFlt-1 production, by neutralizing VEGF might play a role in the occurrence of post-CABG complication (Denizot et al. 2007).

# References

Akila, B. D'Souza, P. Vishwanath, and V. D'Souza. 2007. Oxidative injury and antioxidants in coronary artery bypass graft surgery: off-pump CABG significantly reduces oxidative stress. Clin. Chim. Acta. 375: 147–152.

Amin, A.P., E. Mukhopadhyay, S. Napan, M. Mamtani, R.F. Kelly, and H. Kulkarni. 2009. Value of early cardiac troponin I to predict long-term adverse events after coronary artery bypass graft surgery in patients presenting with acute coronary syndromes. Clin. Cardiol. 32: 386–392.

Anderson, R.J., M. O'Brien, S. MaWhinney, C.B. VillaNueva, T.E. Moritz, G.K. Sethi, W.G. Henderson, K.E. Hammermeister, F.L. Grover, and A.L. Shroyer. 1999. Renal failure predisposes patients to adverse outcome after coronary artery bypass surgery. VA Cooperative Study #5. Kidney Int. 55: 1057–1062.

Anstadt, M.P., D.L. Franga, V. Portik-Dobos, A. Pennathur, M. Bannan, K. Mawulawde, and A. Ergul. 2004. Native matrix metalloproteinase characteristics may influence early stenosis of venous versus arterial coronary artery bypass grafting conduits. Chest 125: 1853–1858.

Arnaoutakis, G.J., A. Bihorac, T.D. Martin, P. J. Hess, Jr., C.T. Klodell, A.A. Ejaz, C. Garvan, C.G. Tribble, and T.M. Beaver. 2007. RIFLE criteria for acute kidney injury in aortic arch surgery. J. Thorac. Cardiovasc. Surg. 134: 1554–1560; discussion 1560–1551.

Aronson, S., M.L. Fontes, Y. Miao, and D.T. Mangano. 2007. Risk index for perioperative renal dysfunction/failure: critical dependence on pulse pressure hypertension. Circulation 115: 733–742.

Ascione, R., C.T. Lloyd, M.J. Underwood, W.J. Gomes, and G.D. Angelini. 1999. On-pump versus off-pump coronary revascularization: evaluation of renal function. Ann. Thorac. Surg. 68: 493–498.

Asimakopoulos, G., A.P. Karagounis, O. Valencia, N. Alexander, M. Howlader, M.A. Sarsam, and V. Chandrasekaran. 2005. Renal function after cardiac surgery off- versus on-pump coronary artery bypass: analysis using the Cockroft-Gault formula for estimating creatinine clearance. Ann. Thorac. Surg. 79: 2024–2031.

Bagshaw, S.M., C. George, and R. Bellomo. 2008. A comparison of the RIFLE and AKIN criteria for acute kidney injury in critically ill patients. Nephrol. Dial. Transplant. 23: 1569–1574.

Bahar, I., A. Akgul, M.A. Ozatik, K.M. Vural, A.E. Demirbag, M. Boran, and O. Tasdemir. 2005. Acute renal failure following open heart surgery: risk factors and prognosis. Perfusion 20: 317–322.

Bansal, S. 2012. Post-surgical acute kidney injury. Clinical Queries: Nephrology 101: 50–57.

Battaloglu, B., E. Kaya, N. Erdil, V. Nisanoglu, F. Kosar, B. Ozgur, B. Yildirim, and H. Karagoz. 2002. Does cardiopulmonary bypass alter plasma level of tumor markers? CA 125 and carcinoembryonic antigen. Thorac. Cardiovasc. Surg. 50: 201–203.

Beiras-Fernandez, A., M.K. Angele, C. Koutang, P. Lohse, B. Reichart, and S. Eifert. 2011. Genetic polymorphisms of TP53 and FAS promoter modulate the progression of coronary artery disease after coronary artery bypass grafting: a gender-specific view. Inflamm. Res. 60: 439–445.

Bellomo, R., C. Ronco, J.A. Kellum, R.L. Mehta, and P. Palevsky. 2004. Acute renal failure-definition, outcome measures, animal models, fluid therapy and information technology needs: the Second International Consensus Conference of the Acute Dialysis Quality Initiative (ADQI) Group. Crit. Care 8: R204–212.

Benedetto, U., E. Angeloni, R. Luciani, S. Refice, M. Stefanelli, C. Comito, A. Roscitano, and R. Sinatra. 2010. Acute kidney injury after coronary artery bypass grafting: does rhabdomyolysis play a role? J. Thorac. Cardiovasc. Surg. 140: 464–470.

Bhamidipati, C.M., D.J. LaPar, G.J. Stukenborg, C.C. Morrison, J.A. Kern, I.L. Kron, and G. Ailawadi. 2011. Superiority of moderate control of hyperglycemia to tight control in patients undergoing coronary artery bypass grafting. J. Thorac. Cardiovasc. Surg. 141: 543–551.

Birdi, I., G.D. Angelini, and A.J. Bryan. 1997. Biochemical markers of myocardial injury during cardiac operations. Ann. Thorac. Surg. 63: 879–884.

Bonacchi, M., E. Prifti, M. Maiani, F. Bartolozzi, M. Di Eusanio, and M. Leacche. 2006. Does off-pump coronary revascularization reduce the release of the cerebral markers, S-100beta and NSE? Heart Lung Circ. 15: 314–319.

Bondiou, M.T., R. Bourbouze, M. Bernard, F. Percheron, N. Perez-Gonzalez, and J.A. Cabezas. 1985. Inhibition of A and B N-acetyl-beta-D-glucosaminidase urinary isoenzymes by urea. Clin. Chim. Acta. 149: 67–73.

Booth, J.V., B. Phillips-Bute, C.B. McCants, M.V. Podgoreanu, P.K. Smith, J.P. Mathew, and M.F. Newman. 2003. Low serum magnesium level predicts major adverse cardiac events after coronary artery bypass graft surgery. Am. Heart J. 145: 1108–1113.

Bove, T., M.G. Calabro, G. Landoni, G. Aletti, G. Marino, G. Crescenzi, C. Rosica, and A. Zangrillo. 2004. The incidence and risk of acute renal failure after cardiac surgery. J. Cardiothorac. Vasc. Anesth. 18: 442–445.

Bronden, B., A. Eyjolfsson, S. Blomquist, A. Dardashti, P. Ederoth, and H. Bjursten. 2011. Evaluation of cystatin C with iohexol clearance in cardiac surgery. Acta. Anaesthesiol. Scand. 55: 196–202.

Brown, J.R., R.S. Kramer, S.G. Coca, and C.R. Parikh. 2010. Duration of acute kidney injury impacts long-term survival after cardiac surgery. Ann. Thorac. Surg. 90: 1142–1148.

Brown, J.R., R.P. Cochran, B.J. Leavitt, L.J. Dacey, C.S. Ross, T.A. MacKenzie, K.S. Kunzelman, R.S. Kramer, F. Hernandez, Jr., R.E. Helm, B.M. Westbrook, R.F. Dunton, D.J. Malenka, and G.T. O'Connor. 2007. Multivariable prediction of renal insufficiency developing after cardiac surgery. Circulation 116: I139–143.

Bulcao, C.F., P.K. Pandalai, K.M. D'Souza, W.H. Merrill, and S.A. Akhter. 2008. Uncoupling of myocardial beta-adrenergic receptor signaling during coronary artery bypass grafting: the role of GRK2. Ann. Thorac. Surg. 86: 1189–1194.

Burton, P.B., V.J. Owen, S. Hafizi, P.J. Barton, G. Carr-White, T. Koh, A. De Souza, M.H. Yacoub, and J.R. Pepper. 2000. Vascular endothelial growth factor release following coronary artery bypass surgery: extracorporeal circulation versus 'beating heart' surgery. Eur. Heart J. 21: 1708–1713.

Cagirci, G., S. Cay, O. Karakurt, N. Yazihan, C. Aydin, S. Acikel, M. Dogan, H. Kilic, S. Topaloglu, D. Aras, and R. Akdemir. 2009. Association between paraoxonase activity and late saphenous vein graft occlusion in patients with coronary artery bypass grafting. Kardiol. Pol. 67: 1063–1068.

Carrier, M., M. Pellerin, L.P. Perrault, B.C. Solymoss, and L.C. Pelletier. 2000. Troponin levels in patients with myocardial infarction after coronary artery bypass grafting. Ann. Thorac. Surg. 69: 435–440.

Castellheim, A., T.N. Hoel, V. Videm, E. Fosse, A. Pharo, J.L. Svennevig, A.E. Fiane, and T.E. Mollnes. 2008. Biomarker profile in off-pump and on-pump coronary artery bypass grafting surgery in low-risk patients. Ann. Thorac. Surg. 85: 1994–2002.

Cata, J.P., B. Abdelmalak, and E. Farag. 2011. Neurological biomarkers in the perioperative period. Br. J. Anaesth. 107: 844–858.

Cavalca, V., E. Sisillo, F. Veglia, E. Tremoli, G. Cighetti, L. Salvi, A. Sola, L. Mussoni, P. Biglioli, G. Folco, A. Sala, and A. Parolari. 2006. Isoprostanes and oxidative stress in off-pump and on-pump coronary bypass surgery. Ann. Thorac. Surg. 81: 562–567.

Cerillo, A.G., S. Bevilacqua, S. Storti, M. Mariani, E. Kallushi, A. Ripoli, A. Clerico, and M. Glauber. 2003. Free triiodothyronine: a novel predictor of postoperative atrial fibrillation. Eur. J. Cardiothorac. Surg. 24: 487–492.

Chandrasena, L.G., H. Peiris, and H.D. Waikar. 2009. Biochemical changes associated with reperfusion after off-pump and on-pump coronary artery bypass graft surgery. Ann Clin. Lab. Sci. 39: 372–377.

Chertow, G.M., J.M. Lazarus, C.L. Christiansen, E.F. Cook, K.E. Hammermeister, F. Grover, and J. Daley. 1997. Preoperative renal risk stratification. Circulation 95: 878–884.

Chertow, G.M., E. Burdick, M. Honour, J.V. Bonventre, and D.W. Bates. 2005. Acute kidney injury, mortality, length of stay, and costs in hospitalized patients. J. Am. Soc. Nephrol. 16: 3365–3370.

Chonchol, M.B., V. Aboyans, P. Lacroix, G. Smits, T. Berl, and M. Laskar. 2007. Long-term outcomes after coronary artery bypass grafting: preoperative kidney function is prognostic. J. Thorac. Cardiovasc. Surg. 134: 683–689.

Cihan, H.B., O. Gulcan, A. Hazar, R. Turkoz, and E. Olmez. 2004. Determination of the pericardial to serum myoglobin ratio for the early diagnosis of perioperative myocardial infarction after coronary artery bypass grafting. Pharmacol. Res. 50: 247–251.

Coca, S.G., B. Yusuf, M.G. Shlipak, A.X. Garg, and C.R. Parikh. 2009. Long-term risk of mortality and other adverse outcomes after acute kidney injury: a systematic review and meta-analysis. Am. J. Kidney Dis. 53: 961–973.

Cooper, W.A., S.M. O'Brien, V.H. Thourani, R.A. Guyton, C.R. Bridges, L.A. Szczech, R. Petersen, and E.D. Peterson. 2006. Impact of renal dysfunction on outcomes of coronary artery bypass surgery: results from the Society of Thoracic Surgeons National Adult Cardiac Database. Circulation 113: 1063–1070.

Cossu, A.P., S. Suelzu, P. Piu, M. Orecchioni, G. Bazzu, G. Padua, M. Portoghese, P.A. Serra, and G. Susini. 2012. Do on- and off-pump coronary bypass surgery differently affect perioperative peripheral tissue metabolism? Minerva Anestesiol 78: 26–33.

Crescenzi, G., G. Landoni, E. Bignami, I. Belloni, C. Biselli, C. Rosica, F. Guarracino, G. Marino, and A. Zangrillo. 2009. N-terminal B-natriuretic Peptide after coronary artery bypass graft surgery. J. Cardiothorac. Vasc. Anest.h 23: 147–150.

Cziraki, A., Z. Ajtay, A. Nemeth, Z. Lenkey, E. Sulyok, S. Szabados, N. Alotti, J. Martens-Lobenhoffer, C. Szabo, and S.M. Bode-Boger. 2011. Effects of coronary revascularization with or without cardiopulmonary bypass on plasma levels of asymmetric dimethylarginine. Coron. Artery Dis. 22: 245–252.

Dabrowski, W., Z. Rzecki, M. Czajkowski, J. Pilat, J. Biernacka, E. Kotlinska, K. Pasternak, K. Stazka, M. Sztanke, and K. Sztanke. 2010. Plasma matrix metalloproteinase 9 correlates with disorders of brain magnesium homeostasis in patients undergoing coronary artery bypass surgery. Magnes. Res. 23: 169–179.

Dacey, L.J., J. DeSimone, J.H. Braxton, B.J. Leavitt, S.J. Lahey, J.D. Klemperer, B.M. Westbrook, E.M. Olmstead, and G.T. O'Connor. 2003. Preoperative white blood cell count and mortality and morbidity after coronary artery bypass grafting. Ann. Thorac. Surg. 76: 760–764.

Dashwood, M.R. 2009. Endothelin-1 and vein graft occlusion in patients undergoing bypass surgery. Eur. J. Clin. Invest. 39 Suppl. 2: 78–87.

Dashwood, M.R., K. Savage, J.C. Tsui, A. Dooley, S.G. Shaw, M.S. Fernandez Alfonso, L. Bodin, and D.S. Souza. 2009. Retaining perivascular tissue of human saphenous vein grafts protects against surgical and distension-induced damage and preserves endothelial nitric oxide synthase and nitric oxide synthase activity. J. Thorac. Cardiovasc. Surg. 138: 334–340.

Dashwood, M.R., A. Dooley, X. Shi-Wen, D.J. Abraham, M. Dreifaldt, and D.S. Souza. 2011. Perivascular fat-derived leptin: a potential role in improved vein graft performance in coronary artery bypass surgery. Interact. Cardiovasc. Thorac. Surg. 12: 170–173.

Denizot, Y., A. Leguyader, E. Cornu, M. Laskar, I. Orsel, C. Vincent, and N. Nathan. 2007. Release of soluble vascular endothelial growth factor receptor-1 (sFlt-1) during coronary artery bypass surgery. J. Cardiothorac. Surg. 2: 38.

Desai, S.P., L.L. Henry, S.D. Holmes, S.L. Hunt, C.T. Martin, S. Hebsur, and N. Ad. 2012. Strict versus liberal target range for perioperative glucose in patients undergoing coronary artery bypass grafting: a prospective randomized controlled trial. J. Thorac. Cardiovasc. Surg. 143: 318–325.

Despotis, G.J. and J.H. Joist. 1999. Anticoagulation and anticoagulation reversal with cardiac surgery involving cardiopulmonary bypass: an update. J. Cardiothorac. Vasc. Anesth. 13: 18–29; discussion 36–17.

Devbhandari, M.P., A.J. Duncan, A.D. Grayson, B.M. Fabri, D.J. Keenan, B. Bridgewater, M.T. Jones, and J. Au. 2006. Effect of risk-adjusted, non-dialysis-dependent renal dysfunction on mortality and morbidity following coronary artery bypass surgery: a multi-centre study. Eur. J. Cardiothorac. Surg. 29: 964–970.

Duncan, L., J. Heathcote, O. Djurdjev, and A. Levin. 2001. Screening for renal disease using serum creatinine: who are we missing? Nephrol. Dial. Transplant. 16: 1042–1046.

Durmaz, I., S. Buket, Y. Atay, T. Yagdi, M. Ozbaran, M. Boga, I. Alat, A. Guzelant, and S. Basarir. 1999. Cardiac surgery with cardiopulmonary bypass in patients with chronic renal failure. J. Thorac. Cardiovasc. Surg. 118: 306–315.

Ek-Von Mentzer, B.A., F. Zhang, and J.A. Hamilton. 2001. Binding of 13-HODE and 15-HETE to phospholipid bilayers, albumin, and intracellular fatty acid binding proteins. implications for transmembrane and intracellular transport and for protection from lipid peroxidation. J. Biol. Chem. 276: 15575–15580.

Eyileten, Z., M.I. Yilmaz, K. Kaya, A.R. Akar, D. Kahraman, S. Bingol, A. Uysalel, and U. Ozyurda. 2007. Coronary artery bypass grafting ameliorates the decreased plasma adiponectin level in atherosclerotic patients. Tohoku J. Exp. Med. 213: 71–77.

Ferguson, M.A., V.S. Vaidya, S.S. Waikar, F.B. Collings, K.E. Sunderland, C.J. Gioules, and J.V. Bonventre. 2010. Urinary liver-type fatty acid-binding protein predicts adverse outcomes in acute kidney injury. Kidney Int. 77: 708–714.

Ferratini, M., V. Ripamonti, S. Masson, P. Grati, V. Racca, I. Cuccovillo, E. Raimondi, S. Capomolla, C. Macchi, P. Coruzzi, T. Vago, M. Calvo, A. Mantovani, and R. Latini. 2012. Pentraxin-3 predicts functional recovery and 1-year major adverse cardiovascular events after rehabilitation of cardiac surgery patients. J. Cardiopulm. Rehabil. Prev. 32: 17–24.

Fontes, M.L., D. Amar, A. Kulak, K. Koval, H. Zhang, W. Shi, and H. Thaler. 2009. Increased preoperative white blood cell count predicts postoperative atrial fibrillation after coronary artery bypass surgery. J. Cardiothorac. Vasc. Anesth. 23: 484–487.

Foody, J.M., F. D. Ferdinand, G.L. Pearce, B.W. Lytle, D.M. Cosgrove, and D.L. Sprecher. 2000. HDL cholesterol level predicts survival in men after coronary artery bypass graft surgery: 20-year experience from The Cleveland Clinic Foundation. Circulation 102: III90–94.

Fox, A.A., E.R. Marcantonio, C.D. Collard, M. Thoma, T.E. Perry, S.K. Shernan, J.D. Muehlschlegel, and S.C. Body. 2011. Increased peak postoperative B-type natriuretic peptide predicts decreased longer-term physical function after primary coronary artery bypass graft surgery. Anesthesiology 114: 807–816.

Frioud, A., S. Comte-Perret, S. Nguyen, M.M. Berger, P. Ruchat, and J. Ruiz. 2010. Blood glucose level on postoperative day 1 is predictive of adverse outcomes after cardiovascular surgery. Diabetes Metab. 36: 36–42.

Gabriel, E.A., R.F. Locali, P.K. Matsuoka, T. Cherbo, and E. Buffolo. 2011. On-pump coronary artery bypass graft surgery: biochemical, hormonal and cellular features. Rev. Bras. Cir. Cardiovasc.26: 525–531.

Girerd, N., P. Pibarot, D. Fournier, P. Daleau, P. Voisine, G. O'Hara, J.P. Despres, and P. Mathieu. 2009. Middle-aged men with increased waist circumference and elevated C-reactive protein level are at higher risk for postoperative atrial fibrillation following coronary artery bypass grafting surgery. Eur. Heart J. 30: 1270–1278.

Gungor, H., M.F. Ayik, B. Kirilmaz, S. Ertugay, I. Gul, B.S. Yildiz, S. Nalbantgil, and M. Zoghi. 2011. Serum resistin level: as a predictor of atrial fibrillation after coronary artery bypass graft surgery. Coron. Artery Dis. 22: 484–490.

Haase-Fielitz, A., R. Bellomo, P. Devarajan, M. Bennett, D. Story, G. Matalanis, U. Frei, D. Dragun, and M. Haase. 2009a. The predictive performance of plasma neutrophil gelatinase-associated lipocalin (NGAL) increases with grade of acute kidney injury. Nephrol. Dial. Transplant.24: 3349–3354.

Haase-Fielitz, A., R. Bellomo, P. Devarajan, D. Story, G. Matalanis, D. Dragun, and M. Haase. 2009b. Novel and conventional serum biomarkers predicting acute kidney injury in adult cardiac surgery—a prospective cohort study. Crit Care Med. 37: 553–560.

Haase, M., R. Bellomo, P. Devarajan, Q. Ma, M.R. Bennett, M. Mockel, G. Matalanis, D. Dragun, and A. Haase-Fielitz. 2009c. Novel biomarkers early predict the severity of acute kidney injury after cardiac surgery in adults. Ann. Thorac. Surg. 88: 124–130.

Haase, M., R. Bellomo, P. Devarajan, P. Schlattmann, and A. Haase-Fielitz. 2009d. Accuracy of neutrophil gelatinase-associated lipocalin (NGAL) in diagnosis and prognosis in acute kidney injury: a systematic review and meta-analysis. Am. J. Kidney Dis. 54: 1012–1024.

Haase, M., R. Bellomo, D. Story, P. Davenport, and A. Haase-Fielitz. 2008. Urinary interleukin-18 does not predict acute kidney injury after adult cardiac surgery: a prospective observational cohort study. Crit. Care 12: R96.

Haase, M., R. Bellomo, and A. Haase-Fielitz. 2011a. Neutrophil gelatinase-associated lipocalin: a superior biomarker for detection of subclinical acute kidney injury and poor prognosis. Biomark. Med. 5: 415–417.

Haase, M., D.A. Story and A. Haase-Fielitz. 2011b. Renal injury in the elderly: diagnosis, biomarkers and prevention. Best Pract. Res. Clin. Anaesthesiol. 25: 401–412.

Haase, M., P. Devarajan, A. Haase-Fielitz, R. Bellomo, D.N. Cruz, G. Wagener, C.D. Krawczeski, J.L. Koyner, P. Murray, M. Zappitelli, S.L. Goldstein, K. Makris, C. Ronco, J. Martensson, C.R. Martling, P. Venge, E. Siew, L.B. Ware, T.A. Ikizler, and P.R. Mertens. 2011c. The outcome of neutrophil gelatinase-associated lipocalin-positive subclinical acute kidney injury: a multicenter pooled analysis of prospective studies. J. Am. Coll. Cardiol. 57: 1752–1761.

Hagg, S., T. Alserius, P. Noori, A. Ruusalepp, T. Ivert, J. Tegner, J. Bjorkegren, and J. Skogsberg. 2011. Blood levels of dual-specificity phosphatase-1 independently predict risk for post-operative morbidities causing prolonged hospitalization after coronary artery bypass grafting. Int. J. Mol. Med. 27: 851–857.

Halonen, J., P. Halonen, O. Jarvinen, P. Taskinen, T. Auvinen, M. Tarkka, M. Hippelainen, T. Juvonen, J. Hartikainen, and T. Hakala. 2007. Corticosteroids for the prevention of atrial fibrillation after cardiac surgery: a randomized controlled trial. JAMA 297: 1562–1567.

Han, W.K., V. Bailly, R. Abichandani, R. Thadhani, and J.V. Bonventre. 2002. Kidney Injury Molecule-1 (KIM-1): a novel biomarker for human renal proximal tubule injury. Kidney Int. 62: 237–244.

Han, W.K., S.S. Waikar, A. Johnson, R.A. Betensky, C.L. Dent, P. Devarajan, and J.V. Bonventre. 2008. Urinary biomarkers in the early diagnosis of acute kidney injury. Kidney Int. 73: 863–869.

Han, W.K., G. Wagener, Y. Zhu, S. Wang, and H.T. Lee. 2009. Urinary biomarkers in the early detection of acute kidney injury after cardiac surgery. Clin. J. Am. Soc. Nephrol. 4: 873–882.

Harris, M., N.W. Shammas, and M. Jerin. 2004. Elevated levels of low-density lipoprotein cholesterol, homocysteine, and lipoprotein(a) are associated with the occurrence of symptomatic bypass graft disease 1 year following coronary artery bypass graft surgery. Prev. Cardiol. 7: 106–108.

Hermann-Arnhof, K.M., T. Kastenbauer, T. Publig, P. Novotny, N. Loho, S. Schwarz, U. Koller, and R. Fitzgerald. 2006. Initially elevated osteoprotegerin serum levels may predict a perioperative myocardial lesion in patients undergoing coronary artery bypass grafting. Crit. Care Med. 34: 76–80.

Herrmann, M., A.D. Ebert, I. Galazky, M.T. Wunderlich, W.S. Kunz, and C. Huth. 2000. Neurobehavioral outcome prediction after cardiac surgery: role of neurobiochemical markers of damage to neuronal and glial brain tissue. Stroke 31: 645–650.

Hillege, H.L., D. Nitsch, M.A. Pfeffer, K. Swedberg, J.J. McMurray, S. Yusuf, C.B. Granger, E.L. Michelson, J. Ostergren, J.H. Cornel, D. de Zeeuw, S. Pocock, and D.J. van Veldhuisen. 2006. Renal function as a predictor of outcome in a broad spectrum of patients with heart failure. Circulation 113: 671–678.

Hobson, C.E., S. Yavas, M.S. Segal, J.D. Schold, C.G. Tribble, A.J. Layon, and A. Bihorac. 2009. Acute kidney injury is associated with increased long-term mortality after cardiothoracic surgery. Circulation 119: 2444–2453.

Hofstra, J.M., J.K. Deegens, E.J. Steenbergen, and J.F. Wetzels. 2008. Urinary excretion of fatty acid-binding proteins in idiopathic membranous nephropathy. Nephrol. Dial. Transplant. 23: 3160–3165.

Holzmann, M.J., S. Ahnve, N. Hammar, L. Jorgensen, K. Klerdal, K. Pehrsson, and T. Ivert. 2005. Creatinine clearance and risk of early mortality in patients undergoing coronary artery bypass grafting. J. Thorac. Cardiovasc. Surg. 130: 746–752.

Hoste, E.A., D.N. Cruz, A. Davenport, R.L. Mehta, P. Piccinni, C. Tetta, G. Viscovo, and C. Ronco. 2008. The epidemiology of cardiac surgery-associated acute kidney injury. Int. J. Artif. Organs 31: 158–165.

Hutfless, R., R. Kazanegra, M. Madani, M.A. Bhalla, A. Tulua-Tata, A. Chen, P. Clopton, C. James, A. Chiu, and A.S. Maisel. 2004. Utility of B-type natriuretic peptide in predicting postoperative complications and outcomes in patients undergoing heart surgery. J. Am. Coll. Cardiol. 43: 1873–1879.

Ichimura, T., J.V. Bonventre, V. Bailly, H. Wei, C.A. Hession, R.L. Cate, and M. Sanicola. 1998. Kidney injury molecule-1 (KIM-1), a putative epithelial cell adhesion molecule containing a novel immunoglobulin domain, is up-regulated in renal cells after injury. J. Biol. Chem. 273: 4135–4142.

Ishani, A., J.L. Xue, J. Himmelfarb, P.W. Eggers, P.L. Kimmel, B.A. Molitoris, and A.J. Collins. 2009. Acute kidney injury increases risk of ESRD among elderly. J. Am. Soc. Nephrol. 20: 223–228.

Ishida, K., F. Kimura, M. Imamaki, A. Ishida, H. Shimura, H. Kohno, M. Sakurai, and M. Miyazaki. 2006. Relation of inflammatory cytokines to atrial fibrillation after off-pump coronary artery bypass grafting. Eur J. Cardiothorac. Surg. 29: 501–505.

Jarvela, K., P. Maaranen, A. Harmoinen, H. Huhtala, and T. Sisto. 2011. Cystatin C in diabetics as a marker of mild renal insufficiency after CABG. Ann. Thorac. Cardiovasc. Surg. 17: 277–282.

Kaireviciute, D., A.D. Blann, B. Balakrishnan, D.A. Lane, J.V. Patel, G. Uzdavinys, G. Norkunas, G. Kalinauskas, V. Sirvydis, A. Aidietis, and G.Y. Lip. 2010. Characterisation and validity of inflammatory biomarkers in the prediction of post-operative atrial fibrillation in coronary artery disease patients. Thromb. Haemost. 104: 122–127.

Kaireviciute, D., G.Y. Lip, B. Balakrishnan, G. Uzdavinys, G. Norkunas, G. Kalinauskas, V. Sirvydis, A. Aidietis, U. Zanetto, H. Sihota, M. Maheshwari, and A.D. Blann. 2011. Intracardiac expression of markers of endothelial damage/dysfunction, inflammation, thrombosis, and tissue remodeling, and the development of postoperative atrial fibrillation. J. Thromb. Haemost. 9: 2345–2352.

Kalman, J., A. Juhasz, G. Bogats, B. Babik, A. Rimanoczy, Z. Janka, B. Penke, and A. Palotas. 2006. Elevated levels of inflammatory biomarkers in the cerebrospinal fluid after coronary artery bypass surgery are predictors of cognitive decline. Neurochem. Int. 48: 177–180.

Kalomenidis, I., G.T. Stathopoulos, R. Barnette, S. Papiris, T.S. Blackwell, C. Roussos, and R.W. Light. 2007. Vascular endothelial growth factor levels in post-CABG pleural effusions are associated with pleural inflammation and permeability. Respir. Med. 101: 223–229.

Kangasniemi, O.P., F. Biancari, J. Luukkonen, S. Vuorisalo, J. Satta, R. Pokela, and T. Juvonen. 2006. Preoperative C-reactive protein is predictive of long-term outcome after coronary artery bypass surgery. Eur J. Cardiothorac. Surg. 29: 983–985.

Karahan, S.C., I. Koramaz, G. Altun, U. Ucar, M. Topbas, A. Mentese, and M. Kopuz. 2010. Ischemia-modified albumin reduction after coronary bypass surgery is associated with the cardioprotective efficacy of cold-blood cardioplegia enriched with N-acetylcysteine: a preliminary study. Eur. Surg. Res. 44: 30–36.

Karkouti, K., D.N. Wijeysundera, T.M. Yau, J. L. Callum, D.C. Cheng, M. Crowther, J.Y. Dupuis, S.E. Fremes, B. Kent, C. Laflamme, A. Lamy, J.F. Legare, C.D. Mazer, S.A. McCluskey, F.D. Rubens, C. Sawchuk, and W.S. Beattie. 2009. Acute kidney injury after cardiac surgery: focus on modifiable risk factors. Circulation 119: 495–502.

Katagiri, D., K. Doi, K. Honda, K. Negishi, T. Fujita, M. Hisagi, M. Ono, T. Matsubara, N. Yahagi, M. Iwagami, T. Ohtake, S. Kobayashi, T. Sugaya, and E. Noiri. 2012. Combination of two urinary biomarkers predicts acute kidney injury after adult cardiac surgery. Ann. Thorac. Surg. 93: 577–583.

Kathiresan, S., S.J. Servoss, J.B. Newell, D. Trani, T.E. MacGillivray, K. Lewandrowski, E. Lee-Lewandrowski, and J.L. Januzzi, Jr. 2004. Cardiac troponin T elevation after coronary artery bypass grafting is associated with increased one-year mortality. Am. J. Cardiol. 94: 879–881.

Kinoshita, T., T. Asai, T. Suzuki, A. Kambara, and K. Matsubayashi. 2012. Preoperative hemoglobin A1c predicts atrial fibrillation after off-pump coronary bypass surgery. Eur J. Cardiothorac. Surg. 41: 102–107.

Knayzer, B., D. Abramov, B. Natalia, D. Tovbin, A. Ganiel, and A. Katz. 2007. Atrial fibrillation and plasma troponin I elevation after cardiac surgery: relation to inflammation-associated parameters. J. Card. Surg. 22: 117–123.

Knight, E.L., J.C. Verhave, D. Spiegelman, H.L. Hillege, D. de Zeeuw, G.C. Curhan, and P.E. de Jong. 2004. Factors influencing serum cystatin C levels other than renal function and the impact on renal function measurement. Kidney Int. 65: 1416–1421.

Kolackova, M., V. Lonsky, M. Trojackova Kudlova, J. Mandak, P. Kunes, V. Svitek, K. Jankovicova, D. Vlaskova, C. Andrys, and J. Krejsek. 2009. Serum level of sCD163, a soluble receptor for hemoglobin, is influenced by cardiac surgery. Perfusion 24: 263–269.

Koyner, J.L., M.R. Bennett, E.M. Worcester, Q. Ma, J. Raman, V. Jeevanandam, K.E. Kasza, M.F. O'Connor, D.J. Konczal, S. Trevino, P. Devarajan, and P.T. Murray. 2008. Urinary cystatin C as an early biomarker of acute kidney injury following adult cardiothoracic surgery. Kidney Int. 74: 1059–1069.

Kulier, A., J. Levin, R. Moser, G. Rumpold-Seitlinger, I.C. Tudor, S.A. Snyder-Ramos, P. Moehnle, and D.T. Mangano. 2007. Impact of preoperative anemia on outcome in patients undergoing coronary artery bypass graft surgery. Circulation 116: 471–479.

Lafrance, J.P. and D.R. Miller. 2010. Acute kidney injury associates with increased long-term mortality. J. Am. Soc. Nephrol. 21: 345–352.

Lane, B.R., E.D. Poggio, B.R. Herts, A.C. Novick, and S.C. Campbell. 2009. Renal function assessment in the era of chronic kidney disease: renewed emphasis on renal function centered patient care. J. Urol. 182: 435–443; discussion 443–434.

Lassnigg, A., D. Schmidlin, M. Mouhieddine, L.M. Bachmann, W. Druml, P. Bauer, and M. Hiesmayr. 2004. Minimal changes of serum creatinine predict prognosis in patients after cardiothoracic surgery: a prospective cohort study. J. Am. Soc. Nephrol. 15: 1597–1605.

Lassnigg, A., E.R. Schmid, M. Hiesmayr, C. Falk, W. Druml, P. Bauer, and D. Schmidlin. 2008. Impact of minimal increases in serum creatinine on outcome in patients after cardiothoracic surgery: do we have to revise current definitions of acute renal failure? Crit. Care Med. 36: 1129–1137.

Leacche, M., J.D. Rawn, T. Mihaljevic, J. Lin, A.N. Karavas, S. Paul, and J.G. Byrne. 2004. Outcomes in patients with normal serum creatinine and with artificial renal support for acute renal failure developing after coronary artery bypass grafting. Am. J. Cardiol. 93: 353–356.

Leal, J.C., O. Petrucci, M.F. Godoy, and D.M. Braile. 2012. Perioperative serum troponin I levels are associated with higher risk for atrial fibrillation in patients undergoing coronary artery bypass graft surgery. Interact. Cardiovasc. Thorac. Surg. 14: 22–25.

Liangos, O. 2006. Urinary kidney injury molecule-1 level is an early and sensitive marker of acute kidney injury J. Am. Soc. Nephrol. 17: 403A.

Liangos, O., H. Tighiouart, M.C. Perianayagam, A. Kolyada, W.K. Han, R. Wald, J.V. Bonventre, and B.L. Jaber. 2009. Comparative analysis of urinary biomarkers for early detection of acute kidney injury following cardiopulmonary bypass. Biomarkers 14: 423–431.

Lim, C.C., F. Cuculi, W.J. van Gaal, L. Testa, J.R. Arnold, T. Karamitsos, J.M. Francis, J.E. Digby, C. Antoniades, R. K. Kharbanda, S. Neubauer, S. Westaby, and A.P. Banning. 2011. Early diagnosis of perioperative myocardial infarction after coronary bypass grafting: a study using biomarkers and cardiac magnetic resonance imaging. Ann. Thorac. Surg. 92: 2046–2053.

Lim, C.C., W.J. van Gaal, L. Testa, F. Cuculi, J.R. Arnold, T. Karamitsos, J.M. Francis, S.E. Petersen, J.E. Digby, S. Westaby, C. Antoniades, R.K. Kharbanda, L.M. Burrell, S. Neubauer, and A.P. Banning. 2011. With the "universal definition," measurement of creatine kinase-myocardial band rather than troponin allows more accurate diagnosis of periprocedural necrosis and infarction after coronary intervention. J. Am. Coll. Cardiol. 57: 653–661.

Loef, B.G., A.H. Epema, T.D. Smilde, R.H. Henning, T. Ebels, G. Navis, and C.A. Stegeman. 2005. Immediate postoperative renal function deterioration in cardiac surgical patients predicts in-hospital mortality and long-term survival. J. Am. Soc. Nephrol. 16: 195–200.

Lok, C.E., P.C. Austin, H. Wang, and J.V. Tu. 2004. Impact of renal insufficiency on short- and long-term outcomes after cardiac surgery. Am. Heart J. 148: 430–438.

Lu, Y.F., H.W. Qi, C.Z. Tang, M.Y. Wu, Y. Wang, and F.H. Zhao. 2007. [Change of N-terminal pro-brain natriuretic peptide and big endothelin in patients undergoing coronary artery bypass grafting]. Zhongguo Wei Zhong Bing Ji Jiu Yi Xue 19: 667–670.

Maatman, R.G., E.M. van de Westerlo, T.H. van Kuppevelt, and J.H. Veerkamp. 1992. Molecular identification of the liver- and the heart-type fatty acid-binding proteins in human and rat kidney. Use of the reverse transcriptase polymerase chain reaction. Biochem. J. 288 (Pt 1): 285–290.

Macedo, E., J. Bouchard, S. H. Soroko, G.M. Chertow, J. Himmelfarb, T.A. Ikizler, E.P. Paganini, and R. L. Mehta. 2010. Fluid accumulation, recognition and staging of acute kidney injury in critically-ill patients. Crit. Care 14: R82.

Maggio, M., F. Nicolini, C. Cattabiani, C. Beghi, T. Gherli, R.S. Schwartz, G. Valenti, and G.P. Ceda. 2012. Effects of testosterone supplementation on clinical and rehabilitative outcomes in older men undergoing on-pump CABG. Contemp. Clin. Trials 33: 730–738.

Makris, K., N. Markou, E. Evodia, E. Dimopoulou, I. Drakopoulos, K. Ntetsika, D. Rizos, G. Baltopoulos, and A. Haliassos. 2009. Urinary neutrophil gelatinase-associated lipocalin (NGAL) as an early marker of acute kidney injury in critically ill multiple trauma patients. Clin. Chem. Lab. Med. 47: 79–82.

Mandal, K., M. Jahangiri, M. Mukhin, J. Poloniecki, A.J. Camm, and Q. Xu. 2004. Association of anti-heat shock protein 65 antibodies with development of postoperative atrial fibrillation. Circulation 110: 2588–2590.

Mandal, K., E. Torsney, J. Poloniecki, A.J. Camm, Q. Xu, and M. Jahangiri. 2005. Association of high intracellular, but not serum, heat shock protein 70 with postoperative atrial fibrillation. Ann. Thorac. Surg. 79: 865–871; discussion 871.

Mangano, C.M., L.S. Diamondstone, J.G. Ramsay, A. Aggarwal, A. Herskowitz, and D.T. Mangano. 1998. Renal dysfunction after myocardial revascularization: risk factors, adverse outcomes, and hospital resource utilization. The Multicenter Study of Perioperative Ischemia Research Group. Ann. Intern. Med. 128: 194–203.

Mariscalco, G., R. Lorusso, C. Dominici, A. Renzulli, and A. Sala. 2011. Acute kidney injury: a relevant complication after cardiac surgery. Ann. Thorac. Surg. 92: 1539–1547.

Martensson, J., M. Bell, A. Oldner, S. Xu, P. Venge, and C.R. Martling. 2010. Neutrophil gelatinase-associated lipocalin in adult septic patients with and without acute kidney injury. Intensive Care Med. 36: 1333–1340.

Mathew, J.P., M.V. Podgoreanu, H.P. Grocott, W.D. White, R.W. Morris, M. Stafford-Smith, G.B. Mackensen, C.S. Rinder, J.A. Blumenthal, D.A. Schwinn, and M.F. Newman. 2007. Genetic variants in P-selectin and C-reactive protein influence susceptibility to cognitive decline after cardiac surgery. J. Am. Coll. Cardiol. 49: 1934–1942.

McIlroy, D.R., G. Wagener, and H.T. Lee. 2010. Neutrophil gelatinase-associated lipocalin and acute kidney injury after cardiac surgery: the effect of baseline renal function on diagnostic performance. Clin. J. Am. Soc. Nephrol. 5: 211–219.

Mehta, R.H., J.D. Grab, S.M. O'Brien, C.R. Bridges, J.S. Gammie, C.K. Haan, T.B. Ferguson, and E.D. Peterson. 2006. Bedside tool for predicting the risk of postoperative dialysis in patients undergoing cardiac surgery. Circulation 114: 2208–2216; quiz 2208.

Mehta, R.L., J.A. Kellum, S.V. Shah, B.A. Molitoris, C. Ronco, D.G. Warnock, and A. Levin. 2007. Acute Kidney Injury Network: report of an initiative to improve outcomes in acute kidney injury. Crit. Care 11: R31.

Mennander, A., T. Angervuori, H. Huhtala, P. Karhunen, M. Tarkka, and P. Kuukasjarvi. 2005. Positive family history of coronary atherosclerosis and serum triglycerides may predict repeated coronary artery bypass surgery. Scand. Cardiovasc. J. 39: 225–228.

Miceli, A., V.D. Bruno, R. Capoun, F. Romeo, G.D. Angelini, and M. Caputo. 2011. Occult renal dysfunction: a mortality and morbidity risk factor in coronary artery bypass grafting surgery. J. Thorac. Cardiovasc. Surg. 141: 771–776.

Mishra, J., C. Dent, R. Tarabishi, M.M. Mitsnefes, Q. Ma, C. Kelly, S.M. Ruff, K. Zahedi, M. Shao, J. Bean, K. Mori, J. Barasch, and P. Devarajan. 2005. Neutrophil gelatinase-associated lipocalin (NGAL) as a biomarker for acute renal injury after cardiac surgery. Lancet 365: 1231–1238.

Mitchell, L.B. 2011. Canadian Cardiovascular Society atrial fibrillation guidelines 2010: prevention and treatment of atrial fibrillation following cardiac surgery. Can. J. Cardiol. 27: 91–97.

Moore, E., R. Bellomo, and A. Nichol. 2010. Biomarkers of acute kidney injury in anesthesia, intensive care and major surgery: from the bench to clinical research to clinical practice. Minerva Anestesiol. 76: 425–440.

Muehlschlegel, J.D., T.E. Perry, K.Y. Liu, A.A. Fox, C.D. Collard, S.K. Shernan, and S.C. Body. 2010. Heart-type fatty acid binding protein is an independent predictor of death and ventricular dysfunction after coronary artery bypass graft surgery. Anesth. Analg. 111: 1101–1109.

Muehlschlegel, J.D., T.E. Perry, K.Y. Liu, L. Nascimben, A.A. Fox, C.D. Collard, E.G. Avery, S.F. Aranki, M.N. D'Ambra, S.K. Shernan, and S.C. Body. 2009. Troponin is superior to electrocardiogram and creatinine kinase MB for predicting clinically significant myocardial injury after coronary artery bypass grafting. Eur. Heart J. 30: 1574–1583.

Negishi, K., E. Noiri, K. Doi, R. Maeda-Mamiya, T. Sugaya, D. Portilla, and T. Fujita. 2009. Monitoring of urinary L-type fatty acid-binding protein predicts histological severity of acute kidney injury. Am. J. Pathol. 174: 1154–1159.

Novella, S., A.C. Martinez, R.M. Pagan, M. Hernandez, A. Garcia-Sacristan, A. Gonzalez-Pinto, J.M. Gonzalez-Santos, and S. Benedito. 2007. Plasma levels and vascular effects of vasopressin in patients undergoing coronary artery bypass grafting. Eur. J. Cardiothorac. Surg. 32: 69–76.

Nowicki, M., M. Misterski, A. Malinska, B. Perek, D. Ostalska-Nowicka, M. Jemielity, W. Witkiewicz, and M. Zabel. 2011. Endothelial integrity of radial artery grafts harvested by minimally invasive surgery—immunohistochemical studies of CD31 and endothelial nitric oxide synthase expressions: a randomized controlled trial. Eur. J. Cardiothorac. Surg. 39: 471–477.

Noyez, L., I. Plesiewicz, and F.W. Verheugt. 2006. Estimated creatinine clearance instead of plasma creatinine level as prognostic test for postoperative renal function in patients undergoing coronary artery bypass surgery. Eur. J. Cardiothorac. Surg. 29: 461–465.

Nozohoor, S., J. Nilsson, L. Algotsson, and J. Sjogren.2011. Postoperative increase in B-type natriuretic peptide levels predicts adverse outcome after cardiac surgery. J. Cardiothorac. Vasc. Anest.h 25: 469–475.

Oc, M., H.I. Ucar, A. Pinar, B. Akbulut, B. Oc, S.B. Akinci, Y. Akyon, M. Kanbak, E. Boke, and R. Dogan. 2007. Heat shock protein 60 antibody. A new marker for subsequent atrial fibrillation development. Saudi Med. J. 28: 844–847.

Ostermann, M.E., D. Taube, C.J. Morgan, and T.W. Evans. 2000. Acute renal failure following cardiopulmonary bypass: a changing picture. Intensive Care Med. 26: 565–571.

Ozkara, C., N. Guler, T. Batyraliev, H. Okut, and M. Agirbasli. 2007. Does off-pump coronary artery bypass surgery reduce secretion of plasminogen activator inhibitor-1? Int. J. Clin. Pract. 61: 763–767.

Palomba, H., I. de Castro, A.L. Neto, S. Lage, and L. Yu. 2007. Acute kidney injury prediction following elective cardiac surgery: AKICS Score. Kidney Int. 72: 624–631.

Paparella, D., G. Scrascia, A. Paramythiotis, P. Guida, V. Magari, P.G. Malvindi, S. Favale, and L. de Luca Tupputi Schinosa. 2010. Preoperative cardiac troponin I to assess midterm risks of coronary bypass grafting operations in patients with recent myocardial infarction. Ann. Thorac. Surg. 89: 696–702.

Parikh, C.R., J. Mishra, H. Thiessen-Philbrook, B. Dursun, Q. Ma, C. Kelly, C. Dent, P. Devarajan, and C.L. Edelstein. 2006. Urinary IL-18 is an early predictive biomarker of acute kidney injury after cardiac surgery. Kidney Int. 70: 199–203.

Parikh, C.R. and P. Devarajan. 2008. New biomarkers of acute kidney injury. Crit. Care Med. 36: S159–165.

Pastuszczak, M., A. Branicka, B. Jakiela, E. Stepien, A.K. Jaworek, A. Wojas-Pelc, B. Kapelak, and J. Sadowski. 2009. The +405 GG variant of vascular endothelial growth factor polymorphism is associated with poor prognosis in patients undergoing coronary artery bypass graft surgery. Pol. Arch. Med. Wewn. 119: 719–725.

Pearce, E.N., Q. Yang, E.J. Benjamin, J. Aragam, and R.S. Vasan. Thyroid function and left ventricular structure and function in the Framingham Heart Study. 2010. Thyroid 20: 369–373.

Peivandi, A.A., M. Dahm, U.T. Opfermann, D. Peetz, F. Doerr, A. Loos, and H. Oelert. 2004. Comparison of cardiac troponin I versus T and creatine kinase MB after coronary artery bypass grafting in patients with and without perioperative myocardial infarction. Herz 29: 658–664.

Pelsers, M.M., W.T. Hermens, and J.F. Glatz. 2005. Fatty acid-binding proteins as plasma markers of tissue injury. Clin. Chim. Acta. 352: 15–35.

Perry, T.E., J.D. Muehlschlegel, K.Y. Liu, A.A. Fox, C.D. Collard, S.C. Body, and S.K. Shernan. 2010. Preoperative C-reactive protein predicts long-term mortality and hospital length of stay after primary, nonemergent coronary artery bypass grafting. Anesthesiology 112: 607–613.

Petaja, L., M. Salmenpera, K. Pulkki, and V. Pettila. 2009. Biochemical injury markers and mortality after coronary artery bypass grafting: a systematic review. Ann. Thorac. Surg. 87: 1981–1992.

Pizon, M.T., T. Gburek, M. Pizon, and K. Sztefko. 2006. Kinetics of plasma heat shock protein HSP-70 release in coronary artery surgery: on-pump versus off-pump. Minerva Chir. 61: 483–491.

Pojar, M., J. Mand'ak, N. Cibicek, V. Lonsky, J. Dominik, V. Palicka, and J. Kubicek. 2008. Peripheral tissue metabolism during off-pump versus on-pump coronary artery bypass graft surgery: the microdialysis study. Eur. J. Cardiothorac. Surg. 33: 899–905.

Portilla, D., C. Dent, T. Sugaya, K.K. Nagothu, I. Kundi, P. Moore, E. Noiri, and P. Devarajan. 2008. Liver fatty acid-binding protein as a biomarker of acute kidney injury after cardiac surgery. Kidney Int. 73: 465–472.

Poston, R., J. Gu, J. Manchio, A. Lee, J. Brown, J. Gammie, C. White, and B.P. Griffith. 2005. Platelet function tests predict bleeding and thrombotic events after off-pump coronary bypass grafting. Eur. J. Cardiothorac. Surg. 27: 584–591.

Provenchere, S., C. Berroeta, C. Reynaud, G. Baron, I. Poirier, J.M. Desmonts, B. Iung, M. Dehoux, I. Philip, and J. Benessiano. 2006. Plasma brain natriuretic peptide and cardiac troponin I concentrations after adult cardiac surgery: association with postoperative cardiac dysfunction and 1-year mortality. Crit. Care Med. 34: 995–1000.

Rabitzsch, G., J. Mair, P. Lechleitner, F. Noll, V. Hofmann, E.G. Krause, F. Dienstl, and B. Puschendorf. 1993. Isoenzyme BB of glycogen phosphorylase b and myocardial infarction. Lancet 341: 1032–1033.

Rade, J.J. and C.W. Hogue, Jr. 2010. Do we really need another biomarker to diagnose myocardial infarction after coronary artery bypass graft surgery? Anesth. Analg. 111: 1086–1087.

Raivio, P., A. Kuitunen, R. Suojaranta-Ylinen, R. Lassila, and J. Petaja. 2006. Thrombin generation during reperfusion after coronary artery bypass surgery associates with postoperative myocardial damage. J. Thromb. Haemost. 4: 1523–1529.

Raivio, P., J.A. Fernandez, A. Kuitunen, J.H. Griffin, R. Lassila, and J. Petaja. 2007. Activation of protein C and hemodynamic recovery after coronary artery bypass surgery. J. Thorac. Cardiovasc. Surg. 133: 44–51.

Raivio, P.M., R. Lassila, A.H. Kuitunen, H. Eriksson, R.T. Suojaranta-Ylinen, and J. Petaja. 2011. Increased preoperative thrombin generation and low protein S level associated with unfavorable postoperative hemodynamics after coronary artery bypass grafting. Perfusion 26: 99–106.

Ranasinghe, A.M., D.W. Quinn, M. Richardson, N. Freemantle, T.R. Graham, J. Mascaro, S.J. Rooney, I.C. Wilson, D. Pagano, and R.S. Bonser. 2011.Which troponometric best predicts midterm outcome after coronary artery bypass graft surgery? Ann. Thorac. Surg. 91: 1860–1867.

Ricci, Z., D.N. Cruz, and C. Ronco. 2011. Classification and staging of acute kidney injury: beyond the RIFLE and AKIN criteria. Nat. Rev. Nephrol. 7: 201–208.

Rifon, J., J.A. Paramo, F. Prosper, M.T. Collados, J. Sarra, and E. Rocha. 1994. Thrombin-antithrombin complexes and prothrombin fragment 1+2 in aorto-coronary bypass surgery: relation to graft occlusion. Hematol. Pathol. 8: 35–42.

Rifon, J., J.A. Paramo, C. Panizo, R. Montes, and E. Rocha. 1997. The increase of plasminogen activator inhibitor activity is associated with graft occlusion in patients undergoing aorto-coronary bypass surgery. Br. J. Haematol. 99: 262–267.

Roques, F., S.A. Nashef, P. Michel, E. Gauducheau, C. de Vincentiis, E. Baudet, J. Cortina, M. David, A. Faichney, F. Gabrielle, E. Gams, A. Harjula, M.T. Jones, P.P. Pintor, R. Salamon, and L. Thulin. 1999. Risk factors and outcome in European cardiac surgery: analysis of the EuroSCORE multinational database of 19030 patients. Eur. J. Cardiothorac. Surg. 15: 816–822; discussion 822–813.

Rosner, M.H. and M.D. Okusa. 2006. Acute kidney injury associated with cardiac surgery. Clin. J. Am. Soc. Nephrol. 1: 19–32.

Rosner, M.H. 2009. Urinary biomarkers for the detection of renal injury. Adv. Clin. Chem. 49: 73–97.

Rothenburger, M., T. Wichter, C. Schmid, J. Stypmann, T.D. Tjan, E. Berendes, C. Etz, A. Pioux, A. Loher, F. Wenzelburger, G. Drees, A. Hoffmeier, G. Breithardt, and H.H. Scheld. 2004. Aminoterminal pro type B natriuretic peptide as a predictive and prognostic marker in patients with chronic heart failure. J. Heart Lung Transplant. 23: 1189–1197.

Sanders, J., E. Hawe, D.J. Brull, C. Hubbart, G.D. Lowe, A. Rumley, S.E. Humphries, and H.E. Montgomery. 2009. Higher IL-6 levels but not IL6 -174G>C or -572G>C genotype are associated with post-operative complication following coronary artery bypass graft (CABG) surgery. Atherosclerosis 204: 196–201.

Santopinto, J.J., K.A. Fox, R.J. Goldberg, A. Budaj, G. Pinero, A. Avezum, D. Gulba, J. Esteban, J.M. Gore, J. Johnson, and E.P. Gurfinkel. 2003. Creatinine clearance and adverse hospital outcomes in patients with acute coronary syndromes: findings from the global registry of acute coronary events (GRACE). Heart 89: 1003–1008.

Sbarouni, E., P. Georgiadou, C. Mihas, A. Chaidaroglou, D. Degiannis, and V. Voudris. 2012. Significant peri-operative reduction in plasma osteopontin levels after coronary artery by-pass grafting. Clin. Biochem.

Schmid, F.X., N. Vudattu, B. Floerchinger, M. Hilker, G. Eissner, M. Hoenicka, E. Holler, and D.E. Birnbaum. 2006. Endothelial apoptosis and circulating endothelial cells after bypass grafting with and without cardiopulmonary bypass. Eur. J. Cardiothorac. Surg. 29: 496–500.

Schmidt-Ott, K.M., K. Mori, A. Kalandadze, J.Y. Li, N. Paragas, T. Nicholas, P. Devarajan, and J. Barasch. 2006. Neutrophil gelatinase-associated lipocalin-mediated iron traffic in kidney epithelia. Curr. Opin. Nephrol. Hypertens. 15: 442–449.

Seabra, V.F., S. Alobaidi, E.M. Balk, A.H. Poon, and B.L. Jaber. 2010. Off-pump coronary artery bypass surgery and acute kidney injury: a meta-analysis of randomized controlled trials. Clin. J. Am. Soc. Nephrol. 5: 1734–1744.

Selvanayagam, J.B., D. Pigott, L. Balacumaraswami, S.E. Petersen, S. Neubauer, and D.P. Taggart. 2005. Relationship of irreversible myocardial injury to troponin I and creatine kinase-MB elevation after coronary artery bypass surgery: insights from cardiovascular magnetic resonance imaging. J. Am. Coll. Cardiol. 45: 629–631.

Serrano, C.V., Jr., J.A. Souza, N.H. Lopes, J.L. Fernandes, J.C. Nicolau, M.H. Blotta, J.A. Ramires, and W.A. Hueb. 2010. Reduced expression of systemic proinflammatory and myocardial biomarkers after off-pump versus on-pump coronary artery bypass surgery: a prospective randomized study. J. Crit. Care 25: 305–312.

Shah, A. A., D.M. Craig, J.K. Sebek, C. Haynes, R. C. Stevens, M.J. Muehlbauer, C.B. Granger, E.R. Hauser, L.K. Newby, C.B. Newgard, W.E. Kraus, G.C. Hughes, and S.H. Shah. 2012. Metabolic profiles predict adverse events after coronary artery bypass grafting. J. Thorac. Cardiovasc. Surg. 143: 873–878.

Shapiro, N.I., S. Trzeciak, J.E. Hollander, R. Birkhahn, R. Otero, T.M. Osborn, E. Moretti, H.B. Nguyen, K. Gunnerson, D. Milzman, D.F. Gaieski, M. Goyal, C.B. Cairns, K. Kupfer, S. W. Lee, and E.P. Rivers. 2010. The diagnostic accuracy of plasma neutrophil gelatinase-associated lipocalin in the prediction of acute kidney injury in emergency department patients with suspected sepsis. Ann. Emerg. Med 56: 52–59 e51.

Shaw, A.D., D.B. Chalfin, and J. Kleintjens. 2011. The economic impact and cost-effectiveness of urinary neutrophil gelatinase-associated lipocalin after cardiac surgery. Clin. Ther. 33: 1713–1725.

Sinha, M.K., D.C. Gaze, J.R. Tippins, P.O. Collinson, and J.C. Kaski. 2003. Ischemia modified albumin is a sensitive marker of myocardial ischemia after percutaneous coronary intervention. Circulation 107: 2403–2405.

Sjoland, H., L. Tengborn, L. Stensdotter, and J. Herlitz. 2007. Lack of very strong association between pre-treatment fibrinogen and PAI-1 with long-term mortality after coronary bypass surgery. Cardiology 108: 82–89.

Sprecher, D.L., G.L. Pearce, E.M. Park, F.J. Pashkow, and B.J. Hoogwerf. 2000. Preoperative triglycerides predict post-coronary artery bypass graft survival in diabetic patients: a sex analysis. Diabetes Care 23: 1648–1653.

St. Rammos, K., G.J. Koullias, M.O. Hassan, N.P. Argyrakis, C.G. Voucharas, S.J. Scarupa, and T.G. Cowte. 2002. Low preoperative HSP70 atrial myocardial levels correlate significantly with high incidence of postoperative atrial fibrillation after cardiac surgery. Cardiovasc. Surg. 10: 228–232.

Svennevig, K., T. Hoel, A. Thiara, S. Kolset, A. Castelheim, T. Mollnes, F. Brosstad, E. Fosse, and J. Svennevig. 2008. Syndecan-1 plasma levels during coronary artery bypass surgery with and without cardiopulmonary bypass. Perfusion 23: 165–171.

Szerafin, T., K. Hoetzenecker, S. Hacker, A. Horvath, A. Pollreisz, P. Arpad, A. Mangold, T. Wliszczak, M. Dworschak, R. Seitelberger, E. Wolner, and H.J. Ankersmit. 2008. Heat shock proteins 27, 60, 70, 90alpha, and 20S proteasome in on-pump versus off-pump coronary artery bypass graft patients. Ann. Thorac. Surg. 85: 80–87.

Szerafin, T., T. Niederpold, A. Mangold, K. Hoetzenecker, S. Hacker, G. Roth, M. Lichtenauer, M. Dworschak, E. Wolner, and H.J. Ankersmit. 2009. Secretion of soluble ST2-possible explanation for systemic immunosuppression after heart surgery. Thorac. Cardiovasc. Surg. 57: 25–29.

Tavil, Y., N. Sen, F. Hizal, S.K. Acikgoz, I. Tasoglu, S. Topal, and M.R. Yalcin. 2008. Relationship between elevated levels of serum uric acid and saphenous vein graft disease. Turk Kardiyol Dern Ars. 36: 14–18.

Tekumit, H., A.R. Cenal, A. Polat, K. Uzun, C. Tataroglu, and E. Akinci. 2010. Diagnostic value of hemoglobin A1c and fasting plasma glucose levels in coronary artery bypass grafting patients with undiagnosed diabetes mellitus. Ann. Thorac. Surg. 89: 1482–1487.

Thakar, C.V., S. Arrigain, S. Worley, J.P. Yared, and E.P. Paganini. 2005. A clinical score to predict acute renal failure after cardiac surgery. J. Am. Soc. Nephrol. 16: 162–168.

Thomas, A.A., S. Demirjian, B.R. Lane, M. N. Simmons, D.A. Goldfarb, V.S. Subramanian, and S.C. Campbell. 2011. Acute kidney injury: novel biomarkers and potential utility for patient care in urology. Urology 77: 5–11.

Thygesen, K., J.S. Alpert, and H.D. White. 2007. Universal definition of myocardial infarction. Eur. Heart J. 28: 2525–2538.

Tolpin, D.A., C.D. Collard, V.V. Lee, S.S. Virani, P.M. Allison, M.A. Elayda, and W. Pan. 2012. Subclinical changes in serum creatinine and mortality after coronary artery bypass grafting. J. Thorac. Cardiovasc. Surg. 143: 682–688 e681.

Tsuruta, R., K. Miyauchi, T. Yamamoto, S. Dohi, K. Tambara, T. Dohi, H. Inaba, K. Kuwaki, H. Daida, and A. Amano. 2011. Effect of preoperative hemoglobin A1c levels on long-term outcomes for diabetic patients after off-pump coronary artery bypass grafting. J. Cardiol. 57: 181–186.

Tuttle, K.R., N.K. Worrall, L.R. Dahlstrom, R. Nandagopal, A.T. Kausz, and C.L. Davis. 2003. Predictors of ARF after cardiac surgical procedures. Am. J. Kidney Dis. 41: 76–83.

Ulus, T., A. Yildirir, L.E. Sade, S. Balta, B. Ozin, A. Sezgin, and H. Muderrisoglu. 2011. [Serum gamma-glutamyltransferase activity: a new marker for coronary artery bypass graft disease]. Turk Kardiyol Dern Ars 39: 205–213.

Vaidya, V.S. and J.V. Bonventre. 2006. Mechanistic biomarkers for cytotoxic acute kidney injury. Expert Opin. Drug Metab. Toxicol. 2: 697–713.

Vallely, M.P., P.G. Bannon, M.S. Bayfield, C.F. Hughes, and L. Kritharides. 2009. Quantitative and temporal differences in coagulation, fibrinolysis and platelet activation after on-pump and off-pump coronary artery bypass surgery. Heart Lung Circ. 18: 123–130.

Vallely, M.P., P.G. Bannon, M.S. Bayfield, C.F. Hughes, and L. Kritharides. 2010. Endothelial activation after coronary artery bypass surgery: comparison between on-pump and off-pump techniques. Heart Lung Circ. 19: 445–452.

van Boven, W.J., W.B. Gerritsen, P. Zanen, J.C. Grutters, H.P. van Dongen, A. Bernard, and L.P. Aarts. 2005. Pneumoproteins as a lung-specific biomarker of alveolar permeability in conventional on-pump coronary artery bypass graft surgery vs mini-extracorporeal circuit: a pilot study. Chest 127: 1190–1195.

van Geene, Y., H.A. van Swieten, and L. Noyez. 2010. Cardiac troponin I levels after cardiac surgery as predictor for in-hospital mortality. Interact. Cardiovasc. Thorac. Surg. 10: 413–416.

Velissaris, T., A.T. Tang, M. Murray, R.L. Mehta, P.J. Wood, D.A. Hett, and S.K. Ohri. 2004. A prospective randomized study to evaluate stress response during beating-heart and conventional coronary revascularization. Ann. Thorac. Surg. 78: 506–512; discussion 506–512.

Verma, S., A. Maitland, R.D. Weisel, P.W. Fedak, S.H. Li, D.A. Mickle, R.K. Li, L. Ko, and V. Rao. 2002. Increased endothelin-1 production in diabetic patients after cardioplegic arrest and reperfusion impairs coronary vascular reactivity: reversal by means of endothelin antagonism. J. Thorac. Cardiovasc. Surg. 123: 1114–1119.

Volkmann, M.A., P.E. Behr, J.E. Burmeister, P.R. Consoni, R.A. Kalil, P.R. Prates, I.A. Nesralla, and J.R. Sant'Anna. 2011. Hidden renal dysfunction causes increased in-hospital mortality risk after coronary artery bypass graft surgery. Rev. Bras. Cir. Cardiovasc. 26: 319–325.

Wagener, G., M. Jan, M. Kim, K. Mori, J.M. Barasch, R.N. Sladen, and H.T. Lee. 2006. Association between increases in urinary neutrophil gelatinase-associated lipocalin and acute renal dysfunction after adult cardiac surgery. Anesthesiology 105: 485–491.

Wagener, G., G. Gubitosa, S. Wang, N. Borregaard, M. Kim, and H.T. Lee. 2008. Urinary neutrophil gelatinase-associated lipocalin and acute kidney injury after cardiac surgery. Am. J. Kidney Dis. 52: 425–433.

Wang, A.Y., Y.V. Bobryshev, S.M. Cherian, H. Liang, D. Tran, S.J. Inder, R.S. Lord, K.W. Ashwell, and A.E. Farnsworth. 2001. Expression of apoptosis-related proteins and structural features of cell death in explanted aortocoronary saphenous vein bypass grafts. Cardiovasc. Surg. 9: 319–328.

Wang, F., J.Y. Dupuis, H. Nathan, and K. Williams. 2003. An analysis of the association between preoperative renal dysfunction and outcome in cardiac surgery: estimated creatinine clearance or plasma creatinine level as measures of renal function. Chest 124: 1852–1862.

Watanabe, M., S. Hasegawa, N. Ohshima, H. Tanaka, T. Sakamoto, and M. Sunamori. 2002. Differential regulation of MMP-2, TIMP-2 and IL-6 in valve replacement versus CABG patients. Perfusion 17: 435–439.

Wenzel, F., T. Gettmann, N. Zimmermann, and G. Giers. 2011. Alterations of serum erythropoietin and thrombopoietin levels in patients undergoing Coronary Artery Bypass Grafting (CABG). Clin. Hemorheol. Microcirc. 49: 399–406.

Westenbrink, B.D., L. Kleijn, R.A. de Boer, J.G. Tijssen, W.J. Warnica, R. Baillot, J.L. Rouleau, and W.H. van Gilst. 2011. Sustained postoperative anaemia is associated with an impaired outcome after coronary artery bypass graft surgery: insights from the IMAGINE trial. Heart 97: 1590–1596.

Wijeysundera, D.N., K. Karkouti, W.S. Beattie, V. Rao, and J. Ivanov. 2006. Improving the identification of patients at risk of postoperative renal failure after cardiac surgery. Anesthesiology 104: 65–72.

Wijeysundera, D.N., K. Karkouti, J.Y. Dupuis, V. Rao, C.T. Chan, J.T. Granton, and W.S. Beattie. 2007. Derivation and validation of a simplified predictive index for renal replacement therapy after cardiac surgery. JAMA 297: 1801–1809.

Witczak, B., A. Hartmann, and J.L. Svennevig. 2005. Multiple risk assessment of cardiovascular surgery in chronic renal failure patients. Ann. Thorac. Surg. 79: 1297–1302.

Yao, B., H.Y. Jiang, Z.Y. Xia, and Z.Y. Huang. 2005. [Change in plasma free 15 - F2t - isoprostane concentrations in smoking patients undergoing coronary artery bypass grafting and its clinical significance]. Zhongguo Wei Zhong Bing Ji Jiu Yi Xue 17: 165–167.

Yin, Y.Q., A.L. Luo, X.Y. Guo, L.H. Li, H.Z. Ren, T.H. Ye, and Y.G. Huang. 2005. [Perioperative cortisol circadian secretion and neuropsychological states in patients undergoing coronary artery bypass grafting surgery]. Zhonghua Wai Ke Za Zhi 43: 463–467.

Zakeri, R., N. Freemantle, V. Barnett, G.W. Lipkin, R.S. Bonser, T.R. Graham, S.J. Rooney, I.C. Wilson, R. Cramb, B.E. Keogh, and D. Pagano. 2005. Relation between mild renal dysfunction and outcomes after coronary artery bypass grafting. Circulation 112: I270–275.

# CHAPTER 9

# Biomarkers of Peripheral Arterial Disease

Marietta Charakida[1,*] and Dimitris Tousoulis[2]

## Introduction

Peripheral arterial disease (PAD) is commonly used to describe atherosclerotic changes affecting the aortic, iliac, and lower limb arteries. The disease is common, affecting 12% to 14% of the general population, with higher preponderance in the elderly (Shanmugasundaram et al. 2011). Its clinical presentation may vary from claudication, changes in limb color, delayed wound healing and diminished hair and nail growth. Musculoskeletal and neurological problems commonly coexist and confound the clinical picture. A small number of patients will present with classic symptoms and 70% to 90% will have atypical leg symptoms or remain asymptomatic and as such, clinical diagnosis may be challenging (Ramos et al. 2009). Structured questionnaires such as the Edinburgh Claudication Questionnaire have improved detection of patients with PAD compared to clinical assessment but these remain valuable for the minority of patients who present with classical symptomatology.

PAD can have adverse implications to the quality of life with reduction in functional capacity, ischemic ulceration and amputation and increased risk for future cardiovascular events, even in asymptomatic individuals (Stoica and Ginghina 2008). As such, identification and optimal management of PAD from the early stages is important to reduce cardiovascular complications and improve life expectancy. In addition, recent data suggest that aggressive treatment of risk factors and initiation of medications such

[1]Vascular Physiology Unit, Institute of Child Health.
[2]1st Cardiology Department, Hippokration Hospital, Athens Greece.
*Corresponding author: charakidadoc@gmail.com

as statins, ACE inhibitors, and anti-platelet therapy can reduce morbidity and mortality in those patients. Currently, screening for PAD is inadequate and as a result, the majority of patients with PAD are under diagnosed with suboptimal and delayed treatment. In this chapter, we will discuss available screening tools for PAD and discuss the need for soluble biomarkers which will allow the identification and stratification of patients with PAD.

## Ankle Brachial Index and Vascular Markers of PAD (Table 1)

Currently, the ankle brachial index (ABI) remains the easiest initial screening test for PAD patients (Widener 2011). It uses a sphygmomanometer and a Doppler ultrasound to accurately detect arterial pulse and define systolic blood pressure in the upper (Parm) and lower (Pleg) limb. By convention, for each limb the higher of two pressures are used for ABI calculation. Results are expressed as a ratio ABI:Pleg/Parm. The ABI has 95% sensitivity and 99% specificity for the detection of PAD. Screening using ABI is currently recommended by the American College of Cardiology and American Heart Association guidelines for patients younger than 50 years with diabetes and one other atherosclerotic risk factor, those between 50–69 years with a history of smoking and finally, individuals older than 70 years with leg symptoms on exertion or ischemic rest pain or with abnormal lower-extremity pulse examination. In addition, the American Diabetes Association recommends annual screening for PAD in diabetics.

Although an ABI of less than 0.9 is the most commonly used index to detect PAD and seems to correlate with disease severity, various studies have demonstrated that even individuals with minimal changes in ABI between 0.90–0.99 are having increased cardiovascular mortality (McDermott et al. 1998). In addition, multiple studies have shown that ABI measurements can be unreliable in patients with calcified arteries or insensitive to mild peripheral arterial disease (Murphy et al. 2012).

Vascular functional and structural abnormalities can now easily be quantified using non-invasive imaging techniques which have been shown to be representative indices of atherosclerotic disease progression in patients with PAD (Coutinho et al. 2011). Endothelial function measured non-invasively in the brachial artery has been shown to be impaired and associated with the functional performance in patients with PAD (Brevetti et al. 2008). Joras et al. studied 30 patients with symptomatic PAD and demonstrated impaired endothelial function compared to matched controls, which was further aggravated following an acute exercise protocol (Joras and Poredos 2008). Higher flow mediated dilatation (FMD) values have been reported in PAD patients with increased daily activity levels, which suggests that regular exercise in this patient population may play a role in improving endothelial dysfunction (Payvandi et al. 2009).

Table 1. Vascular markers for peripheral arterial disease.

| Technique | Arterial sites | Information gained | Screening tool for PAD: Pros and cons |
|---|---|---|---|
| Ankle-brachial index | Brachial-ankle | Difference in the blood pressure in the lower limb to blood pressure in the arm Results are expressed as a ratio ABI:Pleg/Parm. | ABI of less than 0.9 is the most-commonly used index to detect PAD and seems to correlate with disease severity. Measurements are unreliable when the arteries are calcified Insensitive to mild arterial changes. Associated with severity of disease and mortality. Recommended for screening tool in specific populations. |
| Flow mediated dilatation | Brachial artery | Dilatation in brachial artery following 5mins of ischemic stimulus. Measures endothelial function which is mostly nitric oxide dependent. Expressed as ratio PD-BD/BD*100 or as absolute difference PD-BD. | Commonly used to assess early arterial changes relevant to atherosclerosis. Not specific for PAD. Affected by CV risk factors. Associated with functional capacity. Variable method. Requires ultrasound and well trained staff. Not recommended for screening tool. |
| Intima media thickness | Carotid artery | Distance between intima- media. Usually measured in the common carotid artery 1 cm proximal to carotid bifurcation. Cumulative assessment of atherosclerotic progression. | Non specific for PAD. Affected by CV risk factors. Associated with cardiovascular outcome. Variable if not standardized methodology. Not recommended for screening tool |
| Pulse wave velocity | Carotid-radial Carotid and femoral | Velocity of the pulse wave along an artery. Transit time of arterial wave to travel between arterial points of measurement and distance between the two sites. | Not specific for PAD but shown to be increased in PAD. Affected by CV risk factors. Associated with functional capacity in PAD. Variable when measurements of arterial segments are not standardized. Not recommended for screening tool. |
| Augmentation index | Carotid artery Radial artery | Transfer function to calculate aortic AIx increase in BP caused by reflected wave during systole depends on arterial stiffness and vascular tone. | Not specific but increased in PAD. Affected by CV risk factors. Associated with functional capacity in PAD. Not currently recommended for screening tool. |

**Abbreviations:** ABI: ankle-brachial index, CV: cardiovascular, BP: blood pressure, PD: peak diameter, BD: baseline diameter, AIx: augmentation index, SBP: systolic blood pressure.

Indices of structural arterial disease have also been associated with symptomatic and asymptomatic PAD (Tziomalos et al. 2010). Higher carotid intima media thickness values were strongly associated with disease progression and development of subsequent cardiovascular events (Allan et al. 1997b; Allan et al. 1997a). Other markers related to the levels of arterial wall remodeling and stiffness (i.e. pulse wave velocity and augmentation index) have been shown to be impaired in patients with PAD compared to controls (Khandanpour et al. 2009a; Khaleghi and Kullo 2007) and were associated with the functional capacity of these patients. (Brewer et al. 2007; moh-Tonto et al. 2009).

These vascular measurements are very important but the vast majority of primary practitioners lack trained personnel and the necessary equipment for assessment. As such, the scientific interest has focused on the identification of soluble biomarkers which are associated with PAD and can easily be measured in the blood stream (Charakida et al. 2012). These biomarkers should be sensitive and specific to allow stratification of individuals with PAD and to provide prognostic information for disease evolution and cardiovascular morbidity and mortality.

## Inflammatory Biomarkers and PAD (Table 2)

Over the last decades, there has been increasing interest for the role of inflammation in both the development and clinical severity of PAD. Several inflammatory triggers, including classical cardiovascular risk factors, have been shown to promote inflammatory pathways and exert a pro-atherogenic role (Urbonaviciene et al. 2011a; Charakida et al. 2011). Among the different inflammatory markers C-reactive protein has attracted most of the scientific interest.

### *CRP and PAD (Table 3)*

CRP is produced in the liver in response to interleukin -6 and can stimulate endothelial cells to promote adhesion and migration of monocytes and pro-inflammatory cytokine expression (Vainas et al. 2005). Numerous studies have shown that CRP is associated with the development and progression of PAD but can also provide predictive information for later cardiovascular events.

In the Physicians Health study, Ridker et al. demonstrated in a cohort of 14,916 apparently healthy men that the relative risk of developing PAD increased by almost three fold from the lowest to highest quartile of CRP (Ridker et al. 2001). This risk was independent of conventional cardiovascular risk factors and remained the strongest non-lipid predictor

**Table 2.** Key biochemical markers and peripheral arterial disease.

| Biomarkers | PAD vs controls | Link with functional capacity | Link with PAD severity & CV events | Issues for use as screening tools |
|---|---|---|---|---|
| CRP | increased | + | + | Not specific for PAD. Unknown if lowering CRP benefits patients with PAD. |
| Cytokine-IL6 | increased | + | Very few associations | Not specific for PAD. Found in small amounts in plasma. |
| Cellular adhesion molecules | increased | + | Evidence that predict PAD progression | Not specific. Not easily amenable to treatment. |
| Beta 2-microglobulin | increased | + | Limited information | Not specific. Limited discriminating capacity between functional arthritis and PAD. Affected by kidney function. |
| MMPs (MMP9, MMP2) | increased | Limited information | + | Not specific. Limited information about levels in asymptomatic individuals. |
| Fibrinogen | increased | Limited information | + | Not specific. |
| Homocysteine | increased | Limited information | + | Not specific. Reduction in homocysteine levels did not reduce CV risk in PAD patients. |

**Abbreviations:** CRP: C-reactive protein, PAD: peripheral arterial disease, IL6: interleukin 6, MMP: matrix metalloproteinases, CV: cardiovascular.

**Table 3.** Studies investigating the association between CRP and PAD.

| Study | No. of Participants | Study Design | Outcome |
|---|---|---|---|
| Physicians Health study (Ridker et al. 2001) | 14916 | Prospective (9 yrs) | CRP was the strongest non-lipid predictor of PAD |
| Edinburgh Artery Study (Tzoulaki et al. 2005) | 1592 | Prospective (12 yrs) | CRP was associated with progressive atherosclerosis at 12 years, independently of baseline ABI, cardiovascular risk factors, and baseline cardiovascular disease |
| National Health and Nutrition Examination Survey (NHANES) (Pande et al. 2008) | 3242 | Cross-sectional | CRP levels >3 mg/L were strongly associated with PAD, although stratifying patients for levels of insulin resistance removed the significant association between CRP and PAD |
| de Haro et al. 2008 | 3370 | Cross-sectional | CRP plasma levels were increased in association with higher clinical severity of PAD |
| National Health and Nutrition Examination Survey (NHANES) (Vu et al. 2005) | 1600 | Cross-sectional | Elevated CRP levels were associated with increased risk of PAD, with the highest risk being recorded in patients with type 2 diabetes and high levels of CRP |
| van der Meer IM et al./Rotterdam Study (Van et al. 2002) | 773 | Cohort (6.5 yrs) | After adjustment for age, sex and smoking habit, the odds ratio associated with CRP levels in the highest, compared with the lowest quartile, was 2.0 and 1.9 for progression of iliac and lower extremity atherosclerosis, respectively |
| Risk of Arterial Thrombosis in Relation to Oral Contraceptives (RATIO) study (Bloemenkamp et al. 2002) | 687 (212 cases, 475 controls) | Multicenter, population-based, case-control | Elevated CRP levels were associated with an increased likelihood of peripheral arterial disease |
| SPACE Trial (Vainas et al. 2005) | 387 | Prospective (1 yr) | CRP related to the ankle-brachial pressure index (ABPI) at inclusion and at 12 months' follow-up and to future vascular events (death and coronary, cerebral, and peripheral arterial events) |
| Tsurugaya Project (Hozawa et al. 2004) | 946 | Cross-sectional | High levels of CRP related to a low ankle brachial index among Japanese elderly (> 70 years old) |

*Table 3. contd....*

*Table 3. contd.*

| Study | No. of Participants | Study Design | Outcome |
|---|---|---|---|
| Walking and Leg Circulation Study (WALCS) (McDermott et al. 2004) | 188 | Observational | After adjusting for age, sex, race, body mass index, ABI, comorbidities, smoking, total cholesterol/HDL ratio and statin use, a significant inverse linear association was found between physical activity levels and CRP levels in subjects with PAD and age > 55 yrs old |
| The Homocysteine and Progression of Atherosclerosis Study (HPAS) (Musicant et al. 2006) | 384 | Prospective (mean follow up 38.4 months) | No association between CRP levels and markers of atherosclerosis progression |
| Aboyans et al. 2006 | 403 | Prospective (mean follow up 4.6 yrs) | CRP related to large vessel PAD progression but not with small vessel PAD progression |
| Schillinger et al. 2003 | 89 | Prospective (6 months) | High levels of CRP measured before and 48 hours after PTA in patients with PAD correlated with lower ABI and were predictive of re-stenosis after 6 months |
| Schillinger et al. 2002a | 172 | Prospective (6 months) | In patients with PAD, CRP values at baseline and 48 hours after intervention were independently associated with 6-month restenosis after PTA |
| Urbonaviciene et al. 2011a | 463 | Prospective (mean follow up 6.1 yrs) | Levels of CRP predicted cardiovascular mortality in patients with intermittent claudication but not in those with chronic critical limb ischemia |
| Criqui et al. 2010 | 397 | Prospective (mean follow up 2.2 and 6.6 yrs) | Levels of CRP independently predicted mortality at 2.2 but 6.6 yrs in patients PAD |
| Vidula et al. 2010 | 579 | Prospective (mean follow up 3.7 yrs) | Among PAD patients under statin treatment a significant reduction in mortality is recorded only in those with high baseline levels of CRP |
| Schlager et al. 2009 | 447 | Prospective (mean follow up 15.6 months) | In PAD patients, CRP predicts cardiovascular and renal outcome irrespective of the presence of RAS, with levels of CRP above 0.88 mg/dL identifying subjects at particularly high risk for major cardiovascular events |

**Abbreviations:** CRP: C-reactive protein, RAS: renal artery stenosis, PTA: percutaneous transluminal angioplasty, ABPI: ankle-brachial pressure index, HDL: high density lipoprotein.

for the development of PAD. This strong relationship was also reproduced in healthy women (Pradhan et al. 2008). Similar findings were reported from another prospective cohort study, the Edinburg Artery Study, in which 1,592 subjects were followed up over a 12 year period (Tzoulaki et al. 2005).

Data now suggests that inflammatory mechanisms have a key role, not only in the development, but also in the progression of PAD (Brevetti et al. 2010). In the Rotterdam study, CRP and interleukin-6 were inversely associated with ABI after adjusting for classical cardiovascular risk factors (Meijer et al. 2000). Similarly, in the National Health and Nutrition Examination Survey from 1999–2002, inflammatory markers including CRP and leukocyte count were significantly increased in patients with lower ABI category (Wildman et al. 2005). Further evidence for the close link between inflammation and PAD severity is the finding that levels of inflammatory markers increase with advanced Fontaine stage classification, with CRP and IL-6 levels being higher in patients with symptomatic PAD (Andreozzi et al. 2007). McDermott et al. demonstrated that CRP levels were independently associated with six minute walk distance and summary performance score (McDermott et al. 2008). However, when interpreting these results, one has to consider that most of the noted associations are derived from cross-sectional studies and as such the causal link between inflammation and severity of PAD remains to be established.

A number of different studies have demonstrated that inflammatory markers and in particular, CRP have predictive value for the occurrence of cardiovascular events in patients with PAD (Rizzo et al. 2008). In most studies risk stratification based on ABI was significantly improved when inflammatory markers were added to the model (Vainas et al. 2005; Nishida et al. 2011; Owens et al. 2007). Despite these positive associations, the independent role of CRP in risk prediction has recently been questioned. For instance, Hogh et al. demonstrated in 452 prospectively studied patients with symptomatic PAD that CRP was not independently associated with the risk for future cardiovascular events (Hogh et al. 2008).

## *Beta 2 microglobulin and cystatin C in PAD (Table 2)*

In the search of a biomarker for PAD, Wilson et al. performed proteomic analysis and revealed high levels of beta 2 microglobulin in patients with PAD compared to controls (Wilson et al. 2007b). Beta 2 microglobulin (B2M) is a component of the major histocompatibility complex class I molecule. Free beta2 microglobulin circulates in the blood stream from cell surface shedding or intracellular release (Cooper and Plesner 1980; Libby et al. 2002). Increased B2M levels in PAD patients may reflect ischemic reperfusion injury commonly noted in these patients or merely be an index of their inflammatory response. Interestingly, in the same study B2M levels were higher in patients with coronary artery disease and PAD and were associated with lower ABI values and decreased functional capacity, suggesting that measurement of this biomarker may provide information for the severity of the PAD disease. In another smaller study, Kals et al. showed that increased B2M levels are associated with arterial stiffening as assessed by augmentation index in patients with PAD (Kals et al. 2011). The question, though, remains: can this protein be the "unique" biomarker for PAD screening and stratification? Studies so far have demonstrated that increases in B2M are not specific indicators of PAD and has limited discriminating capacity to differentiate between functional arthritis or PAD. The protein is increased in a variety of autoimmune, infectious, and neoplastic conditions and as it is solely excreted by the kidneys, is significantly affected by renal function (Acchiardo et al. 1989).

Another protein which has attracted scientific interest is Cystatic C. Cystatic C, produced by the CST3 gene, is endogenous inhibitors of cathepsine and has been generally used as biomarker of renal function. Recent data however suggest that it may play an important role in vascular remodeling (Koenig et al. 2005). Increased cystatic C levels have been found in patients with PAD compared to controls and provided predictive information for later complications including bypass surgery, angiography or amputation as well as cardiovascular mortality even in patients without renal dysfunction (Urbonaviciene et al. 2011b; Eriksson et al. 2004). As with B2M, the specificity of this protein for PAD is limited. The lack of specificity, however, does not preclude its use as a disease marker. More recently, the biomarker index scores have been introduced to refine diagnosis and improve prognosis of patients with PAD. Fung et al. demonstrated in 540 individuals that a four biomarker panel comprising of B2M, cystatin C, hsCRP, and glucose are associated with PAD independent of traditional risk factors and may provide complementary information for PAD (Fung et al. 2008).

## Other inflammatory biomarkers and PAD

Other inflammatory biomarkers which have been independently associated with increased vascular risk in PAD are adhesion molecules and leukocyte count. Increased inflammation, with production of cytokines and adhesion molecules, endothelial dysfunction, and increased oxidative stress have been associated with vascular risk in patients with PAD at rest and during exercise (Signorelli et al. 2007; Signorelli et al. 2003). Neumann et al. demonstrated that total neutrophil count was higher in the venous blood draining from the affected limb compared to the unaffected contraleteral one following claudication (Neumann et al. 1990).

Blann et al. demonstrated that increased levels of soluble intracellular adhesion molecule-1 predicted adverse cardiovascular events in patients with ultrasound proven atherosclerotic vascular stenosis >70% of the carotid, iliac and femoral artery (Blann and McCollum 1998). Similarly, in patients with intermittent claudication, increased levels of soluble vascular cell adhesion molecule-1 were associated with a four- fold increased cardiovascular risk (Brevetti et al. 2006).

In the Prevention of Atherosclerotic Risk Complication study (including almost 2000 patients with intermittent claudication), baseline leukocyte count was a significant predictor of myocardial infarction, stroke and vascular death (Dormandy and Murray 1991). More recently, Haumer et al. demonstrated that elevated neutrophil count but not leukocyte count were associated with higher composite cardiovascular event (Haumer et al. 2005). On the basis of these findings, it has been suggested that white cell count may not merely reflect acute inflammation but may have pathogenetic role by promoting endothelial and microvascular disturbances, therefore favoring atheroscletotic plaque progression and rupture.

A number of other inflammatory biomarkers have also been tested as prognostic markers for PAD progression and adverse cardiovascular events including monocyte chemo attractant protein-1, myeloperoxidase cytokines i.e. tumor necrosis factor alpha and interleukin 6 and serum osteoprotegerin (Khawaja and Kullo 2009). Despite the positive associations found in various studies, reproduction of the findings in large prospective cohorts is lacking and as such their clinical utility remains as yet unjustified.

## Matrix Metalloproteinases and Peripheral Arterial Disease (Table 2)

Metalloproteinases (MMPs) are a family of enzymes which are produced by inflammatory cells, activated platelets, and degraded extracellular matrix and are thought to be important modulators of atherothrombosis (Hobeika et al. 2007). Platelet-derived growth factor and transforming growth factor-β stimulate MMPs and their activity is regulated by several cytokines, a2 macroglobulin and tissue inhibitor of MMPs. By disrupting and remodeling extracellular matrix, they allow diapedesis of inflammatory cells and the influx of lipoproteins through the vessel wall. MMP2 and 9, in particular, promote smooth muscle cell proliferation promoting atherosclerotic plaque formation. MMP-9 also is strongly associated with plaque instability (Newby 2005). Upregulation of MMP-9 in the atherosclerotic plaque has been associated with increased plaque hemorrhage and rupture in experimental studies (Hobeika et al. 2007). Several studies have demonstrated that MMPs play an important role in the progression of PAD disease. In the initial phases, in response to arterial stenosis and reduction in blood flow, tissue and muscle respond with a series of adaptive changes including capillary growth and remodeling of the arterial wall. Two members of the MMP family, in particular MMP-2 and 9, have been involved in the angiogenic response to ischemia (Hobeika et al. 2007). Higher concentrations of MMP-9 have been noted in coronary artery disease patients and were associated with lipoprotein markers (Tayebjee et al. 2005a; Tayebjee et al. 2005b). In patients with angiographically proven coronary artery disease, MMP-9 levels were independently associated with later cardiovascular mortality in a prospectively followed cohort of 1,127 patients, even after adjustment for cardiovascular risk factors (Blankenberg et al. 2003). Eldrup et al. in a prospectively followed study of patients with severe carotid stenosis, they demonstrated that increased baseline concentrations of MMP-9 were associated with increased risk for stroke or cardiovascular death (Eldrup et al. 2006). Active MMP-9 is elevated in type II diabetics with PAD (Signorelli et al. 2005). MMPs have also been associated with the severity of PAD. Tayebjee et al. demonstrated that MMP-9 levels are significantly increased in patients with peripheral arterial disease and in particular, those with critical limb ischemia (Tayebjee et al. 2005b). Recent findings suggest that MMP 9 is elevated in patients with metabolic syndrome and parallel the rise of inflammatory mediators (Goncalves et al. 2009). The sources of MMP-9 production in the circulation of PAD patients are not well established. It has been suggested that MMP-9 can be derived from ischemic muscle in those patients and by circulating neutrophils or monocytes (Busti et al. 2010). As such, increased MMP circulating concentrations may also reflect increased vessel wall expression and increased inflammatory response. Limited information exists about their concentration in asymptomatic patients with PAD. Further research is needed to clarify the role of specific MMPs in PAD and the mechanisms involved in their dysregulation.

## Hemostatic Factors and Peripheral Arterial Disease (Table 4)

Increased thrombotic tendency and low fibrinolytic capacity has been consistently demonstrated in patients with PAD (Killewich et al. 1998; Cassar et al. 2005). Whether this finding reflects primary thrombogenic derangement or abnormal interaction between the disturbed endothelial surface and blood constituents remains unclarified.

Fibrinogen plays a key role in platelet adhesion and aggregation. Although its plasma levels are at least in part genetically determined common cardiovascular risk factors such as smoking, diabetes and increasing age have consistently been associated with increased fibrinogen levels. Fibrinogen is also involved in inflammatory processes by facilitating leukocyte adherence and migration to the endothelial surface. Higher

**Table 4.** Association between PAD, fibrinogen and fibrinolysis.

| Authors/Study | No. Participants | Study Design | Outcome |
|---|---|---|---|
| Rotterdam Study (Meijer et al. 2000) | 6450 | Cross-sectional | Fibrinogen was a strong and independent determinant associated with PAD. |
| Atherosclerosis Risk in Communities (ARIC) (Wattanakit et al. 2005) | 1651 | Prospective (mean follow up 10.3 yrs) | Elevated fibrinogen levels were independently associated with greater PAD incidence. |
| Bianchi et al. 2007 | 1610 | Cross-sectional | Diabetic patients with PAD had significantly higher fibrinogen levels compared to DM patients without PAD. |
| PLAT Study (Cortellaro et al. 1994) | 108 | Case-control, Cross-sectional | Patients with PAD have high D-dimer levels and low fibrinolytic capacity compared to patients with coronary and/or cerebral atherothrombotic disease, but free of leg atherosclerosis. |
| Roller et al. 1999 | 68 | Case-control, case-control | PAD patients had significantly higher levels of fibrinogen and impaired fibrinolitic activity compared to healthy controls. . |
| Unlu et al. 2006 | 89 | Cross-sectional, case-control | D-Dimer higher in patients with PAD compared to healthy controls and correlate with the severity of PAD. |
| Herren et al. 1994 | 45 | Cross-sectional, case-control | Prothrombin fragments 1+2, thrombin–antithrombin III complexes, fibrinopeptide A were significantly higher in patients with PAD compared to the healthy control subjects and correlated with the severity of PAD. |
| Lee et al. 1995 | 1592 | Cross-sectional | t-PA antigen and fibrin D-dimers levels increased with increasing severity of PAD and fibrin D-dimers correlated independently with the risk of intermittent claudication. |
| Edinburgh Artery Study (Smith et al. 2000) | 1592 | Prospective (12 yrs) | Fibrinogen predicted development of PAD at follow up. |
| Baxter et al. 1988 | 100 | Cross-sectional, case-control | Asymptomatic men with PAD had higher levels of fibrinogen compared to healthy controls. |

*Table 4. contd....*

*Table 4. contd.*

| Authors/Study | No. Participants | Study Design | Outcome |
|---|---|---|---|
| Cardiovascular Heart Study (Newman et al. 1993) | 5084 | Cross-sectional | PAD inversely correlated with fibrinogen levels . |
| Lassila et al. 1993 | 40 | Cross-sectional | Fibrinogen levels associated with PAD severity . |
| Fowkes et al. 1993 | 617 | Prospective | In patients with claudication, plasma fibrinogen was the strongest independent predictor of death from coronary disease. |
| Koksch et al. 1999 | 60 | Cross-sectional | Increased plasma levels of fibrinogen with PAD progression. |
| Woodburn et al. 1995 | 82 | Prospective (follow up 4 months) | Plasma fibrinogen levels decrease significantly following successful resolution of critical limb ischemia in PAD patients. |
| Tschopl et al. 1997 | 71 | Prospective (follow up 6 months) | High plasma fibrinogen levels predicted the development of re-stenosis at the site of PTA. |
| Schillinger et al. 2002b | 246 | Prospective (follow up 6 months) | Patients with high pre-intervention fibrinogen values had increased adjusted risk for re-stenosis following PTA for iliac artery occlusive disease. |
| Lapolla et al. 2003 | 63 | Prospective (follow up 4 yrs) | In diabetic patients without PAD, lower t-PA levels and higher PAI-1 Ag and PAI-1 Act values at baseline were significantly correlated with PAD development. |
| Arterial Disease Multiple Intervention Trial (ADMIT) (Philipp et al. 1997) | 122 | Cross-sectional, case-control | Association between fibrinogen levels and peripheral arterial stenosis and PAD |
| Boneu et al. 1998 | 324 | Prospective (follow up 2 yrs) | Even in patients with only moderate PAD, high levels of TAT complexes and PAI-1 were predictive of vascular complications. |

**Abbreviations:** PAD: peripheral arterial disease, t-PA: tissue plasminogen activator, PAI-1: plasminogen activator inhibitor 1, TAT: thrombin-antithrombin complex, PTA: percutaneous transluminal angioplasty, DM: diabetes mellitus.

baseline fibrinogen levels have been associated with increased risk for development of PAD disease in a number of prospective studies (Price et al. 1999). In the Atherosclerotic Risk in Communities study (ARIC), including 1,651 patients with diabetes who were followed up for 10 years, multivariate analysis showed that elevated fibrinogen and other cardiovascular risk factors (i.e., smoking and diabetes) were independently associated with higher incidence of PAD (Reich et al. 2007). Similarly, in the Edinburgh and the ARIC study, plasma fibrinogen levels showed strong associations with future development of PAD (Lee et al. 1999c; Lee et al. 1999b; Wattanakit et al. 2005).

Patients with PAD have higher levels of D-dimers and prothrombin fragments compared to healthy controls. Higher fibrinogen levels were inversely associated with ABI in a number of cross sectional and prospective studies (Meijer et al. 2000). In the Edinburgh study, plasma fibrinogen levels predict PAD progression over 12 years of follow up (Smith et al. 2000; Tzoulaki et al. 2007). Notably, the association between thrombosis and PAD severity was consistent in different ethnic populations including African Americans and non-Hispanic whites.

Apart from fibrinogen, elevated D-dimers have been associated with lower ABI in several studies. McDermott et al. demonstrated that decline in the functional capacity of patients with PAD as assessed by walking distance and physical activity levels was associated with increased D-dimers (McDermott et al. 2007; McDermott et al. 2006; McDermott et al. 2005). However, this finding was challenged by Vidula et al., where in a group of 384 patients with PAD who were followed up prospectively, D-dimers were not associated with disease progression (Vidula et al. 2008).

Other thrombotic biomarkers including von Willebrand factor, tissue plasminogen activator and plasminogen activator inhibitor 1 have all been associated with increased PAD severity (Lee et al. 1999a; Smith et al. 1995). In particular, vWF levels were highly correlated with the severity of limb ischemia in patients with PAD (Blann et al. 1998b). Interestingly, resolution of ischemia and improvement in endothelial function in these patients was associated with decrease in plasma levels of vWF (Blann et al. 1998a). Increased risk of re-stenosis after percutaneous transluminal angioplasty in patients with elevated PAI-1 levels has also been documented. In the Cardiovascular Heart Study, fibrinogen levels were strongly associated with ABI (Newman et al. 1993). Thrombomodulin, another anticoagulant protein and specific marker of endothelial damage, has been associated with severity of PAD, highlighting the progressive release of thrombomodulin with increasing endothelial damage during PAD progression (Mota et al. 2009).

A number of studies investigated the prognostic value of hemostatic factors in patients with PAD. In a recent meta-analysis, the association between fibrinogen and the risk for coronary heart disease and stroke has been demonstrated (Bosevski et al. 2005). It has been hypothesized that high fibrinogen levels promote platelet aggregation and as such increase the generation of fibrin, therefore promoting atheroma formation. Increased cardiovascular mortality in PAD patients with high fibrinogen levels has also been shown (Vidula et al. 2008). However, as yet, data is not available to suggest that by decreasing fibrinogen levels, cardiovascular risk will decrease.

## Genetic Predisposition and Peripheral Arterial Disease

Recent studies support a genetic predisposition to PAD. The National Heart, Lung and Blood Institute Twin Study assessed the heritability of PAD in 94 monozygotic and 90 dizygotic twins (Carmelli et al. 2000). Low ABI was noted in 33 percent of monozygotic twins and 31 percent of dizygotic twins. In this small study, it appeared that a twin of a patient with PAD was four times more likely to have PAD and that genetic factors determined 48% of the variability in ankle brachial index after adjustment for other cardiovascular risk factors (Murabito et al. 2009; Murabito et al. 2006). In the Framingham Offspring Study, the heritability of low ABI and PAD was estimated as 21% whereas other cardiovascular risk factors contributed by 14% (Murabito et al. 2006). Finally, another family study (GENOA, Genetic Epidemiology Network of Arteriopathy) showed a modest heritability for PAD (Kullo et al. 2006).

Despite these findings, a small number of case-control studies have shown isolated associations with inflammatory and thrombotic polymorphism and increased incidence of PAD (Knowles et al. 2007).

Gaetani et al. demonstrated that the polymorphism of the EE genotype of the intercellular adhesion molecule -1 was associated with increased incidence of PAD (Gaetani et al. 2002). Libra et al. documented that a specific polymorphism of the pro-inflammatory cytokine IL-6 [the G(-174)] influences and promotes the development of PAD in patients with type 2 diabetes (Libra et al. 2006). Associations between polymorphisms in adhesion molecules, monocyte chemoattractant protein-1 and PAD have also been reported (Flex et al. 2007). Polymorphisms at the beta fibrinogen locus (-455G/A polymorphism) were associated with increased risk for PAD, the effect of which was independent of other cardiovascular risk factors. Gene environment interaction has also been shown. In approximately 11,000 Icelandic smokers, a genetic variant on chromosome 15q24 in the nicotine acetylcholine receptor was associated with the smoking habit and presence of PAD.

The available data suggest a modest degree of heritability in PAD and that there is not a consistent genetic marker that could identify individuals at increased risk for PAD.

## Miscellaneous Biomarkers and PAD

Oxidative stress is implicated in the development and progression of atherosclerotic disease (Madamanchi et al. 2005) but only limited information is available for associations between specific markers of oxidative stress and PAD. In small case control studies, glutathione peroxidase-1 activity, 8-iso-prostaglandin F2a, and L ascorbic acid have been associated with PAD (Pipinos et al. 2006; Mueller et al. 2004).

Homocysteine, an amino acid produced during the metabolism of methionine, has also been associated with PAD (Table 2). Darius et al., in a cross sectional study of 6,880 primary care patients, demonstrated that elevated homocysteine plasma levels were associated with reduced ABI (Darius et al. 2003). Similar findings were reported in hypertensive adults and in type II diabetic patients (Ciccarone et al. 2003; Khawaja et al. 2007). In a recent meta-analysis of 14 epidemiological studies, it was demonstrated that patients with PAD have consistently higher levels of homocysteine compared to controls, suggesting that homocysteine may either be a marker of PAD or be etiologically involved in the development of PAD (Khandanpour et al. 2009b). In addition, elevated homocysteine levels predicted all cause mortality in patients with PAD (Taylor et al. 1999; Lange et al. 2005).

Diet supplementation with vitamins B6, B12 and folic acid reduces homocysteine levels. The Heart Outcome Prevention Evaluation (HOPE2) study found that B vitamins do not reduce cardiovascular risk in patients with PAD (Lonn et al. 2006). These results were reproduced by other studies and as such folate supplementation is currently not recommended in patients with PAD (Albert et al. 2008; Khandanpour et al. 2009b).

Markers of endothelial dysfunction like ADMA have also been associated with the development and severity of PAD in some studies but not in others (Boger et al. 1997; Mittermayer et al. 2006; Wilson et al. 2007a). Markers of myocardial injury like troponins and b-natriuretic peptide have been shown to independently predict cardiovascular mortality in symptomatic patients with PAD. Their assays are easily available and reliable. Although the current information suggest that these may provide important diagnostic, therapeutic, and prognostic information in PAD, their clinical use for screening patients with PAD remains limited (Emberson et al. 2007; Feringa et al. 2007; Mueller et al. 2009; Morrow et al. 2001).

## Challenges in Evaluating Biomarkers for PAD

Similar to other cardiovascular diseases, a number of soluble markers (including inflammatory mediators, markers of endothelial injury, lipoproteins and hemostatic factors) have been associated with PAD in population based studies, although none of these biomarkers are specific for PAD. It is thus likely that similar pathophysiological processes are involved in cardiovascular diseases and discrepancies between studies merely reflect differences in the expression of biomarkers in various vascular beds. Another important confounder between studies relates to the effect of medications and other associated diseases including renal or hepatic diseases. Careful phenotypic characterization of patients participating in various studies is necessary for comparison and interpretation of the reported findings.

The current development of proteomic profiling, although still in its infancy, is likely to provide important information for the discovery and characterization of biomarkers for PAD. The discovery process is complicated, considering that the low abundance proteins are often the most important. Technical factors related to blood collection and storage may also create protein degradation and introduce artifacts and these should be carefully standardized when analysis is performed.

For a specific biomarker, there will always be a number of false positive and false negative associations and in this case a biomarker panel might be more useful to refine diagnosis and establish prognosis in patients with PAD. The ideal PAD biomarker panel is likely to incorporate those which can easily be measured in the systemic circulation and reflect pathophysiological processes contributing to inflammation, oxidative stress, coagulation and matrix remodeling.

## Conclusion

Peripheral arterial disease is associated with increased cardiovascular morbidity and mortality in the general population. A number of circulating inflammatory and hemostatic biomarkers have been evaluated as predictors for the development of PAD as well as for the severity and progression of the disease. Currrent data suggests that no single marker is adequate to characterize the complexity of the disease and a multi-marker approach is required to identify patients at increased risk for development and accelerated progression of PAD. However, to be clinically useful, such a biomarker has to be stable, easily measured, and should provide both diagnostic and prognostic incremental information to current biomarkers.

The search for novel and specific biomarkers for PAD is necessary as early recognition of the disease is likely to lead to better clinical care and improved long term outcome.

# References

Aboyans, V., M.H. Criqui, J.O. Denenberg, J.D. Knoke, P.M. Ridker, and A. Fronek. 2006. Risk factors for progression of peripheral arterial disease in large and small vessels. Circulation 113: 2623–2629.

Acchiardo, S., A.P. Kraus, Jr., and B.R. Jennings. 1989. Beta 2-microglobulin levels in patients with renal insufficiency. Am. J. Kidney Dis. 13: 70–74.

Albert, C.M., N.R. Cook, J.M. Gaziano, E. Zaharris, J. MacFadyen, E. Danielson, J.E. Buring, and J.E. Manson. 2008. Effect of folic acid and B vitamins on risk of cardiovascular events and total mortality among women at high risk for cardiovascular disease: a randomized trial. JAMA 299: 2027–2036.

Allan, P.L., P.I. Mowbray, A.J. Lee, and F.G. Fowkes. 1997a. Relationship between carotid intima-media thickness and symptomatic and asymptomatic peripheral arterial disease. The Edinburgh Artery Study. Stroke 28: 348–353.

Allan, P.L., P.I. Mowbray, A.J. Lee, and F.G. Fowkes. 1997b. Relationship between carotid intima-media thickness and symptomatic and asymptomatic peripheral arterial disease. The Edinburgh Artery Study. Stroke 28: 348–3.

Andreozzi, G.M., R. Martini, R. Cordova, A. D'Eri, G. Salmistraro, M. Mussap, and M. Plebani. 2007. Circulating levels of cytokines (IL-6 and IL-1beta) in patients with intermittent claudication, at rest, after maximal exercise treadmill test and during restore phase. Could they be progression markers of the disease? Int. Angiol. 26: 245–252.

Baxter, K., S. Wiseman, J. Powell, and R. Greenhalgh. 1988. Pilot study of a screening test for peripheral arterial disease in middle aged men: fibrinogen as a possible risk factor. Cardiovasc. Res. 22: 300–302.

Bianchi, C., G. Penno, F. Pancani, A. Civitelli, A. Piaggesi, F. Caricato, G. Pellegrini, P.S. Del, and R. Miccoli. 2007. Non-traditional cardiovascular risk factors contribute to peripheral arterial disease in patients with type 2 diabetes. Diabetes Res. Clin. Pract. 78: 246–253.

Blankenberg, S., H.J. Rupprecht, O. Poirier, C. Bickel, M. Smieja, G. Hafner, J. Meyer, F. Cambien, and L. Tiret. 2003. Plasma concentrations and genetic variation of matrix metalloproteinase 9 and prognosis of patients with cardiovascular disease. Circulation 107: 1579–1585.

Blann, A., A. Bignell, and C. McCollum. 1998a. von Willebrand factor, fibrinogen and other plasma proteins as determinants of plasma viscosity. Atherosclerosis 139: 317–322.

Blann, A., A. Bignell, and C. McCollum. 1998b. von Willebrand factor, fibrinogen and other plasma proteins as determinants of plasma viscosity. Atherosclerosis 139: 317–322.

Blann, A.D. and C.N. McCollum. 1998. Circulating ICAM-1 in peripheral arterial disease as a predictor of adverse events. Lancet 351: 1135.

Bloemenkamp, D.G., M.A. Van Den Bosch, W.P. Mali, B.C. Tanis, F.R. Rosendaal, J.M. Kemmeren, A. Algra, F.L. Visseren, and G.Y. van der. 2002. Novel risk factors for peripheral arterial disease in young women. Am. J. Med. 113: 462–467.

Boger, R.H., S.M. Bode-Boger, W. Thiele, W. Junker, K. Alexander, and J.C. Frolich. 1997. Biochemical evidence for impaired nitric oxide synthesis in patients with peripheral arterial occlusive disease. Circulation 95: 2068–2074.

Boneu, B., P. Leger, and C. Arnaud. 1998. Haemostatic system activation and prediction of vascular events in patients presenting with stable peripheral arterial disease of moderate severity. Royat Study Group. Blood Coagul. Fibrinolysis 9: 129–135.

Bosevski, M., S. Kostoska, S. Tosev, and V. Borozanov. 2005. Prognostic importance of haemostatic parameters in polyarterial disease. Prilozi. 26: 81–92.

Brevetti, G., V. Schiano, and M. Chiariello. 2008. Endothelial dysfunction: a key to the pathophysiology and natural history of peripheral arterial disease? Atherosclerosis 197: 1–11.

Brevetti, G., G. Giugliano, L. Brevetti, and W.R. Hiatt. 2010. Inflammation in peripheral artery disease. Circulation 122: 1862–1875.

Brevetti, G., V. Schiano, and M. Chiariello. 2006. Cellular adhesion molecules and peripheral arterial disease. Vasc. Med. 11: 39–47.

Brewer, L.C., H.S. Chai, K.R. Bailey, and I.J. Kullo. 2007. Measures of arterial stiffness and wave reflection are associated with walking distance in patients with peripheral arterial disease. Atherosclerosis 191: 384–390.

Busti, C., E. Falcinelli, S. Momi, and P. Gresele. 2010. Matrix metalloproteinases and peripheral arterial disease. Intern. Emerg. Med. 5: 13–25.

Carmelli, D., R.R. Fabsitz, G.E. Swan, T. Reed, B. Miller, and P.A. Wolf. 2000. Contribution of genetic and environmental influences to ankle-brachial blood pressure index in the NHLBI Twin Study. National Heart, Lung, and Blood Institute. Am. J. Epidemiol. 151: 452–458.

Cassar, K., P. Bachoo, I. Ford, M. Greaves, and J. Brittenden. 2005. Markers of coagulation activation, endothelial stimulation and inflammation in patients with peripheral arterial disease. Eur. J. Vasc. Endovasc. Surg. 29: 171–176.

Charakida, M., F. O'Neil, S. Masi, N. Papageorgiou, and D. Tousoulis. 2011. Inflammatory disorders and atherosclerosis: new therapeutic approaches. Curr. Pharm. Des. 17: 4111–4120.

Charakida, M., S. Masi, and D. Tousoulis. 2012. Functional, genetic and biochemical biomarkers of peripheral arterial disease. Curr. Med. Chem. 19: 2497–2503.

Ciccarone, E., C.A. Di, D. Assanelli, S. Archetti, G. Ruggeri, N. Salcuni, M.B. Donati, F. Capani, and L. Iacoviello. 2003. Homocysteine levels are associated with the severity of peripheral arterial disease in Type 2 diabetic patients. J. Thromb. Haemost. 1: 2540–2547.

Cooper, E.H. and T. Plesner. 1980. Beta-2-microglobulin review: its relevance in clinical oncology. Med. Pediatr. Oncol. 8: 323–334.

Cortellaro, M., E. Cofrancesco, C. Boschetti, L. Mussoni, M.B. Donati, M. Catalano, L. Gabrielli, B. Lombardi, G. Specchia, G. Tavazzi et al. 1994. Association of increased fibrin turnover and defective fibrinolytic capacity with leg atherosclerosis. The PLAT Group. Thromb. Haemost. 72: 292–296.

Coutinho, T., T.W. Rooke, and I.J. Kullo. 2011. Arterial dysfunction and functional performance in patients with peripheral artery disease: a review. Vasc. Med. 16: 203–211.

Criqui, M.H., L.A. Ho, J.O. Denenberg, P.M. Ridker, C.L. Wassel, and M.M. McDermott. 2010. Biomarkers in peripheral arterial disease patients and near- and longer-term mortality. J. Vasc. Surg. 52: 85–90.

Darius, H., D. Pittrow, R. Haberl, H.J. Trampisch, A. Schuster, S. Lange, H.G. Tepohl, J.R. Allenberg, and C. Diehm. 2003. Are elevated homocysteine plasma levels related to peripheral arterial disease? Results from a cross-sectional study of 6880 primary care patients. Eur. J. Clin. Invest 33: 751–757.

de Haro, J., F. Acin, A. Lopez-Quintana, F.J. Medina, E. Martinez-Aguilar, A. Florez, and J.R. March. 2008. Direct association between C-reactive protein serum levels and endothelial dysfunction in patients with claudication. Eur. J. Vasc. Endovasc. Surg. 35: 480–486.

Dormandy, J.A. and G.D. Murray. 1991. The fate of the claudicant—a prospective study of 1969 claudicants. Eur. J. Vasc. Surg. 5: 131–133.

Eldrup, N., M.L. Gronholdt, H. Sillesen, and B.G. Nordestgaard. 2006. Elevated matrix metalloproteinase-9 associated with stroke or cardiovascular death in patients with carotid stenosis. Circulation 114: 1847–1854.

Emberson, J.R., L.L. Ng, J. Armitage, L. Bowman, S. Parish, and R. Collins. 2007. N-terminal Pro-B-type natriuretic peptide, vascular disease risk, and cholesterol reduction among 20,536 patients in the MRC/BHF heart protection study. J. Am. Coll. Cardiol. 49: 311–319.

Eriksson, P., H. Deguchi, A. Samnegard, P. Lundman, S. Boquist, P. Tornvall, C.G. Ericsson, L. Bergstrand, L.O. Hansson, S. Ye, and A. Hamsten. 2004. Human evidence that the cystatin C gene is implicated in focal progression of coronary artery disease. Arterioscler. Thromb. Vasc. Biol. 24: 551–557.

Feringa, H.H., O. Schouten, M. Dunkelgrun, J.J. Bax, E. Boersma, A. Elhendy, J.R. de, S.E. Karagiannis, R. Vidakovic, and D. Poldermans. 2007. Plasma N-terminal pro-B-type natriuretic peptide as long-term prognostic marker after major vascular surgery. Heart 93: 226–231.

Flex, A., E. Gaetani, F. Angelini, A. Sabusco, C. Chilla, G. Straface, F. Biscetti, P. Pola, J.J. Castellot, Jr., and R. Pola. 2007. Pro-inflammatory genetic profiles in subjects with peripheral arterial occlusive disease and critical limb ischemia. J. Intern. Med. 262: 124–130.

Fowkes, F.G., G.D. Lowe, E. Housley, A. Rattray, A. Rumley, R.A. Elton, I.R. MacGregor, and J. Dawes. 1993. Cross-linked fibrin degradation products, progression of peripheral arterial disease, and risk of coronary heart disease. Lancet 342: 84–86.

Fung, E.T., A.M. Wilson, F. Zhang, N. Harris, K.A. Edwards, J.W. Olin, and J.P. Cooke. 2008. A biomarker panel for peripheral arterial disease. Vasc. Med. 13: 217–224.

Gaetani, E., A. Flex, R. Pola, P. Papaleo, M.D. De, E. Pola, F. Aloi, R. Flore, M. Serricchio, A. Gasbarrini, and P. Pola. 2002. The K469E polymorphism of the ICAM-1 gene is a risk factor for peripheral arterial occlusive disease. Blood Coagul. Fibrinolysis 13: 483–488.

Goncalves, F.M., A.L. Jacob-Ferreira, V.A. Gomes, A. Casella-Filho, A.C. Chagas, A.M. Marcaccini, R.F. Gerlach, and J.E. Tanus-Santos. 2009. Increased circulating levels of matrix metalloproteinase (MMP)-8, MMP-9, and pro-inflammatory markers in patients with metabolic syndrome. Clin. Chim. Acta 403: 173–177.

Haumer, M., J. Amighi, M. Exner, W. Mlekusch, S. Sabeti, O. Schlager, I. Schwarzinger, O. Wagner, E. Minar, and M. Schillinger. 2005. Association of neutrophils and future cardiovascular events in patients with peripheral artery disease. J. Vasc. Surg. 41: 610–617.

Herren, T., H. Stricker, A. Haeberli, D.D. Do, and P.W. Straub. 1994. Fibrin formation and degradation in patients with arteriosclerotic disease. Circulation 90: 2679–2686.

Hobeika, M.J., R.W. Thompson, B.E. Muhs, P.C. Brooks, and P.J. Gagne. 2007. Matrix metalloproteinases in peripheral vascular disease. J. Vasc. Surg. 45: 849–857.

Hogh, A.L., J. Joensen, J.S. Lindholt, M.R. Jacobsen, and L. Ostergaard. 2008. C-reactive protein predicts future arterial and cardiovascular events in patients with symptomatic peripheral arterial disease. Vasc. Endovascular. Surg. 42: 341–347.

Hozawa, A., K. Ohmori, S. Kuriyama, T. Shimazu, K. Niu, A. Watando, S. Ebihara, T. Matsui, M. Ichiki, R. Nagatomi, H. Sasaki, and I. Tsuji. 2004. C-reactive protein and peripheral artery disease among Japanese elderly: the Tsurugaya Project. Hypertens. Res. 27: 955–961.

Joras, M. and P. Poredos. 2008. The association of acute exercise-induced ischaemia with systemic vasodilator function in patients with peripheral arterial disease. Vasc. Med. 13: 255–262.

Kals, J., M. Zagura, M. Serg, P. Kampus, K. Zilmer, E. Unt, J. Lieberg, J. Eha, A. Peetsalu, and M. Zilmer. 2011. beta2-microglobulin, a novel biomarker of peripheral arterial disease, independently predicts aortic stiffness in these patients. Scand. J. Clin. Lab Invest 71: 257–263.

Khaleghi, M. and I.J. Kullo. 2007. Aortic augmentation index is associated with the ankle-brachial index: a community-based study. Atherosclerosis 195: 248–253.

Khandanpour, N., M.P. Armon, B. Jennings, A. Clark, and F.J. Meyer. 2009a. The association between ankle brachial pressure index and pulse wave velocity: clinical implication of pulse wave velocity. Angiology 60: 732–738.

Khandanpour, N., Y.K. Loke, F.J. Meyer, B. Jennings, and M.P. Armon. 2009b. Homocysteine and peripheral arterial disease: systematic review and meta-analysis. Eur. J. Vasc. Endovasc. Surg. 38: 316–322.

Khawaja, F.J., K.R. Bailey, S.T. Turner, S.L. Kardia, T.H. Mosley, Jr., and I.J. Kullo. 2007. Association of novel risk factors with the ankle brachial index in African American and non-Hispanic white populations. Mayo Clin. Proc. 82: 709–716.

Khawaja, F.J. and I.J. Kullo. 2009. Novel markers of peripheral arterial disease. Vasc. Med. 14: 381–392.

Killewich, L.A., A.W. Gardner, R.F. Macko, D.J. Hanna, A.P. Goldberg, D.K. Cox, and W.R. Flinn. 1998. Progressive intermittent claudication is associated with impaired fibrinolysis. J. Vasc. Surg. 27: 645–650.

Knowles, J.W., T.L. Assimes, J. Li, T. Quertermous, and J.P. Cooke. 2007. Genetic susceptibility to peripheral arterial disease: a dark corner in vascular biology. Arterioscler. Thromb. Vasc. Biol. 27: 2068–2078.

Koenig, W., D. Twardella, H. Brenner, and D. Rothenbacher. 2005. Plasma concentrations of cystatin C in patients with coronary heart disease and risk for secondary cardiovascular events: more than simply a marker of glomerular filtration rate. Clin. Chem. 51: 321–327.

Koksch, M., F. Zeiger, K. Wittig, D. Pfeiffer, and C. Ruehlmann. 1999. Haemostatic derangement in advanced peripheral occlusive arterial disease. Int. Angiol. 18: 256–262.

Kullo, I.J., S.T. Turner, S.L. Kardia, T.H. Mosley, Jr., E. Boerwinkle, and A.M. de. 2006. A genome-wide linkage scan for ankle-brachial index in African American and non-Hispanic white subjects participating in the GENOA study. Atherosclerosis 187: 433–438.

Lange, S., H.J. Trampisch, R. Haberl, H. Darius, D. Pittrow, A. Schuster, S.B. von, G. Tepohl, J.R. Allenberg, and C. Diehm. 2005. Excess 1-year cardiovascular risk in elderly primary care patients with a low ankle-brachial index (ABI) and high homocysteine level. Atherosclerosis 178: 351–357.

Lapolla, A., F. Piarulli, G. Sartore, C. Rossetti, L. Martano, P. Carraro, P.M. De, and D. Fedele. 2003. Peripheral artery disease in type 2 diabetes: the role of fibrinolysis. Thromb. Haemost. 89: 91–96.

Lassila, R., S. Peltonen, M. Lepantalo, O. Saarinen, P. Kauhanen, and V. Manninen. 1993. Severity of peripheral atherosclerosis is associated with fibrinogen and degradation of cross-linked fibrin. Arterioscler. Thromb. 13: 1738–1742.

Lee, A.J., F.G. Fowkes, G.D. Lowe, and A. Rumley. 1995. Fibrin D-dimer, haemostatic factors and peripheral arterial disease. Thromb. Haemost. 74: 828–832.

Lee, A.J., F.G. Fowkes, G.D. Lowe, J.M. Connor, and A. Rumley. 1999a. Fibrinogen, factor VII and PAI-1 genotypes and the risk of coronary and peripheral atherosclerosis: Edinburgh Artery Study. Thromb. Haemost. 81: 553–560.

Lee, A.J., F.G. Fowkes, G.D. Lowe, J.M. Connor, and A. Rumley. 1999b. Fibrinogen, factor VII and PAI-1 genotypes and the risk of coronary and peripheral atherosclerosis: Edinburgh Artery Study. Thromb. Haemost. 81: 553–560.

Lee, A.J., A.S. MacGregor, C.M. Hau, J.F. Price, A. Rumley, G.D. Lowe, and F.G. Fowkes. 1999c. The role of haematological factors in diabetic peripheral arterial disease: the Edinburgh artery study. Br. J. Haematol. 105: 648–654.

Libby, P., P.M. Ridker, and A. Maseri. 2002. Inflammation and atherosclerosis. Circulation 105: 1135–1143.

Libra, M., S.S. Signorelli, Y. Bevelacqua, P.M. Navolanic, V. Bevelacqua, J. Polesel, R. Talamini, F. Stivala, M.C. Mazzarino, and G. Malaponte. 2006. Analysis of G(-174)C IL-6 polymorphism and plasma concentrations of inflammatory markers in patients with type 2 diabetes and peripheral arterial disease. J. Clin. Pathol. 59: 211–215.

Lonn, E., S. Yusuf, M.J. Arnold, P. Sheridan, J. Pogue, M. Micks, M.J. McQueen, J. Probstfield, G. Fodor, C. Held, and J. Genest, Jr. 2006. Homocysteine lowering with folic acid and B vitamins in vascular disease. N. Engl. J. Med. 354: 1567–1577.

Madamanchi, N.R., A. Vendrov, and M.S. Runge. 2005. Oxidative stress and vascular disease. Arterioscler. Thromb. Vasc. Biol. 25: 29–38.

McDermott, M.M., K. Liu, J.M. Guralnik, S. Mehta, M.H. Criqui, G.J. Martin, and P. Greenland. 1998. The ankle brachial index independently predicts walking velocity and walking endurance in peripheral arterial disease. J. Am. Geriatr. Soc. 46: 1355–1362.

McDermott, M.M., P. Greenland, J.M. Guralnik, L. Ferrucci, D. Green, K. Liu, M.H. Criqui, J.R. Schneider, C. Chan, P. Ridker, W.H. Pearce, G. Martin, E. Clark, and L. Taylor. 2004. Inflammatory markers, D-dimer, pro-thrombotic factors, and physical activity levels in patients with peripheral arterial disease. Vasc. Med. 9: 107–115.

McDermott, M.M., L. Ferrucci, K. Liu, M.H. Criqui, P. Greenland, D. Green, J.M. Guralnik, P.M. Ridker, L.M. Taylor, N. Rifai, L. Tian, J. Zheng, W.H. Pearce, J.R. Schneider, and E. Vonesh. 2005. D-dimer and inflammatory markers as predictors of functional decline in men and women with and without peripheral arterial disease. J. Am. Geriatr. Soc. 53: 1688–1696.

McDermott, M.M., K. Liu, J.M. Guralnik, L. Ferrucci, D. Green, P. Greenland, L. Tian, M.H. Criqui, C. Lo, N. Rifai, P.M. Ridker, J. Zheng, and W. Pearce. 2006. Functional decline in patients with and without peripheral arterial disease: predictive value of annual changes in levels of C-reactive protein and D-dimer. J. Gerontol. A Biol. Sci. Med. Sci. 61: 374–379.

McDermott, M.M., L. Ferrucci, J.M. Guralnik, L. Tian, D. Green, K. Liu, J. Tan, Y. Liao, W.H. Pearce, J.R. Schneider, P. Ridker, N. Rifai, F. Hoff, and M.H. Criqui. 2007. Elevated levels of inflammation, d-dimer, and homocysteine are associated with adverse calf muscle characteristics and reduced calf strength in peripheral arterial disease. J. Am. Coll. Cardiol. 50: 897–905.

McDermott, M.M., K. Liu, L. Ferrucci, L. Tian, J.M. Guralnik, D. Green, J. Tan, Y. Liao, W.H. Pearce, J.R. Schneider, K. McCue, P. Ridker, N. Rifai, and M.H. Criqui. 2008. Circulating blood markers and functional impairment in peripheral arterial disease. J. Am. Geriatr. Soc. 56: 1504–1510.

Meijer, W.T., D.E. Grobbee, M.G. Hunink, A. Hofman, and A.W. Hoes. 2000. Determinants of peripheral arterial disease in the elderly: the Rotterdam study. Arch. Intern. Med. 160: 2934–2938.

Mittermayer, F., K. Krzyzanowska, M. Exner, W. Mlekusch, J. Amighi, S. Sabeti, E. Minar, M. Muller, M. Wolzt, and M. Schillinger. 2006. Asymmetric dimethylarginine predicts major adverse cardiovascular events in patients with advanced peripheral artery disease. Arterioscler. Thromb. Vasc. Biol. 26: 2536–2540.

moh-Tonto, C.A., A.R. Malik, V. Kondragunta, Z. Ali, and I.J. Kullo. 2009. Brachial-ankle pulse wave velocity is associated with walking distance in patients referred for peripheral arterial disease evaluation. Atherosclerosis 206: 173–178.

Morrow, D.A., C.P. Cannon, N. Rifai, M.J. Frey, R. Vicari, N. Lakkis, D.H. Robertson, D.A. Hille, P.T. DeLucca, P.M. DiBattiste, L.A. Demopoulos, W.S. Weintraub, and E. Braunwald. 2001. Ability of minor elevations of troponins I and T to predict benefit from an early invasive strategy in patients with unstable angina and non-ST elevation myocardial infarction: results from a randomized trial. JAMA 286: 2405–2412.

Mota, A.P., M.E. de Castro Santos, F.C. Lima e Silva, N.C. de Carvalho Schachnik, S.M. de Oliveira, and C.M. das Gracas. 2009. Hypercoagulability markers in patients with peripheral arterial disease: association to ankle-brachial index. Angiology 60: 529–535.

Mueller, T., B. Dieplinger, A. Gegenhuber, D. Haidinger, N. Schmid, N. Roth, F. Ebner, M. Landl, W. Poelz, and M. Haltmayer. 2004. Serum total 8-iso-prostaglandin F2alpha: a new and independent predictor of peripheral arterial disease. J. Vasc. Surg. 40: 768–773.

Mueller, T., B. Dieplinger, W. Poelz, G. Endler, O.F. Wagner, and M. Haltmayer. 2009. Amino-terminal pro-B-type natriuretic peptide as predictor of mortality in patients with symptomatic peripheral arterial disease: 5-year follow-up data from the Linz Peripheral Arterial Disease Study. Clin. Chem. 55: 68–77.

Murabito, J.M., C.Y. Guo, C.S. Fox, and R.B. D'Agostino. 2006. Heritability of the ankle-brachial index: the Framingham Offspring study. Am. J. Epidemiol. 164: 963–968.

Murabito, J.M., M.J. Keyes, C.Y. Guo, J.F. Keaney, Jr., R.S. Vasan, R.B. D'Agostino, Sr., and E.J. Benjamin. 2009. Cross-sectional relations of multiple inflammatory biomarkers to peripheral arterial disease: The Framingham Offspring Study. Atherosclerosis 203: 509–514.

Murphy, T.P., R. Dhangana, M.J. Pencina, and R.B. D'Agostino, Sr. 2012. Ankle-brachial index and cardiovascular risk prediction: An analysis of 11,594 individuals with 10-year follow-up. Atherosclerosis 220: 160–167.

Musicant, S.E., L.M. Taylor, Jr., D. Peters, R.A. Schuff, R. Urankar, G.J. Landry, and G.L. Moneta. 2006. Prospective evaluation of the relationship between C-reactive protein, D-dimer and progression of peripheral arterial disease. J. Vasc. Surg. 43: 772–780.

Neumann, F.J., W. Waas, C. Diehm, T. Weiss, H.M. Haupt, R. Zimmermann, H. Tillmanns, and W. Kubler. 1990. Activation and decreased deformability of neutrophils after intermittent claudication. Circulation 82: 922–929.

Newby, A.C. 2005. Dual role of matrix metalloproteinases (matrixins) in intimal thickening and atherosclerotic plaque rupture. Physiol Rev. 85: 1–31.

Newman, A.B., D.S. Siscovick, T.A. Manolio, J. Polak, L.P. Fried, N.O. Borhani, and S.K. Wolfson. 1993. Ankle-arm index as a marker of atherosclerosis in the Cardiovascular Health Study. Cardiovascular Heart Study (CHS) Collaborative Research Group. Circulation 88: 837–845.

Nishida, H., T. Horio, Y. Suzuki, Y. Iwashima, T. Tokudome, F. Yoshihara, S. Nakamura, and Y. Kawano. 2011. Interleukin-6 as an independent predictor of future cardiovascular events in high-risk Japanese patients: comparison with C-reactive protein. Cytokine 53: 342–346.

Owens, C.D., P.M. Ridker, M. Belkin, A.D. Hamdan, F. Pomposelli, F. Logerfo, M.A. Creager, and M.S. Conte. 2007. Elevated C-reactive protein levels are associated with postoperative events in patients undergoing lower extremity vein bypass surgery. J. Vasc. Surg. 45: 2–9.

Pande, R.L., T.S. Perlstein, J.A. Beckman, and M.A. Creager. 2008. Association of insulin resistance and inflammation with peripheral arterial disease: the National Health and Nutrition Examination Survey, 1999 to 2004. Circulation 118: 33–41.

Payvandi, L., A. Dyer, D. McPherson, P. Ades, J. Stein, K. Liu, L. Ferrucci, M.H. Criqui, J.M. Guralnik, D. Lloyd-Jones, M.R. Kibbe, S.T. Liang, B. Kane, W.H. Pearce, M. Verta, W.J. McCarthy, J.R. Schneider, A. Shroff, and M.M. McDermott. 2009. Physical activity during daily life and brachial artery flow-mediated dilation in peripheral arterial disease. Vasc. Med. 14: 193–201.

Philipp, C.S., L.A. Cisar, H.C. Kim, A.C. Wilson, P. Saidi, and J.B. Kostis. 1997. Association of hemostatic factors with peripheral vascular disease. Am. Heart J. 134: 978–984.

Pipinos, I.I., A.R. Judge, Z. Zhu, J.T. Selsby, S.A. Swanson, J.M. Johanning, B.T. Baxter, T.G. Lynch, and S.L. Dodd. 2006. Mitochondrial defects and oxidative damage in patients with peripheral arterial disease. Free Radic. Biol. Med. 41: 262–269.

Pradhan, A.D., S. Shrivastava, N.R. Cook, N. Rifai, M.A. Creager, and P.M. Ridker. 2008. Symptomatic peripheral arterial disease in women: nontraditional biomarkers of elevated risk. Circulation 117: 823–831.

Price, J.F., P.I. Mowbray, A.J. Lee, A. Rumley, G.D. Lowe, and F.G. Fowkes. 1999. Relationship between smoking and cardiovascular risk factors in the development of peripheral arterial disease and coronary artery disease: Edinburgh Artery Study. Eur. Heart J. 20: 344–353.

Ramos, R., M. Quesada, P. Solanas, I. Subirana, J. Sala, J. Vila, R. Masia, C. Cerezo, R. Elosua, M. Grau, F. Cordon, D. Juvinya, M. Fito, C.M. Isabel, A. Clara, M.M. Angel, and J. Marrugat. 2009. Prevalence of symptomatic and asymptomatic peripheral arterial disease and the value of the ankle-brachial index to stratify cardiovascular risk. Eur. J. Vasc. Endovasc. Surg. 38: 305–311.

Reich, L.M., G. Heiss, L.L. Boland, A.T. Hirsch, K. Wu, and A.R. Folsom. 2007. Ankle-brachial index and hemostatic markers in the Atherosclerosis Risk in Communities (ARIC) study cohort. Vasc. Med. 12: 267–273.

Ridker, P.M., M.J. Stampfer, and N. Rifai. 2001. Novel risk factors for systemic atherosclerosis: a comparison of C-reactive protein, fibrinogen, homocysteine, lipoprotein(a), and standard cholesterol screening as predictors of peripheral arterial disease. JAMA 285: 2481–2485.

Rizzo, M., E. Corrado, G. Coppola, I. Muratori, G. Novo, and S. Novo. 2008. Prediction of cardio- and cerebro-vascular events in patients with subclinical carotid atherosclerosis and low HDL-cholesterol. Atherosclerosis 200: 389–395.

Roller, R.E., S. Janisch, V. Carroll, E. Kvas, E. Pilger, B.R. Binder, and J. Wojta. 1999. Changes in the fibrinolytic system in patients with peripheral arterial occlusive disease undergoing percutaneous transluminal angioplasty. Thromb. Res. 94: 241–247.

Schillinger, M., M. Exner, W. Mlekusch, H. Rumpold, R. Ahmadi, S. Sabeti, M. Haumer, O. Wagner, and E. Minar. 2002a. Vascular inflammation and percutaneous transluminal angioplasty of the femoropopliteal artery: association with restenosis. Radiology 225: 21–26.

Schillinger, M., M. Exner, W. Mlekusch, H. Rumpold, R. Ahmadi, S. Sabeti, O. Wagner, and E. Minar. 2002b. Fibrinogen predicts restenosis after endovascular treatment of the iliac arteries. Thromb. Haemost. 87: 959–965.

Schillinger, M., M. Exner, W. Mlekusch, M. Haumer, H. Rumpold, R. Ahmadi, S. Sabeti, O. Wagner, and E. Minar. 2003. Endovascular revascularization below the knee: 6-month results and predictive value of C-reactive protein level. Radiology 227: 419–425.

Schlager, O., J. Amighi, M. Haumer, S. Sabeti, P. Dick, W. Mlekusch, C. Loewe, R. Koppensteiner, E. Minar, and M. Schillinger. 2009. Inflammation and adverse cardiovascular outcome in patients with renal artery stenosis and peripheral artery disease. Atherosclerosis 205: 314–318.

Shanmugasundaram, M., V.K. Ram, U.C. Luft, M. Szerlip, and J.S. Alpert. 2011. Peripheral arterial disease—what do we need to know? Clin. Cardiol. 34: 478–482.

Signorelli, S.S., M.C. Mazzarino, P.L. Di, G. Malaponte, C. Porto, G. Pennisi, G. Marchese, M.P. Costa, D. Digrandi, G. Celotta, and V. Virgilio. 2003. High circulating levels of cytokines (IL-6 and TNFalpha), adhesion molecules (VCAM-1 and ICAM-1) and selectins in patients with peripheral arterial disease at rest and after a treadmill test. Vasc. Med. 8: 15–19.

Signorelli, S.S., G. Malaponte, M. Libra, P.L. Di, G. Celotta, V. Bevelacqua, M. Petrina, G.S. Nicotra, M. Indelicato, P.M. Navolanic, G. Pennisi, and M.C. Mazzarino. 2005. Plasma levels and zymographic activities of matrix metalloproteinases 2 and 9 in type II diabetics with peripheral arterial disease. Vasc. Med. 10: 1–6.

Signorelli, S.S., M.C. Mazzarino, D.A. Spandidos, and G. Malaponte. 2007. Proinflammatory circulating molecules in peripheral arterial disease. Int. J. Mol. Med. 20: 279–286.

Smith, F.B., A.J. Lee, C.M. Hau, A. Rumley, G.D. Lowe, and F.G. Fowkes. 2000. Plasma fibrinogen, haemostatic factors and prediction of peripheral arterial disease in the Edinburgh Artery Study. Blood Coagul. Fibrinolysis 11: 43–50.

Smith, F.B., A.J. Lee, A. Rumley, F.G. Fowkes, and G.D. Lowe. 1995. Tissue-plasminogen activator, plasminogen activator inhibitor and risk of peripheral arterial disease. Atherosclerosis 115: 35–43.

Stoica, A. and C. Ginghina. 2008. Cardiovascular risk in patients with peripheral vascular diseases. Rom. J. Intern. Med. 46: 275–283.

Tayebjee, M.H., G.Y. Lip, K.T. Tan, J.V. Patel, E. A. Hughes, and R.J. MacFadyen. 2005a. Plasma matrix metalloproteinase-9, tissue inhibitor of metalloproteinase-2, and CD40 ligand levels in patients with stable coronary artery disease. Am. J. Cardiol. 96: 339–345.

Tayebjee, M.H., K.T. Tan, R.J. MacFadyen, and G.Y. Lip. 2005b. Abnormal circulating levels of metalloprotease 9 and its tissue inhibitor 1 in angiographically proven peripheral arterial disease: relationship to disease severity. J. Intern. Med. 257: 110–116.

Taylor, L.M., Jr., G.L. Moneta, G.J. Sexton, R.A. Schuff, and J.M. Porter. 1999. Prospective blinded study of the relationship between plasma homocysteine and progression of symptomatic peripheral arterial disease. J. Vasc. Surg. 29: 8–19.

Tschopl, M., D.A. Tsakiris, G.A. Marbet, K.H. Labs, and K. Jager. 1997. Role of hemostatic risk factors for restenosis in peripheral arterial occlusive disease after transluminal angioplasty. Arterioscler. Thromb. Vasc. Biol. 17: 3208–3214.

Tziomalos, K., V.G. Athyros, A. Karagiannis, and D.P. Mikhailidis. 2010. The role of ankle brachial index and carotid intima-media thickness in vascular risk stratification. Curr. Opin. Cardiol. 25: 394–398.

Tzoulaki, I., G.D. Murray, A.J. Lee, A. Rumley, G.D. Lowe, and F.G. Fowkes. 2005. C-reactive protein, interleukin-6, and soluble adhesion molecules as predictors of progressive peripheral atherosclerosis in the general population: Edinburgh Artery Study. Circulation 112: 976–983.

Tzoulaki, I., G.D. Murray, A.J. Lee, A. Rumley, G.D. Lowe, and F.G. Fowkes. 2007. Inflammatory, haemostatic, and rheological markers for incident peripheral arterial disease: Edinburgh Artery Study. Eur. Heart J. 28: 354–362.

Unlu, Y., S. Karapolat, Y. Karaca, and A. Kiziltunc. 2006. Comparison of levels of inflammatory markers and hemostatic factors in the patients with and without peripheral arterial disease. Thromb. Res. 117: 357–364.

Urbonaviciene, G., J. Frystyk, A. Flyvbjerg, S. Urbonavicius, E.W. Henneberg, and J.S. Lindholt. 2011a. Markers of inflammation in relation to long-term cardiovascular mortality in patients with lower-extremity peripheral arterial disease. Int. J. Cardiol. 160: 89–94.

Urbonaviciene, G., G.P. Shi, S. Urbonavicius, E.W. Henneberg, and J.S. Lindholt. 2011b. Higher cystatin C level predicts long-term mortality in patients with peripheral arterial disease. Atherosclerosis 216: 440–445.

Vainas, T., F.R. Stassen, G.R. de, E.L. Twiss, S.B. Herngreen, R.J. Welten, L.H. van den Akker, M.P. van Dieijen-Visser, C.A. Bruggeman, and P.J. Kitslaar. 2005. C-reactive protein in peripheral arterial disease: relation to severity of the disease and to future cardiovascular events. J. Vasc. Surg. 42: 243–251.

Van, D.M., I.M. P. de Maat, A.E. Hak, A.J. Kiliaan, A.I. Del Sol, D.A. van der Kuip, R.L. Nijhuis, A. Hofman, and J.C. Witteman. 2002. C-reactive protein predicts progression of atherosclerosis measured at various sites in the arterial tree: the Rotterdam Study. Stroke 33: 2750–2755.

Vidula, H., L. Tian, K. Liu, M.H. Criqui, L. Ferrucci, J.M. Guralnik, D. Green, P. Ridker, and M.M. McDermott. 2010. Comparison of effects of statin use on mortality in patients with peripheral arterial disease with versus without elevated C-reactive protein and d-dimer levels. Am. J. Cardiol. 105: 1348–1352.

Vidula, H., L. Tian, K. Liu, M.H. Criqui, L. Ferrucci, W.H. Pearce, P. Greenland, D. Green, J. Tan, D.B. Garside, J. Guralnik, P.M. Ridker, N. Rifai, and M.M. McDermott. 2008. Biomarkers of inflammation and thrombosis as predictors of near-term mortality in patients with peripheral arterial disease: a cohort study. Ann. Intern. Med. 148: 85–93.

Vu, J.D., J.B. Vu, J.R. Pio, S. Malik, S.S. Franklin, R.S. Chen, and N.D. Wong. 2005. Impact of C-reactive protein on the likelihood of peripheral arterial disease in United States adults with the metabolic syndrome, diabetes mellitus, and preexisting cardiovascular disease. Am. J. Cardiol. 96: 655–658.

Wattanakit, K., A.R. Folsom, E. Selvin, B.D. Weatherley, J.S. Pankow, F.L. Brancati, and A.T. Hirsch. 2005. Risk factors for peripheral arterial disease incidence in persons with diabetes: the Atherosclerosis Risk in Communities (ARIC) Study. Atherosclerosis 180: 389–397.

Widener, J.M. 2011. Peripheral arterial disease and disability from NHANES 2001–2004 data. J. Vasc. Nurs. 29: 104–112.

Wildman, R.P., P. Muntner, J. Chen, K. Sutton-Tyrrell, and J. He. 2005. Relation of inflammation to peripheral arterial disease in the national health and nutrition examination survey, 1999–2002. Am. J. Cardiol. 96: 1579–1583.

Wilson, A.M., R. Harada, N. Nair, N. Balasubramanian, and J.P. Cooke. 2007a. L-arginine supplementation in peripheral arterial disease: no benefit and possible harm. Circulation 116: 188–195.

Wilson, A.M., E. Kimura, R.K. Harada, N. Nair, B. Narasimhan, X.Y. Meng, F. Zhang, K.R. Beck, J.W. Olin, E.T. Fung, and J.P. Cooke. 2007b. Beta2-microglobulin as a biomarker in peripheral arterial disease: proteomic profiling and clinical studies. Circulation 116: 1396–1403.

Woodburn, K.R., A. Rumley, G.D. Lowe, and J.G. Pollock. 1995. Fibrinogen and markers of fibrinolysis and endothelial damage following resolution of critical limb ischaemia. Eur. J. Vasc. Endovasc. Surg. 10: 272–278.

# CHAPTER 10

# Biomarkers in Cardiorenal Syndrome

Costas Tsioufis,* Themistoklis Katsimichas, Anastasios Milkas
and Dimitris Tousoulis

## Introduction

The cardiorenal syndrome (CRS) is a disorder of the heart and kidneys that comprises five types, depending on which organ's acute or chronic injury or dysfunction manifests first (Ronco et al. 2010). In CRS type 1, acute worsening of heart function leads to acute kidney injury (AKI) and/or dysfunction. In CRS type 2, chronic heart abnormalities lead to kidney injury or dysfunction. CRS type 3 is a disorder where acute worsening of kidney function leads to heart injury and/or dysfunction, whereas in CRS type 4, chronic kidney disease (CKD) leads to heart injury, disease and/or dysfunction. Finally, in CRS type 5, a systemic condition, such as sepsis, leads to a simultaneous injury and/or dysfunction of both the heart and kidneys.

Injury to cells may first produce molecular modifications that will eventually lead to cellular damage and clinical syndromes. An effective biomarker may be a biological substance that will indicate cellular damage even prior to the damage itself, thus providing a time window for early intervention. It is widely accepted, and pursued, that the use of laboratory biomarkers for the early diagnosis of all CRS types will improve prognosis, as the implementation of early therapeutic measures prior to dysfunction onset will become possible. Research data on this matter abound, especially concerning the early diagnosis of AKI. Overall, more than 30 biological substances have been proposed as potential biomarkers in CRS. Table 1 shows relative molecules with respect to each CRS subtype. Research

1st Department of Cardiology, 'Hippokration' Hospital, University of Athens Medical School, Athens, Greece.
*Corresponding author: ktsioufis@hippocratio.gr

**Table 1.** Biomarkers of CRS.

| Syndrome | Acute cardiorenal (type 1) | Chronic cardiorenal (type 2) | Acute renocardiac (type 3) | Chronic renocardiac (type 4) | Secondary CRS (type 5) |
|---|---|---|---|---|---|
| Cardiac Biomarkers | Troponin, CK-MB, BNP, NT-proBNP, MPO, IMA | BNP, NT-proBNP, CRP | BNP, NT-proBNP, MPO, Troponin | BNP, NT-proBNP, CRP, Troponin, assymetricdimethylarginine, plasminogen activator inhibitor type 1, homocysteine, serum amyloid A protein, Hb, IMA | CRP, procalcitonin, BNP |
| Renal Biomarkers | NGAL, Creatinine, KIM 1, IL-6, IL-8, IL-18, Cystatin C, NAG, NHE3, Actin-actin depolymerasing F, α/π-GST, L-FABP, Netrin-1, Keratinocyte-derived chemokine (Gro -α), Fetuin-A, α₁-microglobulin, β₂-microglobulin, retinol binding protein, alaninaminopeptidase | Creatinine, Cystatin C, Urea, Uric Acid, CRP, GFR decrease | NGAL, Creatinine, KIM 1, IL-6, IL-8, IL-18, Cystatin C, NAG, NHE3, Actin-actin Depolymerasing F, α/π-GST, L-FABP, Netrin-1, Keratinocyte- derived chemokine (Gro-α),Fetuin-A, α₁-microglobulin, β₂-microglobulin, retinol binding protein, alaninaminopeptidase | Creatinine, Cystatin C, Urea, Uric Acid, GFR decrease | Creatinine, NGAL, IL-18, KIM-1, NAG |

**Abbreviations:** BNP: B-type Natriuretic Peptide, NT-proBNP: N Terminal proBNP, MPO: Myeloperoxidase, IMA: Ischemia Modified Albumin, CRP: C-Reactive Protein, NGAL: Neutrophil Gelatinase-Associated Lipocalin, KIM-1: Kidney Injury Molecule-1, NAG: N-acetyl-β-D-Glucosaminidase, NHE3: Na⁺/H⁺ Exchanger Isoform 3, α/π-GST: α/πGlutathione S Transferase, IL-: Interleukin-, L-FABP: L-type Fatty Acid Binding Protein, GFR: Glomerulal Filtration Rate, Hb: Hemoglobin.

emphasis has been given to the discovery and clinical utilization of AKI biological markers, which are nowadays considered as a top priority. Focus on patient care surely dictates such a strive, since current classic markers such as serum creatinine, proteinuria, and blood urea nitrogen (BUN) simply indicate damage too late or have serious other limitations, thus making a quest for more sensitive, specific and early biomarkers imperative. Moreover, emphasis has also been given to markers of heart failure (HF), such as the natriuretic peptides.

## Candidate Biomarker Properties

A useful biomarker should be highly sensitive and specific and able to early detect and classify CRS, also properly stratifying patients according to risk. It would also be advantageous if it could be used for treatment or to monitor the effects of treatment, or be combined with other diagnostic modes, such as imaging, for the diagnosis and treatment of CRS. Questions remain as to whether a marker has established reference values, and whether it has been tested in genders, all ages, and multiple ethnic groups. It is also imperative to explore biomarker expression time patterns in several different types of injury, so that relative information is updated and expanded to settings where injury may be of multiple origins. Other questions focus on the gold standard comparison chosen for the marker, and whether it was an appropriate one. Also, can the marker ultimately aid or even change clinical decision making? Finally, concerns arise from the fact that up to now biomarkers have been basically studied in small, single-center contexts and specific clinical settings. Table 2 summarizes ideal biomarker properties.

No matter the predictive strength of a given biomarker, however, chances that it will alone provide all the necessary information on the pathogeny, pathophysiology and prognosis of a certain condition are slim. It is likely that the formation of a panel of several biomarkers will be required, for a concise and thorough understanding of AKI, CKD and CRS effects and progression, and the same is true for other clinical settings, such as heart failure. Efforts must be made so that markers comprising the panel should offer knowledge not only on the type of eminent injury, but on pathophysiology pathways, as well. A biomarker panel may help differentiate between types of injury, verify its duration and severity, predict the outcome and monitor response to therapy. Some promising AKI biomarker panels and combinations have already been suggested, incorporating substances that will be further discussed later in this chapter (Dent et al. 2007; Liangos et al. 2007; Kwon et al. 2003; Devarajan 2007):

**Table 2.** Ideal biomarker properties.

| |
|---|
| Non-invasive |
| Easily detectable in accessible samples (like serum or urine) |
| Highly sensitive and specific for said disorder |
| Rapidly and reliably measurable |
| Capable of early detection |
| Able to give insight into etiology, nature and duration of insult |
| A marker of injury in addition to marker of function |
| Predictor of severity and reversibility |
| Helpful in monitoring course and the response to interventions |
| Unaffected by other biological variables |
| Inexpensive |

Modified from: Soni SS, Ronco C, Katz N, Cruz DN. Early diagnosis of acute kidney injury: the promise of novel biomarkers.BloodPurif. 2009; 28(3):165–74.

### For early AKI diagnosis and differentiation:

1) increased serum cystatin C and increased urinary IL-18, cystatin C, and KIM-1;
2) increased serum cystatin C, and increased urinary IL-18, NGAL, and π-GST.

### For AKI prognosis:

1) increased urinary actin, IL-6, and IL-8.
2) increased urinary actin, IL-6, IL-8, and γ-glutamyl transpeptidase.

### For AKI severity assessment:

1) increased urinary NAG and KIM-1.

### For predicting mortality in AKI:

1) increased urinary KIM-1, IL-18, and NAG levels.

Future multi-center studies and trials are warranted to better evaluate candidate biomarkers and possibly introduce them into clinical practice.

## Phases of Biomarker Clinical Evaluation

The American Heart Association has issued a scientific statement concerning criteria for the evaluation of novel markers of cardiovascular risk, proposing six phases of clinical, candidate biomarker validation (Hlatky et al. 2009; Siew et al. 2011).

Phase 1 concerns proof-of-concept studies, that are expected to show differences in biomarker levels between patients with and without the outcome of interest (e.g., AKI).

Phase 2 relates to prospective studies that determine the association of biomarker levels to the risk of future hard outcomes (such as dialysis requirement or mortality). A biomarker that could convey such information from a patient with as yet minimal injury might really be of great value.

Phase 3 includes studies examining the incremental value of candidate markers to existing predictors. Approaches to this matter have included multivariable regression analyses, discrimination and calibration. Discrimination produces summary characteristics, such as the AUC-ROC (Area Under Curve-Receiver Operating Characteristic) values, by considering such established data as sensitivity, specificity, and predictive value. The curve measures the probability of a marker to correctly rank two patients, one suffering from a given disorder and another who is not. Calibration attempts to quantify the risk for a given outcome and reflects how well the observed risk for an outcome matches the prediction made by using a set of covariates. Newer methods, to assess the way a potential marker improves classification across a wide spectrum of risk, include the integrated discrimination improvement index and the net classification index. The first index determines the extent to which a marker accurately increases or decreases the degree of risk and the latter defines the proportion of subjects reclassified appropriately to another predetermined risk cutoff, which might change treatment choices.

Phase 4 includes studies answering whether a potential marker will change disease management.

Lastly, Phase 5 incorporates studies of potential clininal outcome improvement by use of the biomarker and Phase 6 concerns studies of biomarker cost-effectiveness.

With the exception of established markers like troponins, most potential biomarker studies belong to Phases 1, 2 and 3, meaning there is still a long way ahead before they could claim a place in clinical practice.

## Renal Biomarkers in CRS

AKI is most definitely linked to a high morbidity and mortality, and to prolonged hospitalization and skyrocketing healthcare costs. The delay in the diagnosis of AKI by use of traditional biomarkers, such as serum creatinine, has hindered early, potentially life saving medical interventions. Candidate markers also illuminate the paths of CKD, enhancing knowledge on CKD pathophysiology, prognosis and CKD-linked adverse cardiac outcomes.

In general, desired AKI markers will either directly relate to structural damage, especially at the earliest stage, or characterize disorder pathways, so that the biomarker result will remain directly linked to the clinical outcomes in question. A multitude of research papers in the literature describe more than 30 potential AKI biomarkers and attempt to validate their utility. The most extensively studied potential biomarkers are NGAL, KIM-1, cystatin C, IL-6, IL-18, NAG, L-FABP, a/π GST, NHE3, and Netrin -1 (abbreviations explained in Table 1) (Siew et al. 2011). These markers serve either a functional, inflammatory, or structural role in renal tubular epithelia and are localized in various intracellular locations or on the plasma cell membrane, or they pose as low molecular weight proteins filtered by the glomerulus and/or metabolized by healthy tubular epithelial cells. Impaired tubular reabsorption and metabolism of filtered molecules, release of tubular cell proteins in response to ischemic or nephrotoxic assault, and proteins specifically produced and released in the course of AKI may thus provide urine full of potentially powerful markers. Their discriminating and predictive value is compared to serum creatinine, the reigning reference standard for AKI. Serum creatinine rises following glomerular filtration failure, but this may result even days after initial kidney injury. Creatinine levels have but limited value in the context of AKI and elevate only after about half the kidney function is lost. They are not AKI specific and require differentiation from other prerenal or outrenal causes of azotemia. Furthermore, serum creatinine levels are not specific for renal tubular injury, but seem to reflect the loss of glomerular filtration function that accompanies AKI development. Changes in serum creatinine levels are also influenced by many factors outside the context of renal injury, such as age, sex, body mass, and hydration and nutritional status. Moreover, at lower rates of glomerular filtration, tubular secretion of creatinine leads to overestimation of kidney function. Finally, in acute changes of glomerular filtration, serum creatinine does not accurately define renal function, until a steady-state equilibrium is reached, something that may take several days. Therefore, markers of renal tubular injury, that precedes filtration failure, would constitute a significant breakthrough in nephrology. Although limitations in the use of creatinine exist, the lack of other clinically applicable reference standards make its use as a comparator to new biomarkers reasonable.

In general, data conferred by candidate biomarkers either will agree or disagree with information derived from the interpretation of creatinine levels. When a biomarker suggests injury is present and dysfunction eminent, while creatinine levels have not risen, thorough examination of the case is warranted. The confirmation of increased risk for outcomes such as worsening AKI or adverse events like the need for dialysis will highly potentiate the usefulness of the said marker. Similarly, when biomarker levels suggest no injury but creatinine has risen, confirmation of non-renal causes of creatinine elevation or evidence of rapid recovery will further power biomarker use. When both biomarker and creatinine levels agree, the potential higher or lower risk of adverse outcomes should be confirmed and the correlation between creatinine and candidate marker verified. Hence, biomarkers that aspire to aid AKI diagnosis and prognosis must first be clinically evaluated.

Proof-of-concept studies of markers such as NGAL, NAG, cystatin C, IL-18, and L-FABP generally display an excellent result when it comes to differentiating between patients with established AKI and control subjects without an acutely elevated creatinine (Han et al. 2002; Parikh et al. 2004; Mori et al. 2005; Han et al. 2008; Ferguson et al. 2010; Vaidya et al. 2008). Studies in children undergoing cardiopulmonary bypass showed impressive results in terms of differentiating the levels of biomarkers between AKI and non-AKI patients with a discriminating performance assessed by AUC-ROC values ranging from 0.75 to 0.99 (Mishra et al. 2005; Han et al. 2008; Parikh et al. 2006a; Dent et al. 2007; Bennett et al. 2008; Portilla et al. 2008). Studies in adults in the same setting generally exhibited smaller predictive values, utilizing markers NGAL, IL-18, cystatin C, KIM-1, and NAG (Wagener et al. 2006; Koyner et al. 2008; Haase-Fielitz et al. 2009; Haase et al. 2008; Xin et al. 2008; Haase et al. 2009a; Ristikankare et al. 2010; Han et al. 2008; Liangos et al. 2009). In other clinical settings, the predictive ability of such biomarkers as cystatin C or NGAL appears moderately useful (Zappitelli et al. 2007; Nejat et al. 2010).

## *NGAL*

Neutrophil gelatinase-associated lipocalin (NGAL), or lipocalin-2, is one of the best studied and earliest biomarkers detected in the blood and urine of human AKI subjects (Xu et al. 2000; Mishra et al. 2003; Supavekin et al. 2003; Mori et al. 2005; Mishra et al. 2004a; Schmidt-Ott et al. 2006; Mori et al. 2007). Together with cystatin C, it holds the best chance of introduction into clinical practice in the near future, maybe even becoming "the troponin of the kidney" (Ronco et al. 2010). In humans, NGAL has a 25 kDa polypeptide chain of 178 amino acids and is found mostly as a monomer, though dimer and trimer forms exist. It was originally discovered as a protein covalently bound to matrix metalloproteinase-9 (MMP-9) from neutrophils. Its molecular structure includes eight β-strands that form a β-barrel enclosing a calyx. The calyx binds and transports low molecular weight substances.

NGAL is expressed by neutrophils, proximal collecting epithelia and other epithelial cells. Its gene is also expressed in organs such as the uterus, the trachea, the lungs, the stomach, and the salivary glands. It has been suggested that it operates by binding siderophores, transporting lipophilic substances and acting as a growth/differentiation factor. Its serum levels are known to rise in inflammatory conditions.

Early in AKI, NGAL levels rise within two distinct pools, a systemic and a renal pool. A NGAL rise in the systemic pool during AKI may also come from neutrophils, macrophages and other immune cells. Decreases in glomerular filtration rate (GFR) resulting from AKI should decrease NGAL clearance and result in further accumulation in the systemic pool. Regarding NGAL measurement, clinical studies have used ELIZA-based methods and devices, with cut-off concentrations at two hours being 150 ng/ml for plasma NGAL and 100 ng/ml for urinary NGAL in the cardiac surgery setting (Dent et al. 2007; Bennett et al. 2008). These provided good to excellent AUC, sensitivity, and specificity data and are easy to use in clinical practice, requiring only microliters of blood or urine and providing results in 15 and 35 minutes, respectively. Another study used a cut-off of 50 µg/L and found that at this threshold, NGAL was a powerful independent predictor of AKI (Mishra et al. 2005). Measurement has been commercialized and available for both blood and urine samples. Blood NGAL measurement is easy but invasive and may be confusing, if outrenal diseases affect NGAL concentrations. Urine measurement is non-invasive and there are less potentially interfering proteins present, but may be difficult to collect from oliguric patients. Also, hydration and diuretic therapy affect urinary NGAL concentrations, and it takes longer to analyze samples. Fine tuning cut-off points in the various clinical settings where NGAL has proved useful will be required, to advance its use in everyday practice.

NGAL has been proven useful both in the diagnosis and the prediction of AKI in various clinical settings. In children undergoing cardiac surgery, urinary and plasma NGAL at two hours after cardiopulmonary bypass was found to be a powerful independent predictor of AKI, with an AUC value of 0.998 and 0.91, respectively (Mishra et al. 2005). Urinary NGAL rose from a mean of 1.6 µg/L at baseline to 147 µg/L two hours after cardiopulmonary bypass and serum NGAL increased from a mean of 3.2 µg/L at baseline to 61 µg/L, also two hours after the procedure. Plasma NGAL at two hours following cardiopulmonary bypass in another study of children undergoing cardiac surgery showed excellent results in AKI prediction (AUC-ROC 0.96) and correlated well with postoperative changes in serum creatinine, duration of AKI and length of hospital stay, while 12 hours NGAL was a predictor of mortality (Dent et al. 2007). NGAL levels also mark delayed graft function (Parikh et al. 2006b), contrast media-induced AKI (Bachorzewska-Gajewska et al. 2006) and AKI in critically ill, intensive care child patients, where at 48 hours prior to the development of AKI, urinary NGAL had an AUC-ROC of 0.79 for subsequent AKI (Zappitelli et al. 2007). In another prospective study in children with congenital heart disease undergoing cardiac catheterization, and by using a cut-off of 100 ng/mL, NGAL in both urine (AUC 0.92) and plasma (AUC 0.91) also predicted contrast-induced nephropathy within two hours, when AKI diagnosis by serum creatinine was possible only six to 24 hours after contrast media administration.

In other settings, serum NGAL in critically ill children with systemic inflammatory response syndrome (SIRS) or septic shock was highly sensitive but lacked specificity (84 percent percent as opposed to 39 percent percent) for AKI, also correlating with systemic disease severity (Wheeler et al. 2008). Urinary NGAL in critically ill, mechanically ventilated patients rose more than six-fold two days prior to a 50 percent percent increase in serum creatinine (Zappitelli et al. 2007). NGAL measurement in the emergency room was found to be highly sensitive and specific in diagnosing AKI, and predicted nephrology consultation, admission to ICU and dialysis (Nickolas et al. 2008). Urinary NGAL has been found to be an early AKI biomarker in critically ill polytrauma patients, as well. Contrary to creatinine, NGAL also correlates with established AKI risk factors, such as cardiopulmonary bypass and cardiac surgery aortic cross-clamp time (Wagener et al. 2008).

A recent review and meta-analysis of the diagnostic and prognostic value of NGAL in AKI examined data from studies involving 2,538 patients, of whom 19.2 percent percent developed AKI (Haase et al. 2009b). Overall, NGAL AUC-ROC for AKI prediction was 0.815 (contrast media injury: 0.894, cardiac surgery: 0.775, critically ill patients: 0.728) and diagnostic accuracy of blood NGAL (AUC 0.775) compared well to that of urine NGAL (AUC 0.837). NGAL's prognostic power was better in children (AUC 0.930) than adults (AUC 0.782) and it succeeded in predicting both renal replacement therapy (AUC 0.782) and in-hospital mortality (AUC 0.706).

In chronic settings, urinary and serum NGAL levels were higher in Autosomal Dominal Polycystic Kidney Disease (ADPKD) patients than in controls and strongly correlated with GFR (Bolignano et al. 2007). Higher NGAL levels were also correlated to higher cystic growth, and a possible implication of NGAL in cystic growth was suggested. In a rat model, inflammation and immune responses were involved in CKD pathogenesis and NGAL gene expression was increased, suggesting NGAL as a marker for sustained renal injury after AKI, that could progress to CKD (Ko et al. 2010). In a study of nondiabetic patients with CKD, serum NGAL levels were increased, correlating with disease severity (Malyszko et al. 2009). Serum and urinary NGAL also predicted an estimated GFR (eGFR) fall in multiple etiology CKD patients in mid-term follow up (Bolignano et al. 2009a). In one study of drug-induced, chronic tubulointerstitial nephritis, increased urinary NGAL predicted kidney function worsening (Wu et al. 2010). Thus, NGAL may have an effective future role in CKD prognosis.

An animal study showed that NGAL even exhibited renoprotective results in ischemia reperfusion AKI (Mishra et al. 2004b). Administration of a purified recombinant NGAL before, during, or after ischemia resulted in considerable amelioration of the morphologic and functional sequlae with a reduction in apoptotic tubular cells and an increase in proliferating proximal tubular cells, post ischemic injury. NGAL was one of seven genes highly upregulated in mouse models of renal ischemia reperfusion injury. The protein was detected in urine within two hours after ischemia and its levels correlated to ischemia duration. In the context of ischemic and cisplatin injury, NGAL may decrease injury by reducing apoptosis and enhancing proliferation of kidney tubular cells. It may also increase iron trafficking and cause up-regulation of heme oxygenase-1, further protecting kidney tubular cells (Mishra et al. 2003; Mishra et al. 2004a).

Limitations in NGAL biomarker properties do exist. In one study, NGAL levels were significantly elevated in all patients (with or without AKI), raising the question of whether cardiopulmonary bypass leads to inflammation and activation of neutrophils (Wagener et al. 2008). In another study, *in vitro* testing showed NGAL was ultrafiltered and absorbed by polysulfone membranes, reducing its blood levels. If confirmed *in vivo*, this finding will hinder monitoring the course of AKI using NGAL, as many patients end up in dialysis (Bobek et al. 2008). Age has been found to modify NGAL's performance as an AKI biomarker, with studies on children yielding better predictive results for AKI. Plasma NGAL may also be influenced by confounding factors such as hypertension, inflammation, anemia, and existing cancer, though its rise in blood is generally much less in such conditions than AKI. Lastly, most data have been produced from single-center studies and homogenous patient populations, much of them not including CKD patients. Many studies have failed to report sensitivity, specificity and AUC-ROC values for the diagnosis of AKI. Future studies are needed to fine-tune cut-off levels, better demonstrate an association between NGAL and clinical outcomes such as dialysis, cardiovascular events, and death, and in order to prove that randomization to a treatment for AKI based on increased NGAL levels leads to improvement in kidney function and reduction of adverse clinical outcomes.

## KIM-1

Kidney injury molecule-1 (KIM-1 in humans, Kim-1 in the rat) is a type I transmembrane glycoprotein with two extracellular domains, a six cysteine immunoglobulin-like domain, and a threonine/serine and proline-rich domain characteristic of mucinlike O-glycoslyated proteins, suggesting a role in cell-cell and/or cell-matrix interactions (Bailly et al. 2002). It structurally resembles mucosal addressin cell adhesion molecule-1 (MadCAM-1). Studies have shown that it constitutes the prototype member for a family of molecules called KIMs or TIMs (T-cell immunoglobulin mucin proteins) and it may play a role as a sensor or receptor for adhesion signalling in contexts such as cell-cell and cell-pathogen interaction, including T-cell subtype differentiation (Ichimura et al. 2008). Its gene was found to be the most upregulated in the post ischemic rat kidney (Ichimura et al. 1998). KIM-1 gene and protein are highly expressed in regenerating proximal tubular cells, a finding suggestive of a KIM-1 role in renal repair. It may hold a key part in the dedifferentiation of renal epithelial cells and the remodelling of epithelial architecture during post injury repair procedures, mediating phagocytocis of apoptotic and necrotic cells, among others. It seems to induce neighbouring epithelial cells to phagocytose apoptotic tubular cells, making way for renal repair and moderating the inflammatory response after kidney injury (Bonventre 2010).

KIM-1 is measured only in urine, using ELISA based methods. After proximal tubular kidney injury, its ectodomain is cleaved in a metalloproteinase-dependent fashion and shed from cells into the urine in both rodents and humans. The protein is detectable after both ischemic and nephrotoxic injury and seems highly specific for ischemic AKI and not pre-renal azotemia, CKD, or contrast-induced nephropathy, as levels of urinary KIM-1 were found considerably lower in contrast nephropathy, non-acute tubular necrosis (ATN) AKI, CKD, and were below threshhold in normal controls (Han et al. 2002; Ichimura et al. 2004; Vaidya et al. 2006a; Liangos et al. 2007). KIM-1 was expressed in renal biopsy material of acute tubular necrosis AKI patients (detected by immunohistochemistry assays), specifically on the apical surface of proximal tubular epithelial cells (Han et al. 2002). It was also detected in the urine of similar patients, both by Western Blot protein analysis and ELISA quantification, at least 12 hours post insult, reaching levels of about 2 ng/ml (Han et al. 2002). Adjusted for age, gender, and length of time delay between insult and sampling, a one-unit increase in normalized (to urine creatinine) KIM-1 was associated with a bigger than twelve-fold increase in the presence of ATN (Han et al. 2002). Recently, a fast, direct immunochromatographic lateral flow 15-min assay for the detection of Kim-1 and KIM-1 has also been reported, as a urine dipstick test (Vaidya et al. 2009). The urinary Kim-1 band intensity significantly correlated with Kim-1 levels measured by a microbead-based assay, histopathological damage, and immunohistochemical assessment of renal Kim-1 in a dose- and time-dependent manner. The KIM-1 band intensity was significantly greater in the urine of AKI patients than of normal controls. Also, the KIM-1 dipstick proved useful in evaluating kidney injury and recovery in two patients who developed AKI after surgery for mesothelioma, with local cisplatin administration.

In children undergoing cardiac surgery, urinary KIM-1 levels were markedly increased at 12 hours, with an AUC of 0.83 for predicting AKI (Vaidya et al. 2006a). KIM-1 also predicts mortality and the need for renal replacement therapy in AKI (Liangos et al. 2007). In prospective studies of adult patients undergoing cardiac surgery, KIM-1 AUC-ROC values for AKI prediction at the time of admission were between 0.57 and 0.78 (Han et al. 2008; Liangos et al. 2009; Han et al. 2009). The predictive value of KIM-1 reached an AUC of 0.83 12 hours after operation, however at this time kidney injury might be serious and not reversible (Han et al. 2008). KIM-1 also shows prognostic significance in patients with established AKI, as elevated levels were significantly associated with death or dialysis, even after adjustment for disease severity or comorbidities (Liangos et al. 2007).

In ischemia-reperfusion and cisplatin-induced nephrotoxicity rat models, urinary Kim-1 is a sensitive and specific marker of proximal tubular injury and increased earlier than any of the conventional biomarkers, such as plasma creatinine and blood urea nitrogen, or other markers, such as NAG (Vaidya et al. 2006a). In recent research, Kim-1 had an AUC-ROC of 0.99 for proximal tubular toxicity and of 21 urinary markers studied, it had the highest sensitivity and specificity (Dieterle et al. 2007).

KIM-1 protein expression has been quantified in renal transplant biopsies and correlated with renal functional indices (Zhang et al. 2008). Although biopsies revealed no tubular injury, there was focal positive KIM-1 expression in 28 percent percent of the cases. This might indicate superior KIM-1 expression sensitivity in tracing proximal tubular injury than morphologic examination. In this study, KIM-1 expression was detected in all biopsies from patients with deterioration in kidney function and histological changes suggestive of tubular damage (Zhang et al. 2008). KIM-1 expression substantially correlated with serum creatinine and BUN levels and was inversely correlated with eGFR on the day of biopsy. Furthermore, KIM-1 was expressed focally in affected tubules in 92 percent percent of kidney biopsies from patients with acute cellular rejection.

## *N-acetyl-β-D-Glucosaminidase (NAG)*

NAG is an enzyme of the lysosomes, specifically the most active lysosome glucosidase in proximal tubular cells. It is not filtered due to its high molecular weight of more than 130 kDa, therefore it is a specific urinary marker for the tubular cells. NAG can be measured in urine by fast protein liquid chromatography and can be normalized to units per gram of urinary creatinine (U/g urinary Cr).

During kidney injury, urinary NAG levels remain constantly elevated, compared to normal controls; the increase in urinary NAG activity is highly sensitive for AKI and could indicate tubular cell damage, though it can also reflect increasedlysosomal activity without damage (Liangos et al. 2007; Trof et al. 2006). Increased NAG levels have been reported in studies of nephrotoxicity (Emeigh Hart 2005; D'Amico 2003), delayed kidney graft function (Mukhopadhyay et al. 2004), chronic glomerular disease (Bazzi et al. 2002), diabetic nephropathy (Ikenaga et al. 1993), and after cardiopulmonary bypass (Ascione et al. 1999). Urinary NAG levels were highly sensitive in detecting AKI in critically ill adults, preceding the rise of creatinine levels up to four days (Westhuyzen et al. 2003). Worse outcomes were seen in patients with higher urinary NAG levels on admission to a renal unit, suggesting a dose response to injury (Chew et al. 1993). Another study reported that urinary NAG levels correlated well with the outcomes of death or dialysis (Liangos et al. 2007). Another study reported an impressive NAG AUC-ROC of 0.97 for the diagnosis of AKI (Han et al. 2008). In patients who developed AKI after cardiopulmonary bypass procedures, the rise in urinary NAG levels occurred six hours after surgery and about 18 hours before the increase in serum creatinine levels (Han et al. 2008). In a similar setting, urinary NAG for predicting AKI immediately after surgery and at three hours was 74 percent against 83 percent and 29 percent against 54 percent sensitive, respectively, depending on the values used. NAG specificity was 44 percent against 50 percent and 71 percent against 80 percent, respectively, and the AUC-ROC values 0.61 and 0.63, respectively.

Possible disadvantages of NAG as an AKI biomarker are that NAG activity has been found to be inhibited by endogenous urea (Bondiou et al. 1985) and a number of nephrotoxic agents and heavy metals (Vaidya et al. 2006b), as well as the fact that NAG urinary excretion also increases in diseases of the glomerulus, like diabetic nephropathy (Ikenaga et al. 1993; Marchewka et al. 2001) and in essential hypertension (Harmankaya et al. 2001). These facts suggest a role for NAG in a more chronic setting like CKD, but research in this area is limited.

## *Cystatin C*

Cystatin C is a 13.3 kDa cysteine protease inhibitor produced by all nucleated cells at a constant rate, and serves as an endogenous marker of kidney function. It is freely filtered by the glomerulus, 100 percent reabsorbed by the proximal collecting tubule, and not secreted (Nguyen et al. 2008). Therefore, detection of cystatin C in the urine indicates tubular dysfunction or interstitial disease. Its levels were known to be unaffected by age, gender, race, or muscle mass (ergo, it could be a better predictor of glomerulal function than creatinine in CKD), but recent data indicates higher concentrations in men, those of greater height and weight, those with higher lean body mass and those of older age (Knight et al. 2004). As it is produced constantly, serum cystatin C concentrations may be a marker of glomerular filtration, not affected by infection, liver disease or inflammation (Venkataraman and Kellum 2007; Trof 2006; Royakkers et al. 2007), though its serum levels have been reported by others to be affected by thyroid function, corticosteroids, and inflammation (Friecker et al. 2003). It can be measured in blood or urine samples by various methods, such as nephelometry or ELISA.

Serum cystatin C was found to predict AKI and a need for renal replacement therapy almost two days earlier than serum creatinine (Herget-Rosenthal et al. 2004a). Urinary cystatin C has been shown to be a promising early biomarker of AKI in adult cardiac surgery patients, with an AUC-ROC value of 0.69 to 0.73, depending on cut-off values (Koyner et al. 2008). In this study, urinary cystatin C was a better early predictor of kidney injury than its serum counterpart. Another recent study has shown serum cystatin C to rise considerably at 12 hours after cardiopulmonary bypass in pediatric patients who subsequently developed AKI (Krawczeski et al. 2010). A meta-analysis involving 4,492 patients showed that serum cystatin C has a greater AUC than serum creatinine (0.926 against 0.837) in predicting GFR (Dharnidharka et al. 2002). In ICU patients, elevated urinary cystatin C was associated with AKI, sepsis, and 30-day mortality, with AUC-ROC values of 0.70, 0.80, and 0.64, respectively (Nejat et al. 2010). In this study, urinary cystatin was diagnostic of sepsis, with an optimal cut off point of 0.24 mg/L. Sepsis was eight times more likely in patients above that point. On the contrary, serum cystatin C was not found to be independently associated with sepsis or mortality, although it was mildy diagnostic of sepsis, with an AUC-ROC of 0.60. Cystatin C has been compared to NGAL in cardiac surgery induced AKI and it was observed that both markers predicted AKI at 12 hours, though NGAL performed better at earlier times (VandeVoorde et al. 2006). In patients admitted to hospital with acute heart failure, and by using the 1.4 mg/L as cut-off concentration for normal renal function, serum cystatin C detected a decline in renal function early (within 48 hours) after admission in a considerable percentage of patients with acute heart failure (AHF). A rise in cystatin C levels by 0.3 mg/L or more correlated with longer hospitalization, significantly higher in-hospital mortality and was also found to be an independent predictor of survival during follow-up (Lassus et al. 2010). In adult patients who underwent coronary angiography, cystatin C rapidly detected a falling glomerular filtration rate, early after contrast administration, with an AUC-ROC value of 0.933 (Rickli et al. 2004).

In CKD, cystatin C has been sugggested as a more valuable marker of GFR than serum creatinine (Cruz et al. 2010). In the MMKD study, serum cystatin C also predicted CKD progression (Spanaus et al. 2010). It may also be a marker of cardiovascular risk in CRS types 2 and 4, complementing others cardiac biomarkers such as troponin T and NT-proBNP (Manzano-Fernandez et al. 2009; Cruz et al. 2010). Increased serum cystatin C has been strongly associated with adverse cardiovascular events and cardiovascular mortality (Iwanaga and Miyazaki 2010).

Cystatin C is not without drawbacks. It suffers from a lack of assay standardisation, something that limits the generalizability of many study findings. It is also non specific for AKI, as it is an early marker of impaired GFR, rather than tubular lesions (Trof et al. 2006; Coca et al. 2008). However, it has no circadian variations and is stable in urine, so urinary cystatin C can be quantified in a single sample, with results verified within one hr (Conti et al. 2005).

## *IL-18*

IL-18 is a proinflammatory cytokine and ischemic tissue damage mediator, that is upregulated and detectable in urine after ischemic proximal tubular injury (Melnikov et al. 2001). It is produced as a 24 kDa inactive precursor, cleaved by caspase-1 to its active form, and is constitutively expressed in the intercalated cell of the late distal convoluted tubule, the connecting tubule, and the collecting duct of the human kidney. IL-18 has extensive immunomodulatory actions and seems to hold a role against a number of infections (Gracie et al. 2003). It can be measured by ELISA, and its activity is important in a number of inflammatory diseases, including inflammatory arthritis, multiple sclerosis, chronic hepatitis, and psoriasis (Gracie et al. 2003).

IL-18 has been studied in a variety of AKI clinical settings, displaying promising predictive values. Animal models of ischemia-induced ATN showed that caspase-1 mediated activation of intracellular IL-18 is responsible for kidney injury in the course of AKI (Melnikov et al. 2001; Trof et al. 2006; Edelstein et al. 2007). The first study of IL-18 in human AKI (Parikh et al. 2004) showed good IL-18 sensitivity (85 percent) and specificity (88 percent) in acute tubular necrosis using a cut-off urinary IL-18 to serum creatinine ratio of 500 pg/mg creatinine, with an AUC-ROC of 0.95. Subsequent studies in other clinical settings showed mixed results. For instance, urinary IL-18 from patients with Acute Respiratory Distress Syndrome predicted AKI development and mortality, but AUC-ROC was 0.73 one day prior to AKI, much less than in the previous study (Parikh et al. 2005). In this study, an increased urinary IL-18 to creatinine ratio > 100 pg/mg creatinine had a six and a half-fold increase in odds of AKI within 24 hours. Furthermore, another study found higher IL-18 urinary levels in patients developing AKI after coronary angiography (Ling et al. 2008), whereas another study did not (Bulent et al. 2008). In a study of patients undergoing cardiac surgery, urinary IL-18 levels increased six hours after cardiopulmonary bypass (CPB) and peaked at 12 hours in patients whom creatinine levels diagnosed with AKI 48 hours later (Parikh et al. 2006a). Furthermore, urinary IL-18 is more specific for ischemic AKI and its levels have been shown not to be affected in CKD, urinary tract infections or nephrotoxicity-induced AKI (Parikh et al. 2008a). Conflicting results were yielded by another study, which found IL-18 was not a valuable early predictor of AKI in adults undergoing cardiac surgery (Haase et al. 2008). It showed that urinary IL-18 had poor predictive strength immediately postoperatively (AUC–ROC 0.55) or 24 hours post surgery (AUC–ROC 0.53). IL-18 levels correlated with the duration of cardiopulmonary bypass and their change might be due to cardiopulmonary bypass-associated systemic inflammation. Finally, it is without surprise that a proinflammatory cytokine will rise in sepsis, a result reducing its ability to diagnose AKI in this setting.

Overall though, urinary IL-18 may be a simple, reliable and cost-effective test for the early diagnosis of iscemia or nephrotoxicity-induced kidney injury, which allows exclusion of prerenal azotemia, chronic kidney disease and urinary tract diseases. It is also valuable in determining AKI severity and predicting death in non-septic critically ill patients. IL-18 measurement in urine is highly sensitive and specific for diagnosing human AKI; increased urinary IL-18 levels in a heterogeneous population of critically ill adults and children that underwent cardiac surgery were related to condition severity (Dent et al. 2007; Parikh et al. 2005; Washburn 2008).

## *FABP*

The Fatty Acid-Binding proteins are nine small cytosolic proteins broadly expressed in tissues with an active fatty-acid binding metabolism (Pelsers et al. 2005). FABP main role is facilitating long-chain free fatty-acid transportation from the plasma membrane to stations of oxidation (mitochondria, peroxisomes) (Pelsers et al. 2005; Oyama et al. 2005). FABP may serve as endogenous antioxidant agents, binding polyunsaturated fatty acids (and protecting them from oxidation) and fatty-acid oxidation products (thus minimizing toxic effects of oxidative intermediates on the cell) (Ek-Von Mentzer et al. 2001). In the human kidney, two types of FABP have been reported, namely liver-type FABP (L-FABP) in the proximal tubule and heart-type FABP (H-FABP) in the distal tubule (Ek-Von Mentzer et al. 2001; Maatman et al. 1992).

H-FABP levels are a sensitive marker of nephrotoxic injury in rats (Pelsers et al. 2005). H-FABP as an AKI biomarker has but limited, if any, value. On the contrary, urinary L-FABP has been suggested as a potential biomarker in settings including contrast media-induced nephropathy, IgA nephropathy, CKD and diabetic nephropathy. In mouse transgenic models, renal interstitial injury is associated with increased expression and urinary excretion of human L-FABP (Gok et al. 2003; Kamijo et al. 2004). In the same context, urinary L-FABP levels facilitated an accurate and early detection of both histologic and functional ischemia-induced AKI insults (Negishi et al. 2009). Amelioration of tubulointerstitial damage was observed in transgenic mice, compared to their wild type counterparts, indicating a protective role for L-FABP in the setting of increased tubular stress (Gok et al. 2003; Kamijo et al. 2004). Urinary L-FABP may also be a potential biomarker in the differentiation of IgA nephropathy and thin basement membrane disease (Nakamura et al. 2007), as well as a potential prognostic marker of contrast media-induced nephropathy (Nakamura et al. 2005; Nakamura et al. 2006).

L-FABP also predicted the development of AKI in children undergoing cardiac surgery; a rise in urinary L-FABP within four hours after cardiac surgery predicted the development of AKI by having an AUC-ROC value of 0.81 (Portilla et al. 2008). Urinary L-FABP has been suggested as an early (Bachorzewska-Gajewska et al. 2009; Ohta et al. 2005) and predictive biomarker (Nakamura et al. 2006) of contrast media-induced AKI. In adults undergoing percutaneous coronary intervention urinary L-FABP rose substantially after four hours and remained increased up to two days after cardiac catheterization (Bachorzewska-Gajewska et al. 2009). There was no significant change in the levels of serum creatinine during this period. More research is warranted to assess L-FABP in this and other settings such as ischemia/reperfusion injury, sepsis, and nephrotoxic injury.

In CKD, L-FABP levels correlated with tubulointerstitial damage and urinary protein excretion in one study (Kamijo et al. 2004). It has also been suggested as a more sensitive marker than proteinuria in the prediction of CKD progression (Kamijo et al. 2005) and predicted the development of albuminuria in diabetic patients (Nielsen et al. 2010).

## *Enzymuria*

Many renal enzymes have been researched, albeit less extensively, mostly as potential AKI biomarkers. Such renal enzymes are released in urine after tubular injury and may offer crucial knowledge on the nature, size and site of tubular cell damage and possible cell necrosis or dysfunction (Trof et al. 2006; Westhuyzen et al. 2003). Detectable enzymes mainly derive from the lysosomes, the brush-border membrane, and the cytosole of tubular cells and are measured by different and automated laboratory methods. Alas, increased enzymuria may also be induced by mild and reversible dysfunction of tubular cells and its use is limited by very low sensitivity thresholds. Studies have shown that tubular enzymuria can detect tubular injury up to four days earlier than traditional markers of renal function (Scherberich et al. 1990; Westhuyzen et al. 2003), but enzymes are released into urine in chronic glomerular diseases as well, thus having probably limited use as markers of tubular injury (Scherberich et al. 1990; Westhuyzen et al. 2003; Bosomworth et al. 1999; Marchewka et al. 2001). Optimal cutoff points for enzyme level AKI prediction remain to be determined, their current predictive value for AKI development remaining relatively low. Enzyme detection might however prove useful in the earliest phase of AKI, identifying high risk patients and allowing earlier intervention (Lisowska-Myjak 2010).

## Alkaline phosphatase, γ-glutamyl transpeptidase (γ-GT), and alanine aminopeptidase

Alkaline phosphatase, γ-glutamyl transpeptidase (γ-GT), and alanine aminopeptidase are enzymes located on the brush border. Damage of the brush border membrane, with loss of the microvilli, will lead to their increased excretion in urine. γ-GT is a marker of injury of the proximal tubular cells and in some studies (Westhuyzen et al. 2003; Kwon et al. 2003), aided AKI prediction in critically ill subjects. Brush border enzymes may be highly sensitive for contrast media-induced tubular injury (Hartmann et al. 1985). Unfortunately, these enzymes are unstable and fast analysis is needed (less than four hours from urine collection). Also, interfering substances from the urine sample must be removed prior to analysis (Westhuyzen et al. 2003). It must be noted that γ-GT roles have been suggested in CKD as well. It might be an early predictor of CKD (Ryu et al. 2007), although not all studies agree (Teppala et al. 2010).

## Albuminuria and microalbuminuria

Albuminuria is a long established and traditional diagnostic and prognostic factor of CKD, but evidence suggests a significant role in HF, as well. It represents a unifying sign of common physiological derailments in HF and CKD, such as volume overload, hypertension, diabetes, and inflammation (Damman et al. 2007; Orea-Tejeda et al. 2008; Comper et al. 2008). It marks structural damage, preceding significant decreases of renal function, and dipstick tracing of proteinuria with almost normal renal function prognosticates a higher risk for end-stage kidney disease than even stage 4 CKD but with a negative dipstick (Ruggenenti et al. 1998, Iseki et al. 2004). HF also leads to albuminuria, with the latter prevailing in patients with a lower eGFR. Increased urinary albumin to creatinine ratio is a powerful and independent prognostic factor in HF, correlating with worse outcomes (Jackson et al. 2009). Albuminuria is also known to be associated with worse outcomes in apparently healthy subjects, as well as patients suffering from cardiovascular disease, diabetes, and CKD.

## Low molecular weight proteins

Low molecular weight proteins, such as $\alpha_1$-microglobulin, $\beta_2$-microglobulin, and retinol binding protein (RBP) are normally reabsorbed in the kidney, so that detection in the urine may indicate damage or dysfunction of proximal tubular cells. Tubular proteinuria is more sensitive than enzymuria in detecting proximal tubular injury and may also be superior to enzymuria in predicting the need for renal replacement therapy in AKI (Westhuyzen et al. 2003). However, this finding might be due to the delay in analyzing urinary enzyme levels in related studies, thus missing early peaks in excretion, because most patients are enrolled not at the time of insult, but at a later time (Herget-Rosenthal et al. 2004b).

## $\alpha_1$-microglobulin

The 31 kDa $\alpha_1$-microglobulin is produced in the liver and partly bounds to serum immunoglobulin A. Its free forms are filtered by the glomerulus and reabsorbed in the proximal tubular cells. It is not detected in urine under normal conditions. $\alpha_1$-microglobulin predicted tubular injury and the need for renal replacement therapy after four days, better than other tubular proteins and enzymes including $\gamma$-GST and NAG (Kwon et al. 2003). In one study of neonatal sepsis treated with aminoglycosides, $\alpha_1$-microglobulin was found substantially increased at > three days (Gordjani et al. 1995). Another study showed that early detection of increased urinary $\alpha_1$-microglobulin was an excellent discriminator for future requirement of renal replacement therapy (AUC-ROC 0.86) (Herget-Rosenthal et al. 2004b).

## $\beta_2$-microglobulin

The 12 kDa $\beta_2$-microglobulin is a protein homologous to class I major histocompatibility antigens, freely filtered into urine, reabsorbed and catabolized by proximal renal tubular cells. Normally, it is not traceable in urine. Urinary excretion has been reported in human studies of AKI (Herget-Rosenthal et al. 2004b; Mehta et al. 1997; Cabrera et al. 1982; Gordjani et al. 1995). In one study, 90 percent of critically ill neonates exhibited a significant rise in urinary detection of $\beta_2$-microglobulin, while only 17 percent had evidence of raised serum creatinine levels (Mehta et al. 1997). The highest values were found in septic neonates. However, another study found no difference in $\beta_2$-microglobulin levels between septic and non-septic patients, when tubular injury was defined by values >3,000 µg/l (Cabrera et al. 1982). Another study also found that urinary $\beta_2$-microglobulin was a poor predictor of renal replacement therapy (AUC-ROC 0.51) (Herget-Rosenthal et al. 2004b). Increased urinary levels of $\beta_2$-microglobulin have also been reported in studies examining contrast media nephrotoxicity agents or hypoxia after cardiac surgery or during kidney translapantion (Lisowska-Myjak 2010). Limitations include difficulty in measurements from its instability at pH levels <6. Moreover, $\beta_2$-microglobulin rises in glomerular diseases, so it may not be a specific marker for tubular injury.

### Retinol binding protein (RBP)

The 21 kDa RBP is a protein partly bound to plasma prealbumins, transporting vitamin A. It is freely filtered by the glomerulus, and almost completely reabsorbed in proximal tubules. Even a small reduction in tubular function may lead to RBP increased excretion (Lisowska-Myjak 2010). The detection of RBP in urine has been described in studies including septic patients (du Cheyron et al. 2003; Herget-Rosenthal et al. 2004b). Urinary RBP was significantly increased in AKI compared with non-AKI control subjects, but its levels could not discriminate pre-renal AKI and ATN (du Cheyron et al. 2003). Increased urinary RBP was fairly discriminatory of the need of renal replacement therapy (AUC-ROC 0.80) (Herget-Rosenthal et al. 2004b).

### CYR61

CYR61 is a cysteine-abundant, heparin-binding protein that holds a role as an extracellular growth factor. It exerts local action, and plays a protective part in repair and neovascularization. Post experimental ischemia, CYR61 mRNA is induced in proximal tubular cells and CYR61 is excreted in urine. A rise in urinary CYR61 occurred about three to six hours after a 30 min kidney ischemia, peaking at six to nine hours. Elevated levels were maintained for 24 hours but dropped on follow-up, despite persistence of injury. Urinary CYR61 may therefore be an early marker of renal lesions that could confer early information and help plan early interventions (Trof et al. 2006; Muramatsu et al. 2002; Ferguson 2003). The protein is measured by Western blot analysis.

## *Gro-α*

Gro-a is the human analogue of the mouse keratinocyte-derived chemokine. In a mouse model of kidney injury, increased levels of keratinocyte-derived chemokine both in serum and urine occurred as the earliest, and were maintained the longest, among 18 examined cytokines and chemokines. The levels peaked three hours post ischemia, with values being nine- and 13-fold greater than sham and normal animals, respectively. Histological changes were evident after one hour and serum creatinine rose as late as 24 hours after the initial ischemic insult. This chemokine might thus be a marker for the early diagnosis and prognosis of ischemia-induced AKI. Of note, urinary Gro-α increased substantially in kidney transplant patients requiring dialysis, but not in subjects with good graft function (Molls et al. 2006).

## *Platelet activating factor (PAF)*

Platelet activating factor (PAF) is a phospholipid inflammatory mediator that has been shown to participate in the pathophysiology of septic AKI (Tolins et al. 1989). A significant early urinary PAF rise has been described in critically ill patients with septic shock and AKI (Mariano et al. 1999). Urinary PAF correlated with other serum and urine inflammatory cytokines, such as serum IL-6, serum IL-8 and urine IL-6, whereas serum PAF correlated with serum creatinine levels. PAF may thus contribute to the pathophysiology of AKI in the setting of sepsis and urinary PAF may pose as a predictor of the clinical course and correlate with the severity of AKI.

## *F-Actin*

The apical membrane of proximal tubular cells is very sensitive to ischemia, which induces fast structural modifications, resulting from depolymerization of actin in the microvilli. The decrease in pH levels induced by ischemia leads to generalized protein dephosphorylation. Dephosphorylation of Actin Depolymerizing Factor (ADF) subsequently activates the protein, which leads to attachment, depolymerization and detachment of F-actin in the apical part of proximal tubular cells, resulting in microvilli damage.

Normally, neither ADF nor actin is traced in urine, but they are both detected just 30 min after hypoxia (Schwartz et al. 1999).

## $Na^+/H^+$ Exchanger Isoform 3 (NHE3)

$Na^+/H^+$ Exchanger Isoform 3 (NHE3) is the major sodium transporter in the renal tubules, located in the apical membrane and subapical endosomes of proximal tubules and the thick ascending loop of Henle. This molecular structure is responsible for proximal reabsorption of up to 70 percent of filtered sodium and bicarbonate. Reduced expression of NHE3 in the kidney and significant correlation between urinary excretion of NHE3 and clinical features of AKI in experimental models of ATN suggest that NHE3 is lost in urine during AKI. One study in ICU patients found that urinary NHE3 was undetectable in patients without AKI but significantly elevated in patients with ATN (du Cheyron et al. 2003). Urinary NHE3 was higher in patients with ATN than those with pre-renal acute renal failure and undetectable in control subjects, or patients with other causes of intrinsic acute renal failure. Moreover, NHE3 was superior to RBP in discriminating pre-renal AKI and ATN and correlated positively with serum creatinine levels.

## Netrin-1

Netrin-1, a laminin-related neuronal guidance molecule, is normally not expressed (or expressed very little) in renal tubular cells. In AKI animal models however, this protein is highly expressed and excreted in urine. Netrin-1 levels can be analyzed by sandwich enzyme-linked immunosorbent assay. In one human study, Netrin-1 increased two hours after cardiopulmonary bypass, peaked at six hours, and remained high up to two days, correlating with AKI duration and severity (Ramesh et al. 2010a). In the mouse, urinary Netrin-1 levels rose substantially within three hours of ischemia–reperfusion injury, peaked at six hours, and later dropped near baseline by the third day (Reeves et al. 2008). On the contrary, serum creatinine failed to rise significantly but 24 hours post reperfusion. In the same study, in mice administered cisplatin, folic acid, and lipopolysaccharide, urinary Netrin-1 increased at one hour and peaked at six hours, whereas serum creatinine increased at six, 24, and 72 hours, respectively. Of note, NGAL urinary excretion in folic acid- and lipopolysaccharide-treated animals was detected only 24 hours post drug exposure. Still in the same study, urinary Netrin-1 rose markedly in AKI patients but not in healthy controls. Urinary netrin-1 is found increased in a variety of AKI settings but not in healthy subjects, such as ischemic AKI, contrast media-induced AKI, sepsis, and nephrotoxic substances (Ramesh et al. 2010b).

## Uric acid

Uric acid may be an effective biomarker in CKD, as it is found elevated in CKD patients and may take part in CKD pathophysiology by playing a role in endothelial dysfunction, vascular myocyte proliferation, increased IL-6 production, insulin resistance, and impairment of NO synthesis (Siu et al. 2006). However, more studies are needed to validate uric acid in the context of CKD.

## Asymmetric dimethylarginine (ADMA)

Asymmetric dimethylarginine (ADMA) is the major inhibitor of NO synthase and may be a marker of endothelial dysfunction, which has been implicated in diabetic nephropathy and atherosclerosis. Increased plasma levels of ADMA predicted the development and progression of nephropathy in type 2 diabetics (Hanai et al. 2009). It also predicted progression to end stage renal disease (ESRD) in another CKD study (Ravani et al. 2005). In a cross-sectional study in patients with essential hypertension, ADMA levels were associated with the presence of microalbuminuria and the albumin/creatinine ratio (Tsioufis et al. 2010). ADMA is also a strong independent predictor of cardiovascular events and mortality risk in ESRD (Zoccali et al. 2001) and in one study of CKD patients, plasma ADMA levels independently predicted total mortality (Ravani et al. 2005). Also, plasma ADMA levels in diabetic nephropathy patients predicted cardiovascular outcomes (Lajer et al. 2008).

## Monocyte Chemotactic Peptide-1 (MCP-1)

Monocyte Chemotactic Peptide-1 (MCP-1) mRNA has been found to be up-regulated in ischemia–reperfusion renal injury and has been considered as a biomarker at the setting of monocyte infiltration in the post-ischemic kidney (Rice et al. 2002). MCP-1 is produced by renal cells and mediates acute ischemic and nephrotoxic injury. In a mouse model, MCP-1 protein and mRNA increased in intrarenal injury more than NGAL, whereas in prerenal and postrenal injury, both genes were expressed comparably (Munshi et al. 2011). However, uremia caused only NGAL gene expression with no renal injury, suggesting that MCP-1 is more specific for AKI. Urinary

MCP-1 protein and mRNA levels have also been found increased in human urine samples from AKI patients (Munshi et al. 2011).

## *Adenosine deaminase-binding protein (ABP)*

Adenosine deaminase-binding protein (ABP) is a 120 kDa glycoprotein found also in the brush border of proximal tubular cells. Increased ABP detection in the urine is believed to indicate AKI (Tolkoff-Rubin et al. 1987; Tolkoff-Rubin et al. 1988). ABP has been described in a study of neonatal sepsis and has also been found to be a marker of AKI (Gordjani et al. 1995).

## *α/π-GST*

α-GST is formed in epithelial cells of the proximal tubule and π-GST in the distal tubule. Increases of these GST isoenzymes in urinary excretion (possibly different for each), may indicate the site of the lesion. α/π-GST are predictors of unfavorable AKI development requiring kidney replacement therapy (though utilizing them requires correct storage of urine samples and addition of agents stabilizing enzyme activity) (Trof et al. 2006; Westhuyzen et al. 2003; Herget-Rosenthal et al. 2004B; Branten et al. 2000).

## *Beta trace protein (BTP)*

BTP is a low molecular weight protein member of the lipocalin family, produced mainly in the central nervous system. Because of weight and lack of affinity for protein binding, it may be a useful marker of GFR and kidney function (Bokenkamp et al. 2007; Hoffmann et al. 1997; Priem et al. 1999). Serum BTP levels correlated strongly with GFR in kidney transplant patients and CKD patients (Poge et al. 2005; Bhavsar et al. 2011). Also, in a study of African Americans with hypertensive CKD, BTP levels predicted ESRD more strongly than markers such as creatinine, cystatin C, $eGFR_{SCr}$ and measured GFR (mGFR) by iothalamate clearance (Bhavsar et al. 2011).

## Cardiac Biomarkers in CRS

Acute and chronic heart conditions, such as acute coronary syndromes, heart failure, and cardiovascular medical interventions, are core components of CRS that both affect renal function and pose increased, life threatening risks for patients. Contrary to AKI, accurate and reliable early injury markers of looming cardiac danger have for long been utilized in the context of acute conditions, achieving remarkable risk reduction in heart patients. The clinical use of troponins in the emergency room has changed the history of cardiology in acute coronary syndromes, providing a reliable source of information on ongoing damage that can be effectively addressed. Troponins are excellent biomarkers of ischemic myocardial injury and correlate with outcomes in the general population and in renal patients (Ooi et al. 1999; Needham et al. 2004; Sommerer et al. 2007). They hold a key role in an array of markers that may help physicians treat CRS types 1 and 3 earlier and better. Other promising biomarkers for both acute and chronic heart conditions exist, with research emphasis given mostly to BNP, NT-proBNP, MPO, IMA, and CRP. Troponins, homocysteine, CRP, asymmetric dimethylarginine, plasminogen-activator inhibitor type 1, natriuretic peptides, and ischemia-modified albumin are all biomarkers that also correlate with cardiovascular outcomes in CKD patients (Tonelli et al. 2006; Urquhart and House 2007; Levin et al. 1999). Many candidate markers provide novel information focusing on inflammation, injury, and myocardial stress, and generally processes that precede myocardial necrosis, thus potentially allowing for earlier intervention. The use of such markers may ultimately provide a better and more individualized treatment of cardiovascular and renal disorders, as adverse cardiovascular and renal outcomes associated with them may be succesfully treated.

In the acute clinical setting, biomarkers offer crucial knowledge in the diagnosis, prognosis and management of patients with suspected ACS. Different markers identify different aspects of pathophysiology: For instance, IMA marks ischemia, troponins are markers of cardiomyocyte necrosis, natriuretic peptides mark neurohormonal activation and hemodynamic wall stress, soluble CD40 ligand marks platelet activation, CRP, MPO, MCP-1 and various interleukines reflect inflammation, a major part of atherosclerosis processes. Combined, biomarkers enhance risk stratification and shed brighter light than each one alone.

## *Troponins*

Troponin I (cTnI) and troponin T (cTnT) are highly sensitive and specific protein components for cardiomyocyte injury and necrosis. Cardiac troponins are the ideal biomarkers for the detection of recent myocardial injury and their key role in this context was recently revalidated by the Joint ESC/ACCF/AHA/WHF Task Force for the Redefinition of Myocardial Infarction (Thygesen et al. 2007). Both are subunits of the myocyte troponin regulatory complex that bind to actin. Troponins I and T are not present in the skeletal muscles, thus having an incredibly high specificity for cardiac injury and allowing for a cut-off value very close to zero, which in turn grants them high sensitivity. After myocardial ischemic injury and the loss of myocyte-membrane integrity, cardiac troponins find their way into the blood stream and can be detected in serum within three to six hours after chest pain onset; they peak at around 24 hours and remain elevated up to 10 or 14 days after the initial event. They are currently the reigning gold-standard biomarker for MI diagnosis (Alpert et al. 2000) and allow for the detection of myocardial injury at a much greater sensitivity than creatinine kinase testing. It is recommended that an elevation of troponin significant to detect myocardial necrosis exceed the 99th percentile of troponin values in a reference control group within 24 hours of a relative clinical event, such as chest pain (Alpert et al. 2000). Blood for troponin measurement must be drawn upon admission, at six and 12 hours (Alpert et al. 2000). Laboratory, immunoassay-based testing for cardiac troponins has become a very simple and cost-effective, fast bedside tool to detect or exclude MI in emergency differential diagnosis. Troponins also exhibit a great prognostic value, their levels being strong and independent predictors for adverse events. They mark an almost four-fold increased risk of death or reinfarction (Heidenreich et al. 2001) and ACS patients with increased troponin I have been shown to have a greater atheroma burden at the site of injury and more reference segment atherosclerosis than patients without troponin elevation (Fuchs et al. 2002). The link between a troponin increase and high-risk coronary lesions probably accounts for the benefit of aggressive antiplatelet and antithrombotic therapy in ACS patients with troponin elevation, whereas little, if any such benefit exists in patients with normal troponins (Heeschen et al. 1999; Morrow et al. 2001; Morrow et al. 2000).

Unfortunately, troponins do not elucidate cardiac injury etiology and are found elevated in various disorders of the heart or other organs, or after medical interventions, hence clinical judgement, and the use of other clinical tools are crucial to emergency differential diagnosis. Such conditions include myocarditis, congestive heart failure (CHF), cardiac contusion, cardioversion, chemotherapy, sepsis, pulmonary embolism and ESRD. Multiple other disorders are associated with a rise in cardiac troponins, including left ventricular hypertrophy (LVH), diabetes mellitus, heart transplantation, chronic obstructive pulmonary disease (COPD) exacerbations, hypertensive emergencies and hypertension, aortic dissection, infiltrative diseases (amyloidosis, hemosiderosis, pompe and sarcoidosis), high endurance physical exercise, lobar pneumonia, extensive burns, and rhabdomyolysis (Inbar et al. 2009). However, elevated troponin levels mark patients with cardiac injury that are at high risk for adverse cardiac events, regardless of cardiac injury etiology.

In an analysis of the Acute Decompensated Heart Failure (ADHF) National Registry (ADHERE) of 84,872 patients hospitalized with ADHF, patients with positive cardiac troponin had a higher in-hospital mortality, independent of other predictors (Peacock et al. 2008). Thus troponins seem to extend their uses in the acute worsening of a chronic setting, such as HF. Morever, cTnI I was detected (≥0.04 ng/ml) in advanced HF patients without ischemia and identified as an independent predictor of death (Horwich et al. 2003). Also, cTnT levels ≥ 0.02 ng/ml in CHF patients quadrupled the risk of death in one study (Hudson et al. 2004).

Interpreting troponin levels in patients with renal failure is quite challenging. Persistent increases in troponins is not rare in dialysis patients, and elevated cTnT or cTnI correlate with high cardiac event rates in patients with renal failure (deFilippi et al. 2003; Aviles et al. 2002; Apple et al. 2002). Troponin elevation in CKD is more prominent for troponin T than troponin I (Apple et al. 2002) and CKD related troponin rises reflect ongoing myocardial damage and are strongly associated with diabetes, left ventricular dilatation, and impaired left ventricular systolic and diastolic function, while no severe coronary artery disease may exist (Sharma et al. 2006). Elevated troponin T (>0.1 ng/mL) has been associated with increased all-cause mortality with a relative risk of 2.64 in end-stage renal disease (Khan et al. 2005). Knowing prior troponin values in CKD patients may offer important insight on whether a current troponin elevation is new and potentially of ACS etiology, or is chronic and unrelated to acute injury. If ACS is suspected however, an increase in troponins is to be interpreted similarly in all patients, regardless of their renal function.

## Natriuretic peptides (NP)

Three types of NP, A-type, B-type, and C-type, are found in humans. A-type NP (ANP) is stored in granules mostly in atrial tissue. The stored 126 amino acid (aa) long proANP is cleaved by serine protease corin and yields the active ANP (aa 99–126), and inactive N-ANP (aa 1–98). BNP is produced as pre-proBNP, which is processed intracellularly to pro-BNP (aa 1–108), then cleaved by furin and secreted as inactive N-terminal fragment-BNP (NT-proBNP, aa 1–76) and biologically active, BNP (aa 77–108). B-type NP (BNP) concentration is larger in atrial tissue but more is released from the ventricles, due to the greater ventricular mass. C-type NP (CNP) is primarily released by the central nervous system, vascular endothelium, and only very little by the heart. CNP is synthesized as a pre-prohormone and processed into the 22 amino acid biologically active peptide (CNP) and a larger but inactive pro-form (CNP-53). The role of CNP in the cardiovascular system has not yet been firmly established. Extensive research has suggested the NP system as a compensatory neurohumoral system operating against heart failure.

Natriuretic peptides, particularly BNP and NT-proBNP, are strong predictors of adverse cardiovascular events (Tang et al. 2007) and mortality in critically ill patients (Meyer et al. 2007), ACS (Morrow et al. 2005), and stable heart failure (Latini et al. 2004). They are well accepted diagnostic and prognostic tools both in acute heart failure and acute decompensated heart failure (Maisel et al. 2004) and have been found elevated in patients with CRS type 1, when AKI occurs as a result of ADHF. They also have been of prognostic value in patients with various stages of renal impairment (Carr et al. 2005; Austin et al. 2006), reserving a role in CRS types 2 and 4. CKD patients show higher levels of BNP and NT-proBNP than subjects without renal impairment (after corrections for gender and age), even when clinical CHF is not manifest (McCullough and Sandberg 2003). A reduced renal function may explain this finding, although other explanations are possible.

Natriuretic peptides mark hemodynamic myocardial stress rather than myocyte injury, ergo an increase in cardiac wall stress (by pressure overload or volume expansion) leads to NP release from the heart. Current guidelines refer only to BNP and NT-proBNP, based on their diagnostic power (Dickstein et al. 2008). NP have actions including natriuresis, inhibition of the sympathetic nervous system and the renin-angiotensin-aldosterone axis, and vasodilation, all pathophysiologically linked to counterstrike heart failure (de Lemos et al. 2003a).

BNP has established itself in the diagnosis and prognosis of patients with CHF (Maisel et al. 2002) and the severity of ventricular dysfunction correlates with individual BNP levels (Troughton et al. 2004). Three FDA approved measurement assays have been commercialized (two for BNP and one for NT-proBNP), aiding physicians in gathering crucial information on HF. Higher levels of BNP (and NT-proBNP) in such patients signify a high risk for the progression of heart failure and death. BNP values <100 ng/L essentially rule out HF and have a negative predictive value of 90 percent (Silver et al. 2004). On the contrary, values >500 ng/L indicate HF is very probable, with an associated positive predictive value of 90 percent. BNP levels ranging in between have a smaller predictive value, and their accuracy drops. NT-proBNP levels >450 ng/L in patients younger than 50 years make it both sensitive and specific for HF (Januzzi et al. 2005). Values >900 ng/L in patients older than 50 years again provide a high sensitivity and specificity for HF. NP prognostic value in HF is also promising, higher levels on admission or at discharge generally indicating a worse outcome. Also, patients who have significant reductions in BNP values during treatment have better outcomes.

In HF, BNP has probably the greatest prognostic value among other predictors, with each 100 pg/ml increase of BNP being associated with a 35 percent increase in the relative risk of death (Doust et al. 2005). Patients with BNP <130 pg/ml had a 1 percent risk of sudden cardiac death, whereas a 19 percent risk was predicted by higher concentrations. The Acute

Decompensated Heart Failure National Registry (ADHERE) depicted a positive linear relationship of in-hospital mortality and BNP levels (Fonarow et al. 2008). BNP correlates with death and CHF independently of left ventricle ejection fraction (Omland et al. 2002). Unfortunately, despite its high levels in HF, serum BNP physiological effects fail to prevent disease progression and CRS. In this context, research has suggested a resistance to BNP (Forfia et al. 2007). HF may even be a state of BNP insufficiency.

NP levels also provide guidance in adjusting therapy to hospitalized HF patients. Falling NP levels during treatment have been shown to be proportional to decreasing pulmonary capillary wedge pressure (PCWP) and inversely correlated with symptomatic improvement (Daniels and Maisel 2007). In one study of hospitalized ADHF patients, the pre-discharge BNP level was a strong independent predictor of post-discharge outcomes. It has been proposed that patients with heart failure and BNP levels not falling below ~600 pg/ml should receive intensified treatment before discharge (Logeart et al. 2004). Many trials examined using BNP levels to adjust therapy in the outpatient setting. A recent meta-analysis found that using NP levels to guide HF therapy resulted in a significant reduction in all-cause mortality (Felker et al. 2009). Some studies have shown that NP-guided therapy mostly favors patients less than 75 years old, though this may be due to a lack of treatment aggressiveness in the very elderly. In general, NP-guided therapy in the outpatient setting does seem to improve outcomes.

In the acute setting, BNP measurements in suspected ACS patients have generated interest. It is known that MI increases wall stress and that ischemia can cause BNP release (Toth et al. 1994). However, BNP is not currently considered a useful diagnostic marker in this context and many conditions exist that can lead to an increase in BNP levels analogous to that of ischemia, such as LVH and pulmonary embolism. It should also be noted that an ischemia-induced BNP rise is smaller than that which is seen in the context of HF. It seems though, that the combined use of BNP, CK-MB, and Troponin I increases sensitivity and negative predictive value when compared with CK-MB or troponin I alone (Bassan et al. 2005). In this study, patients with normal troponin I but elevated BNP had a significantly greater risk of MI.

Despite their poor diagnostic value in ACS, NP do have prognostic significance in coronary artery disease. Plasma BNP increases fast in the first 24 hours after MI, proportionately to MI size (Morita et al. 1993; Arakawa et al. 1994), and may remain persistently elevated if severe LV dysfunction and ventricular remodelling develop (Nagaya et al. 1998). BNP has been suggested as a more powerful independent predictor of adverse events than ANP, N-ANP, and ejection fraction (Omland et al. 1996). BNP levels measured in the first days after an acute coronary episode are of important value for the risk stratification of ACS patients, as they predict the long-term risk of death and nonfatal cardiac events even after adjustment for various potentially confounding factors, such as HF co-existence (de Lemos et al. 2001). BNP levels were also positively correlated with the extent of coronary atherosclerosis in the same study. A comparison between troponin and BNP in terms of prognostic value showed that troponin predicts recurrent ischemic events better, but elevated BNP (>80 pg/ml) at admission identifies patients with non ST-elevation ACS who are at higher risk of death and CHF (Morrow et al. 2003). More specifically, in this study troponin-negative patients with increased BNP had a 6.4-fold enhanced risk for death and troponin-positive patients a 3.8-fold increase in such risk. A persistently elevated BNP (>80 pg/ml) during follow-up after discharge also predicts death or CHF. Clinical application of such research data is still limited though, and further studies should evaluate whether incorporating NP levels in decision making optimizes therapeutic strategies in ACS patients.

Natriuretic peptides could also detect cardiac dysfunction early in treatment with cardiotoxic agents, aiding individualized assessment of cancer therapy (Kouloubinis et al. 2007).

In chronic disease (CRS type 4), increased levels of BNP have been associated with an accelerated progression of non-diabetic CKD to ESRD, even after adjustment for other established prognostic factors of CKD progression (Spanaus et al. 2007). BNP and NT-proBNP were significantly higher among the patients who developed ESRD than among those who did not (median value 61 vs 39 ng/L and 320 vs 84 ng/L, respectively). Increased NP levels may reflect a true cardiac dysfunction and may rise due to an increase in atrial pressure, systemic pressure, or ventricular mass. Renal patients often are hypertensive and thus might have significant left ventricular hypertrophy, while other cardiac co-morbidities are also common. Renal impairment affects NT-proBNP levels more than BNP, as its clearance is not mediated by Natriuretic Peptide C Receptor (NPR-C) or neutral endopeptidases and is more renally dependent. It has been suggested that a higher cut-off point of 1,200 pg/ml for the diagnosis of HF is more prudent in the presence of renal dysfunction and a GFR <60 ml/min/1.7 m$^2$ (Anwaruddin et al. 2006). Natriuretic peptide testing for HF may not be suitable in patients on dialysis, as diagnostic value is reduced in this setting (NACB Writing Group 2007). However, BNP is a good prognostic factor for risk stratification in dialysis patients, and serial estimations can help in the identification of cardiac disease (Suresh and Farrington 2005). Despite issues of renal clearance, BNP and NT-proBNP still provide crucial information in patients with renal dysfunction. NP levels are still well correlated with left ventricular wall stress and prognosis. In HF patients that are attended at the emergency room, a NT-proBNP increase was the most powerful independent predictor for 60-day mortality for those with eGFR <60mL/min/1.73m² and independently predicted hospitalization (Anwaruddin et al. 2006).

NP do have limitations that need to be addressed before a wider clinical use is achieved. Healthy women and the elderly have higher values, something more pronounced in those with renal failure, as has already been mentioned (McCullough 2003). On the other hand, obese patients have lower values, especially for NT-proBNP levels. BNP cut-off points to maintain 90 percent sensitivity for HF have been shown to be 170 pg/mL for lean patients and 110 pg/mL for obese patients (Daniels et al. 2006). Moreover, sepsis, volume overload, stroke, cor pulmonale, pulmonary edema, atrial fibrillation, and acute mitral regurgitation are also associated with higher BNP and pro-BNP values. Because of significant variability, NP normal values is a complex issue and it has been proposed that values need to half or double to indicate a definite change (Jaffe et al. 2006). It has also been suggested that NP levels should be used as a continuous variable that takes into account the patient's baseline levels, if available (Maisel et al. 2008). Furthermore, NP cut-off values may differ in patients with chronic heart failure as compared with patients with acute heart failure. In one study, 20 percent of patients with established chronic symptomatic heart failure, had plasma BNP levels < 100 pg/ml (Tang et al. 2003).

The use of BNP in the diagnosis of HF should not take place without proper clinical judgement of the pre-test probability of HF, the degree of elevation of BNP, and the existence of other confounding issues. As mentioned above, a normal BNP value (<100-200 pg/ml depending on age and gender) should urge evaluation of non-cardiac causes of dyspnea as HF is very unlikely. A substantially elevated BNP value (>400 pg/ml) should prompt appropriate testing to confirm the diagnosis and guide the management of HF. If BNP values range in the intermediate, clinical judgement is important and BNP-conferred information value will depend on HF pre-test probability. The role of BNP remains unclear in patients with preserved left ventricular ejection fraction (LVEF), especially for guiding therapy (Daniels and Maisel 2007).

## *Myoglobin*

Myoglobin is one of the earliest markers released into the bloodstream after myocardial necrosis. Myoglobin levels rise as early as one to three hours post ischemic injury and peak at about six hours, returning to baseline 18 to 24 hours after the initial insult. However, myoglobin is not specific for cardiac injury, as it is present in skeletal muscle as well. Measurement of myoglobin is currently recommended only for patients presenting within six hours of chest pain onset, as troponins provide far better and earlier assessment. In subjects with chest pain but no ST segment elevation in the ECG, myoglobin combined to creatine kinase-MB (CK-MB) and troponin may improve sensitivity for MI detection (de Winter et al. 1995; Jernberg et al. 2000; Jurlander et al. 2000).

## CK and CK-MB

Creatinine kinase (CK) was linked to cardiac injury decades ago. It is found in skeletal and cardiac muscle, as well as the gastrointestinal tract, but total CK cannot be used alone for the diagnosis of an acute MI, rather it must be combined with a more sensitive marker such as troponin or CK-MB. After troponins emerged as the key early marker for myocardial injury, CK-MB has been pushed aside. CK-MB remains valuable when troponin testing is not available and for diagnosing recurrent MI, as troponin levels may remain increased for several weeks post MI. Moreover, CK-MB aids in infarct size estimation: CK-MB AUC-ROC value provides a more accurate assessment of infarct size than that of troponin. CK-MB may rise within three to four hours of cardiac injury and drop to baseline levels in 24 to 36 hours. The requirements for CK-MB to detect myocardial necrosis are that values from two successive blood samples exceed the 99th percentile in a control group, and blood tests must be done on admission and at six and 12 hours (Alpert et al. 2000).

## C-reactive protein (CRP)

CRP is a long established, non-specific acute phase reactant and inflammation marker. Inflammation has been increasingly involved in the pathogenesis and pathophysiology of atherosclerosis and heart failure, hence CRP has received great attention as a potential marker of vascular inflammatory processes, acute coronary syndromes, and heart failure that play an important pathogenetic role in CRS types 1 and 2. CRP has been shown to aid adverse events prognosis in healthy individuals (Ridker et al. 2000; Ridker et al. 2002), with high sensitivity CRP (hs-CRP) levels of 1, 1 to 3, and >3 mg/L representing lower, average, or higher vascular risk, respectively, when added to traditional risk factors. Universal CRP cut-off values are, however, a matter of dispute (Khera et al. 2005) and its prognostic value for chronic HF in the general population has been questioned (Danesh et al. 2004). The Reynolds Risk Score added hs-CRP to traditional risk factors and found that risk prediction accuracy was much improved (Ridker 2007).

In the setting of acute or chronic heart failure, CRP levels are an independent predictor of adverse events (Anand et al. 2005). Higher levels were associated with more severe heart failure and were independently associated with mortality and morbidity, adding prognostic value to BNP measurements.

CRP is an independent predictor of mortality at 14 days in ACS, even in patients with negative troponin levels (Morrow et al. 1998). CRP also has direct adverse effects on vascular endothelial cells, reducing NO release, increasing endothelin-1 levels, and promoting the expression of endothelial adhesion molecules (Venugopal et al. 2005). The use of statins after MI has been examined in relation to CRP levels (hence inflammation) and it has been suggested that adjustment of therapy based on CRP could result in fewer future adverse events and enhance regression of atherosclerosis (Nissen et al. 2004; Nissen et al. 2005). The best outcomes were associated with low density lipoprotein cholesterol (LDL-C) levels <70 mg/dL and CRP levels <2 mg/L in 3,813 ACS patients initiating statin therapy.

"Normal" CRP levels have generated much controversy. Some argue that CRP levels vary greatly and fluctuate within individuals. CRP levels vary by gender and ethnicity as well (Khera et al. 2005), and values >10 mg/L are likely due to illness (Rifai et al. 2001).

Compared to other biomarkers such as troponins and natriuretic peptides, CRP is not a strong predictor, and its clinical use for decision making in emergency medicine is still unclear. As it is not a specific inflammation marker, it might soon be substituted by more specific inflammatory biomarkers playing a direct role in the pathophysiology of atherogenesis.

In other contexts, CRP has been associated with the presence of CKD (Fox et al. 2010; Fakhrzadeh et al. 2009) and increased CRP levels were associated with microalbuminuria in patients with untreated essential hypertension (Tisoufis et al. 2010).

## *Monocyte chemoattractant protein-1 (MCP-1)*

Monocytes appear to hold a leading role in atherogenesis, both as the progenitors of foam cells and as a source of growth factors mediating intimal hyperplasia. MCP-1 is the primary chemokine responsible for the recruitment of monocytes into the developing atheroma (Krishnaswamy et al. 1999), thus having a strategic role in atherogenesis and coronary artery disease. MCP-1 is produced by many cells in side the atheroma, including smooth muscle cells, monocytes/macrophages, and endothelial cells (Nelken et al. 1991). MCP-1 does not possess a diagnostic value for ischemia, as increased plasma levels in ACS patients overlap with normal concentrations. It may however possess prognostic value in ACS and might be attractive as a therapy target. In one study, MCP-1 levels above the 75th percentile of concentrations (238 pg/mL) showed a two-fold increased risk of death or MI after adjustment for standard risk markers (de Lemos et al. 2003b).

## Myeloperoxidase (MPO)

MPO is a 150 kDa peroxidase hemoenzyme that is found in lysosomes and is secreted from activated neutrophils, monocytes, and tissue macrophages in inflammatory conditions. It catalyzes the conversion of hydrogen peroxide and chloride to hypochlorite. MPO is an indirect marker of oxidative stress which can cause myocyte apoptosis and necrosis and is associated with endothelial dysfunction, by reducing NO synthase activity and inactivating NO itself. MPO may play a crucial role in low density lipoprotein oxidation, and studies have associated plasma MPO levels with atherosclerosis (Nicholls and Hazen 2005; Lau and Baldus 2006; Zhang et al. 2001). Moreover, MPO levels correlate with HF severity and are independent predictors of death in this setting (Kameda et al. 2003).

Higher MPO concentrations occur in the lesions responsible for unstable angina or acute MI (Apple et al. 2005). MPO is considered a marker of plaque instability and reflects inflammation or activation of hemostasis after plaque rupture. As such, it may provide crucial information before injury is rendered irreversible (Lippi et al. 2006). Increased MPO (>350 µg/L) in patients with chest pain identified subjects without evidence of myocyte necrosis who were at increased risk for MI during their hospital stay or after discharge (Brennan et al. 2003). In this study, MPO, given its inflammatory properties, was suggested as both a marker and mediator of vascular inflammation. Other studies have yielded similar results (Baldus et al. 2003).

However, clinical use of MPO is still a future aspiration as limitations exist, one of them being that MPO lacks specificity because it is associated with other infectious or inflammatory conditions. Much research towards assay development and clinical validation remains to be done.

## Ischemia-modified albumin (IMA)

IMA is currently the only FDA approved ischemia marker in the setting of ACS. It is widely accepted that a cardiac marker for myocardial ischemia in the absence of myocardial injury would be of important value in emergency medicine, directing physicians towards correct diagnoses and allowing for elevated troponins (when injury is present) to account for ischemic-induced necrosis. IMA may play such a role. Myocardial ischemia reduces the

N-terminus of albumin, rendering it unable to bind cobalt. The albumin-cobalt binding test is thus used to measure IMA, with decreased cobalt binding resulting in greater IMA values in ACS. How IMA is produced during ischemia is yet unclear.

One study examined patients with suspected ACS, chest pain, negative troponin on admission, and a non-iagnostic ECG (Roy et al. 2004). Higher IMA levels were measured in those patients who had ACS than those who had not. IMA combined with troponin enhanced the sensitivity and specificity of identifying ACS patients. IMA levels also rise in patients undergoing coronary interventions such as percutaneous transluminal coronary angioplasty (PTCA) (Bar-Or et al. 2001; Sinha et al. 2003).

Challenges in the use of IMA rise from a lack of standard ischemia definition. One study found good sensitivity but low IMA ischemia specificity, at a threshold of 90 U/ml (Anwaruddin et al. 2005). Another study confirmed such results and found a poor prognostic value in the context of ACS (Sinha et al. 2004). More IMA limitations include an influence to IMA levels by albumin. As IMA has low specificity, most testing will be false positive, limiting IMA clinical use to just ruling out ischemia in accordance with ECG and troponin findings.

## *Interleukines*

Interleukines are strategic mediators of inflammation and play important roles in both renal and cardiac pathophysiology in the context of acute and chronic CRS. IL-1 and IL-6 may prove useful as early biomarkers for the diagnosis of CRS, but may also have a pathogenic role, causing myocardial cell injury and apoptosis (Krishnagopalan et al. 2002; Chen et al. 2008) and mediating myocardial damage in ischemic AKI (Kelly 2003). IL-6 induces myocyte hypertrophy (Seta et al. 1996) and may be of value in the prediction of future heart failure development in asymptomatic elderly patients (Lee and Vasan 2005). IL-18 has been associated with atherogenesis, coronary artery disease, plaque rupture leading to ACS (Furtado et al. 2009; Chalikias et al. 2007), myocardial ischemia-reperfusion injury (Mallat et al. 2004) and the pathophysiology of cardiomyocyte hypertrophy and HF (Chandrasekar et al. 2005). HF patients have been found with increased IL-18 levels, higher levels being associated with greater mortality (Mallat et al. 2004). In ACS patients, elevated median plasma IL-18 levels were an independent predictor of adverse cardiac outcomes (Furtado et al. 2009; Chalikias et al. 2007; Youssef et al. 2007), when CRP was of borderline significance for such events. IL-18 could also have a diagnostic value in patients complaining of chest pain. In one study, admission plasma IL-18 elevation preceded CK-MB elevation, and IL-18 concentration rose rapidly after severe myocardial ischemic assault, regardless of evolving myocardial necrosis, marking ischemia as the reason for pain (Kawasaki et al. 2005). In another study, plasma IL-18 levels were significantly increased in ACS patients and also associated with myocardial dysfunction severity (Mallat et al. 2002).

Cardiotrophin-1, a member of the family of IL-6 related cytokines, is elevated in HF in correlation to left ventricular systolic dysfunction (Talwar et al. 2000; Ng et al. 2002). Furthermore, serum soluble ST2, an IL-1 receptor family member, rises early after acute MI. It is induced in patients with HF and has been found to be secreted with myocyte strain (Weinberg et al. 2003). The PRIDE study showed that ST2 levels correlated with HF severity and predicted one-year mortality (Januzzi et al. 2007). ST2 was not as useful as NP in HF diagnosis but their combination might improve prognosis determination. In STEMI dyspnoic patients, ST2 levels were a strong predictor of mortality (Januzzi et al. 2007).

## Soluble CD40 ligand (sCD40L)

Soluble CD40 ligand (sCD40L) is a proinflammatory marker and potential mediator of plaque rupture (Schonbeck and Libby 2001). CD40L has been found on endothelial cells, smooth muscle cells, monocytes/macrophages, and platelets, from where it can trigger an inflammatory endothelial reaction (Henn et al. 1998). After a thrombus is formed, sCD40L, the released and cleaved form of the molecule, acts as a platelet agonist by binding to IIb/IIIa receptors (Andre et al. 2002). Platelets are the main source of released sCD40L. Some studies have suggested an increased release of sCD40L and a potential marker role for this molecule in ACS (Aukrust et al. 1999; Garlichs et al. 2001) and sCD40L may soon find a use as a marker of activated platelets. The CAPTURE trial suggested that sCD40L may both be an independent risk marker of cardiovascular events and help define benefit from therapy with glycoprotein IIb/IIIa inhibitors in ACS (Heeschen et al. 2003). Increased CD40L levels in patients with ACS were associated with risk for recurrent MI in another study (Freedman 2003). Unfortunately, it lacks specificity, as it is related to other inflammatory conditions besides atherosclerosis.

## FABP

Recent studies have proposed roles for both serum H-FABP and urinary L-FABP in the detection of ongoing myocardial damage in ACS, where they could also aid prognosis (Liyan et al. 2009; Fukuda et al. 2009). Combining H-FABP and IMA values on admission is highly sensitive and specific for the risk strafication of ACS patients with negative troponins (Liyan et al. 2009). Serum H-FABP is an independent predictor of adverse cardiac events in

ACS patients (Ishii et al. 2005; Erlikh et al. 2005). It could also be utilized as an effective marker for the evaluation of CHF severity (Goto et al. 2003).

## NGAL

Growing evidence suggests that NGAL may be involved in cell survival, inflammation, and matrix degradation by modulating the activity of matrix metalloproteinase 9, which plays an important role in vascular remodelling and plaque instability in the progress of atherosclerosis (Hemdahl et al. 2006; Yndestad et al. 2009). NGAL is increased in atherosclerotic plaques, weakening plaque structure and facilitating rupture (Hemdahl et al. 2006). In a rat model of post MI HF, NGAL expression was increased in the non ischemic part of the left ventricle mostly located to cardiomyocytes, leading to failing myocardium (Yndestad et al. 2009). In-patients developing HF after an acute MI, elevated baseline NGAL levels were linked to adverse cardiac outcomes and strong NGAL immunostaining was present in cardiomyocytes within the failing myocardium in clinical HF patients (Yndestad et al. 2009). In a study of patients with acute decompensated heart failure, those with worsening renal function had significantly higher levels of serum NGAL on admission and serum NGAL correlated with increased mortality after discharge (Aghel et al. 2010). Admission NGAL $\geq$140 ng/mL was associated with a 7.4-fold increase in the risk of worsening renal function, with a sensitivity and specificity of 86 percent and 54 percent, respectively.

Some studies have evaluated NGAL in the context of chronic heart failure (CHF), as well. Serum and urinary NGAL have been suggested as sensitive early markers of renal dysfunction in CHF patients with normal serum creatinine, but a reduced eGFR (Poniatowski et al. 2009). Urinary NGAL also correlates directly with albumin excretion in the urine and inversely with eGFR in CHF patients (Damman et al. 2008). Higher plasma NGAL levels were also linked directly to CHF severity and associated with higher mortality in another study. A cut-off point of 783 ng/mL was determined for patients to have a higher mortality rate (Bolignano et al. 2009b).

## Pregnancy-associated plasma protein A (PAPP-A)

Pregnancy-associated plasma protein A (PAPP-A) is a high-molecular weight, zinc-binding metalloproteinase, a potentially proatherosclerotic molecule that has been shown to be a specific activator of insulin-like growth factor I (IGF-I), a mediator of atherosclerosis. PAPP-A is typically measured in pregnancy with levels rising to about 100 IU/L at term, but it is present in both men and women (Bayes-Genis et al. 2001). PAPP-A may be a marker of plaque rupture (Apple et al. 2005). PAPP-A was detected in unstable plaques from patients who died a sudden cardiac death (Bayes-Genis et al. 2001). An increase in PAPP-A in suspected ACS patients was an independent predictor of future ischemic events, percutaneous coronary intervention or cardiac bypass surgery (Lund et al. 2003). Elevated PAPP-A levels were also found in patients with negative cardiac troponin I, suggesting a role in detecting high risk patients that are missed by troponins.

However, PAPP-A in the context of ACS cannot be measured by the same assays used in pregnancy and its concentration is affected by laboratory tube additives. More research regarding this molecule is warranted, before a biomarker candidate role is secured.

## Adrenomedullin (ADM)

Adrenomedullin (ADM) is a 52 amino acid vasodilatory peptide released from the adrenal medulla, kidney, and endothelium of large blood vessels, with strong hypotensive effects. It has been shown to be increased in HF patients (Jougasaki et al. 1995). ADM depicts actions very similar to ANP, causing hypotension, vasodilation, and natriuresis. However, it is an unstable molecule, therefore not really useful. Assays detecting MR-proADM, the stable prohormone fragment of proadrenomedullin that reflects ADM levels, have been devised and in one study, MR-proADM was more accurate than BNP or NT-proBNP in predicting HF outcomes at 90 days (Maisel et al. 2010). Such a marker would basically add diagnostic information for patients presenting with intermediate NP levels.

## Cystatin C

Cystatin C has also emerged as a diagnostic and prognostic marker in cardiovascular disease (Patel et al. 2009; Gu et al. 2009; Moran et al. 2008; Ix et al. 2006). Increased serum cystatin C levels were associated with increased left ventricular mass and concentric LVH (Patel et al. 2009) and independently predicted cardiovascular outcomes at six months to one year follow-up of patients with non ST elevation ACS (Taglieri et al. 2010; Kilic et al. 2009). Cystatin C outpowered CRP, NT-proBNP and cTnI in predicting the progression of left ventricular dysfunction in patients appearing with a first ST elevation acute myocardial infarction (STEMI) (Derzhko et al.

2009). In patients with HF exacerbation, cystatin C levels were associated with death or rehospitalization during one year follow-up even after correcting for age, race, sex, comorbidities, and serum creatinine (Campbell et al. 2009). Cystatin C was more predictive of these than creatinine, but their combination was more predictive than either variable alone. In CHF patients, serum cystatin C correlated with more advanced left ventricular diastolic dysfunction and right ventricular systolic dysfunction (Tang et al. 2008). Thus, cystatin C goes more way than just a renal dysfunction marker and may improve cardiovascular risk prognosis in patients without renal impairment (Naruse et al. 2009; Keller et al. 2009; Parikh et al. 2008b; Rifkin et al. 2010). It offers complementary information to other cardiac markers, such as troponin T, hs-CRP, and NT-proBNP, aiding a more accurate risk stratification of patients with AHF or ACS (Manzano-Fernandez et al. 2009; Garcia Acuna et al. 2009; Alehagen et al. 2009).

Additionally, cystatin C is a powerful predictor of CVD mortality in CKD patients. Studies have shown an increased CVD risk associated with increased serum cystatin C levels (Taglieri et al. 2009), which are associated with all-cause mortality, cardiovascular events, and heart failure in CKD patients (Ix et al. 2007).

## *Fas*

Fas or APO-1 is a member of the TNF-α receptor family that is expressed on myocytes. It mediates apoptosis and is of importance in the development of heart failure. HF patients have been found with increased serum levels of a soluble form of Fas, which correlate with disease severity (Okuyama et al. 1997). Administration of pentoxifylline can reduce Fas levels and improve left ventricular function in ischemic or dilated cardiomyopathy (Sliwa et al. 2004).

## Conclusion

The heart and kidneys exhibit closely related functions that maintain cardiac output, systemic pressure, and fluid balance, among others. Accumulative evidence has linked injury or dysfunction of each with adverse or deteriorating operation of the other. The cardiorenal syndrome has thus emerged as a fervently researched clinical setting, incorporating heart and kidney pathophysiology. Laboratory biomarkers that would effectively determine injury or dysfunction early enough for timely and successful medical intervention would greatly enhance CRS treatment. Thorough experimental and clinical research has identified substances like NGAL, KIM-1, cystatin C, NAG, and the natriuretic peptides as powerful markers for both the diagnosis and prognosis of acute or chronic abnormalities of the heart and kidneys. To date, few of all relative biomarkers have already passed into the realm of clinical practice, natriuretic peptides and troponins being such examples. Large, multi-center trials in advanced phases of clinical evaluation and different clinical settings are still needed to further evaluate most others in a number of issues. However, some CRS biomarkers, like NGAL or cystatin C, have quite a promising future and may soon be cleared for practice, aiding or even altering clinical decision making and advancing patient care. Although much remains to be done, research continues to illuminate multiple aspects of the cardiorenal syndrome, bringing better therapies ever closer.

# References

Aghel, A., K. Shrestha, W. Mullens, A. Borowski, and W.H. Tang. 2010. Serum neutrophil gelatinase-associated lipocalin (NGAL) in predicting worsening renal function in acute decompensated heart failure. J. of Cardiac Failure 16: 49–54.

Alehagen, U., U. Dahlstrom, and T.L. Lindahl. Cystatin C and NT-proBNP, a powerful combination of biomarkers for predicting cardiovascular mortality in elderly patients with heart failure: results from a 10-year study in primary care. 2009. Eur. J. Heart Fail. 11: 354–360.

Alpert, J.S., K. Thygesen, E. Antman, and J.P. Bassand. Myocardial infarction redefined. A consensus document of the Joint European Society of Cardiology/American College of Cardiology Committee for the redefinition of myocardial infarction. 2000. J. Am. Coll. Cardiol. 36: 959–969.

Anand, I.S., R. Latini, V.G. Florea, M.A. Kuskowski, T. Rector, S. Masson, S. Signorini, P. Mocarelli, A. Hester, R. Glazer, and J.N.Cohn; Val-HeFT Investigators. 2005. C-reactive protein in heart failure: prognostic value and the effect of valsartan. Circulation 112: 1428–1434.

Andre, P., K.S. Prasad, C.V. Denis, M. He, J.M. Papalia, R.O. Hynes, D.R. Phillips, and D.D. Wagner. 2002. CD40L stabilizes arterial thrombi by a beta3 integrin-dependent mechanism. Nat. Med. 8: 247–252.

Anwaruddin, S., D.M. Lloyd-Jones, A. Baggish, A. Chen, D. Krauser, R. Tung, C. Chae, and J.L. Januzzi Jr. 2006. Renal function, congestive heart failure, and amino-terminal pro-brain natriuretic peptide measurement: results from the ProBNP investigation of dyspnea in the emergency department (PRIDE) study. J. Am. Coll. Cardiol. 47: 91–97.

Anwaruddin, S., J.L. Januzzi Jr., E.L., Lewandrowski, and K.B. Lewandrowski. 2005. Ischemia-modified albumin improves the usefulness of standard cardiac biomarkers for the diagnosis of myocardial ischemia in the emergency department setting. Am. J. Clin. Pathol. 123: 140–145.

Apple, F.S., M.M. Murakami, L.A. Pearce, and C.A. Herzog. 2002. Predictive value of cardiac troponin I and T for subsequent death in end-stage renal disease. Circulation 106: 2941–2945.

Apple, F.S., A.H. Wu, J. Mair, J. Ravkilde, M. Panteghini, J. Tate, F. Pagani, R.H. Christenson, M. Mockel, O. Danne, and A.S. Jaffe; Committee on Standardization of Markers of Cardiac Damage of the IFCC. 2005. Future biomarkers for detection of ischemia and risk stratification in acute coronary syndrome. Clin. Chem. 51: 810–824.

Arakawa, N., M. Nakamura, H. Aoki, and K. Hiramori. 1994. Relationship between plasma level of brain natriuretic peptide and myocardial infarct size. Cardiology 85: 334–340.

Ascione, R., C.T. Lloyd, M.J. Underwood, W.J. Gomes, G.D. Angelini. 1999. On-pump versus off-pump coronary revascularization: evaluation of renal function. Ann. Thorac. Surg. 68: 493–498.

Aukrust, P., F. Muller, T. Ueland, T. Berget, E. Aaser, A. Brunsvig, N.O. Solum, K. Forfang, S.S. Frøland, and L. Gullestad. 1999. Enhanced levels of soluble and membrane-bound CD40 ligand in patients with unstable angina. Possible reflection of T lymphocyte and platelet involvement in the pathogenesis of acute coronary syndromes. Circulation 100: 614–620.

Austin, W.J., V. Bhalla, I. Hernandez-Arce, S.R. Isakson, J. Beede, P. Clopton, A.S. Maisel, and R.L. Fitzgerald. 2006. Correlation and prognostic utility of B-type natriuretic peptide and its amino-terminal fragment in patients with chronic kidney disease. Am. J. Clin. Pathol. 126: 506–512.

Aviles, R.J., A.T. Askari, B. Lindahl, B. Lindahl, L. Wallentin, G. Jia, E.M. Ohman, K.W. Mahaffey, L.K. Newby, R.M. Califf, M.L. Simoons, E.J. Topol, P. Berger, and M.S. Lauer MS. 2002. Troponin T levels in patients with acute coronary syndromes, with or without renal dysfunction. N. Engl. J. Med. 346: 2047–2052.

Bachorzewska-Gajewska, H., J. Malyszko, E. Sitniewska, J.S. Malyszko, and S. Dobrzycki. 2006. Neutrophil-gelatinase-associated lipocalin and renal function after percutaneous coronary interventions. Am. J. Nephrol. 26: 287–292.

Bachorzewska-Gajewska, H., B. Poniatowski, and S. Dobrzycki. 2009. NGAL (neutrophil gelatinaseassociated lipocalin) and L-FABP after percutaneous coronary interventions due to unstable angina in patients with normal serum creatinine. Adv. Med. Sci. 54: 221–224.

Bailly, V., Z. Zhang, W. Meier, R. Cate, M. Sanicola, and J.V. Bonventre. 2002. Shedding of kidney injury molecule-1, a putative adhesion protein involved in renal regeneration. J. Biol. Chem. 277: 39739–39748.

Baldus, S., C. Heeschen, T. Meinertz, A.M. Zeiher, J.P. Eiserich, T. Münzel, M.L. Simoons, and C.W. Hamm; CAPTURE Investigators. 2003. Myeloperoxidase serum levels predict risk in patients with acute coronary syndromes. Circulation 108: 1440–1445.

Bar-Or, D., J.V. Winkler, K. Vanbenthuysen, L. Harris, E. Lau, and F.W. Hetzel. 2001. Reduced albumin-cobalt binding with transient myocardial ischemia after elective percutaneous transluminal coronary angioplasty: a preliminary comparison to creatinine kinase-MB, myoglobin, and troponin I. Am. Heart J. 141: 985–991.

Bassan, R., A. Potsch, A. Maisel, B. Tura, H. Villacorta, M.V. Nogueira, A. Campos, R. Gamarski, A.C. Masetto, and M.A. Moutinho. 2005. B-type natriuretic peptide: a novel early blood marker of acute myocardial infarction in patients with chest pain and no ST-segment elevation. Eur. Heart. J. 26: 234–240.

Bayes-Genis, A., C.A. Conover, M.T. Overgaard, K.R. Bailey, M. Christiansen, D.R. Holmes Jr., R. Virmani, C. Oxvig, and R.S. Schwartz. 2001. Pregnancy-associated plasma protein A as a marker of acute coronary syndromes. N. Engl. J. Med. 345: 1022–1029.

Bazzi, C., C. Petrini, V. Rizza, G. Arrigo, P. Napodano, M. Paparella, and G. D'Amico. 2002. Urinary N-acetyl-beta-glucosaminidase excretion is a marker of tubular cell dysfunction and a predictor of outcome in primary glomerulonephritis, Nephrol. Dial. Transplant. 17: 1890–1896.

Bennett, M., C.L. Dent, Q. Ma, S. Dastrala, F. Grenier, R. Workman, H. Syed, S. Ali, J. Barasch, and P. Devarajan. 2008. Urine NGAL predicts severity of acute kidney injury after cardiac surgery: A prospective study. Clin. J. Am. Soc. Nephrol. 3: 665–673.

Bhavsar, N.A., L.J. Appel, J.W. Kusek, G. Contreras, G. Bakris, J. Coresh, and B.C. Astor; AASK Study Group. 2011. Comparison of Measured GFR, Serum Creatinine, Cystatin C, and Beta-Trace Protein to Predict ESRD in African Americans With Hypertensive CKD. Am. J. Kidney Dis. 58: 886–893.

Bobek, I., M. de Cal, D. Cruz, F. Nalesso, P. Lentini, V. Corradi, M. Happio, and C. Ronco. 2008. *In vitro* removal of NGAL by extracorporeal therapy. J. Am. Soc. Nephrol. 19: 457A.

Bokenkamp, A., I. Franke, M. Schlieber, G. Düker, J. Schmitt, S. Buderus, M.J. Lentze, and B. Stoffel-Wagner. 2007. Beta-trace protein—a marker of kidney function in children: "Original research communication-clinical investigation." Clin. Biochem. 40: 969–975.

Bolignano, D., A. Lacquaniti, G. Coppolino, V. Donato, S. Campo, M.R. Fazio, G. Nicocia, and M. Buemi. 2009a. Neutrophil gelatinaseassociated lipocalin (NGAL) and progression of chronic kidney disease. Clin. J. Am. Soc. Nephrol. 4: 337–344.

Bolignano, D., G. Coppolino, S. Campo, C. Aloisi, G. Nicocia, N. Frisina, and M. Buemi. 2007. Neutrophil gelatinase-associated lipocalin in patients with autosomal dominant polycystic kidney disease. Am. J. Nephrol. 27: 373–378.

Bolignano, D., G. Basile, P. Parisi, G. Coppolino, G. Nicocia, and M. Buemi. 2009b. Increased plasma neutrophil gelatinase-associated lipocalin levels predict mortality in elderly patients with chronic heart failure. Rejuvenation Res. 12: 7–14.

Bondiou, M.T., R. Bourbouze, M. Bernard, F. Percheron, N. Perez-Gonzalez, and J.A. Cabezas. 1985. Inhibition of A and B N-acetyl-beta-D-glucosaminidase urinary isoenzymes by urea. Clin. Chim. Acta 149: 67–73.

Bonventre, J.V. and L. Yang. Kidney injury molecule-1. 2010. Curr. Opin. Crit. Care. 16: 1–6.

Bosomworth, M.P., S.R. Aparicio, and A.W.M. Hay. 1999. Urine N-acetyl-"-D-glucosaminidase-a marker of tubular damage? Nephrol. Dial. Transplant. 14: 620–626.

Branten, A.J.W., T.P.J. Mulder, W.H.M. Peters, K.J.M. Assmann, and J.F.M. Wetzels. 2000. Urinary excretion of glutathione S transferases alpha and pi in patients with proteinuria: reflection of the site of tubular injury. Nephron 85: 120–126.

Brennan, M.L., M.S. Penn, F. Van Lente, V. Nambi, M.H. Shishehbor, R.J. Aviles, M. Goormastic, M.L. Pepoy, E.S. McErlean, E.J. Topol, S.E. Nissen, and S.L. Hazen. 2003. Prognostic value of myeloperoxidase in patients with chest pain. N. Engl. J. Med. 349: 1595–1604.

Bulent Gul, C.B., M. Gullulu, B. Oral, A. Aydinlar, O. Oz, F. Budak, Y. Yilmaz, and M. Yurtkuran. 2008. Urinary IL-18: a marker of contrast-induced nephropathy following percutaneous coronary intervention? Clin. Biochem. 41: 544–547.

Cabrera, J., V. Arroyo, A.M. Ballesta, A. Rimola, J. Gual, M. Elena, and J. Rodes. 1982. Aminoglycoside nephrotoxicity in cirrhosis. Value of urinary beta 2-microglobulin to discriminate functional renal failure from acute tubular damage. Gastroenterology 82: 97–105.

Campbell, C.Y., W. Clarke, H. Park, N. Haq, B.B. Barone, and D.J. Brotman. 2009. Usefulness of cystatin C and prognosis following admission for acute heart failure. Am. J. Cardiol. 104: 389–392.

Carr, S.J., S. Bavanandan, B. Fentum, and L. Ng. 2005. Prognostic potential of brain natriuretic peptide (BNP) in predialysis chronic kidney disease patients. Clin. Sci. (Lond.) 109: 75–82.

Chalikias, G.K., D.N. Tziakas, J.C. Kaski, A. Kekes, E.I. Hatzinikolaou, D.A. Stakos, I.K. Tentes, A.X. Kortsaris, and D.I. Hatseras. 2007. Interleukin-18/interleukin-10 ratio is an independent predictor of recurrent coronary events during a 1-year follow-up in patients with acute coronary syndrome. Int. J. Cardiol. 117: 333–339.

Chandrasekar, B., S. Mummidi, W.C. Claycomb, R. Mestril, and M. Nemer. 2005. Interleukin-18 is a pro-hypertrophic cytokine that acts through a phosphatidylinositol 3-kinase-phosphoinositide-dependent kinase-1-Akt-GATA4 signaling pathway in cardiomyocytes. J. Biol. Chem. 280: 4553–4567.

Chen, D., C. Assad-Kottner, C. Orrego, and G. Torre-Amione. 2008. Cytokines and acute heart failure. Crit. Care. Med. 36: S9–16.

Chew, S.L., R.L. Lins, R. Daelemans, G.D. Nuyts, and M.E. De Broe. 1993. Urinary enzymes in acute renal failure. Nephrol. Dial. Transplant. 8: 507–511.

Coca, S.G., R. Yalavarthy, J. Concato, and C.R. Parikh. 2008. Biomarkers for the diagnosis and risk stratification of acute kidney injury: systematic review. Kidney Int. 73: 1008–1016.

Comper, W.D., L.M. Hilliard, D.J. Nikolic-Paterson, and L.M. Russo. 2008. Disease-dependent mechanisms of albuminuria. Am. J. Physiol. Renal. Physiol. 295: F1589–1600.

Conti, M., M. Zater, K. Lallali, A. Durrbach, S. Mouteraeu, P. Manivet, P. Eschwege, and S. Loric. 2005. Absence of circadian variations in urine cystatin C allows its use on urinary samples. Clin. Chem. 51: 272–273.

Cruz, D.N., S. Soni, L. Slavin, C. Ronco, and A. Maisel. 2010. Biomarkers of cardiac and kidney dysfunction in cardiorenal syndromes. Contrib. Nephrol. 165: 83–92.

D'Amico, G. and C. Bazzi. 2003. Urinary protein and enzyme excretion as markers of tubulardamage, Curr. Opin. Nephrol. Hypertens. 12: 639–643.

Damman, K., G. Navis, A.A. Voors, F.W. Asselbergs, T.D. Smilde, J.G. Cleland, D.J. van Veldhuisen, and H.L. Hillege. 2007. Worsening renal function and prognosis in heart failure: systematic review and meta-analysis. J. Card. Fail. 13: 599–608.

Damman, K., D.J. van Veldhuisen, G. Navis, A.A. Voors, and H.L. Hillege. 2008. Urinary neutrophil gelatinase associated lipocalin (NGAL), a marker of tubular damage, is increased in patients with chronic heart failure. Eur. J. Heart Fail. 10: 997–1000.

Danesh, J., J.G. Wheeler, and G. Hirschfield. 2004. CRP and other circulating markers of inflammation in the prediction of chronic heart disease. N. Engl. J. Med. 350: 1387–1397.

Daniels, L.B., P. Clopton, V. Bhalla, P. Krishnaswamy, R.M. Nowak, J. McCord, J.E. Hollander, P. Duc, T. Omland, A.B. Storrow, W.T. Abraham, A.H. Wu, P.G. Steg, A. Westheim, C.W. Knudsen, A. Perez, R. Kazanegra, H.C. Herrmann, P.A. McCullough, and A.S. Maisel. 2006. How obesity affects the cut-points for B-type natriuretic peptide in the diagnosis of acute heart failure. Results from the Breathing Not Properly Multinational Study. Am. Heart J. 151: 999–1005.

Daniels, L.B. and A.S. Maisel. 2007. Natriuretic peptides. J. Am. Coll. Cardiol. 50: 2357–2368.

de Lemos, J.A., D.A. Morrow, J.H. Bentley, T. Omland, M.S. Sabatine, C.H. McCabe, C. Hall, C.P. Cannon, and E. Braunwald. 2001. The prognostic value of B-type natriuretic peptide in patients with acute coronary syndromes. N. Engl. J. Med. 345: 1014–1021.

de Lemos, J.A., D.K. McGuire, and M.H. Drazner. 2003a. B-type natriuretic peptide in cardiovascular disease. Lancet 362: 316–322.

de Lemos, J.A., D.A. Morrow, M.S. Sabatine, S.A. Murphy, C.M. Gibson, E.M. Antman, C.H. McCabe, C.P. Cannon, and E. Braunwald. 2003b. Association between plasma levels of monocyte chemoattractant protein-1 and long-term clinical outcomes in patients with acute coronary syndromes. Circulation 107: 690–695.

de Winter, R.J., R.W. Koster, A. Sturk, and G.T. Sanders. 1995. Value of myoglobin, troponin T, and CK-MB mass in ruling out an acute myocardial infarction in the emergency room. Circulation 92: 3401–3407.

deFilippi, C., S. Wasserman, S. Rosanio, E. Tiblier, H. Sperger, M. Tocchi, R. Christenson, B. Uretsky, M. Smiley, J. Gold, J. Muniz, J. Badalamenti, C. Herzog, and W. Henrich. 2003. Cardiac troponin T and C-reactive protein for predicting prognosis, coronary atherosclerosis, and cardiomyopathy in patients undergoing long-term hemodialysis. JAMA 290: 353–359.

Dent, C.L., Q. Ma, S. Dastrala, M. Bennett, M.M. Mitsnefes, J. Barasch, and P. Devarajan. 2007. Plasma neutrophil gelatinase-associated lipocalin predicts acute kidney injury, morbidity and mortality after pediatric cardiac surgery: A prospective uncontrolled cohort study. Crit. Care 11: R127.

Derzhko, R., R. Plaksej, M. Przewlocka-Kosmala, and W. Kosmala. 2009. Prediction of left ventricular dysfunction progression in patients with a first ST-elevation myocardial infarction-contribution of cystatin C assessment. Coron. Artery Dis. 20: 453–461.

Devarajan, P. 2007. Emerging biomarkers of acute kidney injury. Contrib. Nephrol. 156: 203–212.

Dharnidharka, V.R., C. Kwon, and G. Stevens. 2002. Serum cystatin C is superior to serum creatinine as a marker of kidney function: a meta-analysis. Am. J. Kidney Dis. 40: 221–226.

Dickstein, K., A. Cohen-Solal, G. Filippatos, J.J. McMurray, P. Ponikowski, P.A. Poole-Wilson, A. Strömberg, D.J. van Veldhuisen, D. Atar, A.W. Hoes, A. Keren, A. Mebazaa, M. Nieminen, S.G. Priori, and K. Swedberg; ESC Committee for Practice Guidelines (CPG). 2008. ESC guidelines for the diagnosis and treatment of acute and chronic heart failure 2008: the Task Force for the Diagnosis and Treatment of Acute and Chronic Heart Failure 2008 of the European Society of Cardiology. Developed in collaboration with the Heart Failure Association of the ESC (HFA) and endorsed by the European Society of Intensive Care Medicine (ESICM). Eur. J. Heart Fail. 10: 933–989.

Dieterle, F., F. Staedtler, O. Grenet, A. Cordier, E. Perentes, D. Roth, A. Mahl, R. Papoian, P. Grass, M. Kammueller, F. Legay, D. Wahl, J. Vonderscher, S. Chibout, and G. Maurer. 2007. Qualification of biomarkers for regulatory decision making: a kidney safety biomarker project (abstract, Society of Toxicology). Toxicologist 96: 383.

Doust, J.A., E. Pietrzak, A. Dobson, and P. Glasziou. 2005. How well does B-type natriuretic peptide predict death and cardiac events in patients with heart failure: systematic review. BMJ 330: 625.

du Cheyron, D., C. Daubin, J. Poggioli, M. Ramakers, P. Houillier, P. Charbonneau, and M. Paillard. 2003. Urinary measurement of Na+/H+ exchanger isoform 3 (NHE3) protein as new marker of tubule injury in critically ill patients with ARF. Am. J. Kidney Dis. 42: 497–506.

Edelstein, C.L., T.S. Hoke, H. Somerset, W. Fang, C.L. Klein, C.A. Dinarello, and S. Faubel. 2007. Proximal tubules from caspase-1-deficient mice are protected against hypoxia-induced membrane. Nephrol. Dial. Transplant. 22: 1052–1061.

Ek-Von Mentzer, B.A., F. Zhang, and J.A. Hamilton. 2001. Binding of 13-HODE and 15-HETE to phospholipid bilayers, albumin, and intracellular fatty acid binding proteins. Implications for transmembrane and intracellular transport and for protection from lipid peroxidation, J. Biol. Chem. 276: 15575–15580.

Emeigh Hart, S.G. 2005. Assessment of renal injury *in vivo*, J. Pharmacol. Toxicol. Methods 52: 30–45.

Erlikh, A.D., A.G. Katrukha, I.R. Trifonov, A.V. Bereznikova, and N.A. Gratsianskiĭ. 2005. Prognostic significance of heart fatty acid binding protein in patients with non-ST elevation acute coronary syndrome: results of follow-up for twelve months. Kardiologiia. 45: 13–21.

Fakhrzadeh, H., M. Ghaderpanahi, F. Sharifi, Z. Badamchizade, M. Mirarefin, and B. Larijani. 2009. Increased risk of chronic kidney disease in elderly with metabolic syndrome and high levels of Creactive protein: Kahrizak Elderly Study. Kidney Blood Press. Res. 32: 457–463.

Felker, G.M., V. Hasselblad, A.F. Hernandez, and C.M. O'Connor. 2009. Biomarker-guided therapy in chronic heart failure: a meta-analysis of randomized controlled trials. Am. Heart J. 158: 422–430.

Ferguson, M.A., V.S. Vaidya, S.S. Waikar, F.B. Collings, K.E. Sunderland, C.J. Gioules, and J.V. Bonventre. 2010. Urinary liver-type fatty acidbinding protein adverse outcomes in acute kidney injury. Kidney Int. 77: 708–714.

Ferguson, S.M. 2003. CYR61 as a marker for acute renal failure. Fed. Regist. 68: 18660.

Fonarow, G.C., W.F. Peacock, T.B. Horwich, C.O. Phillips, M.M. Givertz, M. Lopatin, J. Wynne; ADHERE Scientific Advisory Committee and Investigators. 2008. Usefulness of B-type natriuretic peptide and cardiac troponin levels to predict in-hospital mortality from ADHERE. Am. J. Cardiol. 101: 231–237.

Forfia, P.R., M. Lee, R.S. Tunin, M. Mahmud, H.C. Champion, and D.A. Kass. 2007. Acute phosphodiesterase 5 inhibition mimics hemodynamic effects of B-type natriuretic peptide and potentiates B-type natriuretic peptide effects in failing but not normal canine heart. J. Am. Coll. Cardiol. 49: 1079–1088.

Fox, E.R., E.J. Benjamin, D.F. Sarpong, H. Nagarajarao, J.K. Taylor, M.W. Steffes, A.K. Salahudeen, M.F. Flessner, E.L. Akylbekova, C.S. Fox, R.J. Garrison, and H.A. Taylor Jr. 2010. The relation of C-reactive protein to chronic kidney disease in African Americans: the Jackson Heart Study. BMC Nephrol. 11: 1.

Freedman, J. 2003. SSCD40 ligand: assessing risk instead of damage? N. Engl. J. Med. 348: 1163–1165.

Fuchs, S., E. Stabile, G.S. Mintz, C.K. Pappas, A. Maehara, L. Gruberg, L.F. Satler, A.D. Pichard, K.M. Kent, and N.J. Weissman. 2002. Intravascular ultrasound findings in patients with acute coronary syndromes with and without elevated troponin I level. Am. J. Cardiol. 89: 1111–1113.

Fukuda, Y., S. Miura, B. Zhang, A. Iwata, A. Kawamura, H. Nishikawa, K. Shirai, and K. Saku. 2009. Significance of urinary liver-fatty acid-binding protein in cardiac catheterization in patients with coronary artery disease. Intern. Med. 48: 1731–1737.

Furtado, M.V., A.P. Rossini, R.B. Campani, C. Meotti, M. Segatto, G. Vietta, and C.A. Polanczyk. 2009. Interleukin-18: an independent predictor of cardiovascular events in patients with acute coronary syndrome after 6 months of follow-up. Coron. Artery Dis. 20: 327– 331.

Garcia Acuna, J.M., E. Gonzalez-Babarro, L. Grigorian Shamagian, C. Peña-Gil, R. Vidal Pérez, A.M. López-Lago, M. Gutiérrez Feijoó, and J.R. González-Juanatey. 2009. Cystatin C provides more information than other renal function parameters for stratifying risk in patients with acute coronary syndrome. Rev. Esp. Cardiol. 62: 510–519.

Garlichs, C.D., S. Eskafi, D. Raaz, A. Schmidt, J. Ludwig, M. Herrmann, L. Klinghammer, W.G. Daniel, and A. Schmeisser. 2001. Patients with acute coronary syndromes express enhanced CD40 ligand/CD154 on platelets. Heart 86: 649–655.

Gok, M.A., M. Pelzers, J.F. Glatz, B.K. Shenton, P.E. Buckley, R. Peaston, C. Cornell, D. Mantle, N. Soomro, B.C. Jaques, D.M. Manas, and D. Talbot. 2003. Do tissue damage biomarkers used to assess machine-perfused NHBD kidneys predict long-term renal function post-transplant? Clin. Chim. Acta 338: 33–43.

Gordjani, N., R. Burghard, D. Muller, H. Mathai, G. Mergehenn, J.U. Leititis, and M. Brandis.1995. Urinary excretion of deaminase binding protein in neonates treated with tobramycin. Pediatr. Nephrol. 9: 419–422.

Goto, T., H. Takase, T. Toriyama, T. Sugiura, K. Sato, R. Ueda, and Dohi Y. 2003. Circulating concentrations of cardiac proteins indicate the severity of congestive heart failure. Heart. 89: 1303–1307.

Gracie, J.A., S.E. Robertson, and I.B. McInnes. 2003. Interleukin-18. J. Leukoc. Biol. 73: 213–224.

Gu, F.F., S.Z. Lu, Y.D. Chen, R. Peshock, A. Khera, J.A. de Lemos, J.A. Balko, S. Gupta, P.P. Mammen, M.H. Drazner, and D.W. Markham. 2009. Relationship between plasma cathepsin S and cystatin C levels and coronary plaque morphology of mild to moderate lesions: an *in vivo* study using intravascular ultrasound. Chin. Med. J. 122: 2820–2826.

Haase, M., R. Bellomo, D. Story, P. Davenport, and A. Haase-Fielitz. 2008. Urinary interleukin-18 does not predict acute kidney injury after adult cardiac surgery: A prospective observational cohort study. Crit. Care 12: R96.

Haase, M., R. Bellomo, P. Devarajan, P. Schlattmann, and A. Haase-Fielitz. 2009a. Accuracy of neutrophil gelatinase-associated lipocalin (NGAL) in diagnosis and prognosis in acute kidney injury: a systematic review and meta-analysis. Am. J. Kidney Dis. 54: 1012–24.

Haase, M., R. Bellomo, P. Devarajan, Q. Ma, M.R. Bennett, M. Mockel, G. Matalanis, D. Dragun, and A. Haase-Fielitz. 2009b. Novel biomarkers early predict the severity of acute kidney injury after cardiac surgery in adults. Ann. Thorac. Surg. 88: 124–130.

Haase-Fielitz, A., R. Bellomo, P. Devarajan, D. Story, G. Matalanis, D. Dragun, and M. Haase. 2009c. Novel and conventional serum biomarkers predicting acute kidney injury in adult cardiac surgery—a prospective cohort study. Crit. Care Med. 37: 553–560.

Han, W.K., V. Bailly, R. Abichandani, R. Thadhani, and J.V. Bonventre. 2002. Kidney Injury Molecule-1 (KIM-1): A novel biomarker for human renal proximal tubule injury. Kidney Int. 62: 237–244.

Han, W.K., S.S. Waikar, A. Johnson, R.A. Betensky, C.L. Dent, P. Devarajan and J.V. Bonventre. 2008. Urinary biomarkers in the early diagnosis of acute kidney injury. Kidney Int. 73: 863–869.

Han, W.K., G. Wagener, Y. Zhu, S. Wang, and H.T. Lee. 2009. Urinary biomarkers in the early detection of acute kidney injury after cardiac surgery. Clin. J. Am. Soc. Nephrol. 4: 873–882.

Hanai, K., T. Babazono, I. Nyumura, K. Toya, N. Tanaka, M. Tanaka, A. Ishii, and Y. Iwamoto. 2009. Asymmetric dimethylarginine is closely associated with the development and progression of nephropathy in patients with type 2 diabetes. Nephrol. Dial. Transplant. 24: 1884–1888.

Harmankaya, O., Y. Ozturk, T. Basturk, and A. Obek. 2001. Urinary excretion ofN-acetyl-beta-D-glucosaminidase in newly diagnosed essential hypertensive patients and its changes with effective antihypertensive therapy. Int. Urol. Nephrol. 32: 583–584.

Hartmann, H.G., H.E. Braedel, and G.A. Jutzler. 1985. Detection of renal tubular lesions after abdominal aortography and selective renal arteriography by quantitative measurements of brushborder enzymes in the urine. Nephron 39: 95–101.

Heeschen, C., C.W. Hamm, B. Goldmann, A. Deu, L. Langenbrink, and H.D. White. 1999. Troponin concentrations for stratification of patients with acute coronary syndromes in relation to therapeutic efficacy of tirofiban. PRISM Study Investigators. Platelet Receptor Inhibition in Ischemic Syndrome Management. Lancet 354: 1757–1762.

Heeschen, C., S. Dimmeler, C.W. Hamm, M.J. van den Brand, E. Boersma, A.M. Zeiher, and M.L. Simoons; CAPTURE Study Investigators. 2003. Soluble CD40 ligand in acute coronary syndromes. N. Engl. J. Med. 348: 1104–1111.

Heidenreich, P.A., T. Alloggiamento, K. Melsop, K.M. McDonald, A.S. Go, M.A. Hlatky. 2001. The prognostic value of troponin in patients with non-ST elevation acute coronary syndromes: a meta-analysis. J. Am. Coll. Cardiol. 38: 478–485.

Hemdahl, A.L., A. Gabrielsen, C. Zhu, P. Eriksson, U. Hedin, J. Kastrup, P. Thorén, and G.K. Hansson. 2006. Expression of neutrophil gelatinase-associated lipocalin in atherosclerosis and myocardial infarction. Arterioscler. Thromb. Vasc. Biol. 26: 136–142.

Henn, V., J.R. Slupsky, M. Grafe, I. Anagnostopoulos, R. Förster, G. Müller-Berghaus, and R.A. Kroczek. 1998. CD40 ligand on activated platelets triggers an inflammatory reaction of endothelial cells. Nature 391: 591–594.

Herget-Rosenthal, S., G. Marggraf, J. Husing, F. Goring, F. Pietruck, O. Janssen, T. Philipp, and A. Kribben. 2004a. Early detection of acute renal failure by serum cystatin C. Kidney Int. 66: 1115–1122.

Herget-Rosenthal, S., D. Poppen, J. Hüsing, G. Marggraf, F. Pietruck, H.G. Jakob, T. Philipp, and A. Kribben. 2004b. Prognostic value of tubular proteinuria and enzymuria in nonoliguric acute tubular necrosis. Clin. Chem. 50: 552–558.

Hlatky, M.A., P. Greenland, D.K. Arnett, C.M. Ballantyne, M.H. Criqui, M.S. Elkind, A.S. Go, F.E. Harrell Jr., Y. Hong, B.V. Howard, V.J. Howard, P.Y. Hsue, C.M. Kramer, J.P. McConnell, S.L. Normand, C.J. O'Donnell, S.C. Smith Jr., and P.W. Wilson. 2009. Criteria for evaluation of novel markers of cardiovascular risk: A scientific statement from the American Heart Association. Circulation 119: 2408–2416.

Hoffmann, A., M. Nimtz, and H.S. Conradt. 1997. Molecular characterization of beta-trace protein in human serum and urine: a potential diagnostic marker for renal diseases. Glycobiology 7: 499–506.

Horwich, T.B., J. Patel, W.R. MacLellan, and G.C. Fonarow. 2003. Cardiac troponin I is associated with impaired hemodynamics, progressive left ventricular dysfunction, and increased mortality rates in advanced heart failure. Circulation 108: 833–838.

Hudson, M.P., C.M. O'Connor, W.A. Gattis, G. Tasissa, V. Hasselblad, C.M. Holleman, L.H. Gaulden, F. Sedor, and E.M. Ohman. 2004. Implications of elevated cardiac troponin T in ambulatory patients with heart failure: a prospective analysis. Am. Heart J. 147: 546–552.

Ichimura, T., J.V. Bonventre, V. Bailly, H. Wei, C.A. Hession, R.L. Cate, and M. Sanicola. 1998. Kidney injury molecule-1 (KIM-1), a putative epithelial cell adhesion molecule containing a novel immunoglobulin domain, is up-regulated in renal cells after injury. J. Biol. Chem. 273: 4135–4142.

Ichimura, T., C.C. Hung, S.A. Yang, J.L. Stevens, and J.V. Bonventre. 2004. Kidney injury molecule-1: a tissue and urinary biomarker for nephrotoxicant-induced renal injury. Am. J. Physiol. Renal. Physiol. 286: F552–F563.

Ichimura, T. and S. Mou. 2008. Kidney injury molecule-1 in acute kidney injury and renal repair: a review. J. Chin. Integr. Med. 5: 533–538.

Ikenaga, H., H. Suzuki, N. Ishii, H. Itoh, and T. Saruta. 1993. Enzymuria in non-insulin-dependent diabetic patients: signs of tubular cell dysfunction. Clin. Sci. (Lond.) 84: 469–475.

Inbar, R. and Y. Shoenfeld. 2009. Elevated cardiac troponins: the ultimate marker for myocardial necrosis, but not without a differential diagnosis. Isr. Med. Assoc. J. 1: 50–53.

Iseki, K., K. Kinjo, C. Iseki, and S. Takishita. 2004. Relationship between predicted creatinine clearance and proteinuria and the risk of developing ESRD in Okinawa, Japan. Am. J. Kidney Dis. 44: 806–814.

Ishii, J., Y. Ozaki, J. Lu, F. Kitagawa, T. Kuno, T. Nakano, Y. Nakamura, H. Naruse, Y. Mori, S. Matsui, H. Oshima, M. Nomura, K. Ezaki, and H. Hishida. 2005. Prognostic value of serum concentration of heart-type fatty acid binding protein relative to cardiac troponin T on admission in the early hours of acute coronary syndrome. Clin. Chem. 51: 1397–1404.

Iwanaga, Y. and S. Miyazaki. 2010. Heart failure, chronic kidney disease, and biomarkers—an integrated viewpoint. Circ. J. 74: 1274–1282.

Ix, J.H., M.G. Shlipak, G.M. Chertow, S. Ali, N.B. Schiller, and M.A. Whooley. 2006. Cystatin C, left ventricular hypertrophy, and diastolic dysfunction: data from the Heart and Soul Study. J. Card. Fail. 12: 601– 607.

Ix, J.H., M.G. Shlipak, G.M. Chertow, and M.A. Whooley. 2007. Association of cystatin C with mortality, cardiovascular events, and incident heart failure among persons with coronary heart disease: data from the Heart and Soul Study. Circulation 115: 173–179.

Jackson, C.E., S.D. Solomon, H.C. Gerstein, S. Zetterstrand, B. Olofsson, E.L. Michelson, C.B. Granger, K. Swedberg, M.A. Pfeffer, S. Yusuf, and J.J. McMurray; CHARM Investigators and Committees. 2009. Albuminuria in chronic heart failure: prevalence and prognostic importance. Lancet 374: 543–550.

Jaffe, A.S., L. Babuin, and F.S. Apple. 2006. Biomarkers in acute cardiac disease: the present and the future. J. Am. Coll. Cardiol. 48: 1–11.

Januzzi, J.L., F. Peacock, A.S. Maisel, C.U. Chae, R.L. Jesse, A.L. Baggish, M. O'Donoghue, R. Sakhuja, A.A. Chen, R.R van Kimmenade, K.B. Lewandrowski, D.M. Lloyd-Jones, and A.H. Wu. 2007. Measurement of the interleukin family member ST2 in patients with acute dyspnea: results from the PRIDE (Pro-Brain Natriuretic Peptide Investigation of Dyspnea in the Emergency Department) study. J. Am. Coll. Cardiol. 50: 607–613.

Jernberg, T., B. Lindahl, S. James, G. Ronquist, and L. Wallentin. Comparison between strategies using creatinine kinase-MB (mass), myoglobin, and troponin T in the early detection or exclusion of acute myocardial infarction in patients with chest pain and a nondiagnostic electrocardiogram. 2000. Am. J. Cardiol. 86: 1367–1371.

Jougasaki, M., C.M. Wei, L.J. McKinley, and J.C. Burnett Jr. 1995. Elevation of circulating and ventricular adrenomedullin in human congestive heart failure. Circulation 92: 286–289.

Jurlander, B., P. Clemmensen, G.S. Wagner, and P. Grande. 2000. Very early diagnosis and risk stratification of patients admitted with suspected acute myocardial infarction by the combined evaluation of a single serum value of cardiac troponin-T, myoglobin, and creatinine kinase MB (mass). Eur. Heart J. 21: 382–389.

Kameda, K., T. Matsunaga, N. Abe, H. Hanada, H. Ishizaka, H. Ono, M. Saitoh, K. Fukui, I. Fukuda, T. Osanai, and K. Okumura. 2003. Correlation of oxidative stress with activity of matrix metalloproteinase in patients with coronary artery disease. Eur. Heart J. 24: 2180–2185.

Kamijo, A., T. Sugaya, A. Hikawa, M. Okada, F. Okumura, M. Yamanouchi, A. Honda, M. Okabe, T. Fujino, Y. Hirata, M. Omata, R. Kaneko, H. Fujii, A. Fukamizu, and K. Kimura. 2004. Urinary excretion of fatty acid-binding protein reflects stress overload on the proximal tubules. Am. J. Pathol. 165: 1243–1255.

Kamijo, A., T. Sugaya, A. Hikawa, M. Yamanouchi, Y. Hirata, T. Ishimitsu, A. Numabe, M. Takagi, H. Hayakawa, F. Tabei, T. Sugimoto, N. Mise, and K. Kimura. 2005. Clinical evaluation of urinary excretion of liver-type fatty acid-binding protein as a marker for the monitoring of chronic kidney disease: a multicenter trial. J. Lab. Clin. Med. 145: 125–133.

Kawasaki, D., T. Tsujino, S. Morimoto, M. Masai, M. Masutani, M. Ohyanagi, S. Kashiwamura, H. Okamura, and T. Masuyama 2005.. Plasma interleukin-18 concentration: a novel marker of myocardial ischemia rather than necrosis in humans. Coron. Artery Dis. 16: 437–441.

Keller, T., C.M. Messow, E. Lubos, V. Nicaud, P.S. Wild, H.J. Rupprecht, C. Bickel, S. Tzikas, D. Peetz, K.J. Lackner, L. Tiret, T.F. Münzel, S. Blankenberg, and R.B. Schnabel. 2009. Cystatin C and cardiovascular mortality in patients with coronary artery disease and normal or mildly reduced kidney function: results from the AtheroGene study. Eur. Heart J. 30: 314–320.

Kelly, K.J. 2003. Distant effects of experimental renal ischemia/reperfusion injury. J. Am. Soc. Nephrol. 14: 1549–1558.

Khan, N.A., B.R. Hemmelgarn, M. Tonelli, C.R. Thompson, and A. Levin. 2005. Prognostic value of troponin T and I among asymptomatic patients with end-stage renal disease: a metaanalysis. Circulation 112: 3088–3096.

Khera, A., D. McGuire, S. Murphy, H.G. Stanek, S.R. Das, W. Vongpatanasin, F.H. Wians Jr., S.M. Grundy, and J.A. de Lemos. 2005. Race and gender differences in C-reactive protein levels. J. Am. Coll. Cardiol. 46: 464–9.

Kilic, T., G. Oner, E. Ural, Z. Yumuk, T. Sahin, U. Bildirici, E. Acar, U. Celikyurt, G. Kozdag, and D. Ural. 2009. Comparison of the long-term prognostic value of cystatin C to other indicators of renal function, markers of inflammation and systolic dysfunction among patients with acute coronary syndrome. Atherosclerosis 207: 552–558.

Knight, E.L., J.C. Verhave, D. Spiegelman, H.L. Hillege, D. de Zeeuw, G.C. Curhan, and P.E. De Jong. 2004. Factors influencing serum cystatin C levels other than renal function and the impact on renal function measurement. Kidney Int. 65: 1416–1421.

Ko, G.J., D.N. Grigoryev, D. Linfert, H.R. Jang, T. Watkins, C. Cheadle, L. Racusen, and H. Rabb. 2010. Transcriptional analysis of kidneys during repair from AKI reveals possible roles for NGAL and KIM-1 as biomarkers of AKI to CKD transition. Am. J. Physiol. Renal. Physiol. 298: F1472–1483.

Kouloubinis, A., L. Kaklamanis, N. Ziras, S. Sofroniadou, K. Makaritsis, S. Adamopoulos, I. Revela, A. Athanasiou, D. Mavroudis, and V. Georgoulias. 2007. ProANP and NT-proBNP levels to prospectively assess cardiac function in breast cancer patients treated with cardiotoxic chemotherapy. Int. J. Cardiol. 122: 195–201.

Koyner, J.L., M.R. Bennett, E.M. Worcester, Q. Ma, J. Raman, V. Jeevanandam, K.E. Kasza, M.F. O'Connor, D.J. Konczal, S. Trevino, P. Devarajan, and P.T. Murray. 2008. Urinary cystatin C as an early biomarker of acute kidney injury following adult cardiothoracic surgery. Kidney Int. 74: 1059–1069.

Krauser, D.G., D.M. Lloyd-Jones, C.U. Chae, R. Cameron, S. Anwaruddin, A.L. Baggish, A. Chen, R. Tung, and J.L. Januzzi Jr. 2005. Effect of body mass index on natriuretic peptide levels in patients with acute congestive heart failure: a ProBNP Investigation of Dyspnea in the Emergency Department (PRIDE) substudy. Am. Heart J. 149: 744–750.

Krawczeski, C.D., R.G. Vandevoorde, T. Kathman, M.R. Bennett, J.G. Woo, Y. Wang, R.E. Griffiths, and P. Devarajan. 2010. Serum cystatin C is an early predictive biomarker of acute kidney injury after pediatric cardiopulmonary bypass. Clin. J. Am. Soc. Nephrol. 5: 1552–1557.

Krishnagopalan, S., A. Kumar, J.E. Parrillo, and A. Kumar. 2002. Myocardial dysfunction in the patient with sepsis. Curr. Opin. Crit. Care. 8: 376–388.

Krishnaswamy, G., J. Kelley, L. Yerra, J.K. Smith, and D.S. Chi. 1999. Human endothelium as a source of multifunctional cytokines: molecular regulation and possible role in human disease. J. Interferon. Cytokine. Res. 19: 91–104.

Kwon, O., B.A. Molitoris, M. Pescovitz, and K.J. Kelly. 2003. Urinary actin, interleukin-6, and interleukin-8 may predict sustained ARF after ischemic injury in renal allografts. Am. J. Kidney Dis. 41: 1074–1087.

Lajer, M., L. Tarnow, A. Jorsal, T. Teerlink, H.H. Parving, and P. Rossing. 2008. Plasma concentration of asymmetric dimethylarginine (ADMA) predicts cardiovascular morbidity and mortality in type 1 diabetic patients with diabetic nephropathy. Diabetes Care 31: 747–752.

Lassus, J.P., M.S. Nieminen, K. Peuhkurinen, K. Pulkki, K. Siirilä-Waris, R. Sund, and V.P. Harjola. FINN-AKVA study group. 2010. Markers of renal function and acute kidney injury in acute heart failure: definitions and impact on outcomes of the cardiorenal syndrome. Eur. Heart J. 31: 2791–2798.

Latini, R., S. Masson, I. Anand, M. Salio, A. Hester, D. Judd, S. Barlera, A.P. Maggioni, G. Tognoni, and J.N. Cohn. 2004. The comparative prognostic value of plasma neurohormones at baseline in patients with heart failure enrolled in Val-HeFT. Eur. Heart. J. 25: 292–299.

Lau, D. and S. Baldus. 2006. Myeloperoxidase and its contributory role in inflammatory vascular disease. Pharmacol. Ther. 111: 16–26.

Lee, D.S. and R.S. Vasan. 2005. Novel markers for heart failure diagnosis and prognosis. Curr. Opin. Cardiol. 20: 201–210.

Levin, A., C.R. Thompson, J. Ethier, E.J. Carlisle, S. Tobe, D. Mendelssohn, E. Burgess, K. Jindal, B. Barrett, J. Singer, and O. Djurdjev. 1999. Left ventricular mass index increase in early renal disease: Impact of decline in hemoglobin. Am. J. Kidney. Dis. 34: 125–134.

Liangos, O., M.C. Perianayagam, V.S. Vaidya, W.K. Han, R. Wald, H. Tighiouart, R.W. MacKinnon, L. Li, V.S. Balakrishnan, B.J. Pereira, J.V. Bonventre, and B.L. Jaber. 2007. Urinary NAG activity and KIM-1 level are associated with adverse outcomes in acute renal failure. J. Am. Soc. Nephrol. 18: 904–912.

Liangos, O., H. Tighiouart, M.C. Perianayagam, A. Kolyada, W.K. Han, R. Wald, J.V. Bonventre, and B.L. Jaber. 2009. Comparative analysis of urinary biomarkers for early detection of acute kidney injury following cardiopulmonary bypass. Biomarkers 14: 423–431.

Ling, W., N. Zhaohui, H. Ben, G. Leyi, L. Jianping, D. Huili, and Q. Jiaqi. 2008. Urinary IL-18 and NGAL as early predictive biomarkers in contrastinduced nephropathy after coronary angiography. Nephron. Clin. 108: c176–181.

Lippi, G., M. Montagnana, G.L. Salvagno, and G.C Guidi. 2006. Potential value for new diagnostic markers in the early recognition of acute coronary syndromes. CJEM. 8: 27–31.

Lisowska-Myjak, B. 2010. Serum and urinary biomarkers of acute kidney injury. Blood Purif. 29: 357–365.

Liyan, C., Z. Jie, and H. Xiaozhou. 2009. Prognostic value of combination of heart-type fatty acid-binding protein and ischemia-modified albumin in patients with acute coronary syndromes and normal troponin T values. J. Clin. Lab. Anal. 23: 14–18.

Logeart, D., G. Thabut, P. Jourdain, C. Chavelas, P. Beyne, F. Beauvais, E. Bouvier, and A.C. Solal. 2004. Predischarge B-type natriuretic peptide assay for identifying patients at high risk of re-admission after decompensated heart failure. J. Am. Coll. Cardiol. 43: 635–641.

Lund, J., Q.P. Qin, T. Ilva, K. Pettersson, L.M. Voipio-Pulkki, P. Porela, and K. Pulkki. 2003. Circulating pregnancyassociated plasma protein a predicts outcome in patients with acute coronary syndrome but no troponin I elevation. Circulation 108: 1924–1926.

Maatman, R.G., E.M. van de Westerlo, T.H. van Kuppevelt, and J.H. Veerkamp. 1992. Molecular identification of the liver- and the heart-type fatty acid-binding proteins in human and rat kidney. Use of the reverse transcriptase polymerase chain reaction. Biochem. J. 288: 285–290.

Maisel, A.S., P. Krishnaswamy, R.M. Nowak, J. McCord, J.E. Hollander, P. Duc, T. Omland, A.B. Storrow, W.T. Abraham, A.H. Wu, P. Clopton, P.G. Steg, A. Westheim, C.W. Knudsen, A. Perez, R. Kazanegra, H.C. Herrmann, and P.A. McCullough; Breathing Not Properly Multinational Study Investigators. 2002. Rapid measurement of B-type natriuretic peptide in the emergency diagnosis of heart failure. N. Engl. J. Med. 347: 161–167.

Maisel, A., J.E. Hollander, D. Guss, P. McCullough, R. Nowak, G. Green, M. Saltzberg, S.R. Ellison, M.A. Bhalla, V. Bhalla, P. Clopton, and R. Jesse. 2004. Primary results of the Rapid Emergency Department Heart Failure Outpatient Trial (REDHOT). A multicenter study of B-type natriuretic peptide levels, emergency department decision making, and outcomes in patients presenting with shortness of breath. J. Am. Coll. Cardiol. 44: 1328–1333.

Maisel, A., C. Mueller, K. Adams, S.D. Anker, N. Aspromonte, J.G. Cleland, A. Cohen-Solal, U. Dahlstrom, A. DeMaria, S. Di Somma, G.S. Filippatos, G.C. Fonarow, P. Jourdain, M. Komajda, P.P. Liu, T. McDonagh, K. McDonald, A. Mebazaa, M.S. Nieminen, W.F. Peacock, M. Tubaro, R. Valle, M. Vanderhyden, C.W. Yancy, F. Zannad, and E. Braunwald. 2008. State of the art: using natriuretic peptide levels in clinical practice. Eur. J. Heart Fail. 10: 824–839.

Maisel, A., C. Mueller, R. Nowak, W.F. Peacock, J.W. Landsberg, P. Ponikowski, M. Mockel, C. Hogan, A.H. Wu, M. Richards, P. Clopton, G.S. Filippatos, S. Di Somma, I. Anand, L. Ng, L.B. Daniels, S.X. Neath, R. Christenson, M. Potocki, J. McCord, G. Terracciano, D. Kremastinos, O. Hartmann, S. von Haehling, A. Bergmann, N.G. Morgenthaler, and S.D. Anker. 2010. Mid-region prohormone markers for diagnosis and prognosis in acute dyspnea: results from the BACH (Biomarkers in Acute Heart Failure) trial. J. Am. Coll. Cardiol. 55: 2062–2076.

Mallat, Z., P. Henry, R. Fressonnet, S. Alouani, A. Scoazec, P. Beaufils, Y. Chvatchko, and A. Tedgui. 2002. Increased plasma concentrations of interleukin-18 in acute coronary syndromes. Heart 88: 467–469.

Mallat, Z., C. Heymes, A. Corbaz, D. Logeart, S. Alouani, A. Cohen-Solal, T. Seidler, G. Hasenfuss, Y. Chvatchko, A.M. Shah, and A. Tedgui. 2004. Evidence for altered interleukin 18 (IL)-18 pathway in human heart failure. FASEB J. 18: 1752–1754.

Malyszko, J., J.S. Malyszko, H. Bachorzewska-Gajewska, B. Poniatowski, S. Dobrzycki, and M. Mysliwiec. 2009. Neutrophil gelatinase-associated lipocalin is a new and sensitive marker of kidney function in chronic kidney disease patients and renal allograft recipients. Transplant Proc. 41: 158–161.

Manzano-Fernandez, S., M. Boronat-Garcia, M.D. Albaladejo-Oton, P. Pastor, I.P. Garrido, F.J. Pastor-Pérez, P. Martínez-Hernández, M. Valdés, and D.A. Pascual-Figal. 2009. Complementary prognostic value of cystatin C, N-terminal pro-B-type natriuretic peptide and cardiac troponin T in patients with acute heart failure. Am. J. Cardiol. 103: 1753–1759.

Marchewka, Z., J. Kuzniar, and A. Dlugosz. 2001. Enzymuria and α2-microalbuminuria in the assessment of the influence of proteinuria on the progression of glomerulopathies. Int. Urol. Nephrol. 33: 673–676.

Mariano, F., G. Guida, D. Donati, C. Tetta, P.L. Cavalli, G. Verzetti, G. Piccoli, and G. Camussi. 1999. Production of platelet-activating factor in patients with sepsis-associated acute renal failure. Nephrol. Dial. Transplant. 14: 1150–1157.

McCullough, P.A. and K.R. Sandberg. 2003. Sorting out the evidence on natriuretic peptides. Rev. Cardiovasc. Med. 4: S13–19.

McCullough, P.A., P. Duc, T. Omland, J. McCord, R.M. Nowak, J.E. Hollander, H.C. Herrmann, P.G. Steg, A. Westheim, C.W. Knudsen, A.B. Storrow, W.T. Abraham, S. Lamba, A.H. Wu, A. Perez, P. Clopton, P. Krishnaswamy, R. Kazanegra, and A.S. Maisel; Breathing Not Properly Multinational Study Investigators. 2003. B-type natriuretic peptide and renal function in the diagnosis of heart failure: an analysis from the Breathing Not Properly Multinational Study. Am. J. Kidney Dis. 41: 571–579.

Mehta, K.P., U.S. Ali, L. Shankar, D. Tirthani, and M. Ambadekar. 1997. Renal dysfunction detected by beta-2 microglobulinuria in sick neonates. Indian Pediatr. 34: 107–111.

Melnikov, V.Y., T. Ecder, G. Fantuzzi, B. Siegmund, M.S. Lucia, C.A. Dinarello, R.W. Schrier, and C.L. Edelstein. 2001. Impaired IL-18 processing protects caspase-1-deficient mice from ischemic acute renal failure. J. Clin. Invest. 107: 1145–1152.

Meyer, B., M. Huelsmann, P. Wexberg, G. Delle Karth, R. Berger, D. Moertl, T. Szekeres, R.Pacher, and G. Heinz. 2007. N-terminal pro-B-type natriuretic peptide is an independent predictor of outcome in an unselected cohort of critically ill patients. Crit. Care. Med. 35: 2268–2273.

Mishra, J., Q. Ma, A. Prada, M. Mitsnefes, K. Zahedi, J. Yang, J. Barasch, and P. Devarajan. 2003. Identification of neutrophil gelatinase-associated lipocalin as a novel early urinary biomarker for ischemic renal injury. J. Am. Soc. Nephrol. 14: 2534–43.

Mishra, J., K. Mori, Q. Ma, C. Kelly, J. Barasch, and P. Devarajan. 2004a. Neutrophil gelatinase-associated lipocalin: a novel early urinary biomarker for cisplatin nephrotoxicity. Am. J. Nephrol. 24: 307–315.

Mishra, J., K. Mori, Q. Ma, C. Kelly, J. Yang, M. Mitsnefes, J. Barasch, and P. Devarajan. 2004b. Amelioration of ischemic acute renal injury by neutrophil gelatinase-associated lipocalin. J. Am. Soc. Nephrol. 15: 3073–3082.

Mishra, J., C. Dent, R. Tarabishi, M.M. Mitsnefes, Q. Ma, C. Kelly, S.M. Ruff, K. Zahedi, M. Shao, J. Bean, K. Mori, J. Barasch, and P. Devarajan. 2005. Neutrophil gelatinase-associated lipocalin (NGAL) as a biomarker for acute renal injury after cardiac surgery. Lancet 365: 1231–1238.

Molls, R.R., V. Savransky, M. Liu, S. Bevans, T. Mehta, R.M. Tuder, L.S. King, and H. Rabb. 2006. Keratinocyte- derived chemokine is an early biomarker of ischemic acute kidney injury. Am. J. Physiol. Renal Physiol. 290: F1187–1193.

Moran, A., R. Katz, N.L. Smith, L.F. Fried, M.J. Sarnak, S.L. Seliger, B. Psaty, D.S. Siscovick, J.S. Gottdiener, and M.G. Shlipak. 2008. Cystatin C concentration as a predictor of systolic and diastolic heart failure. J. Card Fail. 14: 19–26.

Mori, K. and K. Nakao. 2007. Neutrophil gelatinase-associated lipocalin as the real-time indicator of active kidney damage. Kidney Int. 71: 967–970.

Mori, K., H.T. Lee, D. Rapoport, I.R. Drexler, K. Foster, J. Yang, K.M. Schmidt-Ott, X. Chen, J.Y. Li, S. Weiss, J. Mishra, F.H. Cheema, G. Markowitz, T. Suganami, K. Sawai, M. Mukoyama, C. Kunis, V. D'Agati, P. Devarajan, and J. Barasch. 2005. Endocytic delivery of lipocalin- siderophore-iron complex rescues the kidney from ischemia-reperfusion injury. J. Clin. Invest. 115: 610–621.

Morita, E., H. Yasue, M. Yoshimura, H. Ogawa, M. Jougasaki, T. Matsumura, M. Mukoyama, and K. Nakao. 1993. Increased plasma levels of brain natriuretic peptide in patients with acute myocardial infarction. Circulation 88: 82–91.

Morrow, D.A., N. Rifai, E.M. Antman, D.L. Weiner, C.H. McCabe, C.P. Cannon, and E. Braunwald. 1998. C-reactive protein is a potent predictor of mortality independently of and in combination with troponin T in acute coronary syndromes: a TIMI 11A substudy. Thrombolysis in Myocardial Infarction. 1998. J. Am. Coll. Cardiol. 31: 1460–1465.

Morrow, D.A., E.M. Antman, M. Tanasijevic, N. Rifai, J.A. de Lemos, C.H. McCabe, C.P. Cannon, and E. Braunwald. 2000. Cardiac troponin I for stratification of early outcomes and the efficacy of enoxaparin in unstable angina: a TIMI-11B substudy. J. Am. Coll. Cardiol. 36: 1812–1817.

Morrow, D.A., C.P. Cannon, N. Rifai, M.J. Frey, R. Vicari, N. Lakkis, D.H. Robertson, D.A. Hille, P.T. DeLucca, P.M. DiBattiste, L.A. Demopoulos, W.S. Weintraub, and E. Braunwald; TACTICS-TIMI 18 Investigators. 2001. Ability of minor elevations of troponins I and T to predict benefit from an early invasive strategy in patients with unstable angina and non-ST elevation myocardial infarction: results from a randomized trial. JAMA 286: 2405–2412.

Morrow, D.A., J.A. de Lemos, M.S. Sabatine, S.A. Murphy, L.A. Demopoulos, P.M. DiBattiste, C.H. McCabe, C.M. Gibson, C.P. Cannon, and E. Braunwald. 2003. Evaluation of B-type natriuretic peptide for risk assessment in unstable angina/non-ST-elevation myocardial infarction: Btype natriuretic peptide and prognosis in TACTICS-TIMI 18. J. Am. Coll. Cardiol. 41: 1264–1272.

Morrow, D.A., J.A. de Lemos, M.A. Blazing, M.S. Sabatine, S.A. Murphy, P. Jarolim, H.D. White, K.A. Fox, R.M. Califf, and E. Braunwald. 2005. Prognostic value of serial B-type natriuretic peptide testing during follow-up of patients with unstable coronary artery disease. J. Am. Med. Assoc. 294: 2866–2871.

Mukhopadhyay, B., S. Chinchole, V. Lobo, S. Gang, and M. Rajapurkar. 2004. Enzymuria pattern in early port renal tranplant period: diagnostic usefulness in graft dysfunction. Indian J. Clin. Biochem. 19: 14–19.

Munshi, R., A. Johnson, E.D. Siew, T.A. Ikizler, L.B.Ware, M.M. Wurfel, J. Himmelfarb, and R.A. Zager. 2011. MCP-1 gene activation marks acute kidney injury. J. Am. Soc. Nephrol. 22: 165–175.

Muramatsu, Y., M. Tsujie, Y. Kohda, B. Pham, A.O. Perantoni, H. Zhao, S.K. Jo, P.S.T. Yuen, L. Craig, X. Hu, and R.A. Star. 2002. Early detection of cysteine rich protein 61 (CYR61, CCN1) in urine following renal ischemic reperfusion injury. Kidney Int. 62: 1601–1610.

NACB Writing Group. 2007. National Academy of Clinical Biochemistry laboratory medicine practice guidelines: Use of cardiac troponin and B-type natriuretic peptide or N-terminal proB-type natriuretic peptide for etiologies other than acute coronary syndromes and heart failure. Clin. Chem. 53: 2086–2096.

Nagaya, N., T. Nishikimi, Y. Goto, Y. Miyao, Y. Kobayashi, I. Morii, S. Daikoku, T. Matsumoto, S. Miyazaki, H. Matsuoka, S. Takishita, K. Kangawa, H. Matsuo, and H. Nonogi. 1998. Plasma brain natriuretic peptide is a biochemical marker for the prediction of progressive ventricular remodeling after acute myocardial infarction. Am. Heart. J. 135: 21–28.

Nakamura, T., T. Sugaya, I. Ebihara, and H. Koide. 2005. Urinary liver-type fatty acid-binding protein:discrimination between IgA nephropathy and thin basement membrane nephropathy. Am. J. Nephrol. 25: 447–450.

Nakamura, T., T. Sugaya, K. Node, Y. Ueda, and H. Koide. 2006. Urinary excretion of liver-type fatty acid-binding protein in contrast medium-induced nephropathy. Am. J. Kidney Dis. 47: 439–444.

Nakamura, T., T. Sugaya, and H. Koide. 2007. Angiotensin II receptor antagonist reduces urinary liver-type fatty acid-binding protein levels in patients with diabetic nephropathy and chronic renal failure. Diabetologia 50: 490–492.

Naruse, H., J. Ishii, T. Kawai, K. Hattori, M. Ishikawa, M. Okumura, S. Kan, T. Nakano, S. Matsui, H. Nomura, H. Hishida, and Y. Ozaki. 2009. Cystatin C in acute heart failure without advanced renal impairment. Am. J. Med. 122: 566–573.

Needham, D.M., K.A. Shufelt, G. Tomlinson, J.W. Scholey, and G.E. Newton. 2004. Troponin I and T levels in renal failure patients without acute coronary syndrome: a systematic review of the literature. Can. J. Cardiol. 20: 1212–1218.

Negishi, K., E. Noiri, K. Doi, R. Maeda-Mamiya, T. Sugaya, D. Portilla, and T. Fujita. 2009. Monitoring of urinary L-type fatty acid-binding protein predicts histological severity of acute kidney. Am. J. Pathol. 174: 1154–1159.

Nejat, M., J.W. Pickering, R.J. Walker, J. Westhuyzen, G.M. Shaw, C.M. Frampton, and Z.H. Endre. 2010. Urinary cystatin C is diagnostic of acute kidney injury and sepsis, and predicts mortality in the intensive care unit. Crit. Care 14: R85.

Nelken, N.A., S.R. Coughlin, D. Gordon, and J.N. Wilcox. 1991. Monocyte chemoattractant protein-1 in human atheromatous plaques. J. Clin. Invest. 88: 1121–1127.

Ng, L.L., R.J. O'Brien, B. Demme, and S. Jennings. 2002. Non-competitive immunochemiluminometric assay for cardiotrophin-1 detects elevated plasma levels in human heart failure. Clin. Sci. 102: 411–416.

Nguyen, M.T. and P. Devarajan. 2008. Biomarkers for the early detection of acute kidney injury. Pediatr. Nephrol. 23: 2151–2157.

Nicholls, S.J. and S.L. Hazen. 2005. Myeloperoxidase and cardiovascular disease. Arterioscler. Thromb. Vasc. Biol. 25: 1102–1111.

Nickolas, T.L., M.J. O'Rourke, J. Yang, M.E. Sise, P.A. Canetta, N. Barasch, C. Buchen, F. Khan, K. Mori, J. Giglio, P. Devarajan, and J. Barasch. 2008. Sensitivity and specificity of a single emergency department measurement of urinary neutrophil gelatinase-associated lipocalin for diagnosing acute kidney injury. Ann. Intern. Med. 148: 810–819.

Nielsen, S.E., T. Sugaya, P. Hovind, T. Baba, H.H. Parving, and P. Rossing. 2010. Urinary liver-type fatty acid-binding protein (u-LFABP) predicts progression to nephropathy in type 1 diabetic patients. Diabetes Care 33: 1320–1324.

Nissen, S.E., E.M. Tuzcu, P. Schoenhagen, B.G. Brown, P. Ganz, R.A. Vogel, T. Crowe, G. Howard, C.J. Cooper, B. Brodie, C.L. Grines, and A.N. DeMaria; REVERSAL Investigators. 2004. Effect of intensive compared with moderate lipid-lowering therapy on progression of coronary atherosclerosis: a randomized controlled trial. JAMA. 291: 1071–1080.

Nissen, S.E., E.M. Tuzcu, P. Schoenhagen, T. Crowe, W.J. Sasiela, J. Tsai, J. Orazem, R.D. Magorien, C. O'Shaughnessy, and P. Ganz; Reversal of Atherosclerosis with Aggressive Lipid Lowering (REVERSAL) Investigators. 2005. Statin therapy, LDL cholesterol, C-reactive protein, and coronary artery disease. N. Engl. J. Med. 352: 29–38.

Ohta, S., T. Ishimitsu, J. Minami, H. Ono, and H. Matsuoka. 2005. Effects of intravascular contrast media on urinary excretion of liver fatty acid-binding protein. Nippon Jinzo Gakkai Shi. 47: 437–444.

Okuyama, M., S. Yamaguchi, N. Nozaki, M. Yamaoka, M. Shirakabe, and H. Tomoike. 1997. Serum levels of soluble form of Fas molecule in patients with congestive heart failure. Am. J. Cardiol. 79: 1698–1701.

Omland, T., A. Aakvaag, V.V. Bonarjee, K. Caidahl, R.T. Lie, D.W. Nilsen, J.A. Sundsfjord, and K. Dickstein. 1996. Plasma brain natriuretic peptide as an indicator of left ventricular systolic function and long-term survival after acute myocardial infarction. Comparison with plasma atrial natriuretic peptide and N-terminal proatrial natriuretic peptide. Circulation 93: 1963–1969.

Omland, T., J.A. de Lemos, D.A. Morrow, E.M. Antman, C.P. Cannon, C. Hall, and E. Braunwald. 2002. Prognostic value of N-terminal pro-atrial and pro-brain natriuretic peptide in patients with acute coronary syndromes. Am. J. Cardiol. 89: 463–465.

Ooi, D.S., J.P. Veinot, G.A. Wells, and A.A. House. 1999. Increased mortality in hemodialyzed patients with elevated serum troponin T: a one-year outcome study. Clin. Biochem. 32: 647–652.

Orea-Tejeda, A., E. Col´ın-Ram´ırez, T. Hern´andez-Gilsoul, L. Castillo-Martínez, M. Abasta-Jiménez, E. Asensio-Lafuente, R. Narváez David, and J. Dorantes-García. 2008. Microalbuminuria in systolic and diastolic chronic heart failure patients. Cardiol J. 15: 143–149.

Oyama, Y., T. Takeda, H. Hama, A. Tanuma, N. Iino, K. Sato, R. Kaseda, M. Ma, T. Yamamoto, H. Fujii, J.J. Kazama, S. Odani, Y. Terada, K. Mizuta, F. Gejyo, and A. Saito. 2005. Lab. Invest. 85: 522–531.

Parikh, C.R., A. Jani, V.Y. Melnikov, S. Faubel, and C.L. Edelstein. 2004. Urinary interleukin-18 is a marker of human acute tubular necrosis. Am. J. Kidney Dis. 43: 405–414.

Parikh, C. R., E. Abraham, M. Ancukiewicz, and C.L. Edelstein. 2005. Urine IL-18 is an early diagnostic marker for acute kidney injury and predicts mortality in the intensive care unit. J. Am. Soc. Nephrol. 16: 3046–3052.

Parikh, C.R., J. Mishra, H. Thiessen-Philbrook, B. Dursun, Q. Ma, C. Kelly, C. Dent, P. Devarajan and C.L. Edelstein. 2006a. Urinary IL-18 is an early predictive biomarker of acute kidney injury after cardiac surgery. Kidney Int. 70: 199–203.

Parikh, C.R., A. Jani, J. Mishra, Q. Ma, C. Kelly, J. Barasch, C.L. Edelstein, and P. Devarajan. 2006b. Urine NGAL and IL-18 are predictive biomarkers for delayed graft function following kidney transplantation. Am. J. Transplant. 6: 1639–1645.

Parikh, C.R. and P. Devarajan. 2008a. New biomarkers of acute kidney injury. Crit. Care Med. 6: S159–165.

Parikh, N.I, S.J. Hwang, Q. Yang, M.G. Larson, C.Y. Guo, S.J. Robins, P. Sutherland, E.J. Benjamin, D. Levy, and C.S. Fox. 2008b. Clinical correlates and heritability of cystatin C (from the Framingham Offspring Study). Am. J. Cardiol. 102: 1194–1198.

Patel, P.C., C.R. Ayers, S.A. Murphy, R. Peshock, A. Khera, J.A. de Lemos, J.A. Balko, S. Gupta, P.P. Mammen, M.H. Drazner, and D.W. Markham. 2009. Association of cystatin C with left ventricular structure and function: the Dallas Heart Study. Circ. Heart Fail. 2: 98–104.

Peacock, W.F., T. De Marco, G.C. Fonarow, D. Diercks, J. Wynne, F.S. Apple, and A.H. Wu; ADHERE Investigators. 2008. Cardiac troponin and outcome in acute heart failure. N. Engl. J. Med. 358: 2117–2126.

Pelsers, M.M., W.T. Hermens, and J.F. Glatz. 2005. Fatty acid-binding proteins as plasma markers of tissue injury. Clin. Chim. Acta. 352: 15–35.

Poge, U., T.M. Gerhardt, B. Stoffel-Wagner, H. Palmedo, H.U. Klehr, T. Sauerbruch, and R.P. Woitas. 2005. Beta-trace protein is an alternative marker for glomerular filtration rate in renal transplantation patients. Clin. Chem. 51: 1531–1533.

Poniatowski, B., J. Malyszko, H. Bachorzewska-Gajewska, J.S. Malyszko, and S. Dobrzycki. 2009. Serum neutrophil gelatinase-associated lipocalin as a marker of renal function in patients with chronic heart failure and coronary artery. Kidney Blood Press. Res. 32: 77–80.

Portilla, D., C. Dent, T. Sugaya, K.K. Nagothu, I. Kundi, P. Moore, E. Noiri, and P. Devarajan. 2008. Liver fatty acid-binding protein as a biomarker of acute kidney injury after cardiac surgery. Kidney Int. 73: 465–472.

Priem, F., H. Althaus, M. Birnbaum, P. Sinha, H.S. Conradt, and K. Jung. 1999. Beta-trace protein in serum: a new marker of glomerular filtration rate in the creatinine-blind range. Clin. Chem. 45: 567–568.

Ramesh, G., C.D. Krawczeski, J.G. Woo, Y. Wang, and P. Devarajan. 2010a. Urinary Netrin-1 is an early predictive biomarker of acute kidney injury after cardiac surgery. Clin. J. Am. Soc. Nephrol. 5: 395–401.

Ramesh, G., O. Kwon, and K. Ahn. 2010b. Netrin-1: a novel universal biomarker of human kidney injury. Transplant Proc. 42: 1519–1522.

Ravani, P., G. Tripepi, F. Malberti, S. Testa, F. Mallamaci, and C. Zoccali. 2005. Asymmetrical dimethylarginine predicts progression to dialysis and death in patients with chronic kidney disease: a competing risks modeling approach. J. Am. Soc. Nephrol. 16: 2449–2455.

Reeves, W.B., O. Kwon, and G. Ramesh. 2008. Netrin-1 and kidney injury. II. Netrin-1 is an early biomarker of acute kidney injury. Am. J. Physiol. Renal Physiol. 294: F731–738.

Rice, J.C., J.S. Spence, D.L. Yetman, and R.L. Safirstein. 2002. Monocyte chemoattractant protein-1 expression correlates with monocyte infiltration in the post-ischemic kidney. Ren. Fail. 24: 703–723.

Rickli, H., K. Benou, P. Ammann, T. Fehr, H.P. Brunner-La Rocca, H. Petridis, W. Riesen, and R.P. Wüthrich. 2004. Time course of serial cystatin C levels in comparison with serum creatinine after application of radiocontrast media. Clin Nephrol. 61: 98–102.

Ridker, P.M., C.H. Hennekens, J.E. Buring, and N. Rifai. 2000. C-reactive protein and other markers of inflammation in the prediction of cardiovascular disease in women. N. Engl. J. Med. 342: 836–843.

Ridker, P.M., N. Rifai, L. Rose, J.E. Buring, and N.R. Cook. 2002. Comparison of C-reactive protein and low-density lipoprotein cholesterol levels in the prediction of first cardiovascular events. N. Engl. J. Med. 347: 1557–1565.

Ridker, P.M. 2007. C-reactive protein and the prediction of cardiovascular events among those at intermediate risk: moving an inflammatory hypothesis toward consensus. J. Am. Coll. Cardiol. 49: 2129–2138.

Rifai, N. and P.M. Ridker. 2001. Proposed cardiovascular risk assessment algorithm using high-sensitivity C-reactive protein and lipid screening. Clin. Chem. 47: 28–30.

Rifkin, D.E., R. Katz, M. Chonchol, L.F. Fried, J. Cao, I.H. de Boer, D.S. Sioviscck, M.G. Shlipak, and M.J. Sarnak. 2010. Albuminuria, impaired kidney function and cardiovascular outcomes or mortality in the elderly. Nephrol. Dial. Transplant. 25: 1560–1567.

Ristikankare, A., R. Poyhia, A. Kuitunen, M. Skrifvars, P. Hammainen, M. Salmenpera, and R. Suojaranta-Ylinen. 2010. Serum cystatin C in elderly cardiac surgery patients. Ann. Thorac. Surg. 89: 689–694.

Ronco, C. 2008. NGAL: an emerging biomarker of acute kidney injury. Int. J. Artif. Organs 31: 199–200.

Ronco, C., P. McCullough, S.D. Anker, I. Anand, N. Aspromonte, S.M. Bagshaw, R. Bellomo, T. Berl, I. Bobek, D.N. Cruz, L. Daliento, A. Davenport, M. Haapio, H. Hillege, A.A. House, N. Katz, A. Maisel, S. Mankad, P. Zanco, A. Mebazaa, A. Palazzuoli, F. Ronco, A. Shaw, G. Sheinfeld, S. Soni, G. Vescovo, N. Zamperetti, and P. Ponikowski; Acute Dialysis Quality Initiative (ADQI) consensus group. 2010. Cardio-renal syndromes: report from the consensus conference of the acute dialysis quality initiative. Eur. Heart. J. 31: 703–711.

Roy, D., J. Quiles, G. Aldama, M. Sinha, P. Avanzas, R. Arroyo-Espliguero, D. Gaze, P. Collinson, and J. Carlos Kaski. 2004. Ischemia modified albumin for the assessment of patients presenting to the emergency department with acute chest pain but normal or non-diagnostic 12-lead electrocardiograms and negative cardiac troponin T. Int. J. Cardiol. 97: 297–301.

Royakkers, A.A., J.D. van Suijlen, L.S. Hofstra, M.A. Kuiper, C.S. Bouman, P.E. Spronk, and M.J. Schultz. 2007. Serum cystatin C—A useful endogenous marker of renal function in intensive care unit patients at risk for or with acute renal failure? Curr. Med. Chem. 14: 2314–2317.

Ruggenenti, P., A. Perna, L. Mosconi, R. Pisoni, and G. Remuzzi. 1998. Urinary protein excretion rate is the best independent predictor of ESRF in non-diabetic proteinuric chronic nephropathies. Kidney Int. 53: 1209–1216.

Ryu, S., Y. Chang, D.I. Kim, W.S. Kim, and B.S. Suh. 2007. gamma-Glutamyltransferase as a predictor of chronic kidney disease in nonhypertensive and nondiabetic Korean men. Clin. Chem. 53: 71–77.

Scherberich, J.E. 1990. Urinary proteins of tubular origin: basic immunological and clinical aspects. Am. J. Nephrol. 10 Suppl. 1: 43–51.

Schmidt-Ott, K.M., K. Mori, A. Kalandadze, J.Y. Li, N. Paragas, T. Nicholas, P. Devarajan, and J. Barasch. 2006. Neutrophil gelatinase-associated lipocalin-mediated iron traffic in kidney epithelia. Curr. Opin. Nephrol. Hypertens. 15: 442–449.

Schonbeck, U. and P. Libby. 2001. CD40 signaling and plaque instability. Circ. Res. 89: 1092–1103.

Schwartz, N., M. Hosford, R.M. Sandoval, M.C. Wagner, S.J. Atkinson, J. Bamburg, and B.A. Molitoris. 1999. Ischemia activates actin depolymerizing factor: role in proximal tubule microvillar actin alterations. Am. J. Physiol. Renal. Physiol. 276: F544–551.

Seta, Y., K. Shan, B. Bozkurt, H. Oral, and D.L. Mann. 1996. Basic mechanisms in heart failure: the cytokine hypothesis. J. Card. Fail. 2: 243–249.

Sharma, R., D.C. Gaze, D. Pellerin, R.L. Mehta, H. Gregson, C.P. Streather, P.O. Collinson, and S.J. Brecker. 2006. Cardiac structural and functional abnormalities in end stage renal disease patients with elevated cardiac troponin T. Heart 92: 804–809.

Siew, E.D., L.B. Ware, and T.A. Ikizler. 2011. Biological markers of acute kidney injury. J. Am. Soc. Nephrol. 22: 810–820.

Silver, M.A., A. Maisel, C.W. Yancy, P.A. McCullough, J.C. Burnett Jr., G.S. Francis, M.R. Mehra, W.F. Peacock 4th, G. Fonarow, W.B. Gibler, D.A. Morrow, and J. Hollander; BNP Consensus Panel. 2004. BNP Consensus Panel 2004: a clinical approach for the diagnostic, prognostic, screening, treatment monitoring, and therapeutic roles of natriuretic peptides in cardiovascular diseases. Congest Heart Fail. 10: 1–30.

Sinha, M.K., D.C. Gaze, J.R. Tippins, P.O. Collinson, and J.C. Kaski. 2003. Ischemia modified albumin is a sensitive marker of myocardial ischemia after percutaneous coronary intervention. Circulation 107: 2403–2405.

Sinha, M.K., D. Roy, D.C. Gaze, P.O. Collinson, and J.C. Kaski. 2004. Role of ischemia modified albumin, a new biochemical marker of myocardial ischemia, in the early diagnosis of acute coronary syndromes. Emerg. Med. J. 21: 29–34.

Siu, Y.P., K.T. Leung, M.K. Tong, and T.H. Kwan. 2006. Use of allopurinol in slowing the progression of renal disease through its ability to lower serum uric acid level. Am. J. Kidney Dis. 47: 51–59.

Sliwa, K., A. Woodiwiss, V.N. Kone, G. Candy, D. Badenhorst, G. Norton, C. Zambakides, F. Peters, and R. Essop. 2004. Therapy of ischemic cardiomyopathy with the immunomodulating agent pentoxifylline. Circulation 109: 750–755.

Sommerer, C., J. Beimler, V. Schwenger, N. Heckele, H.A. Katus, E. Giannitsis, and M. Zeier. 2007. Cardiac biomarkers and survival in haemodialysis patients. Eur. J. Clin. Invest. 37: 350–356.

Spanaus, K.S., F. Kronenberg, E. Ritz, R. Schlapbach, D. Fliser, M. Hersberger, B. Kollerits, P. König, and A. von Eckardstein; Mild-to-Moderate Kidney Disease Study Group. 2007. B-type natriuretic peptide concentrations predict the progression of nondiabetic chronic kidney disease: the mild-to-moderate kidney disease study. Clin. Chem. 53: 1264–1272.

Spanaus, K.S., B. Kollerits, E. Ritz, M. Hersberger, F. Kronenberg, A. von Eckardstein; Mild and Moderate Kidney Disease (MMKD) Study Group. 2010. Serum creatinine, cystatin C, and beta-trace protein in diagnostic staging and predicting progression of primary nondiabetic chronic kidney disease. Clin. Chem. 56: 740–749.

Supavekin, S., W. Zhang, R. Kucherlapati, F.J. Kaskel, L.C. Moore, and P. Devarajan. 2003. Differential gene expression following early renal ischemia/reperfusion. Kidney Int. 63: 1714–1724.

Suresh, M. and K. Farrington. 2005. Natriuretic peptides and the dialysis patient. Semin. Dial. 18: 409–419.

Taglieri, N., W. Koenig, and J.C. Kaski. 2009. Cystatin C and cardiovascular risk. Clin. Chem. 55: 1932–1943.

Taglieri, N., D.J. Fernandez-Berges, W. Koenig, L. Consuegra-Sanchez, J.M. Fernandez, N.R. Robles, P.L. Sánchez, A.C. Beiras, P.M. Orbe, and J.C. Kaski; SIESTA Investigators. 2010. Plasma cystatin C for prediction of 1-year cardiac events in Mediterranean patients with non-ST elevation acute coronary syndrome. Atherosclerosis 209: 300–305.

Talwar, S., I.B. Squire, P.F. Downie, R.J. O'Brien, J.E. Davies, and L.L. Ng. 2000. Elevated circulating cardiotrophin-1 in heart failure: relationship with parameters of left ventricular systolic dysfunction. Clin. Sci. 99: 83–88.

Tang, W.H., F. Van Lente, K. Shrestha, R.W. Troughton, G.S. Francis, W. Tong, M.G. Martin, A.G. Borowski, S. Jasper, R.C. Starling, and A.L. Klein. 2008. Impact of myocardial function on cystatin C measurements in chronic systolic heart failure. J. Card. Fail. 14: 394–399.

Tang, W.H., J.P. Girod, M.J. Lee, R.C. Starling, J.B. Young, F. Van Lente, and G.S. Francis. 2003. Plasma B-type natriuretic peptide levels in ambulatory patients with established chronic symptomatic systolic heart failure. Circulation 108: 2964–2966.

Tang, W.H.W., G.S. Francis, D.A. Morrow, L.K. Newby, C.P. Cannon, R.L. Jesse, A.B. Storrow, R.H. Christenson, F.S. Apple, J. Ravkilde, and A.H. Wu. 2007. NACB Committee. National academy of clinical biochemistry laboratory edicine practice guidelines: clinical utilization of cardiac biomarker testing in heart failure. Circulation 116: e99–109.

Teppala, S., A. Shankar, J. Li, T.Y. Wong, and A. Ducatman. 2010. Association between serum gammaglutamyltransferase and chronic kidney disease among US adults. Kidney Blood Press. Res. 33: 1–6.

Thygesen, K., J.S. Alpert, H.D. White, on behalf of the Joint ESC/ACCF/AHA/ WHF Task Force for the Redefinition of Myocardial Infarction. 2007. Universal definition of myocardial infarction. J. Am. Coll. Cardiol. 50: 2173–2195.

Tolins, J.P., G.M. Vercellotti, M. Wilkowske, B. Ha, H.S. Jacob, and L. Raij. 1989. Role of platelet activating factor endotoxemic acute renal failure in the male rat. J. Lab. Clin. Med. 113: 316–324.

Tolkoff-Rubin, N.E., R.E. Thompson, D.J. Piper, W.P. Hansen, N.H. Bander, C. Cordon-Cardo, C.J. Finstad, L.H. Klotz, L.J. Old, and R.H. Rubin. 1987. Diagnosis of renal proximal tubular injury by urinary immunoassay for a proximal tubular antigen, the adenosine deaminase binding protein. Nephrol. Dial. Transplant. 2: 143–148.

Tolkoff-Rubin, N.E., A.B. Cosimi, F.L. Delmonico, P. S. Russell, R.E. Thompson, and D.J. Piper. 1988. Diagnosis of tubular in renal transplant patients by a urinary assay for a proximal tubular, the adenosine-deaminase binding protein. Transplantation 41: 593–597.

Tonelli, M., N. Wiebe, B. Culleton, A. House, C. Rabbat, M. Fok, F. McAlister, and A.X. Garg. 2006. Chronic kidney disease and mortality risk: a systematic review. J. Am. Soc. Nephrol. 17: 2034–2047.

Toth, M., K.H. Vuorinen, O. Vuolteenaho, Hassinen, P.A. Uusimaa, J. Leppäluoto, and H. Ruskoaho. 1994. Hypoxia stimulates release of ANP and BNP from perfused rat ventricular myocardium. Am. J. Physiol. 266: H1572–1580.

Trof, R.J., F. Di Maggio, J. Leemreis, A.B.J. Groeneveld. 2006. Biomarkers of acute renal injury and renal failure. Shock 26: 245–253.

Troughton, R.W., D.L. Prior, J.J. Pereira, M. Martin, A. Fogarty, A. Morehead, T.G. Yandle, A.M. Richards, R.C. Starling, J.B. Young, J.D. Thomas, and A.L. Klein. 2004. Plasma B-type natriuretic peptide levels in systolic heart failure: importance of left ventricular diastolic function and right ventricular systolic function. J. Am. Coll. Cardiol. 43: 416–422.

Tsioufis, C., K. Dimitriadis, E. Andrikou, C. Thomopoulos, D. Tsiachris, E. Stefanadi, C. Mihas, A. Miliou, V. Papademetriou, and C. Stefanadis. 2010. ADMA, C-reactive protein, and albuminuria in untreated essential hypertension: a cross-sectional study. Am. J. Kidney. Dis. 55: 1050–1059.

Urquhart, B.L. and A.A. House. 2007. Assessing plasma total homocysteine in patients with end-stage renal disease. Perit. Dial. Int. 27: 476–488.

Vaidya, V.S., V. Ramirez, T. Ichimura, N.A. Bobadilla, and J.V. Bonventre. 2006a. Urinary kidney injury molecule-1: a sensitive quantitative biomarker for early detection of kidney tubular injury. Am. J. Physiol. Renal. Physiol. 290: F517–529.

Vaidya, V.S. and J.V. Bonventre. 2006b. Mechanistic biomarkers for cytotoxic acute kidney injury. Expert Opin. Drug Metab. Toxicol. 2: 697–713.

Vaidya, V.S., S.S. Waikar, M.A. Ferguson, F.B. Collings, K. Sunderland, C. Gioules, G. Bradwin, R. Matsouaka, R.A. Betensky, G.C. Curhan, and J.V. Bonventre. 2008. Urinary biomarkers for sensitive and specific detection of acute kidney injury in humans. Clin. Transl. Sci. 1: 200–208.

Vaidya, V.S., G.M. Ford, S.S. Waikar, Y. Wang, M.B. Clement, V. Ramirez, W.E. Glaab, S.P. Troth, F.D. Sistare, W.C. Prozialeck, J.R. Edwards, N.A. Bobadilla, S.C. Mefferd, and J.V. Bonventre. 2009. A rapid urine test for early detection of kidney injury, Kidney Int. 76: 108–114.

VandeVoorde, R.G., T.I. Katlman, and Q. Ma. 2006. Serum NGAL and cystatin C as predictive biomarkers for acute kidney injury. J. Am. Soc. Nephrol. 17: 404A.

Venkataraman, R. and J.A. Kellum. 2007. Defining acute renal failure: the RIFLE criteria. J. Intensive Fcare. Med. 22: 187–193.

Venugopal, S.K., S. Deveraj, and I. Jialal. 2005. Effect of C-reactive protein on vascular cells: evidence for a proinflammatory, proatherogenic role. Curr. Opin. Nephrol. Hypertens. 14: 33–7.

Wagener, G., M. Jan, M. Kim, K. Mori, J.M. Barasch, R.N. Sladen, and H.T. Lee. 2006. Association between increases in urinary neutrophil gelatinase- associated lipocalin and acute renal dysfunction after adult cardiac surgery. Anesthesiology 105: 485–491.

Wagener, G., G. Gubitosa, S. Wang, N. Borregaard, M. Kim, and H.T. Lee. 2008. Urinary neutrophil gelatinase-associated lipocalin and acute kidney injury after cardiac surgery. Am. J. Kidney. Dis. 52: 425–433.

Washburn, K.K., M. Zappitelli, A.A. Arikan, L. Loftis, R. Yalavarthy, C.R. Parikh, C.L. Edelstein, and S.L. Goldstein. 2008. Urinary interleukin-18 is an acute kidney injury biomarker in critically ill children. Nephrol. Dial. Transplant. 23: 566–572.

Weinberg, E.O., M. Shimpo, S. Hurwitz, S. Tominaga, J.L. Rouleau, and R.T. Lee. 2003. Identification of serum soluble ST2 receptor as a novel heart failure biomarker. Circulation 107: 721–726.

Westhuyzen, J., Z.H. Endre, G. Reece, D.M. Reith, D. Saltissi, and T.J. Morgan. 2003. Measurement of tubular enzymuria facilitates early detection of acute renal impairment in the intensive care unit. Nephrol. Dial. Transplant 18: 543–551.

Wheeler, D.S., P. Devarajan, Q. Ma, K. Harmon, M. Monaco, N. Cvijanovich, and H.R. Wong. 2008. Serum neutrophil gelatinase-associated lipocalin (NGAL) as a marker of acute kidney injury in critically ill children with septic shock. Crit. Care Med. 36: 1297–1303.

Wu, Y., T. Su, L. Yang, S.N. Zhu, and X.M. Li. 2010. Urinary neutrophil gelatinase-associated lipocalin: a potential biomarker for predicting rapid progression of druginduced chronic tubulointerstitial nephritis. Am. J. Med. Sci. 339: 537–542.

Xin, C., X. Yulong, C. Yu, C. Changchun, Z. Feng, and M. Xinwei. 2008. Urine neutrophil gelatinase-associated lipocalin and interleukin-18 predict acute kidney injury after cardiac surgery. Ren. Fail. 30: 904–913.

Xu, S. and P. Venge. 2000. Lipocalins as biochemical markers of disease. Biochim. Biophys. Acta 1482: 298–307.

Yndestad, A., L. Landro, T. Ueland, C.P. Dahl, T.H. Flo, L.E. Vinge, T. Espevik, S.S. Frøland, C. Husberg, G. Christensen, K. Dickstein, J. Kjekshus, E. Øie, L. Gullestad, and P. Aukrust. 2009. Increased systemic and myocardial expression of neutrophil gelatinase-associated lipocalin in clinical and experimental heart failure. Eur. Heart. J. 30: 1229– 1236.

Youssef, A.A., L.T. Chang, C.L. Hang, C.J. Wu, C.I. Cheng, C.H. Yang, J.J. Sheu, H.T. Chai, S. Chua, K.H. Yeh, and H.K. Yip. 2007. Level and value of interleukin-18 in patients with acute myocardial infarction undergoing primary coronary angioplasty. Circ. J. 71: 703–708.

Zappitelli, M., K.K. Washburn, A.A. Arikan, L. Loftis, Q. Ma, P. Devarajan, C.R. Parikh, and S.L. Goldstein. 2007. Urine neutrophil gelatinaseassociated lipocalin is an early marker of acute kidney injury in critically ill children: A prospective cohort study. Crit. Care 11: R84.

Zhang, P.L., L.I. Rothblum, W.K. Han, T.M. Blasick, S. Potdar, and J.V. Bonventre. 2008. Kidney injury molecule-1 expression in transplant biopsies is a sensitive measure of cell injury. Kidney Int. 73: 608–614.

Zhang, R., M.L. Brennan, X. Fu, R.J. Aviles, G.L. Pearce, M.S. Penn, E.J. Topol, D.L. Sprecher, and S.L. Hazen. 2001. Association between myeloperoxidase levels and risk of coronary artery disease. JAMA 286: 2136–2142.

Zoccali, C., S. Bode-Boger, F. Mallamaci, F. Benedetto, G. Tripepi, L. Malatino, A. Cataliotti, I. Bellanuova, I. Fermo, J. Frölich, and R. Böger. 2001. Plasma concentration of asymmetrical dimethylarginine and mortality in patients with end-stage renal disease: a prospective study. Lancet 358: 2113–2117.

# CHAPTER 11

# Cardiac Arrhythmias: Is there a Role for Biomarkers?

Polychronis Dilaveris,* Konstantinos Gatzoulis, Dimitris Tousoulis and Christodoulos Stefanadis

## Introduction

Atrial fibrillation (AF) and ventricular tachycardia (VT)/fibrillation (VF) are important causes of cardiovascular morbidity and mortality. To determine the pathophysiology of these disorders and to design clinical trials evaluating pharmacologic therapies and procedural interventions, the use of biological markers (biomarkers) is of increasing importance. A biomarker represents a normal biological or pathogenic process, or pharmacologic response to an intervention. Biomarkers are very important because they can be used in research studies as surrogate endpoints to enable more rapid performance of clinical studies, predict disease risk, monitor disease status, and evaluate the efficacy of implemented therapeutic strategies (Mountantonakis and Deo 2012). This chapter highlights important clinical and translational studies that investigate the role of potential biomarkers to predict the onset of cardiac arrhythmias and their complications.

## Biomarkers in Atrial Fibrillation

Atrial fibrillation (AF) is the most common sustained arrhythmia in clinical practice, affecting approximately 1 percent of persons aged ≤ 65 years and 5 percent of individuals older than 65 years (Kannel 1998). The presence of AF significantly increases mortality and morbidity, as well as adversely

1st University Department of Cardiology, Hippokration Hospital, Athens, Greece.
*Corresponding author: hrodil1@yahoo.com

affects quality of life (Go et al. 2001; Singer et al. 2004; Steger et al. 2004). Previous hallmark studies have established the role of many structural and functional changes within the atria for the perpetuation of AF, a process known as electrical remodelling (Wijffels et al. 1995). AF also confers a prothrombotic or hypercoagulable state, which may contribute to the risk of thromboembolism (Lip 1995). Several classes of biomarkers have been investigated in an attempt to identify those people who are at higher risk of AF and its complications (Table 1).

**Table 1.** Biomarkers in Atrial Fibrillation.

| Categories | Biomarkers |
|---|---|
| Inflammatory biomarkers | CRP, IL-6, IL-1β, IL-8, TNF-α, TGF-β, IFN-γ, sCD40L |
| Biomarkers of oxidative stress | ROS, NADPH oxidase activity, F2-isoprostanes |
| Biomarkers of thrombosis | sP-sel, fibrinogen, vWf, platelet factor 4, thromboglobulin |
| Natriuretic peptide system | BNP, NT-pro BNP, MR-pro ANP |
| Other biomarkers | MAP kinases, bradykinin |

BNP: b-type natriuretic peptide, CRP: C-reactive protein, IFN: interferon, IL: interleukin, MAP: metalloproteinase, MR-pro ANP, mid-region-pro atrial natriuretic peptide, NADPH: nicotinamide adenosine dinucleotide phosphate, NT-pro BNP: N-terminal pro BNP, ROS: reactive oxygen species, sCD40L: soluble CD40 ligand, sP-sel: soluble P-selectin, TGF: transforming growth factor, TNF: tumor necrosis factor, vWF: von Willebrand factor.

## Inflammatory Biomarkers

Histological evidence to support the association between inflammation and AF has been derived from several sources (Nakamura et al. 2003; Verheule et al. 2003; Mihm et al. 2001; Kamiyama 1998; Frustaci et al. 1997). Inflammatory conditions of the heart, such as myocarditis and pericarditis have frequently been associated with AF (Spodick 1976; Morgera et al. 1992). The increased frequency of AF after coronary artery bypass surgery (CABG), with a peak incidence of AF occurring on the second and third post-operative days, which coincided with the peak elevation of CRP level, has earlier been reported (Bruins et al. 1997). The presence of circulating autoantibodies against myosin heavy chain in a significant percentage of patients with idiopathic paroxysmal AF, raising the possibility of an inflammatory autoimmune process in some patients with paroxysmal AF, has also been reported (Maixent et al. 1998). Of paramount importance in the activation of the inflammatory cascade and in the production of acute-phase proteins are cytokines, which are intracellular polypeptides produced by activated cells, usually monocytes and macrophages, in response to inflammatory stimuli. The primary inflammatory-mediated cytokines include IL-6, tumor necrosis factor (TNF)-α, IL-1β, interferon (IFN)-γ, transforming growth factor (TGF)-β, and IL-8. Interleukin-6, however, is the primary stimulator of acute-phase proteins. Measurement of acute-phase proteins, such as CRP, can provide an estimation of the current inflammatory status of a patient (Gabay and Kushner 1999; Ommen et al. 1997; Bruins et al. 1997; Aviles et al. 2003). Many studies have related an increase in CRP and IL-6 in both paroxysmal and persistent AF (Aviles et al. 2003; Chung et al. 2001; Frustaci et al. 1997; Psychari and Apostolou 2005). Elevation of CRP and IL-6 might also contribute to generation and perpetuation of AF, as evidenced by marked inflammatory infiltrates, myocyte necrosis, and fibrosis found in atrial biopsies of patients with lone AF (Ommen et al. 1997; Bruins et al. 1997; Aviles et al. 2003; Chung et al. 2001; Frustaci et al. 1997). Complement activation has also been described in a cohort of patients with AF, without other associated inflammatory diseases (Bruins et al. 1997). The exact mechanism of inflammation leading to tissue remodelling in AF patients is unclear. It is thought that AF leads to myocyte calcium overload, promoting atrial myocyte apoptosis. CRP might then act as an opsonin that binds to atrial myocytes, inducing local inflammation and complement activation. Tissue damage then ensues and fibrosis sets in (Aviles et al. 2003; Dernelis and Panaretou 2001; Abdelhadi et al. 2004; Dernelis and Panaretou 2006).

## Biomarkers of postoperative AF

The inflammatory cascade and catecholamine surge associated with surgery might play a prominent role in initiating atrial tachyarrhythmias after cardiac surgery. It has been reported to occur in up to 40 percent of patients undergoing CABG or up to 50 percent of patients undergoing cardiac valvular surgery (Issac et al. 2007). After cardiac surgery, the complement system is activated and pro-inflammatory cytokines are released. The incidence of atrial arrhythmias follows a similar pattern and peaks on post-operative day 2 or 3 (Gabay and Kushner 1999; Ommen et al. 1997; Bruins et al. 1997; Korantzopoulos et al. 2003). Another study correlated leukocytosis to an increased incidence in AF in post-operative cardiovascular patients (Abdelhadi et al. 2004). At a molecular level, the development of postoperative AF was linked to 174G/C polymorphism of the IL-6 promoter gene (Burzotta et al. 2001). Similarly, a genetic link between inflammation and AF was established and the GG genotype was found to be an independent predictor of postoperative AF (Gaudino et al. 2003). The inflammatory response induced by surgical trauma was also associated with the development of AF after off-pump CABG (Ishida et al. 2006). Finally, elevated preoperative levels of soluble CD40 ligand (sCD40L) were significantly associated with a higher risk of developing AF after off-pump CABG surgery (Fig. 1) (Antoniades et al. 2009). Interestingly, the association of sCD40L with in-hospital AF was independent of systemic

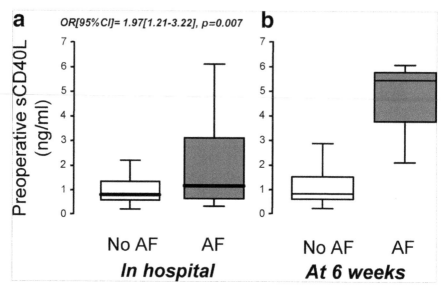

**Figure 1.** a) Preoperative levels of sCD40L were associated with the development of postoperative AF during hospitalization. Probability value was derived from multivariable logistic regression and was adjusted for age, sex, Euroscore, and duration of operation. The OR is expressed per 1-SD change in log(sCD40L). b) Preoperative sCD40L also appeared to be higher in those who were in AF at six weeks, although because of the limited number of patients, formal statistical tests were not performed. Values are expressed as median, [25th-75th percentiles], and range. Reproduction from Antoniades C, Van-Assche T, Shirodaria C, Diesch J, Antonopoulos AS, Lee J, Cunnington C, Tousoulis D, Stefanadis C, Casadei B, Taggart D, Channon KM, and Leeson P. 2009. Preoperative sCD40L levels predict risk of atrial fibrillation after off-pump coronary artery bypass graft surgery. Circulation. 120(11 Suppl.): S170–6, with permission.

endothelial function, markers of systemic inflammation such as CRP and sICAM1, vascular redox state, or clinical markers such as Euroscore. Taken together, these findings suggest that preoperative levels of sCD40L, a marker of platelet activation, may be an important determinant of short-term risk of AF after CABG surgery (Antoniades et al. 2009).

## Biomarkers of post-ablation AF

The relationship of inflammation and outcomes after catheter ablation for AF has been explored in recent past years. It was previously found that preexisting left atrial scarring predicted AF ablation failure and that left atrial scarring was associated with higher CRP level; baseline CRP level, however, was not significantly different between patients with AF recurrence versus no recurrence (Verma et al. 2005). Rotter et al. found no difference in baseline high-sensitive CRP (hsCRP) level between long-lasting persistent-AF patients who had successful versus unsuccessful ablation, but reported a decline of hsCRP level with restoration of sinus rhythm in the non-recurrence group, but not in the recurrence group. This finding suggested that restoration of sinus rhythm by ablation could lead to decrease in the patient's inflammatory state and reverse remodelling of the left atrium (LA) (Rotter et al. 2006). In a later study, it was found that baseline CRP level predicted early, but not late, AF recurrence post-ablation; hsCRP level was not measured in that study (Lellouche et al. 2009). Another study reported no significant difference in baseline hsCRP among patients with paroxysmal AF undergoing AF ablation, who had immediate (1–3 days post-ablation) versus early (4–30 days post-ablation) versus no AF recurrence at 1-month follow-up; however, baseline hsCRP relationship to long-term outcome post-ablation was not evaluated (Koyama et al. 2009). In the same year, it was reported that baseline hsCRP was significantly associated with AF recurrence in univariate, but not multivariate analysis (Letsas et al. 2009). Lin et al. found that high baseline hsCRP level was associated with abnormal LA electrophysiologic substrate and high incidence of non-pulmonary vein AF sources, and was associated with lower single-ablation and final, multiple-ablation success rates (Lin et al. 2010). Kurotobi et al. reported that elevated baseline hsCRP level was associated with LA structural and electrical remodelling maintaining AF, and was associated with increased AF recurrence after ablation (Kurotobi et al. 2010). Finally, it was recently reported that patients with higher hsCRP level before AF ablation had a higher probability of AF recurrence on long-term follow-up (Liu et al. 2011). Furthermore, similar findings were reported in patients with paroxysmal AF and in patients with persistent AF. The results suggest that the patients with high hsCRP level before AF ablation may have heavy LA remodelling and significant systemic inflammatory alterations that adversely affect the clinical outcome of ablation (Liu et al. 2011). The presence of local inflammation assessed by atrial tissue localization of CRP was verified in patients undergoing catheter ablation of AF (Narducci et al. 2011). High plasma homocysteine or YKL-40 levels before ablation have been found in patients with paroxysmal or persistent AF and were associated with recurrence of post-ablation AF (Shimano et al. 2008; Henningsen et al. 2010).

## Biomarkers of new-onset and recurrent AF

Existing evidence suggests that inflammation might also play a prominent role in both the aetiology and maintenance of new-onset and recurrent non-operative AF (Engelmann and Svendsen 2005). Numerous studies have reported specifically on the association of CRP with the development and maintenance of AF. The study by Chung et al. was one of the first to demonstrate an association in elevated CRP levels with the onset of AF in a non-operative setting (Chung et al. 2001). The CRP levels were more than 2-fold higher in patients with AF than in the control subjects. Furthermore, patients with persistent AF had higher CRP levels than those with paroxysmal AF, suggesting that inflammation plays a role in the maintenance of AF. Many studies have drawn similar conclusions, thus validating the notion that inflammation plays a viable role in the perpetuation and maintenance of AF (Aviles et al. 2003; Dernellis and Panaretou 2001). The different levels of inflammatory biomarkers according to the types of AF were verified in newer studies (Gedikli et al. 2007). Moreover, lower CRP levels have also been correlated to increased success rate of electrical cardioversion and subsequent maintenance of normal sinus rhythm (Dernellis and Panaretou 2001; Engelmann and Svendsen 2005; Dernellis and Panaretou 2004; Anderson et al. 2004; Watanabe et al. 2005; Conway et al. 2004; Asselbergs et al. 2005; Sata et al. 2004; Wazni et al. 2005; Loricchio et al. 2007). In one published report (Loricchio et al. 2007), it was found that CRP levels were associated with AF recurrence after electrical cardioversion during long-term follow-up. In particular, patients with CRP < 1.9 mg/L (lowest quartile) had a very low incidence of recurrence after electrical cardioversion, and this low risk was maintained up to one year of follow-up. Although more recent studies are in accordance with these previous findings (Henningsen et al. 2009; Rizos et al. 2010), a rather weak association of inflammatory biomarkers to the risk of first recurrence of AF should be acknowledged in large patient cohorts (Masson et al. 2010).

## Biomarkers of Oxidative Stress in AF

It has been hypothesized that oxidative stress may cause atrial structural and electrical remodelling, which plays an important role in the pathogenesis of AF. The hypothesis is supported by the following aspects: histological studies have demonstrated oxidative damage in both AF patients and animal models of AF. Substantial oxidative damage in right atrial appendages obtained from AF patients who were undergoing the Maze procedure has been reported previously (Mihm et al. 2001). In an experimental model of AF, Carnes et al. revealed that dogs with sustained AF demonstrated increased protein nitration, suggesting enhanced oxidative stress (Carnes et al. 2001). At the genetic level, Kim et al. examined the gene transcriptional profiles in human atrial tissue obtained from 26 patients with permanent AF undergoing the Maze procedure, and found that five genes associated with the production of reactive oxygen species (ROS) increased, whereas two genes related to antioxidants decreased (Kim et al. 2003). More recently, Kim et al. measured nicotinamide adenosine dinucleotide phosphate (NADPH)-stimulated superoxide production in the right atrial appendage samples from 170 consecutive patients undergoing CABG. The authors found that patients who developed AF after surgery had a significant increase in atrial NADPH oxidase activity than patients who remained in sinus rhythm. Furthermore, increased atrial NADPH oxidase activity was independently associated with an increased risk of post-operative AF (Kim et al. 2008). In a cross-sectional case-control design study, it was postulated that oxidative stress markers were significantly elevated in persistent AF patients, and oxidative stress but not inflammatory markers were associated with AF (Neuman et al. 2007). In addition, Ramlawi et al. investigated oxidative stress markers in patients undergoing CABG or valve procedures. They found that patients with post-operative AF had significantly more elevation in total peroxide levels in serum compared to patients in sinus rhythm at six hours after surgery (Ramlawi et al. 2007). A strong correlation was found in one study between inflammatory status and endothelial function, while the only independent predictor for oxidized LDL levels was the maximum diameter of the left atrium in these patients (Tousoulis et al. 2009). However, in a large patient cohort, although inflammatory biomarkers were significantly increased in patients with AF, supporting a strong association between inflammation and AF, urinary F2-isoprostanes, a sensitive index of systemic oxidative stress *in vivo*, were not increased in AF overall or in different subtypes of AF (Li et al. 2010).

Finally, drugs that have antioxidant properties may show beneficial effects on AF development. In chronically instrumented dogs, Carnes et al. found that pre-treatment of dogs with ascorbate attenuated the pacing-induced atrial effective refractory period shortening. In addition, the authors demonstrated that patients undergoing CABG receiving peri-operative treatment with ascorbate were associated with a significant benefit in preventing AF (Carnes et al. 2001). A randomized, double-blind, placebo-controlled trial demonstrated that patients receiving peri-operative treatment with steroids were associated with a significant decrease in post-operative AF (Prasongsukarn et al. 2005). In another study, it was found that patients receiving treatment with vitamin C reduced early recurrence rates after electrical cardioversion of persistent AF (Korantzopoulos et al. 2005).

## Biomarkers of Thrombosis in AF

AF is a major cause of morbidity and mortality through an increased risk of thrombo-embolic stroke (Benjamin et al. 1998; Kannel et al. 1998). The mechanisms behind cerebral thrombo-embolism in AF are incompletely understood, but it is well documented that AF is associated with a prothrombotic state as demonstrated by higher levels of plasma soluble P-selectin (sP-sel; a marker of platelet activation), fibrinogen (the precursor to insoluble fibrin and an important rheological factor), von Willebrand factor (vWf; a marker of endothelial damage/dysfunction), and other haemostatic indices, when compared with healthy control subjects (Benjamin et al. 1998; Kannel et al. 1998; Kumagai et al. 1990; Lip et al. 1995a; Pongratz et al. 1997; Li-Saw-Hee et al. 2000; Watson et al. 2009), leading to suggestions of a generalized prothrombotic state in AF. However, some studies have shown conflicting results (Pongratz et al. 1997; Li-Saw-Hee et al. 2001a; Balleisen et al. 1985; Li-Saw-Hee et al. 2001b; Conway et al. 2003). Among patients with AF who received aspirin, raised levels of vWf were predictive of stroke and vascular events, but raised sP-sel levels were not associated with increased cardiovascular risk (Conway et al. 2003). Recent studies are in accordance with these findings (Kaireviciute et al. 2011; Roldan et al. 2011). Nevertheless, elevations in sP-sel and fibrinogen have both been described in AF cohorts in most of the previous studies (Benjamin et al. 1998; Kannel et al. 1998; Kumagai et al. 1990; Lip et al. 1995b; Pongratz et al. 1997; Minamino et al. 1999; Li-Saw-Hee et al. 2000; Watson et al. 2009). Of note, the AF cohorts in these studies were mostly combined with the underlying cardiovascular diseases. This would certainly disturb the judgement of the association of thrombogenic indices with AF itself. As a result, it has been controversial whether the arrhythmia itself independently leads to platelet activation and abnormal changes in coagulation or not. In some studies, the higher levels of circulating fibrinogen were ascribed to AF per se (Kumagai et al. 1990; Lip et al. 1995b) whereas in another study (Feng et al. 2001), these were presumed to be mainly due to the effect of the confounding traditional risk factors and coexistence of cardiovascular disease. Similarly, in some studies the elevation of sP-sel levels was ascribed to the underlying cardiovascular diseases rather than the AF per se (Choudhury et al. 2007), whereas a more recent study showed that stimulated AF caused local cardiac platelet activation within minutes of AF onset (Akar et al. 2008).

In the subgroup analysis of Framingham Offspring Study (Feng et al. 2001), lone AF patients and controls without cardiovascular disease had no differences in the prothrombotic indices (including fibrinogen). However, Mondillo et al. showed that plasma levels of fibrinogen, platelet factor 4, and thromboglobulin in the 'lone' AF patient were higher than those in the controls, but with the changes of sP-sel level not involved. (Mondillo et al. 2000). Most of the relevant studies found that blood vWf was raised in patients with AF (Freestone and Lip 2008). Because the AF cohorts in these studies were mostly combined with cardiovascular disease, it remains a matter of debate whether the raised vWf levels are due to the AF itself or to the coexistent cardiovascular diseases. In a recently published study, both platelet activation and abnormal changes in coagulation were suggested in idiopathic AF and a platelet activation state in lone AF (Fu et al. 2011). This supports the notion that AF per se contributes to a state of hypercoagulation.

## The Role of the Natriuretic Peptide System in AF

The natriuretic hormone axis is a well-characterized cardiac endocrine pathway comprised of neurohormones that have important physiologic actions including natriuresis, vasodilation, and the modulation of central and peripheral baroreflexes. B-type natriuretic peptide (BNP) is produced primarily in the atrial and ventricular myocardium and the brain (Levin et al. 1998). The precursor protein pro BNP is cleaved to form BNP and the amino terminal N-terminal pro BNP (NT-pro BNP), both of which circulate in the plasma (Levin et al. 1998). Although most widely used as a marker of heart failure, elevated BNP levels have been reported in patients with AF, even in the absence of heart failure or other cardiac pathology (Silvet et al. 2003; Knudsen et al. 2005; Wang et al. 2004). NT-pro BNP is elevated in patients with lone AF (Ellinor et al. 2005). Recently, NT pro-BNP concentrations among the elderly were associated with AF over a 10-year follow-up period (Patton et al. 2009). Of the 5,445 participants, those in the highest NTpro BNP quintile had a four-fold increased incidence of AF compared to those in the lowest quintile. In the setting of decompensated heart failure, elevations in the natriuretic peptides do not appear to have an independent association with AF (Knudsen et al. 2005; Rossi et al. 2000), as ventricular dysfunction is associated with an even greater release in BNP and likely masks the elevation due to AF. Elevations in natriuretic peptides have been associated with post-ablation AF recurrence (den Uijl et al. 2011). In a large patient cohort (Latini et al. 2011), markers of cardiac injury and/or strain (hsTnT, MR-proANP and NT-proBNP) and endothelin showed a modest but statistically significant predictive power of recurrence of AF, whereas the higher the concentration of NT-proBNP, the earlier the onset of AF. Elevations in the natriuretic peptides most likely represent the electrophysiologic and hemodynamic alterations that contribute to the underlying mechanisms in AF. These physiological changes include atrial dysynchrony, increased left ventricular filling pressures, abnormal calcium handling, and activation of the sympathetic nervous system. Atrial natriuretic peptides can also shorten atrial conduction time and the effective refractory period (Crozier et al. 1993). These electrophysiologic changes provide a potential substrate for arrhythmias.

## Other Biomarkers in AF

A recent study associated interleukin-18 with the occurrence of AF (Luan et al. 2010), whereas the significance of elevated metalloproteinase levels in the human atria for AF recurrence has previously been reported (Arndt et al. 2002; Kato et al. 2009). Data from the Framingham Offspring Study (Rienstra et al. 2012) and from a post-CABG patient cohort (Gungor et al. 2011) stressed on the association of elevated resistin levels with the occurrence of AF. Increased expression of the angiotensin converting enzyme (ACE) in fibrillating human atria has been reported previously (Goette et al. 2000). The increased ACE expression was accompanied by an activation of MAP kinases in fibroblasts which was related to the presence of atrial fibrosis. Thus, an activation of the atrial angiotensin system during AF seems to be one mechanism causing the development of structural changes in fibrillating atria. The possible association of the ACE allele with the electrical remodelling in patients with AF has been reported in several studies (Gensini et al. 2003; Ravn et al. 2008; Watanabe et al. 2009). In addition to increasing MAP kinase activity, ACE may also contribute to local degradation of bradykinin (BK). BK has cardioprotective effects by which it diminishes the development of interstitial fibrosis (Kitakaze et al. 2000; Schoemaker and van Heijningen 2000). Thus, changes in BK levels by an altered BK metabolism may influence the structure of fibrillating atria. The observed alterations in ectopeptidase expression/activity may play a role in the structural remodelling of fibrillating atria (Lendeckel et al. 2001). Finally, in a previous publication it was found that factor XIII Val34Leu polymorphism was independently associated with IL6 levels in AF. The Leu34 allele may potentially influence the pro-thrombotic state in these patients by modulating the inflammatory state (Marín et al. 2004).

## Biomarkers in Ventricular Arrhythmias

Sudden cardiac death (SCD) is defined as death that occurs within one hour of symptom onset and is mainly attributed to ventricular tachyarrhythmias including VT and VF (Albert et al. 2003; Zheng et al. 2001). Although patients with structural heart disease including congestive heart failure or coronary heart disease (CAD) are at an increased risk of ventricular tachyarrhythmias and subsequent SCD, two-thirds of SCD events occur in patients without high-risk clinical features. As a result, it is important to evaluate novel risk factors to identify people at risk for SCD (Myerburg et al. 1993). Biomarkers could serve this purpose in improving both the understanding of the pathophysiology underlying SCD and the implementation of preventive measures and appropriate therapeutic strategies (Table 2).

## Inflammatory Biomarkers

Inflammation is known to play a pivotal role in the pathogenesis of atherosclerosis and cardiovascular disease. In terms of predicting SCD, cytokines have gained increasing attention not only in patients with CAD (Fisman et al. 2006) and chronic heart failure (Shehab et al. 2004) but even in

**Table 2.** Biomarkers in Ventricular Arrhythmias.

| Categories | Biomarkers |
|---|---|
| Inflammatory biomarkers | CRP, IL-6, IL-8, sCD40L |
| Natriuretic peptides | BNP, NT-pro BNP |
| Metabolic biomarkers | Trans-18:1 oleic acid, trans-18:2 oleic acid |
| Renal biomarkers | Cystatin-C |
| MicroRNAs | miRNA-1, miRNA-133 |

BNP: b-type natriuretic peptide, CRP: C-reactive protein, IL: interleukin, NT-pro BNP: N-terminal pro BNP, RNA: ribosome nucleic acid, sCD40L: soluble CD40 ligand.

subjects with no apparent heart disease (Sajadieh et al. 2006). The Physicians' Health Study of 22,071 healthy men evaluated whether CRP aids in the identification of participants at an increased risk of SCD. Baseline CRP levels were independently associated with an increased risk of SCD over a 17-year follow-up period (Albert et al. 2002a). Interestingly, the mean time to SCD in this study was 9.2 years. Given this length of time between CRP measures and SCD, the effects of inflammation may be mediated through a chronic, dynamic process such as coronary atherosclerosis. Inflammatory biomarkers have been evaluated in the setting of CAD, where chronic inflammation and thrombosis can transform a stable atherosclerotic plaque to an unstable lesion. This phenotype of acute plaque rupture is known to be associated with ventricular arrhythmias and SCD events (Burke et al. 1997; Myerburg 2001). These findings are consistent with histological studies that demonstrate a strong correlation between CRP and increased numbers of thin cap atheromas in the coronary vasculature (Burke et al. 2002). Elevated CRP may increase the risk of fatal complications from acute coronary syndromes. CRP may also have direct arrhythmogenic properties by locally activating complement and inducing oxidative stress and apoptosis (Lagrand et al. 1999; Deswal et al. 2001). It has been reported that patients with a history of VF have chronically elevated levels of tissue inhibitor of metalloproteinase-1 (TIMP-1) and IL-8 (Elmas et al. 2007). In addition, in a retrospective, cross-sectional study, it was demonstrated that patients with VF during myocardial infarction (MI) display elevated levels of markers of thrombin generation remote from the acute event as compared to MI patients without VF (Elmas et al. 2006). In a later study from the same group, patients with VF during ST-segment elevation MI had supraphysiologic levels of plasma cytokines and markers of thrombin generation (Elmas et al. 2010). Moreover, patients with a history of VF complicating MI show an enhanced expression of CD40L on platelets after *in vitro* lipopolysaccharide-challenge with enhanced platelet activation (Kälsch et al. 2006). The previous findings suggest an activation of inflammatory mechanisms and coagulation activation in patients prone to ischemia-related VF. The mechanisms underlying the association between the occurrence of VF during acute MI and excessive activation of mediators which are known to be linked to plaque rupture and thrombotic coronary occlusion is poorly understood.

With regard to patients wearing an implantable cardioverter defibrillator (ICD), Biasucci et al. presented data describing an association between elevated CRP serum concentrations (>3 mg/L) and malignant ventricular arrhythmias, but these patients were in active ischemia (Biasucci et al. 2006). Other data with patients in NYHA class II–III revealed a significant association between IL-6 serum concentrations and an increased risk of spontaneous VT/VF events in ICD recipients with CAD or dilated cardiomyopathy (DCM) (Streitner et al. 2007). A more recent study describes novel biomarkers related to the occurrence of an electrical storm in ICD-patients with either CAD or DCM. During electrical storm, patients revealed significantly higher serum concentrations of hs-CRP, IL-6 and NT-pro BNP compared to their baseline values and compared to baseline serum levels of ICD-patients with or without single ICD-intervention during a 9-month follow-up (Streitner et al. 2009). The CAMI-GUIDE study, for the first time, tested prospectively the hypothesis that hs-CRP may play a role in sudden death in patients with previous MI and severely depressed left ventricular ejection fraction (LVEF) and may represent a guide for additional risk stratification in patients indicated for ICD implantation according to current guidelines. This study failed to demonstrate an association between CRP, either on a discrete (quartiles, binomial) or continuous scale, and the occurrence of a composite endpoint of SCD or threatening ventricular arrhythmias and, therefore, to validate CRP as a marker of risk in these patients (Biasucci et al. 2012). It might therefore be possible that CRP may emerge as a marker of SCD in low risk population, but that the risk conferred by CRP is completely blunted in the presence of stronger factors, such as depressed LVEF.

## The role of Natriuretic Peptides

As described previously, regulation of the natriuretic peptide systems is important in both cardiovascular physiology and pathology. In a population-based cohort of 121,700 participants from the Nurses' Health Study, increased baseline NT-pro BNP concentrations were independently associated with SCD events after more than 16 years of follow-up (Korngold et al. 2009). These findings are consistent with ones from studies evaluating high-risk patient populations. Among patients with heart failure and left ventricular systolic dysfunction or those post-MI, elevated BNP levels were an independent predictor of SCD events (Berger et al. 2002; Tapanainen et al. 2004). Electrical storm was associated with significantly elevated NT-pro BNP serum concentrations in ICD-patients with structural heart disease

(Streitner et al. 2009). Several small studies also suggest that natriuretic hormones may predict appropriate ICD shocks for VT or VF in ICD recipients (Blangy et al. 2007; Yu et al. 2007). In a relevant meta-analysis, it was demonstrated that BNP predicted both the occurrence of SCD in patients without ICDs and the occurrence of ventricular arrhythmias in patients with devices. In patients without ICDs, a raised BNP predicted nearly a four-fold increase in the risk of SCD compared with patients with a lower BNP. In the subgroup of patients with known heart disease, there was over a four and a half-fold increase in risk. In patients with ICDs, a raised BNP predicted over a two-fold increase in the risk of the occurrence of ventricular arrhythmias (Scott et al. 2009).

Natriuretic peptide dysregulation may represent hemodynamic changes including increased filling pressures that may result in mechanoelectrical feedback or electrophysiologic changes that increase the risk of arrhythmias (Franz et al. 1992). Prior studies have demonstrated increased ventricular ectopy with diastolic stretch in animal models (Hansen et al. 1990). This ectopy may result in sustained, reentrant tachyarrhythmias in patients with cardiac fibrosis and scarring. These conditions may explain the elevated SCD risk associated with natriuretic peptide elevations.

Results of a recent analysis provide evidence that NT-proBNP is not only a significant predictor of death for heart failure, but it is also a predictor of VT/VF and sudden death (Biasucci et al. 2012). The complex pathogenesis of the major ventricular arrhythmias in high risk post-MI patients might explain why NT-proBNP, which is an accurate marker of left ventricular dysfunction, neuro-hormonal activation, and inflammation, is a predictor of life-threatening arrhythmias and sudden death (Verma et al. 2006).

## Metabolic Biomarkers

Dietary fatty acids have been evaluated extensively in clinical and experimental studies as markers of cardiovascular disease. Epidemiologic studies have analyzed both trans-fatty acids and long chain n-3 fatty acids as biomarkers of risk in ventricular arrhythmias and SCD. In population-based studies, higher plasma phospholipids levels of trans-isomers of linoleic acid were associated with a higher risk of SCD events (Lemaitre et al. 2006). In contrast, higher levels of trans-isomers of oleic acid or trans-18:1 were associated with lower risk. Other population based studies reported an independent association between higher levels of trans-18:2 in red blood cell membranes and SCD risk (Lemaitre et al. 2002). The differential associations observed between these fatty acid subtypes and SCD may in part be explained by their biological structures. Other community-based studies have demonstrated an inverse association between n-3 long chain fatty acids and SCD risk (Albert et al. 2002b; Siscovick et al. 1995).

Dietary fatty acids, particularly n-3 polyunsaturated fatty acids can reduce myocardial vulnerability to arrhythmia triggers in experimental settings (Charnock 1994; Billman et al. 1999; Kang and Leaf 2000; Xiao et al. 1997). Despite these epidemiological and translational research findings, clinical trials have not demonstrated a reduction in ventricular tachyarrhythmias in patients receiving omega-3 fatty acid. The study on omega-3 fatty acids and ventricular arrhythmia (SOFA) was a randomized, double-blind, placebo-controlled multicenter study that evaluated the effect of fish oil on the outcome of ICD shocks for VT or VF or all-cause mortality (Brouwer et al. 2006). The study did not indicate any evidence of a strong protective effect of intake of omega-3 polyunsaturated fatty acids from fish oil against ventricular arrhythmia in patients with ICDs.

## Renal Biomarkers

Cystatin C is a novel metabolic marker that appears to correlate closely with glomerular filtration rate. Cystatin-C is a marker of renal function, more accurate than creatinine and glomerular filtration rate, which is an excellent marker of death in ischemic patients (Deo et al. 2009; Ix et al. 2007). Recent studies also demonstrate a strong, independent association between elevated cystatin C concentrations and SCD in population-based studies (Deo et al. 2010). Among 4,465 participants in the Cardiovascular Health Study, baseline cystatin C concentrations were associated with SCD events over a median follow-up of 11.2 years. After multivariate adjustment, the risk of SCD in the highest cystatin C tertile was almost three-fold greater than the risk in tertile 1. Participants with prevalent cardiovascular disease were excluded from this analysis in order to enhance the specificity for identifying arrhythmic events such as VT or VF. The increased risk of SCD observed in this study may have been attributable to a greater degree of cardiac fibrosis or structural heart disease that is known to occur in patients with kidney dysfunction. Finally, rennin-angiotensin-aldosterone system dysregulation may have also mediated the increased arrhythmia risk in this population. In the setting of acute MI, reperfusion ventricular arrhythmias have been associated with angiotensin converting enzyme (ACE) polymorphism; that is, reperfusion ventricular arrhythmias, within 5 h of coronary intervention, have been reported to be significantly more prevalent in the ACE-DD and -ID genotypes than in the ACE-II genotype (Takezako et al. 2001).

## MicroRNAs in Cardiac Arrhythmias

MicroRNAs (miRNAs) are a large sub-group of small non-coding RNAs, which have been demonstrated to post-transcriptionally regulate the expression of protein coding genes in a wide-range biological process. MiRNAs have been shown to be essential for normal heart development and cardiac function (Huang et al. 2010; Papageorgiou et al. 2012). The role of miRNA-1 in arrhythmogenesis has been investigated in a rat model of MI and in human hearts with CAD (Yang et al. 2007). It was found that, the nucleotide miRNA-1 was significantly upregulated in ischemic heart tissue, and its administration exacerbated arrhythmogenesis (Yang et al. 2007). Moreover, it has been shown that mice lacking miRNA-1-2 exhibited a high incidence of electrophysiological abnormalities often resulting in sudden death (Zhao et al. 2007). miRNA-133 is also implicated in arrhythmogenesis, as exogenous administration into diabetic rabbit myocytes resulted in suppression of expression of HERG, a long-QT syndrome gene (encoding a K+ channel) (Krutzfeldt et al. 2006). Moreover, it has been indicated that downregulation of miRNA-1 and miRNA-133 is associated with an increase in protein levels of HCN2 and HCN4, which are two important cardiac pacemaker channel proteins enhancing automaticity and development of arrhythmia (Luo et al. 2008). Those miRNAs have also been implicated in expression changes of KCNQ1 and KCNE1, proteins that form a channel complex when assembled (Xu et al. 2007). Additional evidence suggests the potential involvement of miRNA-133 in the regulation of L-type calcium channel expression (Lu et al. 2009).

## Conclusion

Recent studies have demonstrated the significance of serum biomarkers as risk factors for both atrial and ventricular tachyarrhythmias. Elevations in natriuretic peptides are associated with AF and SCD in population-based studies. Inflammatory biomarkers are also associated with both arrhythmias and may have a significant role in their pathogenesis. RAAS dysregulation has been implicated in structural heart disease and may also be involved in electrophysiologic disturbances. Cystatin C, which is a novel measure of kidney function, is associated with AF and SCD events in community-based studies. Finally, metabolic markers appear to be associated with SCD and cardiovascular disease. Improved diagnostic and risk stratification tools are required to identify patients at high risk of developing cardiac arrhythmias. Serum biomarkers could serve as an inexpensive, safe, and readily accessible tool to accomplish this task. In depth evaluation and a greater understanding, however, is required to determine how these biomarkers will be incorporated into diagnostic and treatment algorithms for patients at risk for AF or SCD.

# References

Abdelhadi, R.H., H.S. Gurm, D.R. Van Wagoner, and M.K. Chung. 2004. Relation of an exaggerated rise in white blood cells after coronary bypass or cardiac valve surgery to development of atrial fibrillation postoperatively. Am. J. Cardiol. 93: 1176–8.

Akar, J.G., W. Jeske, and D.J. Wilber. 2008. Acute onset human atrial fibrillation is associated with local cardiac platelet activation and endothelial dysfunction. J. Am. Coll. Cardiol. 51: 1790–3.

Albert, C.M., J. Ma, N. Rifai, M.J. Stampfer, and P.M. Ridker 2002a. Prospective study of C-Reactive protein, homocysteine, and plasma lipid levels as predictors of sudden cardiac death. Circulation 105: 2595–2599.

Albert, C.M., H. Campos, M.J. Stampfer, P.M. Ridker, J.E. Manson, W.C. Willett, and J. Ma 2002b. Blood levels of long-chain n-3 fatty acids and the risk of sudden death. N. Engl. J. Med. 346: 1113–1118.

Albert, C.M., C.U. Chae, F. Grodstein, L.M. Rose, K.M. Rexrode, J.N. Ruskin, and M.J. Stampfer, J.E. Manson. 2003. Prospective study of sudden cardiac death among women in the United States. Circulation 107: 2096–2101.

Anderson, J.L., C.A. Allen Maycock, D.L. Lappé, B.G. Crandall, B.D. Horne, T.L. Bair, S.R. Morris, Q. Li Q, and J.B. Muhlestein. Intermountain Heart Collaborative Study group. 2004. Frequency of elevation of C-reactive protein in atrial fibrillation. Am. J. Cardiol. 94: 1255– 1259.

Antoniades, C., T. Van-Assche, C. Shirodaria, J. Diesch, A.S. Antonopoulos, J. Lee, C. Cunnington, D. Tousoulis, C. Stefanadis, B. Casadei, D. Taggart, K.M. Channon, and P. Leeson. 2009. Preoperative sCD40L levels predict risk of atrial fibrillation after off-pump coronary artery bypass graft surgery. Circulation. 120(11 Suppl.): S170–176.

Arndt, M.U. Lendeckel, C. Röcken, K. Nepple, C. Wolke, A.S. Huth, S. Ansorge, H.U. Klein, and A. Goette. 2002. Altered Expression of ADAMs (A Disintegrin And Metalloproteinase) in Fibrillating Human Atria. Circulation. 105: 720–725

Asselbergs, F.W., M.P. van den Berg, G.F. Diercks, W.H. van Gilst, and D.J. van Veldhuisen. 2005. C-reactive protein and microalbuminuria are associated with atrial fibrillation. Int. J. Cardiol. 98: 73–77.

Aviles, R., D. Martin, C. Apperson-Hansen, P.L. Houghtaling, P. Rautaharju, R.A. Kronmal, R.P. Tracy, D.R. Van Wagoner, B.M. Psaty, M.S. Lauer, and M.K. Chung. 2003. Inflammation as a risk factor for atrial fibrillation. Circulation 108: 3006–3010.

Balleisen, L., J. Bailey, P.H. Epping, H. Schulte, and J. van de Loo. 1985. Epidemiological study on factor VII, factor VIII and fibrinogen in an industrial population: I. Baseline data on the relation to age, gender, body-weight, smoking, alcohol, pill-using, and menopause. Thromb. Haemost. 54: 475–479.

Benjamin, E.J., P.A.Wolf, R.B. D'Agostino, H. Silbershatz, W.B. Kannel, and D. Levy. 1998. Impact of atrial fibrillation on the risk of death: the Framingham Heart Study. Circulation 98: 946–952.

Berger, R., M. Huelsman, K. Strecker, A. Bojic, P. Moser, B. Stanek, and R. Pacher. 2002. B-type natriuretic peptide predicts sudden death in patients with chronic heart failure. Circulation. 105: 2392–2397.

Biasucci, L.M., G. Giubilato, G. Biondi-Zoccai, T. Sanna, G. Liuzzo, M. Piro, G. De Martino, C. Ierardi, A. dello Russo, G. Pelargonio, F. Bellocci, and F. Crea. 2006. C reactive protein is associated with malignant ventricular arrhythmias in patients with ischaemia with implantable cardioverter-defibrillator. Heart 92: 1147–1148.

Biasucci, L.M., F. Bellocci, M. Landolina, R. Rordorf, A. Vado, E. Menardi, G. Giubilato, S. Orazi, M. Sassara, A. Castro, R. Massa, A. Kheir, G. Zaccone, C. Klersy, F. Accardi, and F. Crea. 2012. Risk stratification of ischaemic patients with implantable cardioverter defibrillators by C-reactive protein and a multi-markers strategy: results of the CAMI-GUIDE study. Eur. Heart J. 33(11): 1344–1350.

Billman, G.E., J.X. Kang, and A. Leaf. 1999. Prevention of sudden cardiac death by dietary pure omega-3 polyunsaturated fatty acids in dogs. Circulation 99: 2452–2457.

Blangy, H., N. Sadoul, B. Dousset, A. Radauceanu, R. Fay, E. Aliot, and F. Zannad. 2007. Serum BNP, hs-C-reactive protein, procollagen to assess the risk of ventricular tachycardia in ICD recipients after myocardial infarction. Europace 9: 724–729.

Brouwer, I.A., P.L. Zock, A.J. Camm, D. Böcker, R.N. Hauer, E.F. Wever, C. Dullemeijer, J.E. Ronden, M.B. Katan, A. Lubinski, H. Buschler, and E.G. Schouten; SOFA Study Group. 2006. Effect of fish oil on ventricular tachyarrhythmia and death in patients with implantable cardioverter defibrillators: The study on omega-3 fatty acids and ventricular arrhythmia [SOFA] randomized trial. JAMA. 295: 2613–2619.

Bruins, P., H. te Velthuis, A.P. Yazdanbakhsh, P.G. Jansen, F.W. van Hardevelt, E.M. de Beaumont, C.R. Wildevuur, L. Eijsman, A. Trouwborst, and C.E. Hack. 1997. Activation of the complement system during and after cardiopulmonary bypass surgery: postsurgery activation involves C-reactive protein and is associated with postoperative arrhythmia. Circulation 96: 3542–3548.

Burke, A.P., A. Farb, G.T. Malcom, Y.H. Liang, J. Smialek, and R. Virmani. 1997. Coronary risk factors and plaque morphology in men with coronary disease who died suddenly. N. Engl. J. Med. 336: 1276–1282.

Burke, A.P., R.P. Tracy, F. Kolodgie, G.T. Malcom, A. Zieske, R. Kutys, J. Pestaner, J. Smialek, and R. Virmani. 2002. Elevated C-reactive protein values and atherosclerosis in sudden coronary death: Association with different pathologies. Circulation 105: 2019–2023.

Burzotta, F., L. Lacoviella, A. Di Castelnuovo, F. Glieca, N. Luciani, R. Zamparelli, R. Schiavello, M.B. Donati, A. Maseri, G. Possati, and F. Andreotti. 2001. Relation of the _174 G/C polymorphism of interleukin-6 to interleukin-6 plasma levels and to length of hospitalization after surgical coronary revascularization. Am. J. Cardiol. 88: 1125–1128.

Carnes, C.A., M.K. Chung, T. Nakayama, H. Nakayama, R.S. Baliga, S. Piao, A. Kanderian S. Pavia, R.L. Hamlin, P.M. McCarthy, J.A. Bauer, and D.R. Van Wagoner. 2001. Ascorbate attenuates atrial pacinginduced peroxy nitrite formation and electrical remodeling and decreases the incidence of postoperative atrial fibrillation. Circ. Res. 89: E32–38.

Charnock, J.S. 1994. Dietary fats and cardiac arrhythmia in primates. Nutrition 10: 161–169.

Choudhury, A., I. Chung, A.D. Blann, and G.Y. Lip. 2007. Platelet surface CD62P and CD63, mean platelet volume, and soluble/platelet P-selectin as indexes of platelet function in atrial fibrillation: a comparison of 'healthy control subjects' and 'disease control subjects' in sinus rhythm. J. Am. Coll. Cardiol. 49: 1957–1964.

Chung, M.K., D.O. Martin, D. Sprecher, O. Wazni, A. Kanderian, C.A. Carnes, J.A. Bauer, P.J. Tchou, M.J. Niebauer, A. Natale, and D.R. Van Wagoner. 2001. C-reactive protein elevation in patients with atrial arrhythmias: inflammatory mechanisms and persistence of atrial fibrillation. Circulation 104(24): 2886–2891.

Conway, D.S., L.A. Pearce, B.S. Chin, R.G. Hart, and G.Y. Lip. 2003. Prognostic value of plasma von Willebrand factor and soluble P-selectin as indices of endothelial damage and platelet activation in 994 patients with nonvalvular atrial fibrillation. Circulation. 107(25): 3141–3145.

Conway, D.S., P. Buggins, E. Hughes, and G.Y. Lip. 2004. Prognostic significance of raised plasma levels of interleukin-6 and C-reactive protein in atrial fibrillation. Am. Heart J. 148: 462–466.

Crozier, I., A.M. Richards, S.G. Foy, and H. Ikram. 1993. Electrophysiological effects of atrial natriuretic peptide on the cardiac conduction system in man. Pacing Clin. Electrophysiol. 16(4 Pt 1): 738–742.

den Uijl, D.W., Delgado V, MD,a Tops LF, MD,a Ng, A, Boersma E, Trines SA, Zeppenfeld K, Schalij MJ, MD, van der Laarse A, and Bax JJ. 2011. Natriuretic peptide levels predict recurrence of atrial fibrillation after radiofrequency catheter ablation. Am. Heart J. 161: 197–203.

Deo, R., M.G. Shlipak, J.H. Ix, S. Ali, N.B. Schiller, and M.A. Whooley. 2009. Association of cystatinC with ischemia in patients with coronary heart disease. Clin. Cardiol. 32: E18–E22.

Deo, R., N. Sotoodehnia, R. Katz, M.J. Sarnak, L.F. Fried, M. Chonchol, B. Kestenbaum, B.M. Psaty, D.S. Siscovick, and M.G. Shlipak. 2010. Cystatin C and sudden cardiac death risk in the elderly. Circ. Cardiovasc. Qual. Outcomes 3: 159–164.

Dernellis, J. and M. Panaretou. 2001. C-reactive protein and paroxysmal atrial fibrillation: evidence of the implication of an inflammatory process in paroxysmal atrial fibrillation. Acta Cardiologica 56: 375– 380.

Dernellis, J. and M. Panaretou. 2004. Relationship between C-reactive protein concentrations during glucocorticoid therapy and recurrent atrial fibrillation. Eur. Heart J. 25: 1100–1107.

Dernellis, J. and M. Panaretou. 2006. Effects of C-reactive protein and the third and fourth components of complement (C3 and C4) on incidence of atrial fibrillation. Am. J. Cardiol. 97: 245–248.

Deswal, A., N.J. Petersen, A.M. Feldman, J.B. Young, B.G. White, and D.L. Mann. 2001. Cytokines and cytokine receptors in advanced heart failure: An analysis of the cytokine database from the vesnarinone trial [VEST]. Circulation 103: 2055–2059.

Ellinor, P.T., A.F. Low, K.K. Patton, M.A. Shea, and C.A. Macrae. 2005. Discordant atrial natriuretic peptide and brain natriuretic peptide levels in lone atrial fibrillation. J. Am. Coll. Cardiol. 45: 82–86.

Elmas, E., T. Kaelsch, C. Wolpert, T. Sueselbeck, T. Bertsch, C.E. Dempfle, and M. Borggrefe. 2006. Assessment of markers of thrombin generation in patients with acute myocardial infarction complicated by ventricular fibrillation. Clin. Cardiol. 29(4): 165–9.

Elmas, E., S. Lang, C.E. Dempfle, T. Kälsch, D. Hannak, T. Sueselbeck, C. Wolpert, M. Borggrefe, and M. Brueckmann. 2007. High plasma levels of tissue inhibitor of metalloproteinase-1 (TIMP-1) and interleukin-8 (IL-8) characterize patients prone to ventricular fibrillation complicating myocardial infarction. Clin. Chem. Lab Med. 45(10): 1360–5.

Elmas, E., T. Popp, S. Lang, C.E. Dempfle, T. Kälsch, and M. Borggrefe. 2010. Sudden death: do cytokines and prothrombotic peptides contribute to the occurrence of ventricular fibrillation during acute myocardial infarction? Int. J. Cardiol. 145(1): 118–9.

Engelmann, M. and H. Svendsen. 2005. Inflammation in the genesis and perpetuation of atrial fibrillation. Eur. Heart J. 26: 2083–92.

Feng, D., R.B. D'Agostino, H. Silbershatz, I. Lipinska, J. Massaro, D. Levy, E.J. Benjamin, P.A. Wolf, and G.H. Tofler. 2001. Hemostatic State and Atrial Fibrillation (The Framingham Offspring Study). Am. J. Cardiol. 87: 168–71.

Fisman, E.Z., M. Benderly, R.J. Esper, S. Behar, V. Boyko, Y. Adler, D. Tanne, Z. Matas, and A. Tenenbaum. 2006 Interleukin-6 and the risk of future cardiovascular events in patients with angina pectoris and/or healed myocardial infarction. Am. J. Cardiol. 98: 14–8.

Franz, M.R., R. Cima, D. Wang, D. Profitt, and R. Kurz. 1992. Electrophysiological effects of myocardial stretch and mechanical determinants of stretch-activated arrhythmias. Circulation 86: 968–978.

Freestone, B. and G.Y. Lip. 2008. The endothelium and atrial fibrillation. The prothrombotic state revisited. Hamostaseologie 28: 207–12.

Frustaci, A., C. Chimenti, F. Bellocci, E. Morgante, M.A. Russo, and A. Maseri. 1997. Histological substrate of atrial biopsies in patients with lone atrial fibrillation. Circulation 96: 1180–1184.

Fu, R., S. Wu, P. Wu, and J. Qiu. 2011. A study of blood soluble P-selectin, fibrinogen, and von Willebrand factor levels in idiopathic and lone atrial fibrillation. Europace 13(1): 31–6.

Gabay, C. and I. Kushner. 1999. Acute phase proteins and other systemic responses to inflammation. N. Engl. J. Med. 340: 448–54.

Gaudino, M., F. Andreotti, R. Zamparelli, A. Di Castelnuovo, G. Nasso, F. Burzotta, L. Iacoviello, M.B. Donati, R. Schiavello, A. Maseri, and G. Possati. 2003. The 174G/C interleukin-6 polymorphism influences postoperative interleukin-6 levels and postoperative atrial fibrillation. Is atrial fibrillation an inflammatory complication? Circulation 108: 195–9.

Gedikli, O., A. Dogan, I. Altuntas, A. Altinbas, M. Ozaydin, O. Akturk, and G. Acar. 2007. Inflammatory markers according to types of atrial fibrillation. Int. J. Cardiol. 120(2): 193–7.

Gensini, F., L. Padeletti, C. Fatini, E. Sticchi, G.F. Gensini, and A. Michelucci. 2003. Angiotensin-converting enzyme and endothelial nitric oxide synthase polymorphisms in patients with atrial fibrillation. Pacing Clin. Electrophysiol. 26(1Pt 2): 295–8.

Go, A.S., E.M. Hylek, K.A. Phillips, Y. Chang, L.E. Henault, J.V. Selby, and D.E. Singer. 2001. Prevalence of diagnosed atrial fibrillation in adults: national implications for rhythm management and stroke prevention: the anticoagulation and risk factors in atrial fibrillation (atria) study. JAMA 285: 2370–5.

Goette, A., T. Staack, C. Rocken, M. Arndy, C. Geller, C. Huth, S. Ansorge, H.U. Klein, and U. Lendeckel. 2000. Increased expression of extracellular signal-regulated kinase and angiotensin-converting enzyme in human atria during atrial fibrillation. J. Am. Coll. Cardiol. 35: 1669–1677.

Gungor, H., M.F. Ayik, B. Kirilmaz, S. Ertugay, I. Gul, B.S. Yildiz, S. Nalbantgil, and M. Zoghi. 2011. Serum resistin level: as a predictor of atrial fibrillation after coronary arterybypass graft surgery. Coron. Artery Dis. 22(7): 484–90.

Hansen, D.E., C.S. Craig, and L.M. Hondeghem. 1990. Stretch-induced arrhythmias in the isolated Canine ventricle. Evidence for the importance of mechanoelectrical feedback. Circulation 81: 1094–1105.

Henningsen, K.M., S.K. Therkelsen, H. Bruunsgaard, K.S. Krabbe, B.K. Pedersen, and J.H. Svendsen. 2009. Prognostic impact of hs-CRP and IL-6 in patients with persistent atrial fibrillation treated with electrical cardioversion. Scand. J. Clin. Lab. Invest. 69(3): 425–32.

Henningsen, K.M., B. Nilsson, J.S. Johansen, X. Chen, S. Pehrson, and J.H. Svendsen. 2010. Plasma YKL-40 is elevated in patients with recurrent atrial fibrillation after catheter ablation. Inflamm. Res. 59(6): 463–9.

Huang, Z.P., R.L. Neppl, and D.Z. Wang. 2010. MicroRNAs in cardiac remodeling and disease. J. Cardiovasc. Transl. Res. 3(3): 212–8. Review.

Issac, T.T., H. Dokainish, and N.M. Lakkis. 2007. Role of inflammation in initiation and perpetuation of atrial fibrillation: a systematic review of the published data. J. Am. Coll. Cardiol. 50(21): 2021–8. Review.

Ishida, K., F. Kimura, M. Imamaki, A. Ishida, H. Shimura, H. Kohno, M. Sakurai, and M. Miyazaki. 2006. Relation of inflammatory cytokines to atrial fibrillation after off-pump coronary artery bypass grafting. Eur. J. Cardiothorac. Surg. 29(4): 501–5.

Ix, J.H., M.G. Shlipak, G.M. Chertow, and M.A. Whooley. 2007. Association of cystatin C with mortality, cardiovascular events, and incident heart failure among persons with coronary heart disease: data from the Heart and Soul Study. Circulation 115: 173–179.

Kaireviciute, D., G.Y. Lip, B. Balakrishnan, G. Uzdavinys, G. Norkunas, G. Kalinauskas, V. Sirvydis, A. Aidietis, U. Zanetto, H. Sihota, M. Maheshwari, and A.D. Blann. 2011. Intracardiac expression of markers of endothelial damage/dysfunction, inflammation, thrombosis, and tissue remodeling, and the development of postoperative atrial fibrillation. J. Thromb. Haemost. 9(12): 2345–52.

Kälsch, T., E. Elmas, X.D. Nguyen, C. Wolpert, H. Klüter, M. Borggrefe, K.K. Haase, and C.E. Dempfle. 2006. Enhanced expression of platelet CD40-ligand by in vitro lipopolysaccharide-challenge in patients with ventricular fibrillation complicating acute myocardial infarction. Int. J. Cardiol. 107(3): 350–5.

Kamiyama, N. 1998. Expression of cell adhesion molecules and the appearance of adherent leukocytes on the left atrial endothelium with atrial fibrillation: rabbit experimental model. Jpn. Circ. J. 62: 837–843.

Kang, J.X. and A. Leaf. 2000. Prevention of fatal cardiac arrhythmias by polyunsaturated fatty acids. Am. J. Clin. Nutr. 71(Suppl. 1): 202S–207S.

Kannel, W.B., P.A. Wolf, E.J. Benjamin, and D. Levy. 1998. Prevalence, incidence, prognosis, and predisposing conditions for atrial fibrillation: population-based estimates. Am. J. Cardiol. 82: 2N–9N.

Kato, K., T. Fujimaki, T. Yoshida, M. Oguri, K. Yajima, T. Hibino, and T. Murohara. 2009. Impact of matrix metalloproteinase-2 levels on long-term outcome following pharmacological or electrical cardioversion in patients with atrial fibrillation. Europace 11: 332–337

Kim, Y.H., D.S. Lim, J.H. Lee, W.J. Shim, Y.M. Ro, G.H. Park, K.G. Becker, Y.S. Cho-Chung, and M.K. Kim. 2003. Gene expression profiling of oxidative stress on atrial fibrillation in humans. Exp. Mo. Med. 35: 336–49.

Kim, Y.M., H. Kattach, C. Ratnatunga, R. Pillai, K.M. Channon, and B. Casadei. 2008. Association of atrial nicotinamide adenine dinucleotide phosphate oxidase activity with the development of atrial fibrillation after cardiac surgery. J. Am. Coll. Cardiol. 51: 68–74.

Kitakaze, M., K. Noue, S. Takashima, T. Minamino, T. Kuzua, and M. Hori. 2000. Cellular mechanisms of cardioprotection afforded by inhibitors of angiotensin-converting enzyme in ischemic hearts: role of bradykinin and nitric oxide. Hypertens. Res. 23: 253–259.

Knudsen, C.W., T. Omland, P. Clopton, A. Westheim, A.H. Wu, P. Duc, J. McCord, R.M. Nowak, J.E. Hollander, A.B. Storrow, W.T. Abraham, P.A. McCullough, and A. Maisel. 2005. Impact of atrial fibrillation on the diagnostic performance of B-type natriuretic peptide concentration in dyspneic patients: An analysis from the breathing not properly multinational study. J. Am. Coll. Cardiol. 46: 838–844.

Korantzopoulos, P., T. Kolettis, K. Siogas, and J. Goudevenos. 2003. Atrial fibrillation and electrical remodelling: the potential role of inflammation and oxidative stress. Med. Sci. Monit. 9: RA225–9.

Korantzopoulos, P., T.M. Kolettis, E. Kountouris, V. Dimitroula, P. Karanikis, E. Pappa, K. Siogas, and J.A. Goudevenos. 2005. Oral vitamin C administration reduces early recurrence rates after electrical cardioversion of persistent atrial fibrillation and attenuates associated inflammation. Int. J. Cardiol. 102: 321–6.

Korngold, E.C., J.L. Januzzi, Jr., M.L. Gantzer, M.V. Moorthy, N.R. Cook, and C.M. Albert. 2009. Amino-terminal pro-B-type natriuretic peptide and high-sensitivity C-reactive protein as predictors of sudden cardiac death among women. Circulation 119: 2868–2876.

Koyama, T., Y. Sekiguchi, H. Tada, T. Arimoto, H. Yamasaki, K. Kuroki, T. Machino, K. Tajiri, X.D. Zhu, M. Kanemoto, A. Sugiyasu, K. Kuga, and K. Aonuma. 2009. Comparison of characteristics and significance of immediate versus early versus no recurrence of atrial fibrillation after catheter ablation. Am. J. Cardiol. 103: 1249–1254.

Krutzfeldt, J., M.N. Poy, and M. Stoffel. 2006. Strategies to determine the biological function of microRNAs. Nat. Genet. 38: S14–S19.

Kumagai, K., M. Fukunami, M. Ohmori, A. Kitabatake, T. Kamada, and N. Hoki. 1990. Increased intracardiovascular clotting in patients with chronic atrial fibrillation. J. Am. Coll. Cardiol. 16: 377–80.

Kurotobi, T., K. Iwakura, K. Inoue, R. Kimura, A. Okamura, Y. Koyama, Y. Toyoshima, N. Ito, and K. Fujii. 2010. A pre-existent elevated C-reactive protein is associated with the recurrence of atrial tachyarrhythmias after catheter ablation in patients with atrial fibrillation. Europace 2: 1213–1218.

Lagrand, W.K., C.A. Visser, W.T. Hermens, H.W. Niessen, F.W. Verheugt, G.J. Wolbink, and C.E. Hack. 1999. C-reactive protein as a cardiovascular risk factor: More than an epiphenomenon? Circulation 100: 96–102.

Latini, R., S. Masson, S. Pirelli, S. Barlera, G. Pulitano, E. Carbonieri, M. Gulizia, T. Vago, C. Favero, D. Zdunek, J. Struck, L. Staszewsky, P.A. Maggioni, M.G. Franzosi, and M. Disertor. 2011. Circulating cardiovascular biomarkers in recurrent atrial fibrillation: data from the GISSI-Atrial Fibrillation Trial. J. Intern. Med. 269: 160–171.

Lellouche, N., F. Sacher, M. Wright, I. Nault, J. Brottier, S. Knecht, S. Matsuo, O. Lomas, M. Hocini, M. Haïssaguerre, and P. Jaïs. 2009. Usefulness of C-reactive protein in predicting early and late recurrences after atrial fibrillation ablation. Europace 11: 662–664.

Lemaitre, R.N., I.B. King, T.E. Raghunathan, R.M. Pearce, S. Weinmann, R.H. Knopp, M.K. Copass, L.A. Cobb, and D.S. Siscovick. 2002. Cell membrane trans-fatty acids and the risk of primary cardiac arrest. Circulation. 105: 697–701.

Lemaitre, R.N., I.B. King, D. Mozaffarian, N. Sotoodehnia, T.D. Rea, L.H. Kuller, R.P. Tracy, and D.S. Siscovick. 2006. Plasma phospholipid trans fatty acids, fatal ischemic heart disease, and sudden cardiac death in older adults: The cardiovascular health study. Circulation 114: 209–215.

Lendeckel, U., M. Arndt, S. Wrenger, K. Nepple, C. Huth, S. Ansorge, H.U. Klein, and A. Goette A. 2001. Expression and activity of ectopeptidases in fibrillating human atria. J. Mol. Cell. Cardiol. 33(6): 1273–81.

Letsas, K.P., R. Weber, G. Burkle, C.C. Mihas, J. Minners, D. Kalusche, and T. Arentz. 2009. Pre-ablative predictors of atrial fibrillation recurrence following pulmonary vein isolation: The potential role of inflammation. Europace 11: 158–163.

Levin, E.R., D.G. Gardner, W.K. and Samson. 1998. Natriuretic peptides. N. Engl. J. Med. 339: 321–328.

Li, J., J. Solus, Q. Chen, Y.H. Rho, G. Milne, C.M. Stein, and D. Darbar. 2010. Role of inflammation and oxidative stress in atrial fibrillation. Heart Rhythm. 7(4): 438–44.

Lin, Y.J., H.M. Tsao, S.L. Chang, L.W. Lo, T.C. Tuan, Y.F. Hu, A.R. Udyavar, W.C. Tsai, C.J. Chang, C.T. Tai, P.C. Lee, K. Suenari, S.Y. Huang, H.T. Nguyen, and S.A. Chen. 2010. Prognostic implications of the high-sensitive C-reactive protein in the catheter ablation of atrial fibrillation. Am. J. Cardiol. 105: 495–501.

Liu, J., P.H. Fang, S. Dibs, Y. Hou, X.F. Li, and S. Zhang. 2011. High-sensitivity C-reactive protein as a predictor of atrial fibrillation recurrence after primary circumferential pulmonary vein isolation. Pacing Clin. Electrophysiol. 34(4): 398–406.

Lip, G.Y. 1995. Does atrial fibrillation confer a hypercoagulable state? Lancet 346: 1313–4.

Lip, G.Y., G.D. Lowe, A. Rumley, and F.G. Dunn. 1995a. Increased markers of thrombogenesis in chronic atrial fibrillation: effects of warfarin treatment. Br. Heart J. 73: 527–33.

Lip, G.Y., G.D. Lowe, M.J. Metcalfe, A. Rumley, and F.G. Dunn. 1995b. Effects of warfarin therapy on plasma fibrinogen, von Willebrand factor, and fibrin D-dimer in left ventricular dysfunction secondary to coronary artery disease with and without aneurysms. Am. J. Cardiol. 76: 453–8.

Li-Saw-Hee, F.L., A.D. Blann, and G.Y. Lip. 2000. A cross-sectional and diurnal study of thrombogenesis among patients with chronic atrial fibrillation. J. Am. Coll. Cardiol. 35: 1926–31.

Li-Saw-Hee, F.L., A.D. Blann, E. Edmunds, C.R. Gibbs, and G.Y. Lip. 2001a. Effect of acute exercise on the raised plasma fibrinogen, soluble P-selectin and von Willebrand factor levels in chronic atrial fibrillation. Clin. Cardiol. 24: 409–14.

Li-Saw-Hee, F.L., A.D. Blann, D. Gurney, and G.Y. Lip. 2001b. Plasma von Willebrand factor, fibrinogen and soluble P-selectin levels in paroxysmal, persistent and permanent atrial fibrillation. Effects of cardioversion and return of left atrial function. Eur. Heart J. 22: 1741–7.

Loricchio, M.L., C. Cianfrocca, V. Pasceri, L. Bianconi, A. Auriti, L. Calo, F. Lamberti, A. Castro, C. Pandozi, A. Palamara, and M. Santini. 2007. Relation of C-reactive protein to long-term risk of recurrence of atrial fibrillation after electrical cardioversion. Am. J. Cardiol. 99(10): 1421–4.

Lu, Y., J. Xiao, H. Lin, Y. Bai, X. Luo, Z. Wang, and B. Yang. 2009. A single antimicroRNA antisense oligodeoxyribonucleotide (AMO) targeting multiple microRNAs offers an improved approach for microRNA interference. Nucleic Acids Res. 37: e24.

Luan, Y., Y. Guo, S. Li, B. Yu, S. Zhu, S. Li, N. Li, Z. Tian, C. Peng, J. Cheng, Q. Li, J. Cui, and Y. Tian. 2010. Interleukin-18 among atrial fibrillation patients in the absence of structural heart disease. Europace 12: 1713–1718.

Luo, X., H. Lin, Z. Pan, J. Xiao, Y. Zhang, Y. Lu, B. Yang, and Z. Wang. 2008. Down-regulation of miR-1/miR-133 contributes to re-expression of pacemaker channel genes HCN2 and HCN4 in hypertrophic heart. J. Biol. Chem. 283: 20045–20052.

Maixent, J.M., F. Paganelli, J. Scaglione, and S. Levy. 1998. Antibodies against myosin in sera of patients with idiopathic paroxysmal atrial fibrillation. J. Cardiovasc. Electrophysiol. 9: 612–617.

Marín, F., J. Corral, V. Roldán, R. González-Conejero, M.L. del Rey, F. Sogorb, G.Y. Lip, and V. Vicente. 2004. Factor XIII Val34Leu polymorphism modulates the prothrombotic and inflammatory state associated with atrial fibrillation. J. Mol. Cell. Cardiol. 37(3): 699–704.

Masson, S., A. Aleksova, C. Favero, L. Staszewsky, M. Bernardinangeli, C. Belvito, G. Cioffi, G. Sinagra, C. Mazzone, F. Bertocchi, T. Vago, G. Peri, I. Cuccovillo, N. Masuda, S. Barlera, A. Mantovani, A.P. Maggioni, M.G. Franzosi, M. Disertori, and R. Latini; GISSI-AF investigators. 2010. Predicting atrial fibrillation recurrence with circulating inflammatory markers in patients in sinus rhythm at high risk for atrial fibrillation: data from the GISSI atrial fibrillation trial. Heart. 96(23): 1909–14.

Mihm, M.J., F. Yu, C.A. Carnes, P.J. Reiser, P.M. McCarthy, D.R. Van Wagoner, and J.A. Bauer. 2001. Impaired myofibrillar energetics and oxidative injury during human atrial fibrillation. Circulation 104: 174–180.

Minamino, T., M. Kitakaze, H. Asanuma, Y. Ueda, Y. Koretsune, T. Kuzuya, and M. Hori. 1999. Plasma adenosine levels and platelet activation in patients with atrial fibrillation. Am. J. Cardiol. 83(2): 194–8.

Mondillo, S., L. Sabatini, E. Agricola, T. Ammaturo, F. Guerrini, R. Barbati, M. Pastore, D. Fineschi, and R. Nami. 2000. Correlation between left atrial size, prothrombotic state and markers of endothelial dysfunction in patients with lone chronic nonrheumatic atrial fibrillation. Int. J. Cardiol. 75: 227–32.

Morgera, T., A. Di Lenarda, L. Dreas, B. Pinamonti, F. Humar, R. Bussani, F. Silvestri, D. Chersevani, and F. Camerini. 1992. Electrocardiography of myocarditis revisited: clinical and prognostic significance of electrocardiographic changes. Am. Heart J. 124: 455–467.

Mountantonakis, S. and Rajat Deo. 2012. Biomarkers in atrial fibrillation, ventricular arrhythmias, and sudden cardiac death. Cardiovascular Therapeutics 30: e74–e80.

Myerburg, R.J., K.M. Kessler, and A. Castellanos. 1993. Sudden cardiac death: Epidemiology, transient risk, and intervention assessment. Ann. Intern. Med. 119: 1187–1197.

Myerburg, R.J. 2001. Sudden cardiac death: Exploring the limits of our knowledge. J. Cardiovasc. Electrophysiol. 12: 369–381.

Nakamura, Y., K. Nakamura, K. Fukushima-Kusano, K. Ohta, H. Matsubara, T. Hamuro, C. Yutani, and T. Ohe. 2003. Tissue factor expression in atrial endothelium associated with nonvalvular atrial fibrillation: possible involvement in intracardiac thrombogenesis. Thromb. Res. 111: 137–142.

Narducci, M.L., G. Pelargonio, A. Dello Russo, M. Casella, L.M. Biasucci, G. La Torre, V. Pazzano, P. Santangeli, A. Baldi, G. Liuzzo, C. Tondo, A. Natale, and F. Crea. 2011. Role of tissue C-reactive protein in atrial cardiomyocytes of patients undergoing catheter ablation of atrial fibrillation: pathogenetic implications. Europace 13(8): 1133–40.

Neuman, R.B., H.L. Bloom, I. Shukrullah, L.A. Darrow, D. Kleinbaum, D.P. Jones, and S.C. Dudley, Jr. 2007. Oxidative stress markers are associated with persistent atrial fibrillation. Clin. Chem. 53: 1652–7.

Ommen, S., J. Odell, and M. Stanton. 1997. Atrial arrhythmias after cardiothoracic surgery. N. Engl. J. Med. 336: 1429–34.

Papageorgiou, N., D. Tousoulis, E. Androulakis, G. Siasos, A. Briasoulis, G. Vogiatzi, A.M. Kampoli, E. Tsiamis, C. Tentolouris, and C. Stefanadis. 2012. The Role of microRNAs in Cardiovascular Disease. Curr. Med. Chem. 19(16): 2605–10.

Patton, K.K., P.T. Ellinor, S.R. Heckbert, R.H. Christenson, C. DeFilippi, J.S. Gottdiener, and R.A. Kronmal. 2009. N-terminal pro-B-type natriuretic peptide is a major predictor of the development of atrial fibrillation: The cardiovascular health study. Circulation 120: 1768–1774.

Pongratz, G., M. Brandt-Pohlmann, K.H. Henneke, C. Pohle, D. Zink, G. Gehling, and K. Bachmann. 1997. Platelet activation in embolic and preembolic status of patients with nonrheumatic atrial fibrillation. Chest 111: 929–33.

Prasongsukarn, K., J.G. Abel, W.R. Jamieson, A. Cheung, J.A. Russell, K.R. Walley, and S.V. Lichtenstein. 2005. The effects of steroids on the occurrence of postoperative atrial fibrillation after coronary artery bypass grafting surgery: a prospective randomized trial. J. Thora.c Cardiovasc. Surg. 130: 93–8.

Psychari, S. and T. Apostolou. 2005. Relation of elevated C-reactive protein and interleukin-6 levels to left atrial size and duration of episodes in patients with atrial fibrillation. Am. J. Cardiol. 95: 764 –7.

Ramlawi, B., H. Otu, S. Mieno, M. Boodhwani, N.R. Sodha, R.T. Clements, C. Bianchi, and F.W. Sellke. 2007. Oxidative stress and atrial fibrillation after cardiac surgery: a case-control study. Ann. Thorac. Surg. 84: 1166–72.

Ravn, L.S., M. Benn, B.G. Nordestgaard, A.A. Sethi, B. Agerholm-Larsen, G.B. Jensen, and A. Tybjaerg-Hansen. 2008. Angiotensinogen and ACE gene polymorphisms and risk of atrial fibrillation in the general population. Pharmacogenet Genomics 18(6): 525–33.

Rienstra, M., J. Sun, S.A. Lubitz, D.S. Frankel, R.S. Vasan, D. Levy, J.W. Magnani, L.M. Sullivan, J.B. Meigs, P.T. Ellinor, and E.J. Benjamin. 2012. Plasma resistin, adiponectin, and risk of incident atrial fibrillation: The Framingham Offspring Study. Am. Heart J. 163: 119–124.

Rizos, I., A.G. Rigopoulos, A.S. Kalogeropoulos, S. Tsiodras, S. Dragomanovits, E.A. Sakadakis E. Faviou, and D.T. Kremastinos. 2010. Hypertension and paroxysmal atrial fibrillation: a novel predictive role of high sensitivity C-reactive protein in cardioversion and long-term recurrence. J. Hum. Hypertens 24(7): 447–57.

Roldán, V., F. Marín, B. Muiña, J.M. Torregrosa, D. Hernández-Romero, M. Valdés, V. Vicente, and G.Y. Lip. 2011. Plasma von Willebrand factor levels are an independent risk factor for adverse events including mortality and major bleeding in anticoagulated atrial fibrillation patients. J. Am. Coll. Cardiol. 57(25): 2496–504.

Rossi, A., M. Enriquez-Sarano, J.C. Burnett, Jr., A. Lerman, M.D. Abel, and J.B. Seward. 2000. Natriuretic peptide levels in atrial fibrillation: A prospective hormonal and doppler-echocardiographic study. J. Am. Coll. Cardiol. 35: 1256–1262.

Rotter, M., P. Jais, M.C. Vergnes, P. Nurden, Y. Takahashi, P. Sanders, T. Rostock, M. Hocini, F. Sacher, and M. Haïssaguerre. 2006. Decline in C-reactive protein after successful ablation of long-lasting persistent atrial fibrillation. J. Am. Coll. Cardiol. 47: 1231–33.

Sajadieh, A., O.W. Nielsen, V. Rasmussen, H. Ole Hein, and J.F. Hansen. 2006. Increased ventricular ectopic activity in relation to C-reactive protein, and NT-pro-brain natriuretic peptide in subjects with no apparent heart disease. Pacing Clin. Electrophysiol. 29: 1188–94.

Sata, N., N. Hamada, T. Horinouchi, S. Amitani, T. Yamashita, Y. Moriyama, and K. Miyahara. 2004. C-reactive protein and atrial fibrillation: Is inflammation a consequence or cause of atrial fibrillation? Jpn. Heart J. 45: 441–5.

Schoemaker, R.G. and C.L. van Heijningen. 2000. Bradykinin mediates cardiac preconditioning at a distance. Am. J. Physiol. Heart Circ. Physiol. 278: H1571–H1576.

Scott, P.A., J. Barry, P.R. Roberts, and J.M. Morgan. 2009. Brain natriuretic peptide for the prediction of sudden cardiac death and ventricular arrhythmias: a meta-analysis. Eur. J. Heart Fail. 11(10): 958–66. Review.

Shehab, A.M., R.J. MacFadyen, M. McLaren, R. Tavendale, J.J. Belch, and A.D. Struthers. 2004. Sudden unexpected death in heart failure may be preceded by short term, intraindividual increases in inflammation and in autonomic dysfunction: a pilot study. Heart. 90: 1263–8.

Shimano, M., Y. Inden, Y. Tsuji, H. Kamiya, T. Uchikawa, R. Shibata, and T. Murohara. 2008. Circulating homocysteine levels in patients with radiofrequency catheter ablation for atrial fibrillation. Europace 10(8): 961–6.

Silvet, H., Y. Young-Xu, D. Walleigh, and S. Ravid. 2003. Brain natriuretic peptide is elevated in outpatients with atrial fibrillation. Am. J. Cardiol. 92: 1124–1127.

Singer, D.E., G.W. Albers, J.E. Dalen, A.S. Go, J.L. Halperin, and W.J. Manning. 2004. Antithrombotic therapy in atrial fibrillation: the Seventh ACCP Conference on Antithrombotic and Thrombolytic Therapy. Chest 126 (Suppl.): 429S–56.

Siscovick, D.S., T.E. Raghunathan, I. King, S. Weinmann, K.G. Wicklund, J. Albright, V. Bovbjerg, P. Arbogast, H. Smith, L.H. Kushi, L.A. Cobb, M.K. Copass, B.M. Psaty, R. Lemaitre, B. Retzlaff, M. Childs, and R.H. Knopp. 1995. Dietary intake and cell membrane levels of long-chain n-3 polyunsaturated fatty acids and the risk of primary cardiac arrest. JAMA 274: 1363–1367.

Spodick, D.H. 1976. Arrhythmias during acute pericarditis: a prospective study of 100 consecutive cases. JAMA 235: 39–41.

Steger, C., A. Pratter, M. Martinek-Bregel, M. Avanzini, A. Valentin, J. Slany, and C. Stöllberger. 2004. Stroke patients with atrial fibrillation have a worse prognosis than patients: data from the Austrian Stroke registry. Eur. Heart J. 25: 1734–40.

Streitner, F., J. Kuschyk, C. Veltmann, M. Brueckmann, I. Streitner, J. Brade, M. Neumaier, T. Bertsch, B. Schumacher, M. Borggrefe, and C. Wolpert. 2007. Prospective study of interleukin-6 and the risk of malignant ventricular tachyarrhythmia in ICD-recipients–A pilot study. Cytokine. 40: 30–4.

Streitner, F., J. Kuschyk, C. Veltmann, D. Ratay, N. Schoene, I. Streitner, M. Brueckmann, B. Schumacher, M. Borggrefe, and C. Wolpert. 2009. Role of proinflammatory markers and NT-proBNP in patients with an implantable cardioverter-defibrillator and an electrical storm. Cytokine. 47(3): 166–72.

Takezako, T., B. Zhang, T. Serikawa, P. Fan, J. Nomoto, and K. Saku. 2001. The D allele of the angiotensin-converting enzyme gene and reperfusion-induced ventricular arrhythmias in patients with acute myocardial infarction. Jpn. Circ. J. 65(7): 603–9.

Tapanainen, J.M., K.S. Lindgren, T.H. Makikallio, O. Vuolteenaho, J. Leppaluoto, and H.V. Huikuri. 2004. Natriuretic peptides as predictors of non-sudden and sudden cardiac death after acute myocardial infarction in the beta-blocking era. J. Am. Coll. Cardiol. 43: 757–763.

Tousoulis, D., K. Zisimos, C. Antoniades, E. Stefanadi, G. Siasos, C. Tsioufis, N. Papageorgiou, E. Vavouranakis, C. Vlachopoulos, and C. Stefanadis. 2009. Oxidative stress and inflammatory process in patients with atrial fibrillation: the role of left atrium distension. Int. J. Cardiol. 136(3): 258–62.

Verma, A., O.M. Wazni, N.F. Marrouche, D.O. Martin, F. Kilicaslan, S. Minor, R.A. Schweikert, W. Saliba, J. Cummings, J.D. Burkhardt, M. Bhargava, W.A. Belden, A. Abdul-Karim, and A. Natale. 2005. Pre-existent left atrial scarring in patients undergoing pulmonary vein antrum isolation: An independent predictor of procedural failure. J. Am. Coll. Cardiol. 45: 285–292.

Verma, A., F. Kilicaslan, D.O. Martin, S. Minor, R. Starling, N.F. Marrouche, S. Almahammed, O.M. Wazni, S. Duggal, R. Zuzek, H. Yamaji, J. Cummings, M.K. Chung, P.J. Tchou, and A. Natale. 2006. Preimplantation B-type natriuretic peptide concentration is an independent predictor of future appropriate implantable defibrillator therapies. Heart. 92: 190–195.

Verheule, S., E. Wilson, T. Everett IV, S. Shanbhag, C. Golden, and J. Olgin. 2003. Alterations in atrial electrophysiology and tissue structure in a canine model of chronic atrial dilatation due to mitral regurgitation. Circulation 107: 2615–2622.

Wang, T.J., M.G. Larson, D. Levy, E.J. Benjamin, E.P. Leip, T. Omland, P.A. Wolf, and R.S. Vasan. 2004. Plasma natriuretic peptide levels and the risk of cardiovascular events and death. N. Engl. J. Med. 350: 655–663.

Watanabe, E., T. Arakawa, T. Uchiyama, I. Kodama, and H. Hishida. 2005. High-sensitivity C-reactive protein is predictive of successful electrical cardioversion for atrial fibrillation and maintenance of sinus rhythm after conversion. Int. J. Cardiol. 108: 346–53.

Watanabe, H., D.W. Kaiser, S. Makino, C.A. MacRae, P.T. Ellinor, B.S. Wasserman, P.J. Kannankeril, B.S. Donahue, D.M. Roden, and D. Darbar. 2009. ACE I/D polymorphism associated with abnormal atrial and atrioventricular conduction in lone atrial fibrillation and structural heart disease: implications for electrical remodeling. Heart Rhythm. 6(9): 1327–32.

Watson, T., E. Shantsila, and G.Y. Lip. 2009. Mechanisms of thrombogenesis in atrial fibrillation: Virchow's triad revisited. Lancet 373: 155–66.

Wazni, O., D. Martin, N.F. Marrouche, M. Shaaraoui, M.K. Chung, S. Almahameed, R.A. Schweikert, W.I. Saliba, and A. Natale. 2005. C-Reactive protein level and recurrence of atrial fibrillation after electrical cardioversion. Heart 91: 1303–5.

Wijffels, M.C., C.J. Kirchhof, R. Dorland, and M.A. Allessie. 1995. Atrial fibrillation begets atrial fibrillation. A study in awake chronically instrumented goats. Circulation 92: 1954–68.

Xiao, Y.F., A.M. Gomez, J.P. Morgan, W.J. Lederer, and A. Leaf. 1997. Suppression of voltage-gated L-type Ca2+ currents by polyunsaturated fatty acids in adult and neonatal rat ventricular myocytes. Proc. Natl. Acad. Sci. USA 94: 4182–4187.

Xu, C., Y. Lu, Z. Pan, W. Chu, X. Luo, H. Lin, J. Xiao, H. Shan, Z. Wang, and B. Yang. 2007. The muscle-specific microRNAs miR-1 and miR-133 produce opposing effects on apoptosis by targeting HSP60, HSP70 and caspase-9 in cardiomyocytes. J. Cell. Sci. 120: 3045–3052.

Yang, B., H. Lin, J. Xiao, Y. Lu, X. Luo, B. Li, Y. Zhang, C. Xu, Y. Bai, H. Wang, G. Chen, and Z. Wang. 2007. The muscle-specific microRNA miR-1 regulates cardiac arrhythmogenic potential by targeting GJA1 and KCNJ2. Nat. Med. 13: 486–491.

Yu, H., H. Oswald, A. Gardiwal, C. Lissel, and G. Klein. 2007. Comparison of N-terminal pro-brain natriuretic peptide versus electrophysiologic study for predicting future outcomes in patients with an implantable cardioverter defibrillator after myocardial infarction. Am. J. Cardiol. 100: 635–639.

Zhao, Y., J.F. Ransom, A. Li, V. Vedantham, M. von Drehle, A.N. Muth, T. Tsuchihashi, M.T. McManus, R.J. Schwartz, and D. Srivastava. 2007. Dysregulation of cardiogenesis, cardiac conduction, and cell cycle in mice lacking miRNA-1-2. Cell. 129: 303–317.

Zheng, Z.J., J.B. Croft, W.H. Giles, and G.A. Mensah. 2001. Sudden cardiac death in the United States, 1989 to 1998. Circulation 104: 2158–2163.

# CHAPTER 12

# The Role of Biomarkers in Antiplatelet Treatment

Gerasimos Siasos,* Dimitris Tousoulis, Marina Zaromitidou,
Evangelos Oikonomou and Christodoulos Stefanadis

## Introduction

Platelets are anuclear cell fragments produced by megakaryocytes in the bone marrow. Platelet activation and aggregation results in the formation of a platelet plug which is the natural hemostatic response following injuries to the vessel wall. However, the protective mechanism of platelets can lead to harmful results when activated by pathological substrates-erosion or rupture of atherosclerotic plaque. Therefore, ideal antiplatelet treatment should balance prevention of pathological coronary thrombi formation with bleeding tendency. Deciphering the platelet cascade is an essential step towards targeted antiplatelet treatment.

Platelet activation, adhesion and aggregation are complex processes. In order for the platelet hemostatic sequence to begin, a disruption of the endothelial layer is demanded. The exposed subendothelial extracellular matrix (ECM) (containing laminin, fibronectin, collagen, vWF) becomes a substrate for platelets. Specifically, platelets bind to ECM via their receptor GPIb, a connection mediated by vWF. The GPIb-vWF-collagen conjunction is not a firm one and basically serves to decelerate platelets long enough for a stronger connection with collagen to form, via GPIV platelet receptor. The latter is an important mediating receptor of platelet adhesion, activation and aggregation. The GPIV-collagen ligation signals two significant pathways,

1st Cardiology Department, University of Athens Medical School, "Hippokration" Hospital, Athens, Greece.
*Corresponding author: gsiasos@med.uoa.gr

the release of secondary mediators such as adenosine diphosphate (ADP) and thromboxane A2 (TX2) and the turn of integrins into high affinity state (Varga-Szabo et al. 2008). ADP is released from the dense-granules of platelet cytoplasm and mediates platelet activation through its linkage with two G protein-coupled receptors (GPCRs), P2Y12 and P2Y1. TX2 is the product of the enzymatic catalysis of arachidonic acid by cyclooxygenase-1 and binds to TX2 receptor (GPCRs), mediating platelet activation. The third important platelet stimuli, thrombin, is produced from tissue factor after exposure to plasma coagulation factors. Thrombin mediates platelet activation through GPCRs protein-activated receptor (PAR-1/PAR4). ADP, TX2 and thrombin through GPCRs activate platelet shape change, degranulation (positive feedback mechanism) and integrin αIIbβ3 mediated aggregation. Apart from the aforementioned stimuli, platelet GPCRs respond to a number of mediators such as epinephrine, prostaglandin E2, serotonin, lysophosphatidic acid and several chemokines. However, only ADP, TX2 and thrombin are responsible for platelet activation whereas the remaining stimuli facilitate and amplify the action of these three mediators (Offermanns 2006).

Integrins are heterodimeric transmembrane receptors with a key role in platelet adhesion and aggregation. Integrins are namely α2β1, αIIbβ3, α5β1, α6β1, αvβ3. They remain in a low affinity state until platelet activation switches them into a high state. The two most important integrins are α2β1, αIIbβ3. α2β1 binds to collagen whereas αIIbβ3 binds to fibrinogen and vWF-collagen. Thus, crossover bridges are created among platelets and between ECP and platelets, leading to thrombi formation (Varga-Szabo et al. 2008).

Thus, platelets are essential elements of thrombi formation. In addition, studies report a crucial role of platelets in the inflammatory, atherosclerotic process (Lindemann et al. 2007). Therefore, the advantages of a targeted antiplatelet treatment are profound and justify the characterization of the specific regimen as the cornerstone in the treatment of coronary artery disease (Fig. 1).

## Antiplatelet Treatment

Numerous studies support that dual antiplatelet treatment with clopidogrel and aspirin is the gold standard in primary and secondary cardiovascular prevention due to the significant reduction of adverse cardiovascular events (Ridker et al. 1991; Antiplatelet Trialists' Collaboration 1994; CAPRIE Steering Committee 1996; Chen et al. 2005).

Aspirin is mainly absorbed in the stomach and is subject to metabolization by esterases present in the liver and blood. Continuing, ASA irreversibly blocks the isoenzyme cyclooxugenase-1 which catalyzes

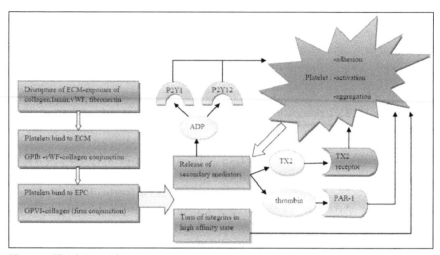

**Figure 1.** Platelet cascade.
**Abbreviations:** ECM: subendothelial extracellular matrix, vWF: von Willebrand factor, ADP: adenosine diphosphate, TX2: thromboxane A2.

the conversion of arachidonic acid into TX2 (Kuliczkowski et al. 2009). On the other hand, clopidogrel after oral intake is absorbed in the intestinal environment, a process mediated by an efflux P-glycoprotein transporter encoded by ABCB1 gene. After absorption, 85 percent of clopidogrel is hydrolyzed by esterases in the blood into an inactive carboxylic acid derivative. The remaining 15 percent of administered prodrug is converted into an active metabolite. This process takes place in the liver and consists of two sequential CYP dependent oxidative steps. The first derivative formed is 2-oxo-clopidogrel. Further oxidation leads to the conversion of the thiophene ring into a carboxyl and a thiol group. The indentified CYP enzymes involved in clopidogrel activation are CYP1A2, CYP2B6, CYP2C9, CYP2C19, CYP3A4 and CYP3A5. The active clopidogrel metabolite exerts its beneficial effects by a selective and irreversible blockage of P2Y12 platelet receptor (Savi et al. 2000; Farid et al. 2007).

However, an important number of patients under dual antiplatelet therapy continue to experience cardiovascular events (Gurbel et al. 2003; O'Donoghue and Wiviott 2006; Angiolillo et al. 2007). This group of patients is categorized as non-responders to antiplatelet treatment and the specific phenomenon as resistance to aspirin and/or clopidogrel. The cardiology research community has dedicated a lot of effort to elucidate the mechanisms responsible for antiplatelet resistance. In general, since platelets are not the only participants in the atherothrombotic process, antiplatelet treatment alone could not intercept this procedure.

As far as aspirin is concerned, resistance is calculated between 6 percent to 60 percent (Lordkipanidze et al. 2007). The mechanisms involved in aspirin resistance could be divided into those concerning bioavailability (non-compliance (Schwartz et al. 2005), under dosing, poor absorption (Sanderson et al. 2005), co-administration of no steroids anti-inflammatory drugs (Catella-Lawson et al. 2001), cigarette smoking (Hung et al. 1995)), platelet activation pathways (increased sensitivity to stimuli such as ADP, epinephrine or collagen (Mason et al. 2005; Sanderson et al. 2005), extra-COX-1 production of TX2 (Cipollone et al. 2000)), and genetic polymorphisms (OX-1,P2Y1,integrins (Cambria-Kiely and Gandhi 2002; Jefferson et al. 2005)).

Regarding clopidogrel resistance, the estimated rate varies from 4 percent to 30 percent. Non-compliance, poor absorption, drug-to-drug interactions, inadequate dosing, elevated body mass index, insulin resistance, and the nature of ACS are clinical factors that may contribute to low clopidogrel response (Angiolillo and Alfonso 2007; Tousoulis et al. 2009). Other parameters that might influence therapy outcomes are accelerated platelet turnover (Michos et al. 2006), platelet number (Giugliano and Braunwald 2005), increased ADP exposure (Wiviott and Antman 2004; Nguyen et al. 2005), upregulation of P2Y12/P2Y1 or non-P2Y dependent pathways (Angiolillo et al. 2007), and post-receptor signaling pathway factors (Lau et al. 2004; Matetzky et al. 2004; Bates and Lau 2005; Guthikonda et al. 2005; Michos et al. 2006; Wang et al. 2006). In addition, genetics polymorphisms encoding for enzymes that participate in the process of clopidogrel's hepatic metabolism, or for receptors involved in intestinal absorption and ADP induced platelet aggregation may affect the percentage of platelet inhibition after clopidogrel's administration (Fontana et al. 2003; Angiolillo et al. 2004; Angiolillo et al. 2006; Angiolillo et al. 2007; Brandt et al. 2007; Giusti et al. 2007; Tousoulis et al. 2009). In addition to efforts aimed at elucidating the mechanisms of resistance, new therapeutic approaches are introduced with the hope that similar drawbacks will not be encountered (Tousoulis et al. 2009).

The new generation of antiplatelet drugs is a thienopyridine, prasugrel and a non-thienopyridine P2Y12 receptor antagonist, ticagrelor. Prasugrel is a third thienopyridine agent whose pharmacokinetic is slightly different than that of clopidogrel. Prasugrel is rapidly hydrolyzed into a thiolactone and consequently coverts into its active form by cytochrome P450 enzymes, namely CYP3A4, CYP2B6 and to a lesser extent by CYP2C19, CYP2C9, in one activation step (Jakubowski et al. 2006). Ticagrelor on the other hand is part of a new class of agents known as cyclopentyltriazolopyrimidines. It does not require transformation and binds directly to P2Y12 receptor.

Consequently, the identification of non-responders is of utmost importance in order to classify high risk patients. Biomarkers monitoring platelet resistance are essential for the selection of the proper individualized therapeutic approach that will ensure secondary prevention.

## Platelet Function Essays

A significant number of studies acknowledge high-on-treatment-platelet reactivity (HTPR) as a biomarker for drug resistance and the occurrence of clinical outcomes. Platelet function assays were developed in order to estimate HTPR (Matetzky et al. 2004; Lev et al. 2006; Buonamici et al. 2007) (Table 1).

### *Light transmittance aggregometry (LTA)*

Light transmittance aggregometry is considered the gold standard method for the evaluation of platelet function. It is based on the change in light transmittance when platelet aggregates form, following the induction of agonists. Depending on the agonist used, different agents can be evaluated (ADP for thienopyridines, arachidonic acid for aspirin, TRAR for change in light transmittance when platelet aggregates in a dose-dependent way. Recent data support the significance of LTA in the prediction of recurrent cardiovascular events (Parodi et al. 2011). However, LTA exhibits certain disadvantages such as the increased cost, the high sample volume and the need for trained personnel. In addition, LTA is a time consuming method with low reproducibility due to lack of standardization and the dependence on technician's skills (Capodanno and Angiolillo 2009; Seidel et al. 2011).

### *Vasodilator-stimulated phosphoprotein (VASP)*

VASP is a flow cytometry-based method, particularly sensitive to P2Y12 inhibitor agents. Studies indicate VASP as a useful predictive tool of adverse cardiovascular events in patients after PCI (El Ghannudi et al. 2010). Nevertheless, VASP remains an expensive, time-consuming method whose operation requires trained technicians, characteristics that deprive the title of point of care assay (Capodanno and Angiolillo 2009; Seidel et al. 2011).

### *Multiple electrode aggregometry (MEA)*

The specific device evaluates platelet aggregation after the addition of agonists by measuring the change in electrical impedance caused by the adhesion of platelets on the sensor surface of two silver-coated copper

**Table 1.** Platelet Function Assays.

| Assay | Sensitivity to agent | Function Principle | Advantages | Disadvantages |
|---|---|---|---|---|
| Light Transmittance Aggregometry (LTA) | Aspirin Thienopyridines GPIIbIIIa-inhibitors | Change in light transmittance due to platelet aggregation | Gold standard in resistance evaluation. Evaluation of a wide spectrum of antiplatelet agents. | Expensive. Time consuming Low reproducibility. Skilled technicians Large sample volume. |
| Vasodilator-stimulated phosphoprotein (VASP) | P2Y12 Inhibitors | Flow cytometry | Low sample volume | Expensive. Time consuming. Need for flow cytometer and skilled personnel. |
| Multiple electrode Aggregometry (MEA) | Aspirin Clopidogrel GPIIbIIIa-inhibitors | Electrical impedance | Relatively rapid and easy measurements. Small sample volume. | Pippeting Expensive. |
| Platelet Function Analyzer (PFA 100) | Aspirin | Closure time | Rapid. Easy. Small sample volume. | Results depend on Hct. |
| INNOVANCE PFA P2Y | Clopidogrel | Closure time | Rapid. Easy. Small sample volume. | Need for studies to validate its predictive value. |
| Verify Now | Aspirin Thienopyridines GPIIbIIIa-inhibitors | Change in light transmittance due to platelet aggregation. | Rapid. Easy. Small sample volume. | Results depend on Hct. |
| Plateletworks | GPIIbIIIa-inhibitors | Difference in platelet number before and after the addition of agonists. | Easy. Rapid. | Time dependent. |
| Impact cone and plate(let) Analyzer (Impact-R) | Aspirin | Shear-induced platelet adhesion | Rapid. Easy. Evaluats Shear-induced platelet adhesion. | Not available for clopidogrel evaluation. |
| Platelet surface P-selectin and GP IIb/IIIa, leukocyte-platelet aggregates | Aspirin Clopidogrel | Flow cytometry | Rapid. Small sample volume. | Expensive. Need for a flow cytometer and skilled personnel. |
| Teg Platelet Mapping System | Aspirin Clopidogrel | Thromboelastography | Provides information about the coagulation process. | Pipeting. |

electrodes (Seidel et al. 2011). Studies indicate that MEA, in addition to having a predictive value for early stent thrombosis, seems a more reliable evaluation method (Sibbing et al. 2009; Siller-Matula et al. 2010). Thus, MEA emerges as a more rapid and easy evaluation method.

### Platelet function analyzer (PFA-100)

The PFA-100 consists of a cartridge whose membrane is coated with collagen and epinephrine or collagen and adenosine diphosphate. The presence of an aperture simulates an injured vessel and after the addition of agonists, the required time for platelets to form a clot closing the aperture is calculated. Unfortunately, PFA-100 was found to have a poor correlation with clopidogrel resistance and thus its use is restricted to aspirin measurements. Recently INNOVANCE PFA P2Y was manufactured to assess clopidogrel resistance. Further studies are needed to verify the credibility of this newly developed assay (Seidel et al. 2011).

### VerifyNow

VerifyNow is probably the most widely used test because its characteristics classify it as a point of care system. It is a rapid and easy test to perform by unskilled personnel. For the measurement, whole blood is used without involving centrifugation or pipeting. It evaluates platelet aggregation by calculating the change in light transmittance. Depending on the agonists used, different antiplatelet agents can be evaluated. Large trials, including GRAVITAS indicate VerifyNow as a predictor marker for recurrent cardiovascular events in patients 30 days after PCI (Price et al. 2011a).

### Plateletworks

Plateletworks is a point of care test, the use of which is restricted in the evaluation of GPIIa-IIIb efficacy. It is designed to work in steps, each step using a separate sample. In both steps, total platelet number is calculated. The only difference is that in the second step an agonist is added to the sample, leading functional platelets to aggregate. Platelets involved in aggregates are not included in the total platelet count. The difference in platelet number between the two samples equals the percent of platelet aggregation. Though Plateworks is an easy and rapid test, its main disadvantage is time-dependency—the sample must be processed in a maximum of ten minutes. As regards Plateworks, more studies should be conducted in order to determine its predictive value (Seidel et al. 2011).

## *Impact cone and plate(let) Analyzer (Impact-R)*

This device was named after the shape and design of its components, a cone and a plate. The manufacturer attempts to simulate Impact-R with vessel flow. Shear-induced platelet adhesion is achieved followed by a washing step. Thereafter, a qualified analyzer calculates the percentage of surface platelet aggregates. Although Impact-R is a rapid and easy evaluation method, studies indicate that it is not suitable for clopidogrel monitoring (van Werkum et al. 2010).

## *Platelet surface P-selectin and GPIIb/IIIa, leukocyte-platelet aggregates*

The quantification of P-selectin receptor, GPIIb/IIa receptor and leukocyte-platelet aggregates is achieved by flow cytometry. The results reflect platelet activation status. Although the technique requires a small sample volume of whole blood, its use is limited by the high cost and the need for a flow cytometre and skilled personnel (Gurbel et al. 2007).

## *Teg platelet mapping system*

The thromboelastograph mapping system measures aspirin and clopidogrel resistance by using arachidonic acid and ADP induced aggregation, respectively. TEG is also equipped with an analytical software providing information on the coagulation process. However, it cannot be considered as a point of care assay due to the pippeting involved (Capodanno and Angiolillo 2009).

The table below summarizes the characteristics of the platelet function assays (Table 1).

## Genetic Biomarkers

Studies have highlighted the need for a more individualized antiplatelet treatment. Research has focused, among others, on the genetic substrate of patients in order to elucidate the interindividual response. The pharmacodynamic and pharmacokinetic of antiplatelet drugs implicates a series of proteins, playing an essential and irreplaceable role either as a transporter, an activator enzyme, or a receptor. These proteins are encoded by certain genes, whose polymorphisms can alter the bioavailability and the effectiveness of the drug. Therefore, the targeted genetic mapping can provide additional information regarding the response to a specific treatment. Every pharmaceutical substance has its unique pharmacokinetic and therefore the genetic monitoring differs according to the regimen administrated (Tousoulis et al. 2011).

## *P2Y12 receptor antagonists and genetics biomarkers*

The P2Y12 receptor antagonists that are used in clinical practice are the thienopyridines clopidogrel and prasugrel and the non-thienopyridine ticagrelor.

A large amount of data regarding the association between clopidogrel resistance and genetic polymorphisms is available, since clopidogrel dominated the field of antiplatelet treatment during the last decade. Clopidogrel efficacy could theoretically be altered if even one of the following key proteins become dysfunctional—the efflux P-glycoprotein transporter, the CYP450 enzymes involved in activation steps, and the P2Y12 receptor.

The efflux P-glycoprotein transporter, responsible for the intestinal absorption of clopidogrel prodrug, is encoded by ABCB1 gene. According to a study by Simon et al. the presence of 3435C > T ABCB1 polymorphism is associated with high risk of adverse cardiovascular events a year after PCI (Simon et al. 2009). In accordance, Mega et al. reported that homozygote carriers of TT ABCB1 polymorphism exhibit reduced platelet inhibition and are at increased risk of adverse cardiovascular event. This effect was more evident when the ABCB1 TT gene was combined with CYP2C19 loss of action polymorphism (Mega et al. 2010). A recent study supports the positive correlation between homozygous carriers of ABCB1 TT polymorphism and the occurrence of early stent thrombosis (Cayla et al. 2011). However, reported results question the contribution of ABCB1 gene in the genetic monitoring of clopidogrel resistance. In the GIFT study, no association was observed between ABCB1 and cardiovascular events (Price et al. 2012).

The P450 cytochrome enzymes are a major oxidative system for endogenous compounds as well as for administrated pharmaceutical substances. The P450CYP enzymes involved in the two oxidation steps of clopidogrel are CYP1A2, CYP2B6, CYP2C9, CYP2C19, CYP3A4 and CYP3A5. The polymorphic CYP2C19 exhibits many variants (1*,2*,3*,4*, 5*,6*,7*,8*,9*,10*,12*13*,14*,17*) coding for either a functional enzyme, a reduced function enzyme or even an enhanced activity enzyme (2C19*17). Polymorphisms in the CYP2C19 gene are encountered in 30 percent of Caucasians, 40 percent of blacks and 55 percent of Asians. Among the loss of function variants, the most frequent is CYP2C19*2 (95%). Regarding clopidogrel's activation process, CYP2C19 participates in both oxidation steps. The CYP2C19 polymorphisms emerge as the main genetic regulator of clopidogrel pharmacodynamic response. Reported data indicate that the antiplatelet effect of clopidogrel *in vivo* is related to the levels of the active metabolite. Therefore, the presence of a reduced variant (CYP2C19*2) could be translated into an inadequate response to clopidogrel due to attenuated formation of the active metabolite (Tousoulis et al. 2011). Laboratory studies conducted on healthy individuals or CAD patients emphasize that the presence of CYP2C19*2 allele is associated with low metabolite levels and high-on-treatment-platelet reactivity (Fontana et al. 2003; Hulot et al. 2006). Consequently, large clinical studies support that the presence of CYP2C19*2 predisposes to clopidogrel resistance, the latter translated in the occurrence of recurrent cardiovascular events including stent thrombosis (Collet et al. 2009; Giusti et al. 2009; Mega et al. 2009a; Simon et al. 2009). However, published studies support opposite results and underline that CYP2C19 polymorphisms have no impact on the rate of adverse cardiovascular outcomes (Pare et al. 2010; Holmes et al. 2011). Therefore, the role of CYP2C19 genetics determinants in the issue of clopidogrel resistance is controversial and warrants further investigation.

The aforementioned studies referred to CYP2C19 loss of action alleles. The CYP2C19*17 polymorphism is alleged to have the exact opposite effect, an enhanced enzyme activity. The conducted studies evaluating CYP2C19*17 effects are limited and rather controversial. Specifically, in a rather small population study, the PRINC Analysis, 2C19*17 allele was interestingly associated with lower platelet inhibition (Gladding et al. 2008). On the contrary, Sibbing et al. reported an enhanced clopidogrel result in carriers of 2c19*17 polymorphism accompanied with increased risk of bleeding, unfortunately without any benefit in stent thrombosis rate (Sibbing et al. 2010a). Additionally, a study conducted in patients with stable CAD emphasizes on the significance of gene to gene interactions and in particular CYP2C19*2 and CYP2C19*17. It is reported that the loss of action effect of 2c19*2 allele is being balanced by the enhanced activity of 2c19*17 when the two alleles co-exist (Sibbing et al. 2010b).

The CYP2C9 polymorphisms, particularly 2c9*3 variant, seem to affect the pathogenesis of stent thrombosis. According to a study by Harmsze et al., carriers of CYP2C9*3 exhibit a 2.4-fold risk of developing stent thrombosis (Harmsze et al. 2010a). The latter study verifies the results of a previous research protocol reporting a four-fold risk of high-on-treatment-platelet reactivity (Harmsze et al. 2010b).

On the other hand, Mega et al. report no correlation between 2C9*3 alleles and stent thrombosis although study results were limited by the small population size (Mega et al. 2009a).

As regards the rest, P450 CYP enzymes involved in the activation process studies document no significant correlation between CYP1A2, CYP3A4, CYP3A5 and CYP2B6 polymorphisms and the laboratory or clinical response to clopidogrel (Fontana et al. 2003; Hulot et al. 2006;

Brandt et al. 2007; Giusti et al. 2007; Mega et al. 2009a; Simon et al. 2009; Varenhorst et al. 2009).

The P2Y12 receptor plays a key role in platelet activation and aggregation. The two haplotypes studied, out of the five polymorphisms discovered, are H1 and H2. Although some data exist connecting P2Y12 genetic variants with clopidogrel resistance[74], the majority of studies report no such correlation (Angiolillo et al. 2005; Giusti et al. 2007; Zuern et al. 2010).

The new P2Y12 receptor antagonists prasugrel and ticagrelor exert their antiplatelet properties independent of genetic polymorphisms. Prasugrel activation is a one step process which involves a different CYP enzymatic group, compared to clopidogrel. Ticagrelor exhibits different pharmacokinetic in comparison to clopidogrel. This novel agent does not require activation and it binds directly to the P2Y12 receptor. Although the data is still limited, researchers report no effect of genetic polymorphisms on the efficacy of prasugrel and ticagrelor (Mega et al. 2009a; Tantry et al. 2010; Wallentin et al. 2010).

## Aspirin genetic biomarkers

The studies designed to decipher the genetic profile of aspirin resistance revolved mainly around the COX-1, P2Y1 receptor and GPIV polymorphisms. The first results presented a positive association of aspirin resistance and the aforementioned gene polymorphisms (Lepantalo et al. 2006). However, further research did not verify the connection of aspirin efficacy and genotype, with perhaps the exception of P2RY1 1622A/G polymorphism, which researchers collate with impaired platelet inhibition and increased risk of recurrent cardiovascular events (Pettinella et al. 2009; Lordkipanidze et al. 2011).

Consequently, the data regarding the role of genetic substrates in antiplatelet treatment response are obscure. There is a respectable number of studies attributing the increased risk of ischemic attacks to genetic factors and encouraging the use of genetic monitoring in order to overcome antiplatelet treatment resistance. On the other hand, studies announcing no genetic impacts to antiplatelet response deprive genetic polymorphism of the properties of a point of care biomarker.

However, due to the considerable amount of data, the FDA announced a 'Box Warning' concerning the possibility of attenuated clopidogrel response in a certain subgroup of patients carrying defective alleles. The 2011 ACCF/ AHA guidelines comment on this controversial issue and remark that "The Class IIb classification suggests that a selective, limited approach to platelet genotype assessment and platelet function testing is the more prudent course until better clinical evidence exists for us to provide a more scientifically derived recommendation" (Wright et al. 2011).

## Serum Biomarkers

Antiplatelet resistance impelled medical research in a quest for novel biochemical markers. Studies concentrated on end-stage products of Thromboxane A2 metabolism and known inflammatory markers (Table 2).

### Thromboxane B2 (TXB2)

Serum Thromboxane B2 is the stable hydrolytic metabolite of Thromboxane A2. Measurement of TXB2 provides information about the pharmacodynamic effect of aspirin and can be used to establish patient compliance or the pharmacodynamic interactions with other non-steroidal anti-inflammatory drugs. Values under 2 ng/ml qualify aspirin efficacy. However, TXB2 levels can be altered by *ex vivo* platelet activation during blood sample collection and processing of results. Unfortunately, according to studies, TBX2 values do not correlate with high-on-treatment-platelet reactivity and so it is considered an indirect marker of platelet activation (Fontana et al. 2010; Seidel et al. 2011).

### 11-dehydro-thromboxane B2

Urinary 11-dehydro-thromboxane B2 is a metabolite of TXB2 and hence its contribution in monitoring aspirin resistance is characterized by approximately the same benefits and limitation as TXB2 measurement. Compared to TXB2, the urinary metabolite has a larger life span and its results are not affected by the technique. However, 11-dehydro-thromboxane B2 levels also include extra platelet production of thromboxane and the admeasured values depend on the renal production of the metabolite. In addition, the interpretation of 11-dehydro-thromboxane B2 results is further complicated by the fact that there are no standardized clinical cut off values (Gasparyan et al. 2008; Kuliczkowski et al. 2009).

### Soluble P-selectin (sP-selectin)

P-selectin is an adhesion glycoprotein stored in platelet α-granules and endothelial cell Weibel-Palade bodies. P-selectin participates in the inflammation processes and thrombi formation via platelet activation and aggregation. Exposure to stimuli, such as thrombin, results in degranulation

**Table 2.** Serum biomarkers.

| Biomarker | Measured substance | Advantages | Disadvantages |
|---|---|---|---|
| Thromboxane B2 (TXB2) | Hydrolytic metabolite of thromboxane A2 | -Pharmacodynamic effect of aspirin<br>-Interactions with other non- steroidals anti inflammatory drugs | -Levels can be alter by *ex vivo* platelet activation during blood sample collection and processing of results.<br>- No correlation with HTPR |
| 11-dehydro-thromboxane B2 | Metabolite of Thromboxane A2 | Pharmacodynamic effect of aspirin | -Extra platelet production of thromboxane<br>-Values depend on the renal production of the metabolite<br>-No standardized clinical cut off values |
| Soluble P-selectin (sP-selectin) | Soluble form of adhesion platelet surface glycoprotein P-selectin | Significant correlation between sp-selectin levels and LTA values | Limited studies |
| Soluble form of CD40 Ligand (sCD40L) | Soluble form of adhesion platelet CD40 Ligand | Platelet activation marker | -Sample procedure<br>-Limited studies |

and finally in the translocation of P-selectin to platelet membrane. P-selectin is also found in a soluble form in the blood circulation after its secretion from activated platelets. Therefore, P-selectin is considered a marker of platelet activation (Barnes et al. 2008).

The connection between P-selectin and platelet activation led to the design of studies assessing the role of P-selectin as a biomarker of antiplatelet resistance. In specific, a small population study conducted in CAD patients reported a significant correlation between sP-selectin levels and LTA values, which is considered the gold standard in evaluating platelet resistance (Ferroni et al. 2009). The same group of investigators verified their former results in a second study on aged-healthy volunteers (Riondino et al. 2010). Nevertheless, the data regarding sP-selectin's role in antiplatelet monitoring are limited and further research is warranted.

## Soluble form of CD40 Ligand (sCD40L)

The CD40 ligand is stored in the α-granules of platelets. The exposure of platelets to stimulating factors (collagen, thrombin) leads to platelet activation and translocation of CD40L to platelet membrane. Continually, CD40L is cleaved and shed from platelet membrane into blood circulation. Thus, sCD40L could be considered as a platelet activation marker since almost 95 percent of circulating sCD40L is secreted by activated platelets (Heeschen et al. 2003; Antoniades et al. 2009). Ge et al. tested the credibility of sCD40L as a biomarker for monitoring clopidogrel resistance in 352 patients with Drug Eluting Stent implantation. According to the results, sCD40L presented a positive association with platelet inhibition rates. In addition, researchers noted that increased elevated sCD40L admission levels worked as a prognostic marker of clopidogrel non-response (Ge et al. 2012). On the contrary, a recent study on PCI patients on clopidogrel treatment recognized no correlation between sCD40L levels and platelet inhibition (Osmancik et al. 2012).

As regards sample procedure, some clarification should be given in order to eliminate false results. Blood collection time should be carefully selected as sCD40L levels are subject to diurnal variation (Dominguez-Rodriguez et al. 2007).The sample should be analyzed without delays and plasma sample should be preferred as it eliminates *in vitro* platelet activation issue (Varo et al. 2006; Rondina et al. 2008).

It is evident that deciphering the role of sCD40L in platelet resistance monitoring requires a larger number of studies.

Recent studies focused on illuminating the association between platelet unresponsiveness and inflammatory status. The data that led to the previous assumption came from studies announcing that the decrease rate of inflammatory markers depended on the increase rate of platelet inhibition

(Gurbel et al. 2006; Vavuranakis et al. 2006). On those grounds, certain inflammatory markers were tested for monitoring antiplatelet treatment with controversial results. Specifically, a study reports positive correlation of IL-10 with high-on-treatment-platelet reactivity and clopidogrel resistance (Osmancik et al. 2012). In a group of stable CAD patients on clopidogrel regimen, CRP levels were associated with high platelet reactivity (Bernlochner et al. 2010).

The aforementioned studies, however, examined only laboratory resistance. The interaction between antiplatelet treatment and inflammatory status and the possible usefulness of known inflammatory markers in resistance monitoring is yet to be discovered.

## Biomarkers and Drug Resistance

Platelet activation and thrombus formation is a complex process involving multiple pathways. Therefore, targeted inhibition of a single pathway is not able to eradicate the risk of adverse cardiovascular events. Resistance to a specific regimen is more properly defined as the failure to inhibit the established target. However, antiplatelet treatment resistance is determined mostly by its effect on platelet activation and not by the pharmacodynamic response to the administrated agent. Up to date biomarkers for platelet monitoring are platelet function assays, biochemical and genetic biomarkers. Platelet function assays based on HTPR evaluation and biochemical markers are the more frequently used methods in designed studies. Certain issues must be clarified before biomarker monitoring enters everyday practice. Firstly, an agreement must be reached considering the method that best evaluates the risk profile of patients. In addition, the validated assay must be easy to perform, with reproducibility, low cost, and rapid results. It is also of extreme importance to determine consensus cut off values. Secondly, a tailored antiplatelet treatment according to monitoring results should decrease the risk of cardiovascular outcomes. Studies report a decrease in cardiovascular events, including stent thrombosis, when an increased loading dose of clopidogrel is administrated in non-responsive patients (Gurbel et al. 2005a; Gurbel et al. 2007).

On the contrary, GRAVITAS trial, conducted on selected HTPR PCI with DES implantation patients, observed no improvement in morbidity and mortality rates irrespective of high daily doses of clopidogrel (Price et al. 2011b).

Genetic monitoring attempted to identify patients in high risk by translating their DNA sequence. ELEVATE TIMI56 study adjusted the clopidogrel daily dose according to CYP2C19*2 carriage. In heterozygous carriers a clopidogrel dose of 225mg/day, decreased platelet reactivity levels to that observed in non-carriers. Unfortunately, elevated clopidogrel doses up to 300mg/day did not alter platelet reactivity in carriers homozygous for 2C19*2 allele (Mega et al. 2011; Siasos et al. 2011; Siasos et al. 2012). The RAPID TEST study conducted in patients after PCI, using a point of care genetic system, reports that carriers of 2c19*2 treated with prasugrel do not exhibit platelet resistance (Roberts et al. 2012). Thus, genetic monitoring is a promising perspective, requiring further studies for validation.

Platelet activation is the only parameter estimated for the evaluation of antiplatelet treatment response. In the PREPARE POST STENTING study, both LTA and TEG were used to evaluate antiplatelet treatment status in patients after PCI. Thromboelastography, by measuring properties of thrombin clot in addition to platelet activation values, was a more accurate marker of cardiovascular events compared to LTA (Gurbel et al. 2005b; Gurbel et al. 2007). Thus, platelet monitoring should be enriched with thrombin related markers.

In conclusion, a precise and documented estimation of antiplatelet treatment response should follow a pattern similar to PREDICT RISK score: a combination of clinical factors, laboratory measurements, and specific gene mapping (Seidel et al. 2011).

## Conclusion

The introduction of antiplatelet treatment in everyday practice is considered a breakthrough in the management of secondary cardiovascular prevention. Aspirin was the first antiplatelet agent used, with encouraging but not desirable effects. The second important step in ACS treatment was the co-administration of aspirin with clopidogrel which improved cardiovascular morbidity and mortality. However, a significant number of patients do not benefit from the dual antiplatelet treatment and are characterized as non-responders. The term "resistance" to antiplatelet treatment entered clinical practice in order to describe the percentage of on-treatment patients who experience an adverse cardiovascular event. The need for identifying non-responders in time is imperative. Increased drug dosages or replacement of standard treatment with novel antiplatelet agents are the alternatives in order to overcome resistance. The contribution of biomarkers monitoring antiplatelet response is valuable. An effective biomarker should gather specific characteristics such as reproducibility, low expense, easy measurements, and rapid results. Unfortunately, none of the available biomarkers are composed of all these desirable properties. Therefore, further research for an ideal biomarker capable of recognizing platelet response is an ongoing process.

# References

Angiolillo, D.J., A. Fernandez-Ortiz, E. Bernardo, F. Alfonso, M. Sabate, C. Fernandez, C. Stranieri, E. Trabetti, P.F. Pignatti, and C. Macaya. 2004. PlA polymorphism and platelet reactivity following clopidogrel loading dose in patients undergoing coronary stent implantation. Blood Coagul. Fibrinolysis 15: 89–93.

Angiolillo, D.J., A. Fernandez-Ortiz, E. Bernardo, C. Ramirez, U. Cavallari, E. Trabetti, M. Sabate, P. Jimenez-Quevedo, R. Hernandez, R. Moreno, J. Escaned, F. Alfonso, C. Banuelos, M.A. Costa, T.A. Bass, P.F. Pignatti, and C. Macaya. 2005. Lack of association between the P2Y12 receptor gene polymorphism and platelet response to clopidogrel in patients with coronary artery disease. Thromb. Res. 116: 491–497.

Angiolillo, D.J., A. Fernandez-Ortiz, E. Bernardo, C. Ramirez, U. Cavallari, E. Trabetti, M. Sabate, R. Hernandez, R. Moreno, J. Escaned, F. Alfonso, C. Banuelos, M.A. Costa, T.A. Bass, P.F. Pignatti, and C. Macaya. 2006. Contribution of gene sequence variations of the hepatic cytochrome P450 3A4 enzyme to variability in individual responsiveness to clopidogrel. Arterioscler. Thromb. Vasc. Biol. 26: 1895–1900.

Angiolillo, D.J. and F. Alfonso. 2007. Platelet function testing and cardiovascular outcomes: steps forward in identifying the best predictive measure. Thromb. Haemost. 98: 707–709.

Angiolillo, D.J., A. Fernandez-Ortiz, E. Bernardo, F. Alfonso, C. Macaya, T.A. Bass, and M.A. Costa. 2007. Variability in individual responsiveness to clopidogrel: clinical implications, management, and future perspectives. J. Am. Coll. Cardiol. 49: 1505–1516.

Antiplatelet Trialists' Collaboration. 1994. Collaborative overview of randomised trials of antiplatelet therapy—I: Prevention of death, myocardial infarction, and stroke by prolonged antiplatelet therapy in various categories of patients. Antiplatelet Trialists' Collaboration. BMJ 308: 81–106.

Antoniades, C., C. Bakogiannis, D. Tousoulis, A.S. Antonopoulos, and C. Stefanadis. 2009. The CD40/CD40 ligand system: linking inflammation with atherothrombosis. J. Am. Coll. Cardiol. 54: 669–677.

Barnes, A., J. Wheat, and C. Milner. 2008. Association between foot type and tibial stress injuries: a systematic review. Br. J. Sports Med. 42: 93–98.

Bates, E.R. and W.C. Lau. 2005. Controversies in antiplatelet therapy for patients with cardiovascular disease. Circulation 111: e267–271.

Bernlochner, I., S. Steinhubl, S. Braun, T. Morath, J. Jaitner, J. Stegherr, J. Mehilli, N. von Beckerath, A. Schomig, A. Kastrati, and D. Sibbing. 2010. Association between inflammatory biomarkers and platelet aggregation in patients under chronic clopidogrel treatment. Thromb. Haemost. 104: 1193–1200.

Brandt, J.T., S.L. Close, S.J. Iturria, C.D. Payne, N.A. Farid, C.S. Ernest, 2nd, D.R. Lachno, D. Salazar, and K.J. Winters. 2007. Common polymorphisms of CYP2C19 and CYP2C9 affect the pharmacokinetic and pharmacodynamic response to clopidogrel but not prasugrel. J. Thromb. Haemost. 5: 2429–2436.

Buonamici, P., R. Marcucci, A. Migliorini, G.F. Gensini, A. Santini, R. Paniccia, G. Moschi, A.M. Gori, R. Abbate, and D. Antoniucci. 2007. Impact of platelet reactivity after clopidogrel administration on drug-eluting stent thrombosis. J. Am. Coll. Cardiol. 49: 2312–2317.

Cambria-Kiely, J.A., and P.J. Gandhi. 2002. Aspirin resistance and genetic polymorphisms. J. Thromb. Thrombolysis 14: 515–518.

Capodanno, D. and D.J. Angiolillo. 2009. Platelet monitoring for PCI: which test is the one to choose? Hamostaseologie 29: 376–380.

CAPRIE Steering Committee. 1996. A randomised, blinded, trial of clopidogrel versus aspirin in patients at risk of ischaemic events (CAPRIE). CAPRIE Steering Committee. Lancet 348: 1329–1339.

Catella-Lawson, F., M.P. Reilly, S.C. Kapoor, A.J. Cucchiara, S. DeMarco, B. Tournier, S.N. Vyas, and G.A. FitzGerald. 2001. Cyclooxygenase inhibitors and the antiplatelet effects of aspirin. N. Engl. J. Med. 345: 1809–1817.

Cayla, G., J.S. Hulot, S.A. O'Connor, A. Pathak, S.A. Scott, Y. Gruel, J. Silvain, J.B. Vignalou, Y. Huerre, A. de la Briolle, F. Allanic, F. Beygui, O. Barthelemy, G. Montalescot, and J.P. Collet. 2011. Clinical, angiographic, and genetic factors associated with early coronary stent thrombosis. JAMA 306: 1765–1774.

Chen, Z.M., L.X. Jiang, Y.P. Chen, J.X. Xie, H.C. Pan, R. Peto, R. Collins, and L.S. Liu. 2005. Addition of clopidogrel to aspirin in 45,852 patients with acute myocardial infarction: randomised placebo-controlled trial. Lancet 366: 1607–1621.

Cipollone, F., G. Ciabattoni, P. Patrignani, M. Pasquale, D. Di Gregorio, T. Bucciarelli, G. Davi, F. Cuccurullo, and C. Patrono. 2000. Oxidant stress and aspirin-insensitive thromboxane biosynthesis in severe unstable angina. Circulation 102: 1007–1013.

Collet, J.P., J.S. Hulot, A. Pena, E. Villard, J.B. Esteve, J. Silvain, L. Payot, D. Brugier, G. Cayla, F. Beygui, G. Bensimon, C. Funck-Brentano, and G. Montalescot. 2009. Cytochrome P450 2C19 polymorphism in young patients treated with clopidogrel after myocardial infarction: a cohort study. Lancet 373: 309–317.

Dominguez-Rodriguez, A., P. Abreu-Gonzalez, M.J. Garcia-Gonzalez, and J.C. Kaski. 2007. Soluble CD40 ligand:interleukin-10 ratio predicts in-hospital adverse events in patients with ST-segment elevation myocardial infarction. Thromb. Res. 121: 293–299.

El Ghannudi, S., P. Ohlmann, N. Meyer, M.L. Wiesel, B. Radulescu, M. Chauvin, P. Bareiss, C. Gachet, and O. Morel. 2010. Impact of P2Y12 inhibition by clopidogrel on cardiovascular mortality in unselected patients treated by percutaneous coronary angioplasty: a prospective registry. JACC Cardiovasc. Interv. 3: 648–656.

Farid, N.A., C.D. Payne, D.S. Small, K.J. Winters, C.S. Ernest, 2nd, J.T. Brandt, C. Darstein, J.A. Jakubowski, and D.E. Salazar. 2007. Cytochrome P450 3A inhibition by ketoconazole affects prasugrel and clopidogrel pharmacokinetics and pharmacodynamics differently. Clin. Pharmacol. Ther. 81: 735–741.

Ferroni, P., F. Martini, S. Riondino, F. La Farina, A. Magnapera, F. Ciatti, and F. Guadagni. 2009. Soluble P-selectin as a marker of *in vivo* platelet activation. Clin. Chim. Acta. 399: 88–91.

Fontana, P., P. Berdague, C. Castelli, S. Nolli, I. Barazer, P. Fabbro-Peray, J.F. Schved, H. Bounameaux, F. Mach, D.E.M.P, and J.L. Reny. 2010. Clinical predictors of dual aspirin and clopidogrel poor responsiveness in stable cardiovascular patients from the ADRIE study. J. Thromb. Haemost. 8: 2614–2623.

Fontana, P., A. Dupont, S. Gandrille, C. Bachelot-Loza, J.L. Reny, M. Aiach, and P. Gaussem. 2003. Adenosine diphosphate-induced platelet aggregation is associated with P2Y12 gene sequence variations in healthy subjects. Circulation 108: 989–995.

Gasparyan, A.Y., T. Watson, and G.Y. Lip. 2008. The role of aspirin in cardiovascular prevention: implications of aspirin resistance. J. Am. Coll. Cardiol. 51: 1829–1843.

Ge, H., Y. Zhou, X. Liu, X. Nie, Z. Wang, Y. Guo, W. Chen, and Q. Yang. 2012. Relationship between plasma inflammatory markers and platelet aggregation in patients with clopidogrel resistance after angioplasty. Angiology 63: 62–66.

Giugliano, R.P. and E. Braunwald. 2005. The year in non-ST-segment elevation acute coronary syndromes. J. Am. Coll. Cardiol. 46: 906–919.

Giusti, B., A.M. Gori, R. Marcucci, C. Saracini, I. Sestini, R. Paniccia, S. Valente, D. Antoniucci, R. Abbate, and G.F. Gensini. 2007. Cytochrome P450 2C19 loss-of-function polymorphism, but not CYP3A4 IVS10 + 12G/A and P2Y12 T744C polymorphisms, is associated with response variability to dual antiplatelet treatment in high-risk vascular patients. Pharmacogenet Genomics 17: 1057–1064.

Giusti, B., A.M. Gori, R. Marcucci, C. Saracini, I. Sestini, R. Paniccia, P. Buonamici, D. Antoniucci, R. Abbate, and G.F. Gensini. 2009. Relation of cytochrome P450 2C19 loss-of-function polymorphism to occurrence of drug-eluting coronary stent thrombosis. Am. J. Cardiol. 103: 806–811.

Gladding, P., M. Webster, I. Zeng, H. Farrell, J. Stewart, P. Ruygrok, J. Ormiston, S. El-Jack, G. Armstrong, P. Kay, D. Scott, A. Gunes, and M.L. Dahl. 2008. The pharmacogenetics and pharmacodynamics of clopidogrel response: an analysis from the PRINC (Plavix Response in Coronary Intervention) trial. JACC Cardiovasc. Interv. 1: 620–627.

Gurbel, P.A., K.P. Bliden, B.L. Hiatt, and C.M. O'Connor. 2003. Clopidogrel for coronary stenting: response variability, drug resistance, and the effect of pretreatment platelet reactivity. Circulation 107: 2908–2913.

Gurbel, P.A., K.P. Bliden, K.M. Hayes, J.A. Yoho, W.R. Herzog, and U.S. Tantry. 2005a. The relation of dosing to clopidogrel responsiveness and the incidence of high post-treatment platelet aggregation in patients undergoing coronary stenting. J. Am. Coll. Cardiol. 45: 1392–1396.

Gurbel, P.A., K.P. Bliden, K. Guyer, P.W. Cho, K.A. Zaman, R.P. Kreutz, A.K. Bassi, and U.S. Tantry. 2005b. Platelet reactivity in patients and recurrent events post-stenting: results of the PREPARE POST-STENTING Study. J. Am. Coll. Cardiol. 46: 1820–1826.

Gurbel, P.A., K.P. Bliden, and U.S. Tantry. 2006. Effect of clopidogrel with and without eptifibatide on tumor necrosis factor-alpha and C-reactive protein release after elective stenting: results from the CLEAR PLATELETS 1b study. J. Am. Coll. Cardiol. 48: 2186–2191.

Gurbel, P.A., R.C. Becker, K.G. Mann, S.R. Steinhubl, and A.D. Michelson. 2007. Platelet function monitoring in patients with coronary artery disease. J. Am. Coll. Cardiol. 50: 1822–1834.

Guthikonda, S., E.I. Lev, and N.S. Kleiman. 2005. Resistance to antiplatelet therapy. Curr. Cardiol. Rep. 7: 242–248.

Harmsze, A.M., J.W. van Werkum, J.M. Ten Berg, B. Zwart, H.J. Bouman, N.J. Breet, A.W. van 't Hof, H.J. Ruven, C.M. Hackeng, O.H. Klungel, A. de Boer, and V.H. Deneer. 2010a. CYP2C19*2 and CYP2C9*3 alleles are associated with stent thrombosis: a case-control study. 31: 3046–3053.

Harmsze, A., J.W. van Werkum, H.J. Bouman, H.J. Ruven, N.J. Breet, J.M. Ten Berg, C.M. Hackeng, M.M. Tjoeng, O.H. Klungel, A. de Boer, and V.H. Deneer. 2010b. Besides CYP2C19*2, the variant allele CYP2C9*3 is associated with higher on-clopidogrel platelet reactivity in patients on dual antiplatelet therapy undergoing elective coronary stent implantation. Pharmacogenet Genomics 20: 18–25.

Heeschen, C., S. Dimmeler, C.W. Hamm, M.J. van den Brand, E. Boersma, A.M. Zeiher, and M.L. Simoons. 2003. Soluble CD40 ligand in acute coronary syndromes. N. Engl. J. Med. 348: 1104–1111.

Holmes, M.V., P. Perel, T. Shah, A.D. Hingorani, and J.P. Casas. 2011. CYP2C19 genotype, clopidogrel metabolism, platelet function, and cardiovascular events: a systematic review and meta-analysis. JAMA 306: 2704–2714.

Hulot, J.S., A. Bura, E. Villard, M. Azizi, V. Remones, C. Goyenvalle, M. Aiach, P. Lechat, and P. Gaussem. 2006. Cytochrome P450 2C19 loss-of-function polymorphism is a major determinant of clopidogrel responsiveness in healthy subjects. Blood 108: 2244–2247.

Hung, J., J.Y. Lam, L. Lacoste, and G. Letchacovski. 1995. Cigarette smoking acutely increases platelet thrombus formation in patients with coronary artery disease taking aspirin. Circulation 92: 2432–2436.

Jakubowski, J.A., C.D. Payne, J.T. Brandt, G.J. Weerakkody, N.A. Farid, D.S. Small, H. Naganuma, G.Y. Li, and K.J. Winters. 2006. The platelet inhibitory effects and pharmacokinetics of prasugrel after administration of loading and maintenance doses in healthy subjects. J. Cardiovasc. Pharmacol. 47: 377–384.

Jefferson, B.K., J.H. Foster, J.J. McCarthy, G. Ginsburg, A. Parker, K. Kottke-Marchant, and E.J. Topol. 2005. Aspirin resistance and a single gene. Am. J. Cardiol. 95: 805–808.

Kuliczkowski, W., A. Witkowski, L. Polonski, C. Watala, K. Filipiak, A. Budaj, J. Golanski, D. Sitkiewicz, J. Pregowski, J. Gorski, M. Zembala, G. Opolski, K. Huber, H. Arnesen, S.D. Kristensen, and R. De Caterina. 2009. Interindividual variability in the response to oral antiplatelet drugs: a position paper of the Working Group on antiplatelet drugs resistance appointed by the Section of Cardiovascular Interventions of the Polish Cardiac Society, endorsed by the Working Group on Thrombosis of the European Society of Cardiology. Eur. Heart J. 30: 426–435.

Lau, W.C., P.A. Gurbel, P.B. Watkins, C.J. Neer, A.S. Hopp, D.G. Carville, K.E. Guyer, A.R. Tait, and E.R. Bates. 2004. Contribution of hepatic cytochrome P450 3A4 metabolic activity to the phenomenon of clopidogrel resistance. Circulation 109: 166–171.

Lepantalo, A., J. Mikkelsson, J.C. Resendiz, L. Viiri, J.T. Backman, E. Kankuri, P.J. Karhunen, and R. Lassila. 2006. Polymorphisms of COX-1 and GPVI associate with the antiplatelet effect of aspirin in coronary artery disease patients. Thromb. Haemost. 95: 253–259.

Lev, E.I., R.T. Patel, K.J. Maresh, S. Guthikonda, J. Granada, T. DeLao, P.F. Bray, and N.S. Kleiman. 2006. Aspirin and clopidogrel drug response in patients undergoing percutaneous coronary intervention: the role of dual drug resistance. J. Am. Coll. Cardiol. 47: 27–33.

Lindemann, S., B. Kramer, P. Seizer, and M. Gawaz. 2007. Platelets, inflammation and atherosclerosis. J. Thromb. Haemost. 5 Suppl. 1: 203–211.

Lordkipanidze, M., C. Pharand, E. Schampaert, J. Turgeon, D.A. Palisaitis, and J.G. Diodati. 2007. A comparison of six major platelet function tests to determine the prevalence of aspirin resistance in patients with stable coronary artery disease. Eur. Heart J. 28: 1702–1708.

Lordkipanidze, M., J.G. Diodati, D.A. Palisaitis, E. Schampaert, J. Turgeon, and C. Pharand. 2011. Genetic determinants of response to aspirin: appraisal of 4 candidate genes. Thromb. Res. 128: 47–53.

Mason, P.J., A.K. Jacobs, and J.E. Freedman. 2005. Aspirin resistance and atherothrombotic disease. J. Am. Coll. Cardiol. 46: 986–993.

Matetzky, S., B. Shenkman, V. Guetta, M. Shechter, R. Beinart, I. Goldenberg, I. Novikov, H. Pres, N. Savion, D. Varon, and H. Hod. 2004. Clopidogrel resistance is associated with increased risk of recurrent atherothrombotic events in patients with acute myocardial infarction. Circulation 109: 3171–3175.

Mega, J.L., S.L. Close, S.D. Wiviott, L. Shen, R.D. Hockett, J.T. Brandt, J.R. Walker, E.M. Antman, W. Macias, E. Braunwald, and M.S. Sabatine. 2009a. Cytochrome p-450 polymorphisms, and response to clopidogrel. N. Engl. J. Med. 360: 354–362.

Mega, J.L., S.L. Close, S.D. Wiviott, L. Shen, R.D. Hockett, J.T. Brandt, J.R. Walker, E.M. Antman, W.L. Macias, E. Braunwald, and M.S. Sabatine. 2009b. Cytochrome P450 genetic polymorphisms and the response to prasugrel: relationship to pharmacokinetic, pharmacodynamic, and clinical outcomes. Circulation 119: 2553–2560.

Mega, J.L., S.L. Close, S.D. Wiviott, L. Shen, J.R. Walker, T. Simon, E.M. Antman, E. Braunwald, and M.S. Sabatine. 2010. Genetic variants in ABCB1 and CYP2C19 and cardiovascular outcomes after treatment with clopidogrel and prasugrel in the TRITON-TIMI 38 trial: a pharmacogenetic analysis. Lancet 376: 1312–1319.

Mega, J.L., W. Hochholzer, A.L. Frelinger, 3rd, M.J. Kluk, D.J. Angiolillo, D.J. Kereiakes, S. Isserman, W.J. Rogers, C.T. Ruff, C. Contant, M.J. Pencina, B.M. Scirica, J.A. Longtine, A.D. Michelson, and M.S. Sabatine. 2011. Dosing clopidogrel based on CYP2C19 genotype and the effect on platelet reactivity in patients with stable cardiovascular disease. JAMA 306: 2221–2228.

Michos, E.D., R. Ardehali, R.S. Blumenthal, R.A. Lange, and H. Ardehali. 2006. Aspirin and clopidogrel resistance. Mayo Clin. Proc. 81: 518–526.

Nguyen, T.A., J.G. Diodati, and C. Pharand. 2005. Resistance to clopidogrel: a review of the evidence. J. Am. Coll. Cardiol. 45: 1157–1164.

O'Donoghue, M. and S.D. Wiviott. 2006. Clopidogrel response variability and future therapies: clopidogrel: does one size fit all? Circulation 114: e600–606.

Offermanns, S. 2006. Activation of platelet function through G protein-coupled receptors. Circ. Res. 99: 1293–1304.

Osmancik, P., P. Paulu, P. Tousek, V. Kocka, and P. Widimsky. 2012. High leukocyte count and interleukin-10 predict high on-treatment-platelet-reactivity in patients treated with clopidogrel. J. Thromb. Thrombolysis 33: 349–354.

Pare, G., S.R. Mehta, S. Yusuf, S.S. Anand, S.J. Connolly, J. Hirsh, K. Simonsen, D.L. Bhatt, K.A. Fox, and J.W. Eikelboom. 2010. Effects of CYP2C19 genotype on outcomes of clopidogrel treatment. N. Engl. J. Med. 363: 1704–1714.

Parodi, G., R. Marcucci, R. Valenti, A.M. Gori, A. Migliorini, B. Giusti, P. Buonamici, G.F. Gensini, R. Abbate and D. Antoniucci. 2011. High residual platelet reactivity after clopidogrel loading and long-term cardiovascular events among patients with acute coronary syndromes undergoing PCI. JAMA 306: 1215–1223.

Pettinella, C., M. Romano, L. Stuppia, F. Santilli, R. Liani, and G. Davi. 2009. Cyclooxygenase-1 haplotype C50T/A-842G does not affect platelet response to aspirin. Thromb. Haemost. 101: 687–690.

Price, M.J., D.J. Angiolillo, P.S. Teirstein, E. Lillie, S.V. Manoukian, P.B. Berger, J.F. Tanguay, C.P. Cannon and E.J. Topol. 2011a. Platelet reactivity and cardiovascular outcomes after percutaneous coronary intervention: a time-dependent analysis of the Gauging Responsiveness with a VerifyNow P2Y12 assay: Impact on Thrombosis and Safety (GRAVITAS) trial. Circulation 124: 1132–1137.

Price, M.J., P.B. Berger, P.S. Teirstein, J.F. Tanguay, D.J. Angiolillo, D. Spriggs, S. Puri, M. Robbins, K.N. Garratt, O.F. Bertrand, M.E. Stillabower, J.R. Aragon, D.E. Kandzari, C.T. Stinis, M.S. Lee, S.V. Manoukian, C.P. Cannon, N.J. Schork, and E.J. Topol. 2011b. Standard- vs high-dose clopidogrel based on platelet function testing after percutaneous coronary intervention: the GRAVITAS randomized trial. JAMA 305: 1097–1105.

Price, M.J., S.S. Murray, D.J. Angiolillo, E. Lillie, E.N. Smith, R.L. Tisch, N.J. Schork, P.S. Teirstein, and E.J. Topol. 2012. Influence of genetic polymorphisms on the effect of high- and standard-dose clopidogrel after percutaneous coronary intervention: the GIFT (Genotype Information and Functional Testing) study. J. Am. Coll. Cardiol. 59: 1928–1937.

Ridker, P.M., J.E. Manson, J.E. Buring, S.Z. Goldhaber, and C.H. Hennekens. 1991. The effect of chronic platelet inhibition with low-dose aspirin on atherosclerotic progression and acute thrombosis: clinical evidence from the Physicians' Health Study. Am. Heart J. 122: 1588–1592.

Riondino, S., F. Martini, F. La Farina, A. Spila, F. Guadagni, and P. Ferroni. 2010. Increased plasma levels of soluble CD40 ligand correlate with platelet activation markers and underline the need for standardized pre-analytical conditions. Clin. Biochem. 43: 666–670.

Roberts, J.D., G.A. Wells, M.R. Le May, M. Labinaz, C. Glover, M. Froeschl, A. Dick, J.F. Marquis, E. O'Brien, S. Goncalves, I. Druce, A. Stewart, M.H. Gollob, and D.Y. So. 2012. Point-of-care genetic testing for personalisation of antiplatelet treatment (RAPID GENE): a prospective, randomised, proof-of-concept trial. Lancet 379: 1705–1711.

Rondina, M.T., J.M. Lappe, J.F. Carlquist, J.B. Muhlestein, M.J. Kolek, B.D. Horne, R.R. Pearson, and J.L. Anderson. 2008. Soluble CD40 ligand as a predictor of coronary artery disease and long-term clinical outcomes in stable patients undergoing coronary angiography. Cardiology 109: 196–201.

Sanderson, S., J. Emery, T. Baglin, and A.L. Kinmonth. 2005. Narrative review: aspirin resistance and its clinical implications. Ann. Intern. Med. 142: 370–380.

Savi, P., J.M. Pereillo, M.F. Uzabiaga, J. Combalbert, C. Picard, J.P. Maffrand, M. Pascal, and J.M. Herbert. 2000. Identification and biological activity of the active metabolite of clopidogrel. Thromb. Haemost. 84: 891–896.

Schwartz, K.A., D.E. Schwartz, K. Ghosheh, M.J. Reeves, K. Barber, and A. DeFranco. 2005. Compliance as a critical consideration in patients who appear to be resistant to aspirin after healing of myocardial infarction. Am. J. Cardiol. 95: 973–975.

Seidel, H., M.M. Rahman and R.E. Scharf. 2011. Monitoring of antiplatelet therapy. Current limitations, challenges, and perspectives. Hamostaseologie 31: 41–51.

Siasos, G., D. Tousoulis, and C. Stefanadis. 2011. CYP2C19 genotype and outcomes of clopidogrel treatment. N. Engl. J. Med. 364: 481–482.

Siasos, G., D. Tousoulis, and C. Stefanadis. 2012. CYP2C19 genotype and cardiovascular events. JAMA 307: 1484–1485.

Sibbing, D., S. Braun, T. Morath, J. Mehilli, W. Vogt, A. Schomig, A. Kastrati, and N. von Beckerath. 2009. Platelet reactivity after clopidogrel treatment assessed with point-of-care analysis and early drug-eluting stent thrombosis. J. Am. Coll. Cardiol. 53: 849–856.

Sibbing, D., W. Koch, D. Gebhard, T. Schuster, S. Braun, J. Stegherr, T. Morath, A. Schomig, N. von Beckerath, and A. Kastrati. 2010a. Cytochrome 2C19*17 allelic variant, platelet aggregation, bleeding events, and stent thrombosis in clopidogrel-treated patients with coronary stent placement. Circulation 121: 512–518.

Sibbing, D., D. Gebhard, W. Koch, S. Braun, J. Stegherr, T. Morath, N. Von Beckerath, J. Mehilli, A. Schomig, T. Schuster, and A. Kastrati. 2010b. Isolated and interactive impact of common CYP2C19 genetic variants on the antiplatelet effect of chronic clopidogrel therapy. J. Thromb. Haemost. 8: 1685–1693.

Siller-Matula, J.M., G. Christ, I.M. Lang, G. Delle-Karth, K. Huber, and B. Jilma. 2010. Multiple electrode aggregometry predicts stent thrombosis better than the vasodilator-stimulated phosphoprotein phosphorylation assay. J. Thromb. Haemost. 8: 351–359.

Simon, T., C. Verstuyft, M. Mary-Krause, L. Quteineh, E. Drouet, N. Meneveau, P.G. Steg, J. Ferrieres, N. Danchin, and L. Becquemont. 2009. Genetic determinants of response to clopidogrel and cardiovascular events. N. Engl. J. Med. 360: 363–375.

Tantry, U.S., K.P. Bliden, C. Wei, R.F. Storey, M. Armstrong, K. Butler, and P.A. Gurbel. 2010. First analysis of the relation between CYP2C19 genotype and pharmacodynamics in patients treated with ticagrelor versus clopidogrel: the ONSET/OFFSET and RESPOND genotype studies. Circ. Cardiovasc. Genet. 3: 556–566.

Tousoulis, D., A. Briasoulis, S.S. Dhamrait, C. Antoniades, and C. Stefanadis. 2009. Effective platelet inhibition by aspirin and clopidogrel: where are we now? Heart 95: 850–858.

Tousoulis, D., G. Siasos, M. Zaromytidou, N. Papageorgiou, E. Stefanadi, E. Oikonomou, and C. Stefanadis. 2011. The role of the cytochrome P450 polymorphisms in clopidogrel efficacy and clinical utility. Curr. Med. Chem. 18: 427–438.

van Werkum, J.W., H.J. Bouman, N.J. Breet, J.M. ten Berg, and C.M. Hackeng. 2010. The Cone-and-Plate(let) analyzer is not suitable to monitor clopidogrel therapy: a comparison with the flowcytometric VASP assay and optical aggregometry. Thromb. Res. 126: 44–49.

Varenhorst, C., S. James, D. Erlinge, J.T. Brandt, O.O. Braun, M. Man, A. Siegbahn, J. Walker, L. Wallentin, K.J. Winters, and S.L. Close. 2009. Genetic variation of CYP2C19 affects both pharmacokinetic and pharmacodynamic responses to clopidogrel but not prasugrel in aspirin-treated patients with coronary artery disease. Eur. Heart J. 30: 1744–1752.

Varga-Szabo, D., I. Pleines, and B. Nieswandt. 2008. Cell adhesion mechanisms in platelets. Arterioscler. Thromb. Vasc. Biol. 28: 403–412.

Varo, N., R. Nuzzo, C. Natal, P. Libby, and U. Schonbeck. 2006. Influence of pre-analytical and analytical factors on soluble CD40L measurements. Clin. Sci. (Lond.) 111: 341–347.

Vavuranakis, M., G. Latsios, D. Aggelis, I. Bosinakou, I. Karambelas, D. Tousoulis, K. Toutouzas, and C. Stefanadis. 2006. Randomized comparison of the effects of ASA plus clopidogrel versus ASA alone on early platelet activation in acute coronary syndromes with elevated high-sensitivity C-reactive protein and soluble CD40 ligand levels. Clin. Ther. 28: 860–871.

Wallentin, L., S. James, R.F. Storey, M. Armstrong, B.J. Barratt, J. Horrow, S. Husted, H. Katus, P.G. Steg, S.H. Shah, and R.C. Becker. 2010. Effect of CYP2C19 and ABCB1 single nucleotide polymorphisms on outcomes of treatment with ticagrelor versus clopidogrel for acute coronary syndromes: a genetic substudy of the PLATO trial. Lancet 376: 1320–1328.

Wang, T.H., D.L. Bhatt, and E.J. Topol. 2006. Aspirin and clopidogrel resistance: an emerging clinical entity. Eur. Heart J. 27: 647–654.

Wiviott, S.D. and E.M. Antman. 2004. Clopidogrel resistance: a new chapter in a fast-moving story. Circulation 109: 3064–3067.

Wright, R.S., J.L. Anderson, C.D. Adams, C.R. Bridges, D.E. Casey, Jr., S.M. Ettinger, F.M. Fesmire, T.G. Ganiats, H. Jneid, A.M. Lincoff, E.D. Peterson, G.J. Philippides, P. Theroux, N.K. Wenger, J.P. Zidar, E.M. Antman, R.M. Califf, W.E. Chavey, 2nd, J.S. Hochman, and T.N. Levin. 2011. ACCF/AHA focused update of the Guidelines for the Management of Patients with Unstable Angina/Non-ST-Elevation Myocardial Infarction (updating the 2007 guideline): a report of the American College of Cardiology Foundation/American Heart Association Task Force on Practice Guidelines developed in collaboration with the American College of Emergency Physicians, Society for Cardiovascular Angiography and Interventions, and Society of Thoracic Surgeons. J. Am. Coll. Cardiol. 57: 1920–1959.

Zuern, C.S., M. Schwab, M. Gawaz, and T. Geisler. 2010. Platelet pharmacogenomics. J. Thromb. Haemost. 8: 1147–1158.

# Part II
# Novel Biomarkers

# CHAPTER 13

# Novel Biomarkers Used in the Assessment of Acute Coronary Syndrome

Dimitris Tousoulis,* Emmanuel Androulakis,
Nikolaos Papageorgiou and Christodoulos Stefanadis

## Introduction

It is well-known that cardiovascular disease is the leading cause of mortality, while coronary heart disease (CAD) accounts for more than 50 percent of all cases (Lloyd-Jones et al. 2010; Fonarow 2007; Steg et al. 2007). Anginal chest pain or other symptoms typical for an acute coronary syndrome (ACS) is the main cause of admission to an emergency department and remains the major clinical condition responsible for hospitalization (Braunwald et al. 2002; Bertrand et al. 2002; Antman et al. 2004). ACS covers the spectrum of clinical conditions ranging from unstable angina to non-Q-wave myocardial infarction (MI) and Q-wave MI. In particular, an acute reduction in coronary blood flow and hence myocardial oxygen supply is usually the mechanism of ACS. In this case, recent events including rupture, erosion and hemorrhage often superimposed on thrombosis and/or microembolism reduce coronary blood flow and lead to acute ischemic myocyte injury (Libby and Theroux 2005). The main pathophysiological mechanism contributing to the expression of the ACS is the progression of atherosclerosis, whereas the central process to the initiation of an ACS is disruption of an atheromatous plaque (Kampoli et al. 2012a).

1st Department of Cardiology, 'Hippokration' Hospital, University of Athens Medical School, Athens, Greece.
*Corresponding author: drtousoulis@hotmail.com

Many novel biomarkers have been discovered and used to predict cardiovascular risk. In order to be integrated as a complementary diagnostic tool of the cardiovascular disease (CVD), a biomarker must meet several criteria, such as specificity and cost-effectiveness which are considered as top priority. The ideal biomarker should also fulfil the following properties: reliability, accessibility, cost effectiveness, and easy interpretation by clinicians. Since there is no individual biomarker exhibiting all the necessary requirements, the use of a combination of biomarkers appears to be the most likely and effective solution (Tousoulis et al. 2006; Tousoulis et al. 2008).

In the present article, we will focus on the presentation and evaluation of the most promising biochemical markers used in acute myocardial ischemic states and will review the underlying pathophysiological characteristics of each biomarker, as well as potential clinical implications in daily practice.

## General Considerations

Cardiac ischemia is the result of an imbalance between myocardial oxygen supply and demand. In general, the reduction in coronary blood flow results in development of an ACS. Regarding the sensation of pain, several mediators such as lactate are produced, as a consequence of hypoxia which disrupts oxidative metabolic pathways. Cardiac ischemia seems to be subdivided into two clinical phenotypes; ACS and chronic coronary syndrome (CCS). In particular, an acute reduction in coronary blood flow and in myocardial oxygen supply is usually the mechanism of ACS. In the latter case, recent events including rupture, erosion and hemorrhage often superimposed on thrombosis and/or microembolism reduce coronary blood flow and lead to acute ischemic myocyte injury. On the contrary, a sudden increase in oxygen demand, in the ground of disability of the myocardium to meet that demand, is usually the mechanism of ischemia in CCS. In that case, the ischemia is transient and without evidence of myocyte injury. Life-threatening arrhythmias may occur in both cases but more frequently after reperfusion in ACS (Kampoli et al. 2012a; Foreman 2004).

## Diagnosis

Interestingly, approximately half the cases of admission to an emergency department are presented with non-cardiac chest pain (Kachintorn 2005). Diagnosis relies on history, patient characteristics, and identification of classical risk factors. It has been indicated that high risk patients would benefit more from coronary angiography rather than non-invasive testing (Fraker et al. 2007). The electrocardiogram is a convenient but insensitive method of diagnosing myocardial ischemia. It is often normal in stable angina in the absence of a previous MI, while during anginal pain, it could show ST-segment depression and/or T wave inversion. Also, exercise treadmill testing is recommended for the initial non-invasive risk stratification and diagnosis of CAD, though only in patients who are capable of exercising. The gold standard for definitive diagnosis of CAD remains invasive coronary angiography which indicates the site and severity of a coronary lesion, especially in cases of absolute contraindications to stress testing, or medically refractory angina (Agarwal et al. 2010).

## Myocyte Necrosis—The Role of Troponins

Cardiac troponin I or troponin T (cTnI and cTnT) are the most widely used markers of myocyte necrosis since they allow sensitive detection of myocardial damage. cTn has been established as the 'gold standard' in the diagnosis of ACS (Thygesen et al. 2007). Given the specificity of cTn for myocardial damage, elevated levels of cTn, along with clinical evidence, is indicative of myocardial injury. According to European and American guidelines, the use of a decision limit is recommended for myocardial injury at the 99th percentile of the reference population for cTn and cTnI. However, these traditionally used biomarkers are currently not thought to be the true 'gold' standard in the evaluation of this group of patients. Elevated troponin can be detected four to six hours after the onset of symptoms and is found positive only in 33 percent of patients presenting with an ACS (Tousoulis et al. 2012; Jesse and Kontos 1997).

cTnI and cTnT are products of specific genes and therefore have the potential to be unique for the heart. Regarding pathophysiology, they are cardiac regulatory proteins that control the calcium-mediated interaction of actin and myosin; thus their use in the diagnosis of acute myocardial infarction is essential. Cardiac troponin concentrations usually rise within three hours after the onset of acute myocardial infarction (AMI). The appearance of troponin four to six hours after the onset of chest pain, a peak in troponin at 12–16 hours, and a subsequent fall in troponin level would be consistent with the diagnosis of ACS. However, other conditions which result in myocardial cell damage, such as myocarditis, cardiac surgery, and sepsis can increase cTn levels. Current assays measure concentrations of troponin in nanograms per milliliter, whereas assays in development can detect concentrations in picograms per milliliter. In order to avoid false positive results (when troponin will be measured in picograms per milliliter), new reference values should be defined after close collaboration between cardiologists and pathologist (Tousoulis et al. 2008).

According to GUSTO-IIA trial, which included 855 patients within 12 hours of the onset of symptoms, cardiac troponin T levels have been proved a powerful, independent risk marker and they have allowed further stratification of risk when combined with standard measures such as electrocardiography and the serum creatine kinase MB level (Ohman et al. 1996). In turn, a GUSTO-IV substudy has demonstrated in over 7,000 patients that in ACS, baseline levels of TnT and C-reactive protein (CRP) are independently related to 30-day mortality. Furthermore, any detectable elevation of TnT, but not of CRP, is also associated with an increased risk of subsequent MI while the combination of both markers provides a better risk stratification than either one alone with regard to mortality (James et al. 2003). An elevated cTnT was an independent predictor of mortality at 30 days in a multivariate regression analysis. A pooled analysis of 21 studies including 18,982 patients with ACS found that elevated serum cTnI or cTnT was associated with an increased risk of cardiac death or reinfarction at 30 days and similarly, these patients showed a substantial increase in risk during long-term follow-up (five months to three years) (Ottani et al. 2000).

## Novel Biomarkers in ACS

### *Markers of inflammation*

As a result of chronic inflammation, various biomarkers such CRP, interleukin-6 (IL-6) and -18, tumor necrosis factor-α (TNF-α), intercellular adhesion molecule 1 (ICAM-1) and E-selectin are increased in plasma and may serve as possible risk factors for atherosclerosis (Stefanadi et al. 2010; Rebuzzi et al. 1998). More specifically, a growing body of evidence has focused on the predictive value of CRP in subjects both with and without cardiovascular disease (Stefanadi et al. 2010; Rebuzzi et al. 1998). Even though the most commonly used inflammatory markers in ACS are CRP, myoglobin and CD40 lingand (CD40L), all markers which have been proposed to be useful in ACS are presented in Table 1.

CRP is an acute-phase protein with a homopentameric structure and Ca-binding specificity for phosphocholine (Goodman et al. 1996). The predictive value of CRP for disease severity and prognosis has been studied in a wide variety of settings. As an acute phase reactant and a potent inflammatory marker, CRP has raised a debate over its role in the pathogenesis of atherothrombosis. It is considered a mediator rather than marker and may contribute directly to atherogenesis. For example, it has been demonstrated that CRP may be detected in the atherosclerotic tissue (Torzewski et al. 1998), while mRNAs encoding CRP may be increased in atherosclerotic plaque (Yasojima et al. 2001). The association between ACS and elevated serum concentrations of acute phase reactants, such as CRP (Liuzzo et al. 1994; Tomoda et al. 2000), serum amyloid A (SAA) (Liuzzo et al. 1994; Tomoda et al. 2000), and IL-6 (Biasucci et al. 1996) suggests that

**Table 1.** Proposed biomarkers in acute coronary syndrome.

| **Markers of necrosis-Heart specific Markers** |
| --- |
| Myoglobin<br>Troponin<br>Creatine phosphokinase MB<br>Heart Fatty-Acid Binding Protein |
| **Markers of Myocardial dysfunction** |
| Atrial natriuretic peptides<br>Brain natriuretic peptides |
| **Markers of Stress** |
| Copeptin<br>Proadremedulin |
| **Markers of inflammation and plaque rupture/destabilization** |
| Adiponectin<br>C-reactive protein<br>Interleukin-6<br>Tumor necrosis factor-α<br>Matrix metalloproteinase-9<br>Myeloperoxidase<br>Placental growth factor<br>Pregnancy-associated plasma protein-A<br>Phospholipase A2<br>Soluble tyrosine kinase 1<br>Soluble intracellular adhesion molecule 1<br>Lipoprotein associated phospholipase A2 |
| **Markers of platelet activation** |
| Soluble CD40 ligand<br>Soluble P-selectin |

chronic inflammation of the coronary arterial wall may play an important role. The elevated levels of CRP in ACS appear to be a marker of extensive vascular inflammation and hyperesponsiveness of the inflammatory system (Antoniades et al. 2007). In patients who have survived AMI, it has been proposed that serum CRP and SAA potentially predict the risk of a recurrent in-hospital cardiac event (Tomoda et al. 2000) or 30-day or long-term mortality after an MI (Suleiman et al. 2003; Suleiman et al. 2006). The value of such a delay was assessed in the THROMBO trial in which serum CRP and SAA were measured in 1,045 patients two months after an AMI (Harb et al. 2002). Even though patients in the highest quartile of CRP and SAA concentrations had an increased rate of coronary events compared to those in the lower quartiles, neither CRP nor SAA were independent markers for recurrent events on multivariate analysis that included gender, diabetes, prior MI, heart failure, ejection fraction, and location of the MI. It is well known that elevations in serum CRP early after AMI may reflect the extent and reaction to myocardial injury. However, other reports have not confirmed such a relationship (Androulakis et al. 2009; Androulakis et al. 2011; Srinivasan et al. 1994; Volanakis 2001).

In patients with unstable angina, CRP levels were useful in prediction of the long-term prognosis, while TnT levels were useful in prediction of in-hospital prognosis (Tanaka et al. 2006). In 447 patients with unstable angina enrolled in CAPTURE trial, Troponin T, but not CRP, was predictive of cardiac risk during the initial 72-hour period. However, CRP was an independent predictor of both cardiac risk and repeated coronary revascularization (coronary artery bypass graft surgery and percutaneous transluminal coronary angioplasty) during a six month follow-up (Heeschen et al. 2000). Moreover, a growing body of literature has suggested that high sensitivity CRP (hs-CRP) could be of major importance for clinical practice. A value >3 mg/L is useful for predicting outcomes in patients with stable CAD, while a threshold >10 mg/L may be more predictive in patients with an ACS. When used for early risk stratification in combination with troponins, values obtained as soon as possible after presentation are less likely to be influenced by the extent of necrosis (Tousoulis et al. 2012).

## CD40 ligand (CD40L)

The potential role of CD40/CD40L interactions in ACS and CAD at a clinical level has been examined by several studies. Elevation of soluble CD40 (sCD40L) may identify a subgroup of patients at high risk of unstable angina and ACS (Tousoulis et al. 2010). Furthermore, given that interaction between CD40L and its membrane receptor participates in the proinflammatory and prothrombotic milieu for aggravating the development of atherosclerosis and instability of atherosclerotic plaques, this may be a valuable marker in the aim of predicting the severity of ACS (Yan et al. 2002; Wang et al. 2007; Aukrust et al. 1999). More specifically, it has been suggested that high sCD40L levels were associated with increased risk of major cardiovascular events, including AMI, sudden death and recurrent angina in unstable CAD (Heeschen et al. 2003; Yan et al. 2004).

Notably, patients with AMI exhibited elevated levels of sCD40L and IL-6 compared to both CAD and controls. Even though IL-6 was independently associated with sCD40L in healthy individuals, in advanced disease states (CAD or AMI), only diabetes mellitus and smoking were independently associated with sCD40L levels (Tousoulis et al. 2007a). Similarly, blood coagulation and platelet activation, as shown by platelet-derived sCD40L release at the microvascular injury site, were enhanced compared with that in stable CAD patients (Undas et al. 2009). Notably, in patients with coronary syndromes undergoing percutaneous coronary intervention, intracoronary CD40L and IL-1 receptor antagonist levels are higher in the culprit artery compared with the levels from a peripheral artery. Therefore, it is possible that these cytokines contribute directly to inflammatory status in response to coronary intervention (Aggarwal et al. 2004; Tousoulis et al. 2007b).

Regarding prognostic information derived from sCD40L, it has been indicated that the predictive value may be independent of cTnI concentrations. Interestingly, assessment of sCD40L in combination with cTnI potentially yields independent and complementary prognostic information in the prediction of adverse cardiac outcomes (Martín-Ventura et al. 2009). For example, it has been shown that elevation of sCD40L levels one month after PCI predicts angiographic restenosis (L'Allier et al. 2005).

Despite these important findings, a growing body of evidence supports the notion that the clinical significance of circulating sCD40L in ACS is still controversial. In more detail, according to a large population-based study investigating the prognostic performance of sCD40L, it was not associated with higher risk of non-fatal MI or rehospitalization for ACS (Morrow et al. 2008). Also, elevated admission levels of sCD40 were not associated with the onset of STEMI, and provided no prognostic information for future cardiac mortality (Tan et al. 2008). Moreover, according to several reports in patients with ASC, elevated levels of sCD40L were not predictive of long-term adverse cardiovascular event, including death and AMI (Rondina et al. 2008; Olenchock et al. 2008).

## *Myoglobin*

Myoglobin is another biomarker of cardiac injury. It is a ubiquitous heme protein that is rapidly released from damaged tissue because of its small size. Its half-life in plasma is in the range of nine minutes (Klocke et al. 1982). According to data derived from TIMI 11B and TACTICS-TIMI 18 studies, serum myoglobin above the MI detection threshold (>110 microg/l) is associated with an increased risk of six-month mortality, independent of baseline clinical characteristics, electrocardiographic changes and elevation in CK-MB and cTnI. These findings were independent of other prognostic variables such as ECG changes and serum cTnI and CK-MB and may suggest that myoglobin could be a useful addition to cardiac biomarker panels for early risk-stratification in ACS (de Lemos et al. 2002b). The same investigators had previously shown that a low 12-hour myoglobin level (≤ 239 ng/mL) may discriminate patients at low risk and could potentially contribute to early risk stratification and triage after ST-segment elevation MI (Srinivas et al. 2001).

## *Microalbuminuria*

According to international consensus, microalbuminuria is defined as an elevated urinary albumin excretion rate of 20–200 micrograms/min, which is below the proteinuric range. Nephropathy is a major complication in insulin-dependent diabetes mellitus, seen in about 30 percent of patients after many years of the disease and increasing microalbuminuria is an excellent marker of subsequent nephropathy in these patients (Mogensen et al. 1992). Microalbuminuria in non-diabetics appears to be related to a dysfunctional endothelium. Establishing the diagnosis of microalbuminuria requires the demonstration of a persistent elevation in albumin excretion with a 24-hour urine collection (Mogensen et al. 1992). An increasing number of studies in different patient populations have suggested that, in addition to its relation to renal disease, microalbuminuria is an important risk factor for CVD and early cardiovascular mortality in patients with and without diabetes and/or hypertension (Gerstein et al. 2001; Wachtell et al. 2003; Hillege et al. 2002). According to the HOPE trial, microalbuminuria is associated with an increased relative risk of the primary aggregate endpoint MI, stroke, or cardiovascular death in those with and without diabetes (Gerstein et al. 2001). Moreover, an increased risk was also demonstrated, in an analysis from the LIFE trial, of patients with hypertension and electrocardiographic evidence of left ventricular hypertrophy. The urine albumin-to-creatinine ratio was measured in 7,143 nondiabetic subjects and 1,063 subjects with diabetes (Wachtell et al. 2003). In an analysis from the PREVEND study including 40,548 participants who were followed for a median of 2.6 years, albumin excretion was associated with increased relative risk of cardiovascular mortality of 1.35 for each doubling of urinary albumin excretion (Hillege et al. 2002). In hypertensives, we have previously demonstrated that increasing levels of microalbuminuria in tandem with decreasing levels of adiponectin are associated with aortic stiffness (Tsioufis et al. 2007), and that molecules participating to the perpetuation of sub-clinical inflammation such CRP, IL-18 and CD-40L are interrelated with microalbuminuria (Tsioufis et al. 2006).

## *Cystatin-C*

Cystatin-C is a protein that is a member of cysteine protease inhibitors, filtered at the glomerulus and not reabsorbed, though metabolized in the tubules, which prevents use of cystatin-C to directly measure clearance. It has been supposed to be unaffected by gender, age, or muscle mass. Of note, it is affected by hyper- and hypothyroidism, and has been correlated with markers of inflammation (CRP) (Manetti et al. 2005; Knight 2004). Recent data have indicated that cystatin-C potentially predicts an increased risk of all-cause mortality and cardiovascular events, even though the underlying mechanisms remain to be clarified (Toft et al. 2012). Of note, in a recent study, cystatin-C was significantly correlated with cardio-ankle vascular index, suggesting a significant role of arterial stiffness in renal insufficiency (Nakamura et al. 2009). In addition, Health ABC study has demonstrated that medium and high cystatin-C groups were associated with a 5.3 percent and 8.0 percent higher PWV than the low cystatin-C group in older adults; also, recent evidence has suggested that cystatin-C levels had no independent association with carotid IMT (Madero et al. 2009; Han et al. 2010). Increased plasma levels of cystatin-C are associated with adverse cardiovascular outcomes, cardiovascular and non-cardiovascular death, and possibly cerebrovascular disease (Shlipak et al. 2006). For example, the risk of death relative to cystatin-C concentration increased in a dose-response fashion in one study (Shlipak et al. 2005). These results suggest that elevated plasma cystatin-C concentration may be a sensitive marker of cardiovascular risk, while increased inflammation and other cardiovascular factors, may underlie this association.

## Metalloproteinases (MMPs)

It is well-known that extracellular matrix (ECM) are produced by vascular cells, provide structural support to vessels and define critical mechanical properties of arteries. The ECM is a complex structure composed of collagen, laminin, fibronectin and proteoglykanes. Degradation of ECM by the MMPs system plays an important role in many physiological and pathological processes (Androulakis et al. 2012). MMPs are a family of zinc metallo-endopeptidases secreted by cells and are responsible for much of the turnover of matrix components. Microscopic studies of human plaques have demonstrated a net excess of matrix degrading activity in regions of human atherosclerotic plaque. A growing body of evidence has suggested that MMPs are involved in all stages of the atherosclerotic process, from the initial lesion to plaque rupture. Recent data have suggested that MMP activity may facilitate atherosclerosis, plaque destabilization, and platelet aggregation, while matrix metalloproteinases participate in vascular remodeling, plaque instability, and ventricular remodeling after cardiac injury (Kampoli et al. 2012b). For example, serum MMP-2 and MMP-9 concentrations on the day of admission were two to three fold higher in patients with an ACS compared to controls or those with stable angina (Jones et al. 2003). There is an expanding body of evidence reporting that ACS may also be influenced by MMPs through degradation of the fibrous cap of vulnerable atherosclerotic lesions. Moreover, the accumulation of macrophage-derived foam cells in atherosclerotic lesions has been associated with local release of MMPs and a thin fibrous cap (Moreau et al. 1999). Statin therapy is associated with decreased macrophage infiltration and MMP expression and with plaque stabilization. The inhibitory effect on macrophages and MMPs may be mediated in part by statin-induced inhibition of prostaglandin synthesis (Cipollone et al. 2003).

## Brain-type natriuretic peptide (BNP)

BNP is a 32 amino acid peptide which is synthesized and released mainly from ventricular myocardium in response to ventricular dilatation and pressure overload. Similar to atrial natriuretic peptide (ANP), BNP exhibits almost solely beneficial physiological properties, including balanced vasodilation, natriuresis, and inhibition of sympathetic nervous system and renin-angiotensin-aldosterone axis (Gopal et al. 2011). BNP and the amino terminal fragment of the prohormone BNP (NT-proBNP) appear to provide prognostic information in individuals following hospital admission for ACS. The predictive value of plasma BNP was best illustrated in a study which included 2,525 and BNP was measured at a mean of 40 hours after the onset of symptoms. After adjusting for other predictors of risk, the odds ratios for death at 10 months were 3.8, 4, and 5.8 for concentrations in the second, third, or fourth quartiles compared to those in the lowest quartile. Higher plasma BNP was also associated with an increased risk of new or recurrent MI and new or worsening heart failure (de Lemos et al. 2001; de Lemos et al. 2002a). However, its use for the diagnosis and risk stratification of an ACS is not currently thought to be the true "gold" standard in the evaluation of this group of patients (Tousoulis et al. 2012).

## Pregnancy associated plasma protein A (PAPP-A)

Pregnancy-associated plasma protein-a (PAPP-A) is a metalloproteinase that is widely used as a prenatal screening test for the detection of trisomy 21 during the first trimester of pregnancy. It is a 200 kDa protein, circulating in a form of a heterotetrameric complex, which consists of two subunits covalently bound to their inhibitor. It is an insulin-like growth factor dependent that, apart from its expression in the reproductive system, participates in local proliferative processes and may also have additional impacts in the function and pathophysiology of various human systems (Tousoulis et al. 2012). It also seems to be a biomarker of plaque rupture. In a study of patients with an angiographically confirmed ACS, elevated serum PAPP-A was a strong independent predictor of death or nonfatal MI, even in patients with normal serum troponin T, while this association was prospectively validated in patients with chest pain (Heeschen et al. 2005). PAPP-A has emerged as a promising biomarker useful in the clinical management of ACS patients. However more prospective and interventional studies with carefully established immunoassays are required to validate its clinical utility. The use and the contribution of all the aforementioned markers have been summarized in Table 2.

**Table 2.** The use of new biochemical markers in ACS.

| Biochemical marker | Characteristics | Use in ACS |
|---|---|---|
| CRP and hs-CRP | Acute-phase protein with a homopentameric structure | A marker of widespread underlying vascular inflammation hs-CRP is considered to be the most useful inflammatory for clinical practice May be useful as an independent marker of prognosis in patients with ACS |
| Myoglobin | Heme protein | Biomarker of cardiac injury Serum myoglobin above 110 µg/L is associated with increased likelihood of thrombosis and six-month mortality |
| CD40 ligand | Expressed on platelets and released from them on activation | May identify patients with ACS at heightened risk for recurrent cardiac events |
| MMPs | Metallo-endopeptidases secreted by cells and are responsible for much of the turnover of matrix components | MMP activity may facilitate atherosclerosis, plaque destabilization, and platelet aggregation |
| BNP and NT-proBNP | Neurohormone synthesized in the left ventricular myocardium | Associations with ACS N-pro-BNP had a stronger correlation with mortality than any other marker studied |
| Microalbuminuria | Urinary albumin excretion rate of 20-200 micrograms/min | The presence of microalbuminuria is associated with an increased relative risk of myocardial infarction, stroke, or cardiovascular death |
| Cystatin-C | Biochemical marker of vascular damage produced by all nucleated cells | Levels of cystatin-C are associated with cardiovascular and noncardiovascular death, and possibly cerebrovascular disease |
| PAPP-A | Metalloproteinase used as a prenatal screening test for the detection of trisomy 21 | Associations with adverse cardiovascular outcomes, cardiovascular death, even with normal troponins |

**Abbreviations**: cTn: cardiac troponin, ACS: acute coronary syndrome, CVD: cardiovascular disease, hs-CRP: high-sensitivity CRP, MMP: matrix metalloproteinase, BNP: Brain-type natriuretic peptide, PAPP-A pregnancy associated plasma protein A.

## Conclusion

The treatment of a patient presenting in an emergency department with symptoms of myocardial ischemia remains a challenging procedure for the clinician. In most cases, a potential ACS demands rapid evaluation and decision making. The aforementioned data suggest that there are several promising biomarkers that may be capable of complementing current routine assessment in this group of patients. However, an establishment of such a biomarker in clinical practice demands several properties, such as reliability, accessibility, cost effectiveness, and easy interpretation by clinicians. Troponins, CRP, BNP and its fragment are well-established in the prognosis and the risk assessment of ACS. Cystatin-C, myoglobin, CD40L, MMPs microalbuminuria and PAPP-A are newer biomarkers, not regularly used in everyday clinical practice, but with important contributions in this setting. A combination of biomarkers, rather than individual ones, may provide more important prognostic information and contribute to a better understanding of the pathophysiological mechanisms in ACS. Further studies are needed in order to elucidate and resolve all controversies in these areas.

## References

Agarwal, M., P.K. Mehta, and C.N. Bairey Merz. 2010. Nonacute coronary syndrome anginal chest pain. Med. Clin. North Am. 94: 201–216.

Aggarwal, A., D.J. Schneider, E.F. Terrien, B.E. Sobel, and H.L. Dauerman. 2004. Increased Coronary arterial release of interleukin-1 receptor antagonist and soluble CD40 ligand indicative of inflammation associated with culprit coronary plaques. Am. J. Cardiol. 93: 6–9.

Androulakis, E., D. Tousoulis, N. Papageorgiou, G. Latsios, G. Siasos, C. Tsioufis, A. Giolis, and C. Stefanadis. 2011. Inflammation in hypertension: current therapeutic approaches. Curr. Pharm. Des. 17: 4121–4131.

Androulakis, E., D. Tousoulis, N. Papageorgiou, G. Latsios, G. Siasos, and C. Stefanadis. 2012. The role of matrix metalloproteinases in essential hypertension. Curr. Top. Med. Chem. 12: 1149–1158.

Androulakis, E.S., D. Tousoulis, N. Papageorgiou, C. Tsioufis, I. Kallikazaros, and C. Stefanadis. 2009. Essential hypertension: is there a role for inflammatory mechanisms? Cardiol. Rev. 17: 216–221.

Antman, E.M., D.T. Anbe, P.W. Armstrong, E.R. Bates, L.A. Green, M. Hand, J.S. Hochman, H.M. Krumholz, F.G. Kushner, G.A. Lamas, C.J. Mullany, J.P. Ornato, D.L. Pearle, M.A. Sloan, S.C. Jr. Smith, J.S. Alpert, J.L. Anderson, D.P. Faxon, V. Fuster, R.J. Gibbons, G. Gregoratos, J.L. Halperin, L.F. Hiratzka, S.A. Hunt, A.K. Jacobs, and J.P. Ornato. 2004. ACC/AHA guidelines for the management of patients with ST-elevation myocardial infarction; A report of the American College of Cardiology/American Heart Association Task Force on Practice Guidelines (Committee to Revise the 1999 Guidelines for the Management of patients with acute myocardial infarction). J. Am. Coll. Cardiol. 44: E1–E211.

Antoniades, C., D. Tousoulis, K. Marinou, N. Papageorgiou, E. Bosinakou, C. Tsioufis, E. Stefanadi, G. Latsios, C. Tentolouris, G. Siasos, and C. Stefanadis. 2007. Effects of insulin dependence on inflammatory process, thrombotic mechanisms and endothelial function, in patients with type 2 diabetes mellitus and coronary atherosclerosis. Clin. Cardiol. 30: 295–300.

Aukrust, P., F. Müller, T. Ueland, T. Berget, E. Aaser, A. Brunsvig, N.O. Solum, K. Forfang, S.S. Frøland, and L. Gullestad. 1999. Enhanced levels of soluble and membrane-bound CD40 ligand in patients with unstable angina. Possible reflection of T lymphocyte and platelet involvement in the pathogenesis of acute coronary syndromes. Circulation 100: 614–620.

Bertrand, M.E., M.L. Simoons, K.A. Fox, L.C. Wallentin, C.W. Hamm, E. McFadden, P.J. De Feyter, G. Specchia, and W. Ruzyllo. 2002. Task Force on the Management of Acute Coronary Syndromes of the European Society of Cardiology. Management of acute coronary syndromes in patients presenting without persistent ST-segment elevation. Eur. Heart J. 23: 1809–1840.

Biasucci, L.M., A. Vitelli, G. Liuzzo, S. Altamura, G. Caligiuri, C. Monaco, A.G. Rebuzzi, G. Ciliberto, and A. Maseri. 1996. Elevated levels of interleukin-6 in unstable angina. Circulation 94: 874–877.

Braunwald, E., E.M. Antman, J.W. Beasley, R.M. Califf, M.D. Cheitlin, J.S. Hochman, R.H. Jones, D. Kereiakes, J. Kupersmith, T.N. Levin, C.J. Pepine, J.W. Schaeffer, E.E. 3rd. Smith, D.E. Steward, P. Theroux, R.J. Gibbons, J.S. Alpert, D.P. Faxon, V. Fuster, G. Gregoratos, L.F. Hiratzka, A.K. Jacobs, and S.C. Jr. Smith; American College of Cardiology; American Heart Association. Committee on the Management of Patients with Unstable Angina. 2002. ACC/AHA 2002 guideline update for the management of patients with unstable angina and non-ST-segment elevation myocardial infarction—summary article: a report of the American College of Cardiology/American Heart Association task force on practice guidelines (Committee on the Management of Patients with Unstable Angina). J. Am. Coll. Cardiol. 40: 1366–1374.

Cipollone, F., M. Fazia, A. Iezzi, M. Zucchelli, B. Pini, D. De Cesare, S. Ucchino, F. Spigonardo, G. Bajocchi, R. Bei, R. Muraro, L. Artese, A. Piattelli, F. Chiarelli, F. Cuccurullo, and A. Mezzetti. 2003. Suppression of the functionally coupled cyclooxygenase-2/prostaglandin E synthase as a basis of simvastatin dependent plaque stabilization in humans. Circulation 107: 1479–1485.

de Lemos, J.A., D.A. Morrow, J.H. Bentley, T. Omland, M.S. Sabatine, C.H. McCabe, C. Hall, C.P. Cannon, and E. Braunwald. 2001. The prognostic value of B-type natriuretic peptide in patients with acute coronary syndromes. N. Engl. J. Med 345: 1014–1021.

de Lemos, J.A. and D.A. Morrow. 2002a. Brain natriuretic peptide measurement in acute coronary syndromes: ready for clinical application? Circulation 106: 2868–2870.

de Lemos, J.A., D.A. Morrow, C.M. Gibson, S.A. Murphy, M.S. Sabatine, N. Rifai, C.H. McCabe, E.M. Antman, C.P. Cannon, and E. Braunwald. 2002b. The prognostic value of serum myoglobin in patients with non-ST-segment elevation acute coronary syndromes. Results from the TIMI 11B and TACTICS-TIMI 18 studies. J. Am. Coll. Cardiol. 40: 238–244.

Fonarow, G.C. 2007. The global burden of atherosclerotic vascular disease. Nat. Clin. Pract. Cardiovasc. Med. 4: 530–531.

Foreman, R.D. 2004. Mechanisms of visceral pain: from nociception to targets. Drug Discovery Today: Disease Mechanisms 1: 457–463.

Fraker, T.D. Jr. and S.D. Fihn. 2002. Chronic Stable Angina Writing Committee; American College of Cardiology; American Heart Association, R.J. Gibbons, J. Abrams, K. Chatterjee, J. Daley, P.C. Deedwania, J.S. Douglas, T.B. Jr. Ferguson, J.M. Gardin, R.A. O'Rourke, S.V. Williams, S.C. Jr. Smith, A.K. Jacobs, C.D. Adams, J.L. Anderson, C.E. Buller, M.A. Creager, S.M. Ettinger, J.L. Halperin, S.A. Hunt, H.M. Krumholz, F.G. Kushner, B.W. Lytle, R. Nishimura, R.L. Page, B. Riegel, L.G. Tarkington, and C.W. Yancy. 2007. Chronic angina focused update of the ACC/AHA 2002 guidelines for the management of patients with chronic stable angina: a report of the American College of Cardiology/

American Heart Association Task Force on Practice Guidelines Writing Group to develop the focused update of the 2002 guidelines for the management of patients with chronic stable angina. J. Am. Coll. Cardiol. 50: 2264–2274.

Gerstein, H.C., J.F. Mann, Q. Yi, B. Zinman, S.F. Dinneen, B. Hoogwerf, J.P. Hallé, J. Young, A. Rashkow, C. Joyce, S. Nawaz, and S. Yusuf; HOPE Study Investigators. 2001. Albuminuria and risk of cardiovascular events, death, and heart failure in diabetic and nondiabetic individuals. JAMA 286: 421–426.

Goodman, A.R., T. Cardozo, R. Abagyan, A. Altmeyer, H.G. Wisniewski, and J. Vilcek. 1996. Long pentraxins: an emerging group of proteins with diverse functions. Cytokine Growth Factor Rev. 7: 191–202.

Gopal, D.J., M.N. Iqbal, and A. Maisel. 2011. Updating the role of natriuretic peptide levels in cardiovascular disease. Postgrad. Med. 123: 102–13.

Han, L., X. Bai, H. Lin, X. Sun, and X.M. Chen. 2010. Lack of independent relationship between age-related kidney function decline and carotid intima-media thickness in a healthy Chinese population. Nephrol. Dial. Transplant. 25: 1859–1865.

Harb, T.S., W. Zareba, A.J. Moss, P.M. Ridker, V.J. Marder, N. Rifai, L.F. Miller Watelet, R. Arora, M.W. Brown, R.B. Case, E.M. Jr. Dwyer, J.A. Gillespie, R.E. Goldstein, H. Greenberg, J. Hochman, R.J. Krone, C.S. Liang, E. Lichstein, W. Little, F.I. Marcus, D. Oakes, C.E. Sparks, and L. VanVoorhees; THROMBO Investigators. 2002. Association of C-reactive protein and serum amyloid A with recurrent coronary events in stable patients after healing of acute myocardial infarction. Am. J. Cardiol. 89: 216–221.

Heeschen, C., S. Dimmeler, C.W. Hamm, M.J. van den Brand, E. Boersma, A.M. Zeiher, and M.L. Simoons. CAPTURE Study Investigators. 2003. Soluble CD40 Ligand in Acute Coronary Syndromes. N. Engl. J. Med. 348: 1104–1111.

Heeschen, C., C.W. Hamm, J. Bruemmer, and M.L. Simoons. 2000. Predictive value of C-reactive protein and troponin T in patients with unstable angina: a comparative analysis. CAPTURE Investigators. Chimeric c7E3 AntiPlatelet Therapy in Unstable angina REfractory to standard treatment trial. J. Am. Coll. Cardiol. 35: 1535–1542.

Heeschen, C., S. Dimmeler, C.W. Hamm, S. Fichtlscherer, M.L. Simoons, A.M. Zeiher. CAPTURE Study Investigators. 2005. Pregnancy-associated plasma protein-A levels in patients with acute coronary syndromes Comparison with markers of systemic inflammation, platelet activation, and myocardial necrosis. J. Am. Coll. Cardiol. 45: 229–237.

Hillege, H.L., V. Fidler, G.F. Diercks, W.H. van Gilst, D. de Zeeuw, D.J. van Veldhuisen, G.O. Gans, W.M. Janssen, D.E. Grobbee, and P.E. de Jong. 2002. Prevention of Renal and Vascular End Stage Disease (PREVEND) Study Group. Urinary albumin excretion predicts cardiovascular and noncardiovascular mortality in general population. Circulation 106: 1777–1782.

James, S.K., P. Armstrong, E. Barnathan, R. Califf, B. Lindahl, A. Siegbahn, M.L. Simoons, E.J. Topol, P. Venge, and L. Wallentin; GUSTO-IV-ACS Investigators. 2003. Troponin and C-reactive protein have different relations to subsequent mortality and myocardial infarction after acute coronary syndrome: a GUSTO-IV substudy. J. Am. Coll. Cardiol. 41: 916–924.

Jesse, R.L. and M.C. Kontos. 1997. Evaluation of chest pain in the emergency department. Curr. Probl. Cardiol. 22: 149–236.

Jones, C.B., D.C. Sane, and D.M. Herrington. 2003. Matrix metalloproteinases: a review of their structure and role in acute coronary syndrome. Cardiovasc. Res. 59: 812–823.

Kachintorn, U. 2005. How do we define non-cardiac chest pain? J. Gastroenterol. Hepatol. 20(Suppl.): S2–S5.

Kampoli, A.M., D. Tousoulis, N. Papageorgiou, Z. Pallatza, G. Vogiatzi, A. Briasoulis, E. Androulakis, C. Toutouzas, P. Stougianos, C. Tentolouris, and C. Stefanadis. 2012a. Clinical utility of biomarkers in premature atherosclerosis. Curr. Med. Chem. 19: 2521–2533.

Kampoli, A.M., D. Tousoulis, N. Papageorgiou, C. Antoniades, E. Androulakis, E. Tsiamis, G. Latsios, and C. Stefanadis. 2012b. Matrix metalloproteinases in acute coronary syndromes: current perspectives. Curr. Top. Med. Chem. 12: 1192–1205.

Klocke, F.J., D.P. Copley, J.A. Krawczyk, and M. Reichlin. 1982. Rapid renal clearance of immunoreactive canine plasma myoglobin. Circulation 65: 1522–1528.

Knight, E.L., J.C. Verhave, D. Spiegelman, H.L. Hillege, D. de Zeeuw, G.C. Curhan, and P.E. de Jong. 2004. Factors influencing serum cystatin C levels other than renal function and the impact on renal function measurement. Kidney Int. 65: 1416–1421.

L'Allier, P.L., J.C. Tardif, J. Grégoire, M. Joyal, J. Lespérance, A. Fortier, and M.C. Guertin. 2005. Sustained elevation of serum CD40 ligand levels one month after coronary angioplasty predicts angiographic restenosis. Can. J. Cardiol. 21: 495–500.

Libby, P. and P. Theroux. 2005. Pathophysiology of coronary artery disease. Circulation 111: 3481–3488.

Liuzzo, G., L.M. Biasucci, J.R. Gallimore, R.L. Grillo, A.G. Rebuzzi, M.B. Pepys, and A. Maseri. 1994. The prognostic value of C-reactive protein and serum amyloid a protein in severe unstable angina. N. Engl. J. Med. 331: 417–424.

Lloyd-Jones, D., R.J. Adams, T.M. Brown, M. Carnethon, S. Dai, G. De Simone, T.B. Ferguson, E. Ford, K. Furie, C. Gillespie, A. Go, K. Greenlund, N. Haase, S. Hailpern, P.M. Ho, V. Howard, B. Kissela, S. Kittner, D. Lackland, L. Lisabeth, A. Marelli, M.M. McDermott, J. Meigs, D. Mozaffarian, M. Mussolino, G. Nichol, V.L. Roger, W. Rosamond, R. Sacco, P. Sorlie, R. Stafford, T. Thom, S. Wasserthiel-Smoller, N.D. Wong, and J. Wylie-Rosett. American Heart Association Statistics Committee and Stroke Statistics Subcommittee. 2010. Executive summary: heart disease and stroke statistics—2010 update: a report from the American Heart Association. Circulation 121: 948–954.

Madero, M., C.L. Wassel, C.A. Peralta, S.S. Najjar, K. Sutton-Tyrrell, L. Fried, R. Canada, A. Newman, M.G. Shlipak, and M.J. Sarnak. Health ABC Study. 2009. Cystatin C associates with arterial stiffness in older adults. J. Am. Soc. Nephrol. 20: 1086–1093.

Manetti, L., E. Pardini, M. Genovesi, A. Campomori, L. Grasso, L.L. Morselli, I. Lupi, G. Pellegrini, L. Bartalena, F. Bogazzi, and E. Martino. 2005. Thyroid function differently affects serum cystatin C and creatinine concentrations. J. Endocrinol. Invest. 28: 346–349.

Martín-Ventura, J.L., L.M. Blanco-Colio, J. Tuñón, B. Muñoz-García, J. Madrigal-Matute, J.A. Moreno, M. Vega de Céniga, and J. Egido. 2009. Biomarkers in Cardiovascular Medicine. Rev. Esp. Cardiol. 62: 677–688.

Mogensen, C.E., E.M. Damsgaard, A. Frøland, S. Nielsen, N. de Fine Olivarius, and A. Schmitz. 1992. Microalbuminuria in non-insulin-dependent diabetes. Clin. Nephrol. 38 Suppl. 1:S28–39.

Moreau, M., I. Brocheriou, L. Petit, E. Ninio, M.J. Chapman, and M. Rouis. 1999. Interleukin-8 mediates down regulation of tissue inhibitor of metalloproteinase-1 expression in cholesterol-loaded human macrophages: relevance to stability of atherosclerotic plaque. Circulation 99: 420–426.

Morrow, D.A., M.S. Sabatine, M.L. Brennan, J.A. de Lemos, S.A. Murphy, C.T. Ruff, N. Rifai, C.P. Cannon, and S.L. Hazen. 2008. Concurrent evaluation of novel cardiac biomarkers in acute coronary syndrome: myeloperoxidase and soluble CD40 ligand and the risk of recurrent ischaemic events in TACTICS-TIMI 18. Eur. Heart J. 29: 1096–1102.

Nakamura, K., T. Iizuka, M. Takahashi, K. Shimizu, H. Mikamo, T. Nakagami, M. Suzuki, K. Hirano, Y. Sugiyama, T. Tomaru, Y. Miyashita, K. Shirai, and H. Noike. 2009. Association between cardio-ankle vascular index and serum cystatin C levels in patients with cardiovascular risk factor. J. Atheroscler. Thromb. 16: 371–379.

Ohman, E.M., P.W. Armstrong, R.H. Christenson, C.B. Granger, H.A. Katus, C.W. Hamm, M.A. O'Hanesian, G.S. Wagner, N.S. Kleiman, F.E. Jr. Harrell, R.M. Califf, and E.J. Topol. 1996. Cardiac troponin T levels for risk stratification in acute myocardial ischemia. GUSTO IIA Investigators. N. Engl. J. Med. 335: 1333–1341.

Olenchock, B.A., S.D. Wiviott, S.A. Murphy, C.P. Cannon, N. Rifai, E. Braunwald, and D.A. Morrow. 2008. Lack of association between soluble CD40L and risk in a large cohort of patients with acute coronary syndrome in OPUS TIMI-16. J. Thromb. Thrombolysis. 26: 79–84.

Ottani, F., M. Galvani, F.A. Nicolini, D. Ferrini, G. Pozzati, Di Pasquale, and A.S. Jaffe. 2000. Elevated cardiac troponin levels predict the risk of adverse outcome in patients with acute coronary syndromes. Am. Heart J. 140: 917–927.

Rebuzzi, A.G., G. Quaranta, G. Liuzzo, G. Caligiuri, G.A. Lanza, J.R. Gallimore, R.L. Grillo, D. Cianflone, L.M. Biasucci, and A. Maseri. 1998. Incremental prognostic value of serum levels of troponin T and C-reactive protein on admission in patients with unstable angina pectoris. Am. J. Cardiol. 82: 715–719.

Rondina, M.T., J.M. Lappé, J.F. Carlquist, J.B. Muhlestein, M.J. Kolek, B.D. Horne, R.R. Pearson, and J.L. Anderson. 2008. Soluble CD40 ligand as a predictor of coronary artery disease and long-term clinical outcomes in stable patients undergoing coronary angiography. Cardiology 109: 196–201.

Shlipak, M.G., M.J. Sarnak, R. Katz, L.F. Fried, S.L. Seliger, A.B. Newman, D.S. Siscovick, and C. Stehman-Breen. 2005. Cystatin C and the risk of death and cardiovascular events among elderly persons. N. Engl. J. Med. 352: 2049–2060.

Shlipak, M.G., R. Katz, M.J. Sarnak, L.F. Fried, A.B. Newman, C. Stehman-Breen, S.L. Seliger, B. Kestenbaum, B. Psaty, R.P. Tracy, and D.S. Siscovick. 2006. Cystatin C and prognosis for cardiovascular and kidney outcomes in elderly persons without chronic kidney disease. Ann. Intern. Med. 145: 237–246.

Srinivas V.S., C.P. Cannon, C.M. Gibson, E.M. Antman, M.A. Greenberg, M.J. Tanasijevic, S. Murphy, J.A. de Lemos, S. Sokol, E. Braunwald, and H.S. Mueller. 2001. Myoglobin levels at 12 hours identify patients at low risk for 30-day mortality after thrombolysis in acute myocardial infarction: a Thrombolysis in Myocardial Infarction 10B substudy. Am. Heart J. 142: 29–36.

Srinivasan, N., H.E. White, and J. Emsley. 1994. Comparative analyses of pentraxins: implications for promoter assembly and ligand binding. Structure 2: 1017–1027.

Stefanadi, E., D. Tousoulis, N. Papageorgiou, A. Briasoulis, and C. Stefanadis. 2010. Inflammatory biomarkers predicting events in atherosclerosis. Curr Med Chem 17: 1690–1707.

Steg. P.G., D.L. Bhatt, P.W. Wilson, R. Sr. D'Agostino, E.M. Ohman, J. Röther, C.S. Liau, A.T. Hirsch, J.L. Mas, Y. Ikeda, M.J. Pencina, S. Goto; REACH Registry Investigators. 2007. One-year cardiovascular event rates in outpatients with atherothrombosis. JAMA 297: 1197–1206.

Suleiman, M., D. Aronson, S.A. Reisner, M.R. Kapeliovich, W. Markiewicz, Y. Levy, and H. Hammerman. 2003. Admission C-reactive protein levels and 30-day mortality in patients with acute myocardial infarction. Am. J. Med. 115: 695–701.

Suleiman, M., R. Khatib, Y. Agmon, R. Mahamid, M. Boulos, M. Kapeliovich, Y. Levy, R. Beyar, W. Markiewicz, H. Hammerman, and D. Aronson. 2006. Early inflammation and risk of long-term development of heart failure and mortality in survivors of acute myocardial infarction predictive role of C-reactive protein. J. Am. Coll. Cardiol. 47: 962–8.

Tan, J., Q. Hua, J. Gao, and Z.X. Fan. 2008. Clinical implications of elevated serum interleukin-6, soluble CD40 ligand, metalloproteinase-9, and tissue inhibitor of metalloproteinase-1 in patients with acute ST-segment elevation myocardial infarction. Clin. Cardiol. 31: 413–418.

Tanaka, H., Y. Tsurumi, and H. Kasanuki. 2006. Prognostic value of C-reactive protein and troponin T level in patients with unstable angina pectoris. J. Cardiol. 47: 173–179.

Thygesen, K., J.S. Alpert, and H.D. White; Joint ESC/ACCF/AHA/WHF Task Force for the Redefinition of Myocardial Infarction, A.S. Jaffe, F.S. Apple, M. Galvani, H.A. Katus, L.K. Newby, J. Ravkilde, B. Chaitman, P.M. Clemmensen, M. Dellborg, H. Hod, P. Porela, R. Underwood, J.J. Bax, G.A. Beller, R. Bonow, E.E. Van der Wall, J.P. Bassand, W. Wijns, T.B. Ferguson, P.G. Steg, B.F. Uretsky, D.O. Williams, P.W. Armstrong, E.M. Antman,

K.A. Fox, C.W. Hamm, E.M. Ohman, M.L. Simoons, P.A. Poole-Wilson, E.P. Gurfinkel, J.L. Lopez-Sendon, P. Pais, S. Mendis, J.R. Zhu, L.C. Wallentin, F. Fernández-Avilés, K.M. Fox, A.N. Parkhomenko, S.G. Priori, M. Tendera, L.M. Voipio-Pulkki, A. Vahanian, A.J. Camm, R. De Caterina, V. Dean, K. Dickstein, G. Filippatos, C. Funck-Brentano, I. Hellemans, S.D. Kristensen, K. McGregor K, U. Sechtem, S. Silber, M. Tendera, P. Widimsky, J.L Zamorano, J. Morais, S. Brener, R. Harrington, D. Morrow, M. Lim, M.A. Martinez-Rios, S. Steinhubl, G.N. Levine, W.B. Gibler, D. Goff, M. Tubaro, D. Dudek, and N. Al-Attar. 2007. Universal definition of myocardial infarction. Circulation 116: 2634–2653.

Toft, I., M. Solbu, J. Kronborg, U.D. Mathisen, B.O. Eriksen, H. Storhaug, T. Melsom, M.L. Løchen, E.B. Mathiesen, I. Njølstad, T. Wilsgaard, and J. Brox. 2012. Cystatin C as risk factor for cardiovascular events and all-cause mortality in the general population. The Tromso Study. Nephrol. Dial. Transplant. 27: 2780–2787.

Tomoda, H. and N. Aoki. 2000. Prognostic value of C-reactive protein levels within six hours after the onset of acute myocardial infarction. Am. Heart J. 140: 324–328.

Torzewski, J., M. Torzewski, D.E. Bowyer, M. Fröhlich, W. Koenig, J. Waltenberger, C. Fitzsimmons, and V. Hombach. 1998. C-reactive protein frequently colocalizes with the terminal complement complex in the intima of early atherosclerotic lesions of human coronary arteries. Arterioscler. Thromb. Vasc. Biol. 18: 1386–1392.

Tousoulis, D., C. Antoniades, and C. Stefanadis. 2007a. Assessing inflammatory status in cardiovascular disease. Heart 93: 1001–1007.

Tousoulis, D., C. Antoniades, A. Nikolopoulou, K. Koniari, C. Vasiliadou, K. Marinou, N. Koumallos, N. Papageorgiou, E. Stefanadi, G. Siasos, and C. Stefanadis. 2007b. Interaction between cytokines and sCD40L in patients with stable and unstable coronary syndromes. Eur. J. Clin. Invest. 37: 623–628.

Tousoulis D., E. Androulakis, N. Papageorgiou, A. Briasoulis, G. Siasos, C. Antoniades, and C. Stefanadis. 2010. From atherosclerosis to acute coronary syndromes: the role of soluble CD40 ligand. Trends Cardiovasc. Med. 20: 153–64.

Tousoulis, D., G. Hatzis, N. Papageorgiou, E. Androulakis, G. Bouras, A. Giolis, C. Bakogiannis, G. Siasos, G. Latsios, C. Antoniades, and C. Stefanadis. 2012. Assessment of acute coronary syndromes: focus on novel biomarkers. Curr. Med. Chem. 19: 2572–2587.

Tousoulis D., A.M. Kampoli, E. Stefanadi, C. Antoniades, G. Siasos, A.G. Papavassiliou, and C. Stefanadis. 2008. New biochemical markers in acute coronary syndromes. Curr. Med. Chem. 15: 1288–1296.

Tousoulis, D., M. Charakida, and C. Stefanadis. 2006. Endothelial function and inflammation in coronary artery disease. Heart 92: 441–444.

Tsioufis, C., K. Dimitriadis, E. Taxiarchou, C. Vasiliadou, G. Chartzoulakis, D. Tousoulis, A. Manolis, C. Stefanadis, and I. Kallikazaros. 2006. Diverse associations of microalbuminuria with C reactive protein, interleukin 18, and soluble CD 40 ligand in male essential hypertensive subjects. Am. J. Hypertens. 19: 462–466.

Tsioufis, C., K. Dimitriadis, M. Selima, C. Thomopoulos, C. Mihas, I. Skiadas, D. Tousoulis, C. Stefanadis, and I. Kallikazaros. 2007. Low-grade inflammation and hypoadiponectinemia have an additive detrimental effect on aortic stiffness in essential hypertensive patients. Eur. Heart J. 28: 1162–1169.

Undas, A., K. Szułdrzyński, K.E. Brummel-Ziedins, W. Tracz, K. Zmudka, and K.G. Mann. 2009. Systemic blood coagulation activation in acute coronary syndromes. Blood 113: 2070–2078.

Volanakis, J.E. 2001. Human C-reactive protein: expression, structure, and function. Mol. Immunol. 38: 189–197.

Wachtell, K., H. Ibsen, M.H. Olsen, K. Borch-Johnsen, L.H. Lindholm, C.E. Mogensen, B. Dahlöf, R.B. Devereux, G. Beevers, U. de Faire, F. Fyhrquist, S. Julius, S.E. Kjeldsen, K. Kristianson, O. Lederballe-Pedersen, M.S. Nieminen, P.M. Okin, P. Omvik, S. Oparil, H. Wedel, S.M. Snapinn, and P. Aurup. 2003. Albuminuria and cardiovascular risk in hypertensive patients with left ventricular hypertrophy: the LIFE study. Ann. Intern. Med. 139: 901–906.

Wang, Y., L. Li, H.W. Tan, G.S. Yu, Z.Y. Ma, Y.X. Zhao, and Y. Zhang. 2007. Transcoronary concentration gradient of sCD40L and hsCRP in patients with coronary heart disease. Clin. Cardiol. 30: 86–91.

Yan, J., Z. Wu, Z. Huang, L. Li, R. Zhong, and X. Kong. 2002. Clinical implications of increased expression of CD40L in patients with acute coronary syndromes. Chin. Med. J. (Engl.) 115: 491–493.

Yan, J.C., J. Zhu, L. Gao, Z.G. Wu, X.T. Kong, R.Q. Zong, and L.Z. Zhan. 2004. The effect of elevated serum soluble CD40 ligand on the prognostic value in patients with acute coronary syndromes. Clinica Chimica Acta 343: 155–159.

Yasojima, K., C. Schwab, E.G. McGeer, and P.L. McGeer. 2001. Generation of C-reactive protein and complement components in atherosclerotic plaques. Am. J. Pathol. 158: 1039–1051.

# CHAPTER 14

# Novel Cardiac Biomarkers and Application to Heart Failure

Evangelos Oikonomou,* Dimitris Tousoulis, Gerasimos Siasos, Nikolaos Papageorgiou and Christodoulos Stefanadis

## Introduction

Heart failure (HF) is defined as an abnormality of cardiac structure or function impairing oxygen delivery from the heart at a rate proportionate to the requirements of metabolizing tissues. The prevalence of HF in the western world is calculated at 1–2 percent; in the population over 70 years old, the prevalence is rising to 10 percent. Hence, it is estimated that 10–15 million Europeans suffer from HF (Mosterd and Hoes 2007). Despite new treatments, exacerbations of HF and readmission to hospital are common and the overall prognosis remains poor, equivalent or even worse than some malignant neoplasm, posing significant costs to healthcare systems (Levy et al. 2002).

Several functional or structural abnormalities of the myocardium, endocardium, pericardium, heart valves, and conduction system lead to HF development—hence, the term "syndrome" is applied to this condition. The multiplicity of this syndrome is reflected in its categorization in HF with preserved and reduced ejection fraction, in acute and chronic and in mild, moderate, or severe HF according to the severity of symptoms. HF is also categorized based on its pathophysiology as ischemic, hypertensive, valvular, alcoholic, idiopathic, dilated, etc. (McMurray et al. 2012).

---

1st Cardiology Department, University of Athens Medical School, "Hippokration" Hospital, Athens, Greece.
*Corresponding author: boikono@gmail.com

Consequently, the diagnosis of HF cannot be established by a simple test, but—in line with recently published European guidelines—is based on clinical signs and symptoms and on the demonstration of an underlying cardiac cause. Nevertheless, the diagnosis especially in the early stages, in elderly and in subjects with other co-morbidities such as obesity and chronic pulmonary disease, can be difficult (Kelder et al. 2011). In addition, symptoms of HF are not discriminating and thus of inadequate diagnostic value. Moreover, signs may be absent, equivocal or less reproducible depending on the physician's expertise (Fonseca 2006; McMurray et al. 2012).

In addition, although considerable variability exists concerning the pathophysiology, clinical course, severity and prognosis, most HF patients are prescribed the same medications and until now there have been no practical ways to definitively identify subjects likely to benefit from special treatments or interventions. For example, some patients fulfilling the criteria for left ventricle function and QRS width are not benefited from cardiac resynchronization therapy and the "one size fits all" approach seems ineffectual.

The considerable variability of HF syndrome, in combination with the increased prevalence of the disease and the lack of individualized therapy, make attractive the use of biomarkers—peptides that can be objectively measured in the blood as indicators of normal or pathogenic processes. Except from the natriuretic peptides and mainly B-type natriuretic peptide (BNP) and N terminal-pro-BNP (NT-pro-BNP) which are incorporated in the diagnostic algorithm of acute and chronic HF in the recently published European guidelines (McMurray et al. 2012), a series of novel biomarkers with different pathophysiologic roles have been identified (Table 1).

In the setting of HF, biomarkers can be used to establish the diagnosis, to identify patients at risk for HF, to differentiate between different pathophysiologic categories, to identify patients at risk for adverse outcome, to guide therapy and to assess treatment outcome. In the present review, we will present the role of novel biomarkers in HF.

## Neurohormonal Peptides

### *Cardiac natriuretic peptides*

Several natriuretic peptides have been identified and have been used for the diagnosis, monitoring and prognosis of HF patients. They regulate the body's intravascular volume and facilitate sodium excretion, diuresis, vasodilation, inhibit the rennin-angiotensin-aldosterone system (RAAS), downregulate cardiac and vascular myocyte growth and fibrosis, and facilitate myocardial relaxation (Daniels and Maisel 2007). The best studied natriuretic peptides are BNP and its counterpart NT-pro-BNP which are

**Table 1.** Biomarkers associated with heart failure according to their pathophysiologic role.

| Neurohormonal markers | |
|---|---|
| Cardiac natriuretic peptides | BNP (Fonarow et al. 2007) |
| | NT-pro-BNP (Hartmann et al. 2004) |
| | MR-pro-ANP (Masson et al. 2010) |
| | Chromogranin A (Dieplinger et al. 2010) |
| Renin angiotensin aldosterone system | Aldosterone (Palmer et al. 2008) |
| | Renin (Latini et al. 2004b) |
| Adrenergic nervous system | Norepinephrine (Yan et al. 2005) |
| | Epinephrine (Latini et al. 2004b) |
| | Arginine vasopressin (Rouleau et al. 1994) |
| **Endothelial derived peptides** | |
| | Endothelin-1 (Milo-Cotter et al. 2011) |
| | C-terminal pro-endothelin-1 (Jankowska et al. 2011) |
| | MR-proADM (Maisel et al. 2010) |
| **Inflammatory markers** | |
| | TNFα (Deswal et al. 2001) |
| | sTNFR1 and sTNFR2 (Deswal et al. 2001) |
| | IL-6 (Maeda et al. 2000) |
| | CRP (Kaneko et al. 1999) |
| | Lp-PLA2 (Gerber et al. 2009) |
| **Oxidative stress markers** | |
| | MPO (Reichlin et al. 2010) |
| | oxLDL antibodies (Charach et al. 2009) |
| | F2-isoprostanes (LeLeiko et al. 2009) |
| | Uric acid (Anker et al. 2003) |
| **Extracellular matrix remodeling markers** | |
| | MMP-2 (George et al. 2005) |
| | MMP-9 (Buralli et al. 2010) |
| | TIMP-1(Frantz et al. 2008) |
| | TIMP-4 (Felkin et al. 2009) |
| | Type I collagen telopeptide (Iraqi et al. 2009) |
| **Cardiorenal markers** | |
| | Cystatin-C (Shlipak et al. 2005) |
| | NGAL (Bolignano et al. 2009) |
| | NAG and KIM-1(Damman et al. 2010) |
| | Sodium (Gheorghiade et al. 2007b) |
| | Blood urea nitrogen (Filippatos et al. 2007) |
| | Creatinine (Abraham et al. 2008) |

*Table 1. contd....*

*...Table 1. contd.*

| Myocyte injury markers | |
|---|---|
| | Cardiac Troponin T (Latini et al. 2007) |
| | Cardiac Troponin I (Kawahara et al. 2011) |
| | Heart fatty acid binding protein (Niizeki et al. 2008) |
| **Other biomarkers** | |
| | Osteoprotogerin (Ueland et al. 2011) |
| | ST2 (Januzzi et al. 2007) |
| | Adiponectin (Kistorp et al. 2005) |
| | Resistin (Butler et al. 2009) |
| | CA125 (Nunez et al. 2010) |
| | Galectin 3 (van Kimmenade et al. 2006) |
| | Co-enzyme Q10 (Molyneux et al. 2008) |
| | Endoglin (Kapur et al. 2010) |
| | Hemoglobin and red cell Distribution (Felker et al. 2007) |
| | GDF-15 (Kempf et al. 2007) |

**Abbreviations:** BNP: B-type natriuretic peptide, NT-pro-BNP: amino-terminal BNP fragment, MR-pro-ANP: midregional pro-atrial natriuretic peptide, MR-proADM: mid-regional pro-adrenomedullin, TNFα: Tumor necrosis factor alpha, sTNFR: soluble TNFα receptor, IL-6: intereukin 6, CRP: C reactive protein, Lp-PLA2: Lipoprotein-associated phosholipase A2, MPO: Myeloperoxidase, oxLDL: oxidised low density lipoproteins, MMP-2: Matrix metalloproteinase 2, MMP-9: Matrix metalloproteinase 9, TIMP-1: Tissue inhibitor of matrix metalloproteinase-1, TIMP-4: Tissue inhibitor of matrix metalloproteinase-4, NGAL: Neutrophil gelatinase–associated lipocalin, NAG: N-acetyl-β-D-glucosaminidase, KIM-1: Kidney injury molecule 1, ST2: the interleukin 1 receptor family ST2, CA125: Tumor marker antigen carbohydrate 125, GDF-15: member of the transforming grow factor-β.

produced after cleavage of the pro-hormone (108 amino acids) in the biologically active BNP (32 amino acids) and the biologically inactive NT-pro-BNP (76 amino acids). Atrial natriuretic peptide (ANP), urodilantin and C-type natriuretic peptide (CNP) are also categorized in the natriuretic peptides (Daniels and Maisel 2007) (Fig. 1). Recent studies have recognized the role of the pro-atrial natriuretic peptide (MR-pro-ANP) (Masson et al. 2010) and Chromogranin A (Dieplinger et al. 2010).

## *BNP and NT-pro-BNP*

Although BNP and its amino-terminal counterpart were discovered a few decades ago and after the expansion of our knowledge regarding their clinical utility during the past decade, they still remain the most studied and useful biomarkers in HF subjects. Importantly, they are incorporated in the diagnostic algorithm of suspected HF in the recent European guidelines. Hence, in the acute settings, a value of BNP>100pg/ml and of

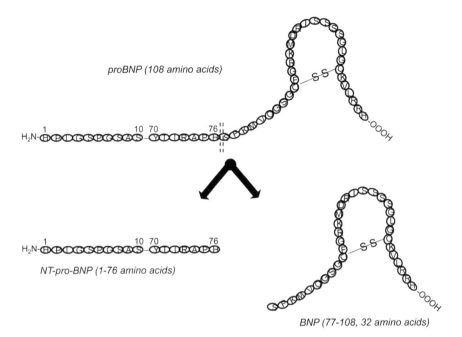

**Figure 1.** Structure of pro-BNP, NT-pro-BNB and BNP.

NT-pro-BNP>300pg/ml are considered indicative of HF. The cut-off values in chronic conditions are 35pg/ml for BNP and 125pg/ml for NT-pro-BNP (McMurray et al. 2012).

The Breathing Not Properly (BNP) study, in 1,586 subjects presenting in the emergency department with dyspnea, was the first study in 2002 that established a cut-off BNP value of 100pg/ml as 76 percent specific and 90 percent sensitive for the diagnosis of heart failure (Maisel et al. 2002). The clinical utility of BNP was confirmed by the Acute Shortness of Breath Evaluation (BASEL) study which concluded that used in conjunction with other clinical information, rapid measurement of BNP in the emergency department improved the evaluation and treatment of patients with acute dyspnea and thereby reduced the discharge time and the total cost of treatment (Mueller et al. 2004).

Since then several studies have confirmed the clinical utility of BNP. McCullough et al. concluded that the use of BNP reduces clinical indecision by 74 percent (McCullough et al. 2002). Moreover, there is an inverse correlation of BNP with left ventricle ejection fraction (Davis et al. 1994) and a positive association with New York Heart Association (NYHA) functional classification (Maisel et al. 2001). Importantly, monitoring of natriuretic peptides levels can be used to decide treatment strategies and to predict the response to specific treatment. Thus, a reduction of more than

58 percent in BNP levels after initiation of levosimendan can predict a low rate of events incidence in the following six months, with a sensitivity of 87 percent and a specificity of 83 percent (Farmakis et al. 2010). In addition, in non-ischemic HF patients treated with ventricular assist device, a faster decrease in BNP levels can recognize those patients who could be weaned from the device (Sodian et al. 2001).

NT-pro-BNP is also increased in HF, renders diagnostic and prognostic significance and is a valuable tool in the emergency department when there is clinical indecision about the causes of shortness of breath (Clerico et al. 2007; Hartmann et al. 2004).

## BNP and NT-pro-BNP guided therapy

Natriuretic peptides and especially BNP and NT-pro-BNP have also been used to guide HF patients' treatment. Several studies have addressed this issue (Table 2). The results are somewhat conflicting. STARBITE (Shah et al. 2011) and STARS-BNP (Jourdain et al. 2007), based on BNP value, didn't report a significant reduction in all cause mortality. TIME-CHF (Pfisterer et al. 2009) and BATTLESCARRED (Lainchbury et al. 2009) based on NT-pro-BNP levels concluded that only patients younger than 75 years old are benefited from a natriuretic peptide guided treatment in terms of all cause mortality. Recently, PROTECT trial which included only systolic HF subjects with a NT-pro-BNP target less than 1000 pg/ml concluded that such an approach was superior to standard care and reduced event rates, improved quality of life, and favorably affected cardiac remodeling (Januzzi et al. 2011). Based on these results, we can conclude that when using BNP or NT-pro-BNP to guide treatment, a target concentration for BNP at 125 pg/ml and for NT-pro-BNP at 1000 pg/ml must be set. Importantly, a rise or fall of 25 percent for NT-pro-BNP and to 40 percent for BNP is biologically meaningful. Finally, physicians must know that not all HF patients are benefited from a natriuretic peptide guided treatment strategy.

## MR-pro-ANP

ANP as BNP promotes natriuresis, diuresis and has vasodilatory properties. Recently, MR-pro-ANP has gained research interest as midregional epitopes of pro-hormones may be more stable to degradation by exoproteases and its concentration in the circulation is 10 to 50-fold higher than that of BNP (Potocki et al. 2010). Thus, in 1,237 patients with chronic stable HF enrolled in the GISSI-HF trial, the addition of MR-pro-ANP improved net reclassification for mortality when added to multivariable models based on clinical risk factors alone or together with NT-pro-BNP

**Table 2.** Summary of studies with natriuretic peptides guided heart failure treatment.

| Study | Natriuretic peptide | Subjects-Design-Strategy | Comments |
|---|---|---|---|
| **Negative results** | | | |
| STARBRITE (Shah et al. 2011) | BNP | 130 recently hospitalized HF patients. Randomized. Medications were titrated if BNP was 2x the discharge values. | There was no difference in all cause mortality. |
| TIME-CHF (Pfisterer et al. 2009) | NT-pro-BNP | 499 elderly (>60 years) HF patients. Randomized. NT-pro-BNP goal was 400pg/ml for subjects <75y and 800pg/ml for subjects≥75y. | There was only a trend (p=0.06) for NP guided treatment to improve all cause survival. In subjects <75y NP treatment improve survival. |
| BATTLESCARRD (Lainchbury et al. 2009) | NT-pro-BNP | 364 systolic and diastolic HF subjects. Randomized. Three treatment group: SC vs. NP guided (<1270pg/ml) vs. intensive care. | 1 year mortality was the same between NP guided treatment and intensive care. Subjects <75y had less event with NP guided treatment. |
| PRIMA (Eurlings et al. 2010) | NT-pro-BNP | 345 systolic and diastolic HF subjects. Randomized. NT-pro-BNP goal was set at the lowest level post discharge. | In the NT-pro-BNP guided group mortality was lower but not statistically significant (p = 0.206). |
| SINGAL-HF (Persson et al. 2010) | NT-pro-BNP | 252 HF subjects. Randomized simple blinded. | There were no differences between the groups concerning days alive, days out of hospital, and symptom score (P=0.28). |
| **Positive results** | | | |
| STARS-BNP (Jourdain et al. 2007) | BNP | 220 HF subjects. Randomized. BNP goal was (<100pg/ml). | In the BNP guided group the risk of HF related death and hospital stay was reduced (p<0.001). |
| PROTECT (Januzzi et al. 2011) | NT-pro-BNP | 151 systolic HF patients. Randomized. NT-pro-BNP goal was (<1,000 pg/ml). | NT-pro-BNP guided treatment has significantly reduced total cardiovascular events (p=0.02). |

**Abbreviations:** BNP: B-type natriuretic peptide; HF: heart failure, NT-pro-BNP: N terminal pro BNP, NP: natriuretic peptide.

(Masson et al. 2010). Moreover, changes in MR-pro-ANP concentrations were related to mortality. Thus, MR-pro-ANP is as accurate in the diagnosis of heart failure as NT-pro-BNP and provides incremental diagnostic information to BNP and NT-pro-BNP in patients presenting with BNP levels in the grey zone between 100 and 500pg/ml (Potocki et al. 2010). It has also been documented that MR-pro-ANP levels increased with NYHA functional class in chronic HF patients. Increased MR-pro-ANP levels were a predictor of poor outcome independently of NT-pro-BNP levels, age, creatinine, left ventricle ejection fraction, and body mass index adding prognostic value to NT-pro-BNP (von Haehling et al. 2007). Recently, it was also documented that in patients who had suffered acute myocardial infarction, there was a lower readmission rate for HF in those in the lower three quartiles of MR-pro-ANP (p<0.0001) and a higher HF readmission rate in those with both biomarkers (MR-pro-ANP and NT-pro-BNP) elevated in the highest quartile (p<0.0005) (Khan et al. 2008). The diagnostic accuracy of MR-pro-ANP is summarized in the recent guidelines of the European community which recommend that a cut of value of MR-pro-ANP of 120pmol/l can be used to distinguish patients presented in the emergency department with dyspnea to those with HF or other reasons for shortness of breath (McMurray et al. 2012).

## Chromogranin A and Chromogranin B

Chromogranin A (CgA) is a protein produced throughout the neuroendocrine system. CgA levels have been found to reflect sympathetic activity in both the adrenal medulla and the peripheral nervous system, indicating that circulating CgA levels may represent an index of overall neuroendocrine activity (Takiyyuddin et al. 1990). Although increased activity of the sympathetic nervous system is a recognized risk factor for poor outcome in HF, direct measurement of catecholamine levels is difficult and impractical, making CgA an interesting surrogate marker of sympathetic nervous system activity. Indeed, some early studies documented that serum CgA levels are elevated in relation to severity of HF (Gupta et al. 2009) and independently predict mortality in acute HF (Dieplinger et al. 2009). Nevertheless, recently GISSI-HF trial in 1,233 stable HF subjects concluded that circulating CgA levels do not provide incremental prognostic information to that obtained from physical examination, routine biochemical analysis, and contemporary HF biomarkers (Rosjo et al. 2010b).

Chromogranin B (CgB), another member of the granin protein family has recently been studied in 80 patients with systolic HF and it was found that CgB production is increased and regulated in proportion to disease severity in the left ventricle and circulation during HF development (Rosjo et al. 2010a). Nevertheless, until now there have been no studies evaluating the diagnostic and prognostic significance of CgB.

## Other Neurohormonal Markers

### *Aldosterone*

Aldosterone acts as a regulator of blood volume, electrolyte, and blood pressure through its direct action on the kidney. The important prognostic significance of aldosterone in HF is highlighted by three studies—the Randomized Aldactone Evaluation Study (RALES), the Eplerenone Post-Acute Myocardial Infarction Heart Failure Efficacy and Survival Study (EPHESUS), and the Eplerenone in patients with systolic heart failure and mild symptoms (EMPHASIS-HF) study (Pitt et al. 1999; Zannad et al. 2011). All three studies documented that blockade of aldosterone after acute myocardial infarction or in patients with mild or moderate systolic heart failure reduce both morbidity and mortality, emphasizing the role of aldosterone blockade in the standard care of HF patients.

### *Renin activity*

Renin activity is used as a surrogate of renin-angiotensin-aldosterone system activation. In Val-HeFT study, plasma renin activity was measured in 4,300 patients with stable, moderate to severe HF. The study concluded that plasma renin activity in this setting of subjects was an independent predictor of death with a hazard ratio of 1.27 (Latini et al. 2004b). Moreover, active renin activity levels are an independent predictor of survival in HF patients already receiving angiotensin-converting enzyme inhibitors or angiotensin-receptor blockers (Tsutamoto et al. 2007).

### *Adrenergic hormones*

Increased activity of the sympathetic nervous system is a recognized risk factor for poor outcome in HF and epinephrine and norepinephrine are increased according to the severity of HF (Peng et al. 2003). Additionally, in Val-HeFT study in 4,300 patients with HF, norepinephrine has emerged as an independent predictor of morbidity and mortality (Latini et al. 2004b).

### *Vasopressin*

Arginine vasopressin is increased in HF and promotes hyponatremia and congestion. Nevertheless, few studies have evaluated the prognostic significance of arginine vasopressin in HF mortality and morbidity. In the Survival and Ventricular Enlargement (SAVE) study which recruited 534 HF patients, arginine vasopressin levels were related to subsequent cardiovascular events and were independent predictors of cardiovascular mortality, development of severe HF or recurrent myocardial infarction (Rouleau et al. 1994). Despite the theoretical advantages of a treatment based on a vasopressin antagonism the Efficacy of Vasopressin Antagonism in Heart Failure Outcome Study With Tolvaptan (EVEREST) trial in 4,133 HF patients failed to improve the primary endpoints of all cause mortality, the composite of cardiovascular death and HF hospitalization or overall quality of life scores (Gheorghiade et al. 2007a).

## Endothelial Derived Peptides

### *Endothelin-1*

Endothelin-1 (ET-1) is a potent, endothelial cell-derived venous and arterial vasoconstrictor peptide that functions as both a circulating hormone and as a paracrine factor in the regulation of vascular tone (Ohmae 2011). Plasma ET-1 concentrations are generally increased in patients with HF (Tousoulis et al. 2005). It also appears that ET-1 levels are correlated with the severity of HF, the NYHA functional classification and inversely associated with the exercise capacity and left ventricle ejection fraction of HF patients (Yancy et al. 2004). Interestingly, the VERITAS study in 112 acute HF subjects concluded that plasma levels of ET-1 were predictive of worsening HF (Milo-Cotter et al. 2011). Recently, Minoru Ohmae suggested that ET-1 plays an important role in chronic congestive HF with preserved ejection fraction (Ohmae 2011). Importantly, a recent study in 236 post acute myocardial infarction patients and evidence of HF documented that the cluster of biomarkers, with proven independent predictive value for the outcome, included pro-ET-1, MR-pro-ADM, tumor necrosis factor alpha (TNFα), growth differentiation factor 15, C-terminal pro arginine vasopressin, uric acid, CgA, and procollagen type III N-terminal. Interesting the cluster of biomarkers with the natriuretic peptides does not contribute significantly to the outcome of post acute myocardial infarction HF patients (Manhenke et al. 2011). In accordance, in patients with systolic congestive HF, plasma C-terminal pro-endothelin-1 (CT-proET-1) constitutes a novel predictor of increased one year cardiovascular mortality. High CT-proET-1, in combination with high NT-pro-BNP enables the identification of patients with congestive HF and particularly unfavorable outcomes (Jankowska et al. 2011). Despite this evidence concerning the association of ET-1 with HF and prognosis, to date there are no data to establish the diagnostic accuracy of ET-1 or its use to guide and monitor treatment, limiting its use in the clinical practice.

## *Adrenomedullin*

Adrenomedullin (ADM) is a peptide hormone with hypertensive, vasodilatory, positive isotropic and natriuretic effects and its levels are found elevated in fluid-overloaded HF patients who have elevated filling pressures. Pressure and volume overload act as stimuli to multiple tissue, including kidney and medulla to release this hormone. Midregional pro-adrenomedullin (MR-pro-ADM) is a stable prohormone fragment of adrenomedullin that is easier to measure. The BACH (Biomarkers in Acute Heart Failure) trial, a prospective study that recruited 1,641 patients presenting to the emergency department with dyspnea, concluded that MR-pro-ADM levels added to the utility of BNP levels in patients with intermediate BNP values and with obesity. More precisely, the accuracy to predict 90-day survival of HF patients was 73 percent for MR-pro-ADM and 62 percent for BNP (difference p<0.001). In adjusted multivariable Cox regression, MR-pro-ADM, but not BNP, carried independent prognostic value (p<0.001). Results were consistent using NT-pro-BNP instead of BNP (p<0.001) (Maisel et al. 2010). Recently, another prospective study of 560 subjects presenting with dyspnea in the emergency department concluded that MR-pro-ADM was an independent predictor of 4-year mortality (Shah et al. 2012). In addition, MR-pro-ADM is not only associated with HF mortality but also with the clinical status of the patient as it is expressed with NYHA functional classification (von Haehling et al. 2010).

Despite the diagnostic and prognostic utility of ADM in HF to date there are no data on how we can use it in clinical practice. A recent study in patients with acute decompensated HF presenting in the emergency department follows the levels of MR-pro-ADM the first 24 hours after the initiation of treatment. This study revealed that in the first two to four hours, patients with elevations of MR-pro-ADM exhibited jugular venous distension and, later in the 12- to 24-hour time period, had increased peripheral edema and respiratory rates, and relatively worse hyponatremia. Similarly, patients who had decreased MR-pro-ADM levels over 12 to 24 hours showed deceleration in their respiratory rate and improvement in pulse oximetry. These patients also had decreases in BNP and increases in serum sodium, suggesting that decreases in MR-pro-ADM may be associated with decreased congestion. These preliminary results imply the hypothesis that changes in MR-pro-ADM levels can be used to stratify risk of acute decompensated HF subjects and to decide safely, the time of discharge from the emergency department (Boyer et al. 2012).

## Myocardial Injury Markers

During the past decades, diagnosis of myocardial injury was based on less specific markers of myocardial necrosis such as glutamineoxaloacetic transaminase (GOT), lactate dehydrogenase (LDH), creatine kinase (CK) and the MB fraction of CK, i.e., CKMB activity and CKMB mass (Thygesen et al. 2007). At present, more specific biomarkers of myocyte injury are in use such as cardiac troponins I and T, myosin light-chain kinase I, heart fatty acid binding protein and ischemia modified albumin.

## *The role of cardiac troponins in heart failure*

Cardiac troponins have long been used in the diagnosis of myocardial injury secondary to ischemia. Although three types of troponins (T, I, and C) have been identified, only cardiac troponin T and cardiac troponin I have been utilized as diagnostic assay targets. The elevation of cardiac troponins with acute or chronic HF (of ischemic or non ischemic etiology) is well recognized (Kawahara et al. 2011).

The precise explanation for such "troponin leak" in the non-acute coronary syndrome setting of HF is unclear. It is estimated that 1 g of myocardial mass corresponding to 64 million cells is lost per year in the human heart while it is believed that additional myocardial damage occurs from necrosis or apoptosis due to acute or chronic HF (Olivetti et al. 1995). Ongoing apoptosis and myocardial necrosis is believed to be the result of increased myocardial wall stress, while other potential contributors include diminished global (rather than regional) myocardial perfusion and oxygen delivery and/or diminished renal clearance.

Several studies have confirmed the prognostic significance of elevated cardiac troponins in the setting of HF. PROTECT pilot study in patients with acute HF concluded that positive troponin at baseline, or conversion to positive levels during hospitalization, were associated with worse outcomes at 60 days (O'Connor et al. 2011). In addition, the Acute Decompensated Heart Failure National Registry (ADHERE) have described a 2.6-fold increased risk of in-hospital mortality for patients with positive cardiac troponin levels at the time of admission (Peacock et al. 2008). Importantly, in patients with acute HF, elevated cardiac troponin levels were associated with lower left ventricle ejection fraction (La Vecchia et al. 2000) and mortality prediction was improved with the association of cardiac troponin level and BNP level (Horwich et al. 2003).

Unlike the consistent association of cardiac troponins with the prognosis of HF subjects, they cannot be used for the diagnosis of HF as they have limited specificity and sensitivity. Moreover, there are no studies evaluating how cardiac troponins can be used to guide therapy and monitor treatment's results. Thus their clinical applicability is limited.

## *Heart fatty acid binding proteins in heart failure*

Heart fatty acid binding protein is a cytoplasmic and non-enzymatic protein transporting long-chain fatty acids in cardiomyocytes that the damaged myocardium rapidly releases into the circulation (Tanaka et al. 1991). Heart fatty acid binding protein is constantly elevated in systolic HF, is correlated positively with NYHA class and is also associated with heart failure morbidity (Arimoto et al. 2005). In accordance, in elderly chronic HF individuals followed-up for 421 days, heart fatty acid binding proteins effectively risk stratified these patients for cardiac events (Niizeki et al. 2005). Moreover, the combination of BNP, pentrexin-3 and heart fatty acid binding protein has shown an additive predictive value for cardiac events in a study with 164 chronic HF individuals (Ishino et al. 2008). Interestingly, a recent study in 238 consecutive patients with congenital heart disease concluded that heart fatty acid binding protein could serve as a new monitoring tool to provide information that will guide the optimal therapy and management of these patients (Hayabuchi et al. 2011). Importantly, a study based on 130 patients with diastolic dysfunction revealed elevated levels of heart fatty acid binding protein, suggesting that ongoing myocardial damage plays a critical role in the pathophysiology of this entity and concluded that the combination of heart fatty acid binding protein with echocardiographic parameters might improve diagnostic accuracy and risk stratification of patients with diastolic HF (Dinh et al. 2011).

Taken together, we can conclude that heart fatty acid binding protein, a marker of myocardial damage, can definitely used as a prognostic marker in HF and in some cases can also enhance diagnostic ability and monitoring of HF patients.

## Inflammatory and Oxidative Stress Biomarkers in Heart Failure

### *The role of inflammatory cytokines*

Imbalance between pro-inflammatory and anti-inflammatory mediators is a well established pathogenetic mechanism which contributes significantly to the progression of HF (Tousoulis et al. 2012). Thus, according to the "cytokine hypothesis", enhanced levels of inflammatory mediators such as TNF-a, IL-6, IL-1 and the corresponding decrease in anti-inflammatory cytokines (Adamopoulos et al. 2001; Levine et al. 1990; Ohtsuka et al. 2005; Tousoulis et al. 2007a) has a central role in the maladaptive responses seen in advanced HF (Fig. 2). Nevertheless, the cause of this immune activation is largely unknown and several mechanisms have been proposed. According to these hypotheses, microbial antigens may cause myocardial damage through molecular mimicry and autoimmune responses (Becker et al. 2001). Toll-like receptors (TLRs) 2 and 4 are also involved in the preservation of immune activation after the initial insult (Dunzendorfer et al. 2004). In addition, the "endotoxin hypothesis" advocates that impairment of the gastrointestinal tract function leads to leakage of endotoxins from the gastrointestinal tract into the circulation (Niebauer et al. 1999).

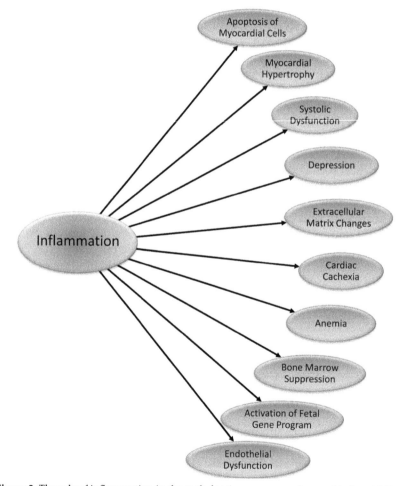

**Figure 2.** The role of inflammation in the maladaptive responses observed in heart failure.

## Prognostic significance of inflammatory cytokines in heart failure

As we have already mentioned, several inflammatory cytokines are implicated in the progression of HF (Oikonomou et al. 2011) and data have shown that they are also implicated in the prognosis of this condition. Indeed, recently it was shown that levels of TNF-α and IL-6 are related with echocardiographic indexes of both systolic and diastolic left ventricular function (Chrysohoou et al. 2009). Cytokines are also responsible for the endothelial dysfunction observed in HF patients (Antoniades et al. 2007; Tousoulis et al. 2007b; Tousoulis et al. 2005). Importantly, the Vesnarinone trial (VEST) which recruited 384 patients with moderate to severe HF documented a decline in survival with increasing TNF-α levels, with the worst survival in patients with TNFα>75th percentile (Deswal et al. 2001). Moreover, high plasma BNP and IL-6 levels three months after HF optimized treatment are independent risk factors for mortality despite improvements in left ventricle ejection fraction and symptoms (Maeda et al. 2000). There are also other biomarkers less well established with significant prognostic ability in HF. Thus, levels of lipoprotein-associated phospholipase A2 (Lp-PLA2) were proved to be independently associated with an increased risk of mortality in 646 consecutive HF patients (Gerber et al. 2009).

## The role of C-reactive protein in heart failure

C-reactive protein (CRP) is a widely known acute phase protein and has a central role in the inflammatory cataract. Although not an ideal index, the fact that it is produced by the liver in response to stimulation by proinflammatory cytokines such as IL-6, TNF-α, etc., in combination with its ease of measurement and the availability of the method make it attractive as a biomarker. Interestingly, the Framingham Heart Study showed that patients free from ischemic heart disease and increased levels of CRP (serum CRP level ≥5 mg/dL) had 2.8 times increased risk of HF development during a follow-up period of approximately 5 years compared to subjects with normal CRP levels (Vasan et al. 2003). Moreover, data from Val-HeFT confirmed that higher CRP levels are associated with severe symptoms and adverse outcome (Anand et al. 2005). Some studies have also found an association between CRP levels and left ventricle ejection fraction (Yin et al. 2004).

Importantly, as it was shown from GISSI-HF and CORONA trial, statin, with its anti-inflammatory and pleiotropic properties, cannot modulate the physical history of HF (Kjekshus et al. 2007; Tavazzi et al. 2008). Nevertheless,

CRP levels can guide treatment, as a post-hoc analysis of the CORONA trial concluded that rosuvastatin reduces cardiovascular mortality in those HF subjects with CRP levels above 2mg/l (McMurray et al. 2009).

## Pentraxin 3

Pentraxin-3 is a newly indentified biomarker which is produced in response to inflammatory stimulation. It is associated with the development of incident HF after coronary heart disease (Dubin et al. 2012). Moreover, plasma levels of pentraxin-3 are associated with NYHA functional class and with the prognosis of HF (Latini et al. 2004a; Suzuki et al. 2008). Importantly, pentraxin-3 can be used in the clinical settings to distinguish normal subjects from subjects with left ventricle diastolic dysfunction and heart failure with preserved ejection fraction (Matsubara et al. 2011).

## Anti-inflammatory strategies in heart failure

The growing recognition of the importance of immune and inflammatory mediators in the pathogenesis of HF has given rise to the hypothesis that treatment capable of modulating the inflammatory status of HF patients can also modify the natural history of the disease and improve the patient's clinical status. Nevertheless, the results were not as expected (Table 3).

Anti-inflammatory treatments are divided into three categories. The first includes conventional cardiovascular drugs such as statin, beta-blockers angiotensin converting enzyme inhibitors, and angiotensin II receptor antagonists etc., which have been proved to have anti-inflammatory properties. The second category includes drugs with specific anti-inflammatory properties which inhibit specific mediators and the third one includes treatment interventions with the potential to modulate the immune and inflammatory system as a whole.

## Established cardiovascular treatments with anti-inflammatory properties

Except for statin, data from studies with angiotensin converting enzyme inhibitors or angiotensin receptor blockers have shown a decrease in inflammatory markers, with a parallel decrease in indices of left ventricle function (Gullestad et al. 1999). Moreover, short term inotropic therapy with dobutamine or milrinone resulted in improved clinical status with a parallel reduction in inflammatory indices such as IL-6, IL-18, sICAM-1 and CRP (White et al. 2006).

**Table 3.** Studies examining the impact of immunomodulatory and anti-inflammatory strategies in HF patients.

| Study/(year) | Treatment | Patients population | Comments |
|---|---|---|---|
| ATTACH/(2003)(Chung et al. 2003) | Infliximab | 150 patients with NYHA III and IV and LVEF<35% | There was no improvement in clinical status and in LVEF |
| RECOVER/(2004) (Mann et al. 2004) | Etanercept | 1123 patients with ischemic and non ischemic CHF and NYHA II-IV | There was no clinical improvement and mortality was reduced only in patients treated with 25mg twice weekly |
| RENAISSANCE/(2004) (Mann et al. 2004) | Etanercept | 925 patients with ischemic and non ischemic CHF and NYHA II-IV | There was no improvement in clinical status and in overall mortality |
| RENEWAL/(2004) | Etanercept | 2,048 patients with ischemic and non ischemic CHF and NYHA II-IV | There was no improvement in the overall mortality and in the rates of hospitalization |
| ACCLAIM/(2008) (Torre-Amione et al. 2008) | Immunomodulation (Celacede) | 2,426 patients with ischemic and non ischemic CHF and LVEF<30% | There were no changes in mortality, hospitalization, LVEF, and NYHA class. Only in patients with no history of MI and NYHA class II, there was a reduction in rates of hospitalization and death |
| CORONA/(2007) (Cleland et al. 2009; McMurray et al. 2009) | Rosuvastatin | 5011 patients with ischemic CHF and a LVEF<40% | There was a reduction in CV hospitalizations and no improvement in cardiovascular events and in overall mortality. In patients with CRP>2mg/L there was a reduction in overall mortality and patients with BNP < 103pmol/L had lower rates of CV events |
| GISSI-HF/(2008) (Tavazzi et al. 2008) | Rosuvastatin | 4,574 with CHF of any cause and NYHA class II-IV | There was no change in time to death and hospitalization for CV reasons. |
| Intravenous immunoglobulin/(2001) (Gullestad et al. 2001) | IVIG | 40 patients with ischemic and dilated cardiomyopathy | Resulted in increase LVEF, in increased levels of anti-inflammatory mediators and in a reduction in NT-proBNP levels |
| Therapeutic plasma exchange/(2010)(Torre-Amione et al. 2010) | Plasma exchange and IVIG | 10 patients with LVEF<30% and NYHA II or IV | Resulted in improved LVEF and in improvements in quality of life as measured by the Minnesota Living with Heart Failure Questionnaire |
| Physical training/ (2002) (Adamopoulos et al. 2002) | 12 week program of physical training | 24 patients with CHF and 20 normal control subjects | Resulted in a significant improvement in VO2max and in functional status |

IVIG: intravenous immunoglobulin; LVEF: left ventricle ejection fraction; NYHA: New York Heart Association classification; MI: Myocardial infarction; CV: cardio-vascular; VO2max: peak oxygen consumption

## Specific anti-inflammatory and immune modulation treatments

Three studies—RENAISSANCE, RECOVER and ATTACH—with anti-TNFα anti-bodies failed to improve outcome in HF (Mann et al. 2004). ACCLAIM, a trial, enrolled 2,426 subjects with HF to study the effect of Celacade therapy, which induces apoptosis in a sample of the patient's blood exposed *in vivo* to oxidative stress, and concluded that there was no difference in the total population in all cause mortality and cardiovascular hospitalization (Torre-Amione et al. 2008). Finally, autoantibodies against cardiac cell proteins are thought to have an integral role in the pathogenesis of myocarditis and idiopathic dilated cardiomyopathy. Small studies have shown that removal of autoantibodies by immunoabsortion had beneficial effects in left ventricle function of those patients although there was no improvement in inflammatory biomarkers (Felix et al. 2002). Interestingly, immunoabsorption, when combined with intravenous immunoglobulin infusion offers a potential additive effect in the action of the last treatment.

## Other interventions modulating inflammatory status

Results concerning cardiac resynchronization (CRT) therapy are conflicting and some studies have shown a neutral effect of CRT in CRP levels while others documented a reduction in IL-6, IL-8, MCP-1 levels (Glick et al. 2006; Lappegard and Bjornstad 2006). Furthermore, a number of studies examine the impact of cardiac rehabilitation programs in the inflammatory status of chronic HF patients. Indeed, it was found that in HF patients, a physical training program results in a reduction of inflammatory markers such as TNF-α, IL-6, soluble intracellular adhesion molecule-1 (sICAM-1), etc. More importantly, it was found that training-induced improvement in exercise tolerance, in maximum oxygen consumption and in six minutes walk test was correlated with the attenuation of the inflammatory process (Adamopoulos et al. 2002; Bjornstad et al. 2008).

## Markers of oxidative stress

Myeloperoxidase (MPO), oxidised low density lipoproteins (oxLDL), lipid peroxides, F2-isoprostane, and malondialdehyde are oxidative stress markers found elevated in HF and for some of them, there are data connecting their levels with the prognosis of the disease.

## *Myeloperoxidase*

Myeloperoxidase (MPO) is a leukocyte derived enzyme that has multiple mechanistic links with vulnerable plaque development and progression of HF. MPO levels are also associated with NYHA functional class (Michowitz et al. 2008) and although it has a limited diagnostic accuracy to determine the cause of acute dyspnea in the emergency department, it adds predictive value when it is combined with BNP (Reichlin et al. 2010). Moreover rosuvastatin treatment decreases plasma myeloperoxidase in HF patients (Andreou et al. 2010).

## *Uric acid*

Uric acid is a commonly used biomarker that reflects underlying oxidative stress. It is considered as an independent predictor of poor prognosis in HF (Anker et al. 2003). Treatment with allopurinol, a xanthine osidase inhibitor, can benefit endothelial function in heart failure (Tousoulis et al. 2011). However, strategies to reduce uric acid by oxypurinol, another xanthine oxidase inhibitor, have demonstrated improvement in clinical outcomes restricted only to those with elevated serum uric acid levels >9.5pg/mL (Hare et al. 2008).

## Biomarkers of Cardiorenal Syndrome

Cardiorenal syndrome is defined as worsening renal function in the setting of worsening HF clinical status. Renal dysfunction not only co-exists with HF but also predicts poor outcome (Smilde et al. 2009). Except from classic indices of renal function such as serum creatinine, blood urea nitrogen, and glomerular filtration rate (GFR), recently novel biomarkers of renal function have been associated with the prognosis of HF subjects (Siasos et al. 2012).

## *Cystatin-C*

Cystatin-C is a cysteine protease inhibitor associated with early renal failure (Abu-Omar et al. 2005) and can predict renal function independently of age, sex, or muscle mass. In HF patients, Cystatin-C is a stronger predictor of mortality than creatinine (Shlipak et al. 2005) and for each standard deviation increase in cystatin-C, there was a 31 percent anticipated increase in adjusted mortality risk. Importantly, the combination of NT-pro-BNP and Cystatin-C was better in predicting 10-year mortality in HF patients than each individual measure (Alehagen et al. 2009). Despite these findings, the recent ARIC study concluded that the addition of cystatin-C in Framingham-recalibrated score does not add incremental risk prediction for the development of HF in the follow up of 13,555 subjects during 15.5 years (Agarwal et al. 2012). Whether Cystatin-C provides true incremental insight in the prognosis of HF remains to be determined in the future.

## Neutrophil gelatinase–associated lipocalin

Neutrophil gelatinase-associated lipocalin (NGAL) is a glycoprotein released by multiple cells including renal tubular cells as a response to inflammation or injury. In clinical settings such as cardiac surgery, contrast procedures and in the emergency department NGAL allow early identification of acute kidney injury (Siasos et al. 2012).

A prospective study concerning HF demonstrated that NGAL values increased in parallel with the clinical severity of chronic HF according to NYHA classification. Moreover, a two-year follow up revealed that patients with baseline NGAL > 783 ng/mL had a significantly higher mortality than those with lower levels of NGAL (Bolignano et al. 2009). Interestingly, NGAL can also be used for the prediction of acute decompensated HF patients prone to develop cardiorenal syndrome. Indeed, in patients with acute HF, a cutoff NGAL value of 170ng/L predicts 48- to 72-hour development of type 1 cardio-renal syndrome (Damman et al. 2008).

## Other Biomarkers in Heart Failure

### Growth differentiation factor 15

Growth differentiation factor 15 (GDF-15) is a member of the transforming factor-β cytokine group that plays a role in the development and regulation of responses to stress and inflammation. Although, it is hardly detectable in the healthy myocardium it becomes detectable in the human myocardium after myocardial infarction (Khan et al. 2009).

Interestingly, in chronic HF patients GDF-15 levels are associated with NYHA functional class (Kempf et al. 2007). In addition, GDF-15 has been identified as an independent marker of left ventricle remodeling in patients with ST-segment elevation myocardial infarction (Dominguez-Rodriguez et al. 2011). Moreover, in post myocardial infarction patients, levels of GDF-15 are associated with NT-pro-BNP levels and are a useful independent predictor of HF and death (Khan et al. 2009). Interestingly, in adults with operated congenital heart defects, GDF-15 might be used as a surrogate marker for latent heart failure and could help to identify patients who are at risk for developing HF, even if they are clinically asymptomatic (Norozi et al. 2011).

The previous studies highlight the potential role of GDF-15 as a monitoring tool or as a predictive biomarker in HF. Nevertheless, to date there are no studies evaluating the response of GDF-15 to HF treatment and its potential use to guide therapy.

## Galectin-3

Galectin-3 is a b-galactoside-binding protein that was identified in macrophages and other tissues, such as heart, liver, and kidney. Galectin-3 exhibits profibrotic properties, and experimental studies have shown that it is involved in cardiac fibrosis and HF.

In clinical settings, Galectin-3 appeared elevated in the plasma of HF patients presenting in the emergency department. Nevertheless, as it was documented from the PRIDE study, levels of Galectin-3 were not correlated with the NYHA functional class and were less sensitive and specific from NT-pro-BNP for the diagnosis of HF (Januzzi et al. 2005). Interestingly, the same study revealed that Galectin-3 was better than NT-pro-BNP in predicting 60 days mortality. In accordance with the PRIDE study, the COACH trial highlighted the ability of Galectin-3 in predicting mortality of patients with diastolic HF (de Boer et al. 2011). Of clinical importance is also the evidence that higher Galectin-3 levels are associated with worse right ventricular systolic function and increased right ventricle systolic pressure (Shah et al. 2010).

Other than its role in the diagnosis and prediction of HF outcome, Galectin-3 is also associated with the risk of developing HF following acute coronary syndrome, supporting its role as a biomarker of adverse remodeling that contributes to HF as well as a potential therapeutic target (Grandin et al. 2012). In accordance, a post hoc analysis of the CORONA trial concluded that patients with systolic HF of ischemic etiology who have Galectin-3 values <19.0ng/ml may benefit from rosuvastatin treatment. Importantly, high Galerctin-3 levels at the time of ventricular assist device implantation predict worse survival and lower rates of transplantation (Milting et al. 2008).

## ST2

ST2 protein is a member of the interleukin-1 receptor family and exists in two forms, soluble and membrane bound. ST2 levels are elevated in HF and the PRIDE study involving 593 patients with HF concluded that an ST2 concentration >0.20ng/ml strongly predicts death at one year among dyspneic patients as a whole as well as in those with acute HF (Januzzi et al. 2007). Moreover, in 1,449 patients with ischemic HF from the CORONA study, ST2 levels are associated with worsening HF (Broch et al. 2012). Importantly, in a cohort of 995 presenting in the emergency department, ST2 was more specific for acute HF than BNP and is independently predictive of future death and/or HF (Aldous et al. 2012). In accordance, the combined use of ST2 and NT-pro-BNP substantially improved the risk stratification for death beyond that of a model based only on established mortality risk factors (Bayes-Genis et al. 2012).

Previous data confirm the diagnostic and prognostic utility of ST2 in HF. Nevertheless there are no studies evaluating the impact of treatment in ST2 levels and how this biomarker can be used to guide treatment, limiting its clinical value.

## *The role of proteomics*

The term proteomics is used to describe all the proteins expressed by the genome that defines an organism. By studying proteomics we evaluate the large scale expression, function, and interaction of the complement of proteins in an organism in health and disease. They differ from genomics in complexity and dynamic variability. Thus, the genome is relative constant while, the proteome changes according to interactions between the genome and environment. As HF is a complex syndrome with varied appearances, proteomics seem attractive for the diagnosis, differentiation, and staging of HF.

To date most evidence has been based on experimental studies. Indeed, analysis of ventricular proteomic changes during the initial inception, development, and progression to heart failure revealed large-scale pattern differences (Gramolini et al. 2007). In accordance, in canine models of acute ischemic HF, nicotinamide adenine dinucleotide, isocitrate dehydrogenase and mitochondrial adenosine triphosphate (ATP) synthase D chain were increased while creatine kinase M chain and myosin light chain-1 were decreased (Sawicki and Jugdutt 2004).

Recently, proteomics have been used in human HF models. Accordingly, changes in several heat shock proteins which are thought to stabilize components of the sarcomere in pathological conditions have been established (Lu et al. 2008). Moreover, based on proteomics it was identified that in post acute myocardial infarction, plasma level of haptoglobin below 1.4g/L predicted the occurrence of HF (NYHA 2, 3, 4) at one year with 100 percent sensitivity (Haas et al. 2011). Importantly, proteomics can help understand the remodeling process in the failing heart. Thus, from specimens in failing and non-failing hearts it was recognized that changes in expression of at least 25 proteins in failing human heart are associated with the metabolic function, cytoskeleton, and stress response (Li et al. 2012).

Although, several studies have evaluated the significance of proteomics in understanding the pathophysiology and in the diagnosis and progression of HF, limited studies exist concerning change in proteomics by established or novel HF treatment. Furthermore, there are no data concerning their usefulness to guide treatment. Further studies have to address these issues for the establishment of proteomics as a useful tool in the management of HF.

## *The role of microRNAs*

MicroRNAs is a class of evolutionarily conserved small non-coding RNAs of 19 to 25 nucleotides in length, playing a role in most cellular processes as modulators of gene-expression (Papageorgiou et al. 2012). It is estimated that more than 1,000 deferent microRNAs exist in humans and can be measured in plasma, serum, and urine where they remain remarkably stable.

Their ability to govern adaptive and maladaptive cardiac remodeling seems to play a key role in the pathogenesis of HF. Thum et al. reported that microRNAs expression in the failing human heart results in reactivation of a fetal program (Thum et al. 2007). In addition, plasma microRNA-126 levels were increased following improvement of NYHA functional classification from IV to III (Fukushima et al. 2011). Moreover, Ikeda et al. concluded that according to microRNAs, pattern patients can accurately be grouped by their etiology of HF (Ikeda et al. 2007). In accordance, Greco et al. concluded that different microRNAs are expressed in diabetic and non-diabetic HF (Greco et al. 2012). Importantly, microRNA-423-5p is correlated with BNP levels and with left ventricular ejection fraction and has the potential to be used as a diagnostic predictor of HF in patients presenting with shortness of breath (Tijsen et al. 2010).

On the whole, we can conclude that microRNAs can offer insights in the pathopsysiology of HF and diagnostic utilities, but further studies are needed to conclude on their prognostic and therapeutic utilities.

## Combined Approach

As we already have mentioned, HF is a complex entity which can be categorized according to its severity, pathophysiology, chronicity, etc. The wide heterogeneity of the syndrome is probably responsible for the investigators' inability to indentify a single biomarker capable of providing accurate diagnostic and prognostic information in all circumstances.

These difficulties can possibly be overcome by a combined approach adding the diagnostic and prognostic value of several biomarkers and providing a useful clinical algorithm, capable of identifying accurately, the different aspects and presentations of HF syndrome. Accordingly, the combination of biomarkers associated with different pathophysiologic processes of HF, such as myocardial stress, injury, and remodeling appear interesting and may improve diagnostic power.

BNP, as the best studied biomarker in HF, has been used in conjunction with cardiac troponin I, and resulted in improvement of prognostic ability. Thus, patients with detectable cardiac troponin I and BNP >485 pg/ml had a 12-fold increased risk of death compared to those with undetectable cardiac Troponin I and BNP <485pg/ml (Horwich et al. 2003). Moreover, the combined elevation of BNP, heart-type fatty acid-binding protein and pentraxin 3 in chronic HF can predict a significantly higher cardiac event rate than that of patients with an elevation in one or two of the examined biomarkers (Ishino et al. 2008). Importantly, a study in 876 consecutive HF patients concluded that simultaneous addition of high sensitivity cardiac troponin T and NT-pro-BNP into a model that includes established risk factors improves mortality risk stratification (de Antonio et al. 2012). In line, a recent study concluded that clustering of multiple biomarkers might prove useful in exploring the biological interactions between different biomarkers in cardiovascular disease and, thus increase our understanding of the complicated interplay at the molecular level (Manhenke et al. 2011).

## Conclusion

Heart failure is a complex entity with increasing incidence, high morbidity, and mortality despite recent advances in treatment strategies and diagnostic techniques. Nowadays, a variety of biomarkers are being tested for their ability to predict outcome and to facilitate diagnosis and treatment. Nevertheless, only natriuretic peptides and cardiac troponins are widely used in every day clinical practice as the majority of the remaining biomarkers are not easily available, are expensive, or their predictive value is low. These remaining biomarkers, despite their lack of clinical usefulness in prognosis and treatment, are still valuable for our further understanding of the complicated pathophysiology of HF. Moreover, recent studies with several biomarkers give rise to the expectation of a combined panel of biomarkers being able to differentiate and to diagnose HF, to predict outcome and to guide treatment with high accuracy.

## References

Abraham, W.T., G.C. Fonarow, N.M. Albert, W.G. Stough, M. Gheorghiade, B.H. Greenberg, C.M. O'Connor, J.L. Sun, C.W. Yancy, and J.B. Young. 2008. Predictors of in-hospital mortality in patients hospitalized for heart failure: insights from the Organized Program to Initiate Lifesaving Treatment in Hospitalized Patients with Heart Failure (OPTIMIZE-HF). J. Am. Coll. Cardiol. 52: 347–356.

Abu-Omar, Y., S. Mussa, M.J. Naik, N. MacCarthy, S. Standing, and D.P. Taggart. 2005. Evaluation of Cystatin C as a marker of renal injury following on-pump and off-pump coronary surgery. Eur. J. Cardiothorac. Surg. 27: 893–898.

Adamopoulos, S., J.T. Parissis, and D.T. Kremastinos. 2001. A glossary of circulating cytokines in chronic heart failure. Eur. J. Heart Fail. 3: 517–526.

Adamopoulos, S., J. Parissis, D. Karatzas, C. Kroupis, M. Georgiadis, G. Karavolias, J. Paraskevaidis, K. Koniavitou, A.J. Coats, and D.T. Kremastinos. 2002. Physical training modulates proinflammatory cytokines and the soluble Fas/soluble Fas ligand system in patients with chronic heart failure. J. Am. Coll. Cardiol. 39: 653–663.

Agarwal, S.K., L.E. Chambless, C.M. Ballantyne, B. Astor, A.G. Bertoni, P.P. Chang, A.R. Folsom, M. He, R.C. Hoogeveen, H. Ni, M. Quibrera, W.D. Rosamond, S.D. Russell, E. Shahar, and G. Heiss. 2012. Prediction of Incident Heart Failure in General Practice: The ARIC Study. Circ. Heart Fail. 5: 422–429.

Aldous, S.J., A.M. Richards, R. Troughton, and M. Than. 2012. ST2 has diagnostic and prognostic utility for all-cause mortality and heart failure in patients presenting to the emergency department with chest pain. J. Card. Fail. 18: 304–310.

Alehagen, U., U. Dahlstrom, and T.L. Lindahl. 2009. Cystatin C and NT-proBNP, a powerful combination of biomarkers for predicting cardiovascular mortality in elderly patients with heart failure: results from a 10-year study in primary care. Eur. J. Heart Fail. 11: 354–360.

Anand, I.S., R. Latini, V.G. Florea, M.A. Kuskowski, T. Rector, S. Masson, S. Signorini, P. Mocarelli, A. Hester, R. Glazer, and J.N. Cohn. 2005. C-reactive protein in heart failure: prognostic value and the effect of valsartan. Circulation 112: 1428–1434.

Andreou, I., D. Tousoulis, A. Miliou, C. Tentolouris, K. Zisimos, P. Gounari, G. Siasos, N. Papageorgiou, C.A. Papadimitriou, M.A. Dimopoulos, and C. Stefanadis. 2010. Effects of rosuvastatin on myeloperoxidase levels in patients with chronic heart failure: a randomized placebo-controlled study. Atherosclerosis 210: 194–198.

Anker, S.D., W. Doehner, M. Rauchhaus, R. Sharma, D. Francis, C. Knosalla, C.H. Davos, M. Cicoira, W. Shamim, M. Kemp, R. Segal, K.J. Osterziel, F. Leyva, R. Hetzer, P. Ponikowski, and A.J. Coats. 2003. Uric acid and survival in chronic heart failure: validation and application in metabolic, functional, and hemodynamic staging. Circulation 107: 1991–1997.

Antoniades, C., D. Tousoulis, K. Marinou, N. Papageorgiou, E. Bosinakou, C. Tsioufis, E. Stefanadi, G. Latsios, C. Tentolouris, G. Siasos, and C. Stefanadis. 2007. Effects of insulin dependence on inflammatory process, thrombotic mechanisms and endothelial function, in patients with type 2 diabetes mellitus and coronary atherosclerosis. Clin. Cardiol. 30: 295–300.

Arimoto, T., Y. Takeishi, R. Shiga, A. Fukui, H. Tachibana, N. Nozaki, O. Hirono, J. Nitobe, T. Miyamoto, B.D. Hoit, and I. Kubota. 2005. Prognostic value of elevated circulating heart-type fatty acid binding protein in patients with congestive heart failure. J. Card. Fail. 11: 56–60.

Bayes-Genis, A., M. de Antonio, A. Galan, H. Sanz, A. Urrutia, R. Cabanes, L. Cano, B. Gonzalez, C. Diez, T. Pascual, R. Elosua, and J. Lupon. 2012. Combined use of high-sensitivity ST2 and NTproBNP to improve the prediction of death in heart failure. Eur. J. Heart Fail. 14: 32–38.

Becker, A.E., O.J. de Boer, and A.C. van Der Wal. 2001. The role of inflammation and infection in coronary artery disease. Annu. Rev. Med. 52: 289–297.

Bjornstad, H.H., J. Bruvik, A.B. Bjornstad, B.L. Hjellestad, J.K. Damas, and P. Aukrust. 2008. Exercise training decreases plasma levels of soluble CD40 ligand and P-selectin in patients with chronic heart failure. Eur. J. Cardiovasc. Prev. Rehabil. 15: 43–48.

Bolignano, D., G. Basile, P. Parisi, G. Coppolino, G. Nicocia, and M. Buemi. 2009. Increased plasma neutrophil gelatinase-associated lipocalin levels predict mortality in elderly patients with chronic heart failure. Rejuvenation Res. 12: 7–14.

Boyer, B., K.W. Hart, M.I. Sperling, C.J. Lindsell, and S.P. Collins. 2012. Biomarker changes during acute heart failure treatment. Congest. Heart Fail. 18: 91–97.

Broch, K., T. Ueland, S.H. Nymo, J. Kjekshus, J. Hulthe, P. Muntendam, J.J. McMurray, J. Wikstrand, J.G. Cleland, P. Aukrust, and L. Gullestad. 2012. Soluble ST2 is associated with adverse outcome in patients with heart failure of ischaemic aetiology. Eur. J. Heart Fail. 14: 268–277.

Buralli, S., F.L. Dini, P. Ballo, U. Conti, P. Fontanive, E. Duranti, M.R. Metelli, M. Marzilli, and S. Taddei. 2010. Circulating matrix metalloproteinase-3 and metalloproteinase-9 and tissue Doppler measures of diastolic dysfunction to risk stratify patients with systolic heart failure. Am. J. Cardiol. 105: 853–856.

Butler, J., A. Kalogeropoulos, V. Georgiopoulou, N. de Rekeneire, N. Rodondi, A.L. Smith, U. Hoffmann, A. Kanaya, A.B. Newman, S.B. Kritchevsky, R.S. Vasan, P.W. Wilson, and T.B. Harris. 2009. Serum resistin concentrations and risk of new onset heart failure in older persons: the health, aging, and body composition (Health ABC) study. Arterioscler. Thromb. Vasc. Biol. 29: 1144–1149.

Charach, G., J. George, A. Afek, D. Wexler, D. Sheps, G. Keren, and A. Rubinstein. 2009. Antibodies to oxidized LDL as predictors of morbidity and mortality in patients with chronic heart failure. J. Card. Fail. 15: 770–774.

Chrysohoou, C., C. Pitsavos, J. Barbetseas, I. Kotroyiannis, S. Brili, K. Vasiliadou, L. Papadimitriou, and C. Stefanadis. 2009. Chronic systemic inflammation accompanies impaired ventricular diastolic function, detected by Doppler imaging, in patients with newly diagnosed systolic heart failure (Hellenic Heart Failure Study). Heart Vessels 24: 22–26.

Chung, E.S., M. Packer, K.H. Lo, A.A. Fasanmade, and J.T. Willerson. 2003. Randomized, double-blind, placebo-controlled, pilot trial of infliximab, a chimeric monoclonal antibody to tumor necrosis factor-alpha, in patients with moderate-to-severe heart failure: results of the anti-TNF Therapy Against Congestive Heart Failure (ATTACH) trial. Circulation 107: 3133–3140.

Cleland, J.G., J.J. McMurray, J. Kjekshus, J.H. Cornel, P. Dunselman, C. Fonseca, A. Hjalmarson, J. Korewicki, M. Lindberg, N. Ranjith, D.J. van Veldhuisen, F. Waagstein, H. Wedel, and J. Wikstrand. 2009. Plasma concentration of amino-terminal pro-brain natriuretic peptide in chronic heart failure: prediction of cardiovascular events and interaction with the effects of rosuvastatin: a report from CORONA (Controlled Rosuvastatin Multinational Trial in Heart Failure). J. Am. Coll. Cardiol. 54: 1850–1859.

Clerico, A., M. Fontana, L. Zyw, C. Passino, and M. Emdin. 2007. Comparison of the diagnostic accuracy of brain natriuretic peptide (BNP) and the N-terminal part of the propeptide of BNP immunoassays in chronic and acute heart failure: a systematic review. Clin. Chem. 53: 813–822.

Damman, K., D.J. van Veldhuisen, G. Navis, A.A. Voors, and H.L. Hillege. 2008. Urinary neutrophil gelatinase associated lipocalin (NGAL), a marker of tubular damage, is increased in patients with chronic heart failure. Eur. J. Heart Fail. 10: 997–1000.

Damman, K., D.J. Van Veldhuisen, G. Navis, V.S. Vaidya, T.D. Smilde, B.D. Westenbrink, J.V. Bonventre, A.A. Voors, and H.L. Hillege. 2010. Tubular damage in chronic systolic heart failure is associated with reduced survival independent of glomerular filtration rate. Heart 96: 1297–1302.

Daniels, L.B. and A.S. Maisel. 2007. Natriuretic peptides. J. Am. Coll. Cardiol. 50: 2357–2368.

Davis, M., E. Espiner, G. Richards, J. Billings, I. Town, A. Neill, C. Drennan, M. Richards, J. Turner, and T. Yandle. 1994. Plasma brain natriuretic peptide in assessment of acute dyspnoea. Lancet 343: 440–444.

de Antonio, M., J. Lupon, A. Galan, J. Vila, A. Urrutia, and A. Bayes-Genis. 2012. Combined use of high-sensitivity cardiac troponin T and N-terminal pro-B type natriuretic peptide improves measurements of performance over established mortality risk factors in chronic heart failure. Am. Heart J. 163: 821–828.

de Boer, R.A., D.J. Lok, T. Jaarsma, P. van der Meer, A.A. Voors, H.L. Hillege, and D.J. van Veldhuisen. 2011. Predictive value of plasma galectin-3 levels in heart failure with reduced and preserved ejection fraction. Ann. Med. 43: 60–68.

Deswal, A., N.J. Petersen, A.M. Feldman, J.B. Young, B.G. White, and D.L. Mann. 2001. Cytokines and cytokine receptors in advanced heart failure: an analysis of the cytokine database from the Vesnarinone trial (VEST). Circulation 103: 2055–2059.

Dieplinger, B., A. Gegenhuber, J. Struck, W. Poelz, W. Langsteger, M. Haltmayer, and T. Mueller. 2009. Chromogranin A and C-terminal endothelin-1 precursor fragment add independent prognostic information to amino-terminal proBNP in patients with acute destabilized heart failure. Clin. Chim. Acta. 400: 91–96.

Dieplinger, B., A. Gegenhuber, G. Kaar, W. Poelz, M. Haltmayer, and T. Mueller. 2010. Prognostic value of established and novel biomarkers in patients with shortness of breath attending an emergency department. Clin. Biochem. 43: 714–719.

Dinh, W., W. Nickl, R. Futh, M. Lankisch, G. Hess, D. Zdunek, T. Scheffold, M.C. Barroso, K. Tiroch, D. Ziegler, and M. Seyfarth. 2011. High sensitive troponin T and heart fatty acid binding protein: novel biomarker in heart failure with normal ejection fraction? A cross-sectional study. BMC Cardiovasc. Disord. 11: 41.

Dominguez-Rodriguez, A., P. Abreu-Gonzalez, and P. Avanzas. 2011. Relation of growth-differentiation factor 15 to left ventricular remodeling in ST-segment elevation myocardial infarction. Am. J. Cardiol. 108: 955–958.

Dubin, R., Y. Li, J.H. Ix, M.G. Shlipak, M. Whooley, and C.A. Peralta. 2012. Associations of pentraxin-3 with cardiovascular events, incident heart failure, and mortality among persons with coronary heart disease: data from the Heart and Soul Study. Am. Heart J. 163: 274–279.

Dunzendorfer, S., H.K. Lee, K. Soldau, and P.S. Tobias. 2004. Toll-like receptor 4 functions intracellularly in human coronary artery endothelial cells: roles of LBP and sCD14 in mediating LPS responses. FASEB J. 18: 1117–1119.

Eurlings, L.W., P.E. van Pol, W.E. Kok, S. van Wijk, C. Lodewijks-van der Bolt, A.H. Balk, D.J. Lok, H.J. Crijns, D.J. van Kraaij, N. de Jonge, J.G. Meeder, M. Prins, and Y.M. Pinto. 2010. Management of chronic heart failure guided by individual N-terminal pro-B-type natriuretic peptide targets: results of the PRIMA (Can PRo-brain-natriuretic peptide guided therapy of chronic heart failure IMprove heart fAilure morbidity and mortality?) study. J. Am. Coll. Cardiol. 56: 2090–2100.

Farmakis, D., J.T. Parissis, V. Bistola, I.A. Paraskevaidis, E.K. Iliodromitis, G. Filippatos, and D.T. Kremastinos. 2010. Plasma B-type natriuretic peptide reduction predicts long-term response to levosimendan therapy in acutely decompensated chronic heart failure. Int. J. Cardiol. 139: 75–79.

Felix, S.B., A. Staudt, M. Landsberger, Y. Grosse, V. Stangl, T. Spielhagen, G. Wallukat, K.D. Wernecke, G. Baumann, and K. Stangl. 2002. Removal of cardiodepressant antibodies in dilated cardiomyopathy by immunoadsorption. J. Am. Coll. Cardiol. 39: 646–652.

Felker, G.M., L.A. Allen, S.J. Pocock, L.K. Shaw, J.J. McMurray, M.A. Pfeffer, K. Swedberg, D. Wang, S. Yusuf, E.L. Michelson, and C.B. Granger. 2007. Red cell distribution width as a novel prognostic marker in heart failure: data from the CHARM Program and the Duke Databank. J. Am. Coll. Cardiol. 50: 40–47.

Felkin, L.E., E. Lara-Pezzi, R. George, M.H. Yacoub, E.J. Birks, and P.J. Barton. 2009. Expression of extracellular matrix genes during myocardial recovery from heart failure after left ventricular assist device support. J. Heart Lung Transplant 28: 117–122.

Filippatos, G., J. Rossi, D.M. Lloyd-Jones, W.G. Stough, J. Ouyang, D.D. Shin, C. O'Connor, K.F. Adams, C. Orlandi, and M. Gheorghiade. 2007. Prognostic value of blood urea nitrogen in patients hospitalized with worsening heart failure: insights from the Acute and Chronic Therapeutic Impact of a Vasopressin Antagonist in Chronic Heart Failure (ACTIV in CHF) study. J. Card. Fail. 13: 360–364.

Fonarow, G.C., W.F. Peacock, C.O. Phillips, M.M. Givertz, and M. Lopatin. 2007. Admission B-type natriuretic peptide levels and in-hospital mortality in acute decompensated heart failure. J. Am. Coll. Cardiol. 49: 1943–1950.

Fonseca, C. 2006. Diagnosis of heart failure in primary care. Heart Fail. Rev. 11: 95–107.

Frantz, S., S. Stork, K. Michels, M. Eigenthaler, G. Ertl, J. Bauersachs, and C.E. Angermann. 2008. Tissue inhibitor of metalloproteinases levels in patients with chronic heart failure: an independent predictor of mortality. Eur. J. Heart Fail. 10: 388–395.

Fukushima, Y., M. Nakanishi, H. Nonogi, Y. Goto, and N. Iwai. 2011. Assessment of plasma miRNAs in congestive heart failure. Circ. J. 75: 336–340.

George, J., S. Patal, D. Wexler, A. Roth, D. Sheps, and G. Keren. 2005. Circulating matrix metalloproteinase-2 but not matrix metalloproteinase-3, matrix metalloproteinase-9, or tissue inhibitor of metalloproteinase-1 predicts outcome in patients with congestive heart failure. Am. Heart J. 150: 484–487.

Gerber, Y., S.M. Dunlay, A.S. Jaffe, J.P. McConnell, S.A. Weston, J.M. Killian, and V.L. Roger. 2009. Plasma lipoprotein-associated phospholipase A2 levels in heart failure: association with mortality in the community. Atherosclerosis 203: 593–598.

Gheorghiade, M., M.A. Konstam, J.C. Burnett, Jr., L. Grinfeld, A.P. Maggioni, K. Swedberg, J.E. Udelson, F. Zannad, T. Cook, J. Ouyang, C. Zimmer, and C. Orlandi. 2007a. Short-term clinical effects of tolvaptan, an oral vasopressin antagonist, in patients hospitalized for heart failure: the EVEREST Clinical Status Trials. JAMA 297: 1332–1343.

Gheorghiade, M., J.S. Rossi, W. Cotts, D.D. Shin, A.S. Hellkamp, I.L. Pina, G.C. Fonarow, T. DeMarco, D.F. Pauly, J. Rogers, T.G. DiSalvo, J. Butler, J.M. Hare, G.S. Francis, W.G. Stough, and C.M. O'Connor. 2007b. Characterization and prognostic value of persistent hyponatremia in patients with severe heart failure in the ESCAPE Trial. Arch. Intern. Med. 167: 1998–2005.

Glick, A., Y. Michowitz, G. Keren, and J. George. 2006. Neurohormonal and inflammatory markers as predictors of short-term outcome in patients with heart failure and cardiac resynchronization therapy. Isr Med. Assoc. J. 8: 391–395.

Gramolini, A.O., T. Kislinger, P. Liu, D.H. MacLennan, and A. Emili. 2007. Analyzing the cardiac muscle proteome by liquid chromatography-mass spectrometry-based expression proteomics. Methods Mol. Biol. 357: 15–31.

Grandin, E.W., P. Jarolim, S.A. Murphy, L. Ritterova, C.P. Cannon, E. Braunwald, and D.A. Morrow. 2012. Galectin-3 and the development of heart failure after acute coronary syndrome: pilot experience from PROVE IT-TIMI 22. Clin. Chem. 58: 267–273.

Greco, S., P. Fasanaro, S. Castelvecchio, Y. D'Alessandra, D. Arcelli, M. Di Donato, A. Malavazos, M.C. Capogrossi, L. Menicanti, and F. Martelli. 2012. MicroRNA Dysregulation in Diabetic Ischemic Heart Failure Patients. Diabetes 61: 1633–1641.

Gullestad, L., P. Aukrust, T. Ueland, T. Espevik, G. Yee, R. Vagelos, S.S. Froland, and M. Fowler. 1999. Effect of high-versus low-dose angiotensin converting enzyme inhibition on cytokine levels in chronic heart failure. J. Am. Coll. Cardiol. 34: 2061–2067.

Gullestad, L., H. Aass, J.G. Fjeld, L. Wikeby, A.K. Andreassen, H. Ihlen, S. Simonsen, J. Kjekshus, S. Nitter-Hauge, T. Ueland, E. Lien, S.S. Froland, and P. Aukrust. 2001. Immunomodulating therapy with intravenous immunoglobulin in patients with chronic heart failure. Circulation 103: 220–225.

Gupta, S., M.H. Drazner, and J.A. de Lemos. 2009. Newer biomarkers in heart failure. Heart Fail. Clin. 5: 579–588.

Haas, B., T. Serchi, D.R. Wagner, G. Gilson, S. Planchon, J. Renaut, L. Hoffmann, T. Bohn, and Y. Devaux. 2011. Proteomic analysis of plasma samples from patients with acute myocardial infarction identifies haptoglobin as a potential prognostic biomarker. J. Proteomics 75: 229–236.

Hare, J.M., B. Mangal, J. Brown, C. Fisher, Jr., R. Freudenberger, W.S. Colucci, D.L. Mann, P. Liu, M.M. Givertz, and R.P. Schwarz. 2008. Impact of oxypurinol in patients with symptomatic heart failure. Results of the OPT-CHF study. J. Am. Coll. Cardiol. 51: 2301–2309.

Hartmann, F., M. Packer, A.J. Coats, M.B. Fowler, H. Krum, P. Mohacsi, J.L. Rouleau, M. Tendera, A. Castaigne, S.D. Anker, I. Amann-Zalan, S. Hoersch, and H.A. Katus. 2004. Prognostic impact of plasma N-terminal pro-brain natriuretic peptide in severe chronic congestive heart failure: a substudy of the Carvedilol Prospective Randomized Cumulative Survival (COPERNICUS) trial. Circulation 110: 1780–1786.

Hayabuchi, Y., M. Inoue, N. Watanabe, M. Sakata, T. Ohnishi and S. Kagami. 2011. Serum concentration of heart-type fatty acid-binding protein in children and adolescents with congenital heart disease. Circ. J. 75: 1992–1997.

Horwich, T.B., J. Patel, W.R. MacLellan, and G.C. Fonarow. 2003. Cardiac troponin I is associated with impaired hemodynamics, progressive left ventricular dysfunction, and increased mortality rates in advanced heart failure. Circulation 108: 833–838.

Ikeda, S., S.W. Kong, J. Lu, E. Bisping, H. Zhang, P.D. Allen, T.R. Golub, B. Pieske, and W.T. Pu. 2007. Altered microRNA expression in human heart disease. Physiol. Genomics 31: 367–373.

Iraqi, W., P. Rossignol, M. Angioi, R. Fay, J. Nuee, J.M. Ketelslegers, J. Vincent, B. Pitt, and F. Zannad. 2009. Extracellular cardiac matrix biomarkers in patients with acute myocardial infarction complicated by left ventricular dysfunction and heart failure: insights from the Eplerenone Post-Acute Myocardial Infarction Heart Failure Efficacy and Survival Study (EPHESUS) study. Circulation 119: 2471–2479.

Ishino, M., Y. Takeishi, T. Niizeki, T. Watanabe, J. Nitobe, T. Miyamoto, T. Miyashita, T. Kitahara, S. Suzuki, T. Sasaki, O. Bilim, and I. Kubota. 2008. Risk stratification of chronic heart failure patients by multiple biomarkers: implications of BNP, H-FABP, and PTX3. Circ. J. 72: 1800–1805.

Jankowska, E.A., G.S. Filippatos, S. von Haehling, J. Papassotiriou, N.G. Morgenthaler, M. Cicoira, J.C. Schefold, P. Rozentryt, B. Ponikowska, W. Doehner, W. Banasiak, O. Hartmann, J. Struck, A. Bergmann, S.D. Anker, and P. Ponikowski. 2011. Identification of chronic heart failure patients with a high 12-month mortality risk using biomarkers including plasma C-terminal pro-endothelin-1. PLoS One 6: e14506.

Januzzi, J.L., Jr., C.A. Camargo, S. Anwaruddin, A.L. Baggish, A.A. Chen, D.G. Krauser, R. Tung, R. Cameron, J.T. Nagurney, C.U. Chae, D.M. Lloyd-Jones, D.F. Brown, S. Foran-Melanson, P.M. Sluss, E. Lee-Lewandrowski, and K.B. Lewandrowski. 2005. The N-terminal Pro-BNP investigation of dyspnea in the emergency department (PRIDE) study. Am. J. Cardiol. 95: 948–954.

Januzzi, J.L., Jr., W.F. Peacock, A.S. Maisel, C.U. Chae, R.L. Jesse, A.L. Baggish, M. O'Donoghue, R. Sakhuja, A.A. Chen, R.R. van Kimmenade, K.B. Lewandrowski, D.M. Lloyd-Jones, and A.H. Wu. 2007. Measurement of the interleukin family member ST2 in patients with acute dyspnea: results from the PRIDE (Pro-Brain Natriuretic Peptide Investigation of Dyspnea in the Emergency Department) study. J. Am. Coll. Cardiol. 50: 607–613.

Januzzi, J.L., Jr., S.U. Rehman, A.A. Mohammed, A. Bhardwaj, L. Barajas, J. Barajas, H.N. Kim, A.L. Baggish, R.B. Weiner, A. Chen-Tournoux, J.E. Marshall, S.A. Moore, W.D. Carlson, G.D. Lewis, J. Shin, D. Sullivan, K. Parks, T.J. Wang, S.A. Gregory, S. Uthamalingam, and M.J. Semigran. 2011. Use of amino-terminal pro-B-type natriuretic peptide to guide outpatient therapy of patients with chronic left ventricular systolic dysfunction. J. Am. Coll. Cardiol. 58: 1881–1889.

Jourdain, P., G. Jondeau, F. Funck, P. Gueffet, A. Le Helloco, E. Donal, J.F. Aupetit, M.C. Aumont, M. Galinier, J.C. Eicher, A. Cohen-Solal, and Y. Juilliere. 2007. Plasma brain natriuretic peptide-guided therapy to improve outcome in heart failure: the STARS-BNP Multicenter Study. J. Am. Coll. Cardiol. 49: 1733–1739.

Kaneko, K., T. Kanda, Y. Yamauchi, A. Hasegawa, T. Iwasaki, M. Arai, T. Suzuki, I. Kobayashi, and R. Nagai. 1999. C-Reactive protein in dilated cardiomyopathy. Cardiology 91: 215–219.

Kapur, N.K., K.S. Heffernan, A.A. Yunis, P. Parpos, M.S. Kiernan, N.A. Sahasrabudhe, C.D. Kimmelstiel, D.A. Kass, R.H. Karas, and M.E. Mendelsohn. 2010. Usefulness of soluble endoglin as a noninvasive measure of left ventricular filling pressure in heart failure. Am. J. Cardiol. 106: 1770–1776.

Kawahara, C., T. Tsutamoto, H. Sakai, K. Nishiyama, M. Yamaji, M. Fujii, T. Yamamoto, and M. Horie. 2011. Prognostic value of serial measurements of highly sensitive cardiac troponin I in stable outpatients with nonischemic chronic heart failure. Am. Heart J. 162: 639–645.

Kelder, J.C., M.J. Cramer, J. van Wijngaarden, R. van Tooren, A. Mosterd, K.G. Moons, J.W. Lammers, M.R. Cowie, D.E. Grobbee, and A.W. Hoes. 2011. The diagnostic value of physical examination and additional testing in primary care patients with suspected heart failure. Circulation 124: 2865–2873.

Kempf, T., S. von Haehling, T. Peter, T. Allhoff, M. Cicoira, W. Doehner, P. Ponikowski, G.S. Filippatos, P. Rozentryt, H. Drexler, S.D. Anker, and K.C. Wollert. 2007. Prognostic utility of growth differentiation factor-15 in patients with chronic heart failure. J. Am. Coll. Cardiol. 50: 1054–1060.

Khan, S.Q., O. Dhillon, D. Kelly, I.B. Squire, J. Struck, P. Quinn, N.G. Morgenthaler, A. Bergmann, J.E. Davies, and L.L. Ng. 2008. Plasma N-terminal B-Type natriuretic peptide as an indicator of long-term survival after acute myocardial infarction: comparison with plasma midregional pro-atrial natriuretic peptide: the LAMP (Leicester Acute Myocardial Infarction Peptide) study. J. Am. Coll. Cardiol. 51: 1857–1864.

Khan, S.Q., K. Ng, O. Dhillon, D. Kelly, P. Quinn, I.B. Squire, J.E. Davies, and L.L. Ng. 2009. Growth differentiation factor-15 as a prognostic marker in patients with acute myocardial infarction. Eur. Heart J. 30: 1057–1065.

Kistorp, C., J. Faber, S. Galatius, F. Gustafsson, J. Frystyk, A. Flyvbjerg, and P. Hildebrandt. 2005. Plasma adiponectin, body mass index, and mortality in patients with chronic heart failure. Circulation 112: 1756–1762.

Kjekshus, J., E. Apetrei, V. Barrios, M. Bohm, J.G. Cleland, J.H. Cornel, P. Dunselman, C. Fonseca, A. Goudev, P. Grande, L. Gullestad, A. Hjalmarson, J. Hradec, A. Janosi, G. Kamensky, M. Komajda, J. Korewicki, T. Kuusi, F. Mach, V. Mareev, J.J. McMurray, N. Ranjith, M. Schaufelberger, J. Vanhaecke, D.J. van Veldhuisen, F. Waagstein, H. Wedel, and J. Wikstrand. 2007. Rosuvastatin in older patients with systolic heart failure. N. Engl. J. Med. 357: 2248–2261.

La Vecchia, L., G. Mezzena, L. Zanolla, M. Paccanaro, L. Varotto, C. Bonanno, and R. Ometto. 2000. Cardiac troponin I as diagnostic and prognostic marker in severe heart failure. J. Heart Lung Transplant 19: 644–652.

Lainchbury, J.G., R.W. Troughton, K.M. Strangman, C.M. Frampton, A. Pilbrow, T.G. Yandle, A.K. Hamid, M.G. Nicholls, and A.M. Richards. 2009. N-terminal pro-B-type natriuretic peptide-guided treatment for chronic heart failure: results from the BATTLESCARRED (NT-proBNP-Assisted Treatment To Lessen Serial Cardiac Readmissions and Death) trial. J. Am. Coll. Cardiol. 55: 53–60.

Lappegard, K.T. and H. Bjornstad. 2006. Anti-inflammatory effect of cardiac resynchronization therapy. Pacing Clin. Electrophysiol. 29: 753–758.

Latini, R., A.P. Maggioni, G. Peri, L. Gonzini, D. Lucci, P. Mocarelli, L. Vago, F. Pasqualini, S. Signorini, D. Soldateschi, L. Tarli, C. Schweiger, C. Fresco, R. Cecere, G. Tognoni, and A. Mantovani. 2004a. Prognostic significance of the long pentraxin PTX3 in acute myocardial infarction. Circulation 110: 2349–2354.

Latini, R., S. Masson, I. Anand, M. Salio, A. Hester, D. Judd, S. Barlera, A.P. Maggioni, G. Tognoni, and J.N. Cohn. 2004b. The comparative prognostic value of plasma neurohormones at baseline in patients with heart failure enrolled in Val-HeFT. Eur. Heart J. 25: 292–299.

Latini, R., S. Masson, I.S. Anand, E. Missov, M. Carlson, T. Vago, L. Angelici, S. Barlera, G. Parrinello, A.P. Maggioni, G. Tognoni, and J. N. Cohn. 2007. Prognostic value of very low plasma concentrations of troponin T in patients with stable chronic heart failure. Circulation 116: 1242–1249.

LeLeiko, R.M., C.S. Vaccari, S. Sola, N. Merchant, S.H. Nagamia, M. Thoenes, and B.V. Khan. 2009. Usefulness of elevations in serum choline and free F2)-isoprostane to predict 30-day cardiovascular outcomes in patients with acute coronary syndrome. Am. J. Cardiol. 104: 638–643.

Levine, B., J. Kalman, L. Mayer, H.M. Fillit, and M. Packer. 1990. Elevated circulating levels of tumor necrosis factor in severe chronic heart failure. N. Engl. J. Med. 323: 236–241.

Levy, D., S. Kenchaiah, M.G. Larson, E.J. Benjamin, M.J. Kupka, K.K. Ho, J.M. Murabito, and R.S. Vasan. 2002. Long-term trends in the incidence of and survival with heart failure. N. Engl. J. Med. 347: 1397–1402.

Li, W., R. Rong, S. Zhao, X. Zhu, K. Zhang, X. Xiong, X. Yu, Q. Cui, S. Li, L. Chen, J. Cai, and J. Du. 2012. Proteomic analysis of metabolic, cytoskeletal and stress response proteins in human heart failure. J. Cell. Mol. Med. 16: 59–71.

Lu, X.Y., L. Chen, X.L. Cai, and H.T. Yang. 2008. Overexpression of heat shock protein 27 protects against ischaemia/reperfusion-induced cardiac dysfunction via stabilization of troponin I and T. Cardiovasc. Res. 79: 500–508.

Maeda, K., T. Tsutamoto, A. Wada, N. Mabuchi, M. Hayashi, T. Tsutsui, M. Ohnishi, M. Sawaki, M. Fujii, T. Matsumoto, and M. Kinoshita. 2000. High levels of plasma brain natriuretic peptide and interleukin-6 after optimized treatment for heart failure are independent risk factors for morbidity and mortality in patients with congestive heart failure. J. Am. Coll. Cardiol. 36: 1587–1593.

Maisel, A., C. Mueller, R. Nowak, W.F. Peacock, J.W. Landsberg, P. Ponikowski, M. Mockel, C. Hogan, A.H. Wu, M. Richards, P. Clopton, G.S. Filippatos, S. Di Somma, I. Anand, L. Ng, L.B. Daniels, S.X. Neath, R. Christenson, M. Potocki, J. McCord, G. Terracciano, D. Kremastinos, O. Hartmann, S. von Haehling, A. Bergmann, N.G. Morgenthaler, and S.D. Anker. 2010. Mid-region pro-hormone markers for diagnosis and prognosis in acute dyspnea: results from the BACH (Biomarkers in Acute Heart Failure) trial. J. Am. Coll. Cardiol. 55: 2062–2076.

Maisel, A.S., J. Koon, P. Krishnaswamy, R. Kazenegra, P. Clopton, N. Gardetto, R. Morrisey, A. Garcia, A. Chiu, and A. De Maria. 2001. Utility of B-natriuretic peptide as a rapid, point-of-care test for screening patients undergoing echocardiography to determine left ventricular dysfunction. Am. Heart J. 141: 367–374.

Maisel, A.S., P. Krishnaswamy, R.M. Nowak, J. McCord, J.E. Hollander, P. Duc, T. Omland, A.B. Storrow, W.T. Abraham, A.H. Wu, P. Clopton, P.G. Steg, A. Westheim, C.W. Knudsen, A. Perez, R. Kazanegra, H.C. Herrmann, and P. A. McCullough. 2002. Rapid measurement of B-type natriuretic peptide in the emergency diagnosis of heart failure. N. Engl. J. Med. 347: 161–167.

Manhenke, C., S. Orn, S. von Haehling, K.C. Wollert, T. Ueland, P. Aukrust, A.A. Voors, I. Squire, F. Zannad, S.D. Anker and K. Dickstein. 2011. Clustering of 37 circulating biomarkers by exploratory factor analysis in patients following complicated acute myocardial infarction. Int. J. Cardiol.

Mann, D.L., J.J. McMurray, M. Packer, K. Swedberg, J.S. Borer, W.S. Colucci, J. Djian, H. Drexler, A. Feldman, L. Kober, H. Krum, P. Liu, M. Nieminen, L. Tavazzi, D.J. van Veldhuisen, A. Waldenstrom, M. Warren, A. Westheim, F. Zannad, and T. Fleming. 2004. Targeted anticytokine therapy in patients with chronic heart failure: results of the Randomized Etanercept Worldwide Evaluation (RENEWAL). Circulation 109: 1594–1602.

Masson, S., R. Latini, E. Carbonieri, L. Moretti, M.G. Rossi, S. Ciricugno, V. Milani, R. Marchioli, J. Struck, A. Bergmann, A.P. Maggioni, G. Tognoni, and L. Tavazzi. 2010. The predictive value of stable precursor fragments of vasoactive peptides in patients with chronic heart failure: data from the GISSI-heart failure (GISSI-HF) trial. Eur. J. Heart Fail. 12: 338–347.

Matsubara, J., S. Sugiyama, T. Nozaki, K. Sugamura, M. Konishi, K. Ohba, Y. Matsuzawa, E. Akiyama, E. Yamamoto, K. Sakamoto, Y. Nagayoshi, K. Kaikita, H. Sumida, S. Kim-Mitsuyama, and H. Ogawa. 2011. Pentraxin 3 is a new inflammatory marker correlated with left ventricular diastolic dysfunction and heart failure with normal ejection fraction. J. Am. Coll. Cardiol. 57: 861–869.

McCullough, P.A., R.M. Nowak, J. McCord, J.E. Hollander, H.C. Herrmann, P.G. Steg, P. Duc, A. Westheim, T. Omland, C. W. Knudsen, A.B. Storrow, W.T. Abraham, S. Lamba, A.H. Wu, A. Perez, P. Clopton, P. Krishnaswamy, R. Kazanegra, and A.S. Maisel. 2002. B-type natriuretic peptide and clinical judgment in emergency diagnosis of heart failure: analysis from Breathing Not Properly (BNP) Multinational Study. Circulation 106: 416–422.

McMurray, J.J., J. Kjekshus, L. Gullestad, P. Dunselman, A. Hjalmarson, H. Wedel, M. Lindberg, F. Waagstein, P. Grande, J. Hradec, G. Kamensky, J. Korewicki, T. Kuusi, F. Mach, N. Ranjith, and J. Wikstrand. 2009. Effects of statin therapy according to plasma high-sensitivity C-reactive protein concentration in the Controlled Rosuvastatin Multinational Trial in Heart Failure (CORONA): a retrospective analysis. Circulation 120: 2188–2196.

McMurray, J.J., S. Adamopoulos, S.D. Anker, A. Auricchio, M. Bohm, K. Dickstein, V. Falk, G. Filippatos, C. Fonseca, M.A. Sanchez, T. Jaarsma, L. Kober, G.Y. Lip, A.P. Maggioni, A. Parkhomenko, B.M. Pieske, B.A. Popescu, P.K. Ronnevik, F.H. Rutten, J. Schwitter, P. Seferovic, J. Stepinska, P.T. Trindade, A.A. Voors, F. Zannad, A. Zeiher, J.J. Bax, H. Baumgartner, C. Ceconi, V. Dean, C. Deaton, R. Fagard, C. Funck-Brentano, D. Hasdai, A. Hoes, P. Kirchhof, J. Knuuti, P. Kolh, T. McDonagh, C. Moulin, Z. Reiner, U. Sechtem, P.A. Sirnes, M. Tendera, A. Torbicki, A. Vahanian, S. Windecker, L.A. Bonet, P. Avraamides, H.A. Ben Lamin, M. Brignole, A. Coca, P. Cowburn, H. Dargie, P. Elliott, F.A. Flachskampf, G.F. Guida, S. Hardman, B. Iung, B. Merkely, C. Mueller, J.N. Nanas, O.W. Nielsen, S. Orn, J.T. Parissis, and P. Ponikowski. 2012. ESC Guidelines for the diagnosis and treatment of acute and chronic heart failure 2012: The Task Force for the Diagnosis and Treatment of Acute and Chronic Heart Failure 2012 of the European Society of Cardiology. Developed in collaboration with the Heart Failure Association (HFA) of the ESC. Eur. Heart J.

Michowitz, Y., S. Kisil, H. Guzner-Gur, A. Rubinstein, D. Wexler, D. Sheps, G. Keren, and J. George. 2008. Usefulness of serum myeloperoxidase in prediction of mortality in patients with severe heart failure. Isr Med. Assoc. J. 10: 884–888.

Milo-Cotter, O., B. Cotter-Davison, C. Lombardi, H. Sun, L. Bettari, S. Bugatti, M. Rund, M. Metra, E. Kaluski, I. Kobrin, A. Frey, M. Rainisio, J.J. McMurray, J.R. Teerlink, and G. Cotter-Davison. 2011. Neurohormonal Activation in Acute Heart Failure: Results from VERITAS. Cardiology 119: 96–105.

Milting, H., P. Ellinghaus, M. Seewald, H. Cakar, B. Bohms, A. Kassner, R. Korfer, M. Klein, T. Krahn, L. Kruska, A. El Banayosy, and F. Kramer. 2008. Plasma biomarkers of myocardial fibrosis and remodeling in terminal heart failure patients supported by mechanical circulatory support devices. J. Heart Lung Transplant 27: 589–596.

Molyneux, S.L., C.M. Florkowski, P.M. George, A.P. Pilbrow, C.M. Frampton, M. Lever, and A.M. Richards. 2008. Coenzyme Q10: an independent predictor of mortality in chronic heart failure. J. Am. Coll. Cardiol. 52: 1435–1441.

Mosterd, A. and A.W. Hoes. 2007. Clinical epidemiology of heart failure. Heart 93: 1137–1146.

Mueller, C., A. Scholer, K. Laule-Kilian, B. Martina, C. Schindler, P. Buser, M. Pfisterer, and A.P. Perruchoud. 2004. Use of B-type natriuretic peptide in the evaluation and management of acute dyspnea. N. Engl. J. Med. 350: 647–654.

Niebauer, J., H.D. Volk, M. Kemp, M. Dominguez, R.R. Schumann, M. Rauchhaus, P.A. Poole-Wilson, A.J. Coats, and S.D. Anker. 1999. Endotoxin and immune activation in chronic heart failure: a prospective cohort study. Lancet 353: 1838–1842.

Niizeki, T., Y. Takeishi, T. Arimoto, H. Okuyama, N. Takabatake, H. Tachibana, N. Nozaki, O. Hirono, Y. Tsunoda, T. Miyashita, A. Fukui, H. Takahashi, Y. Koyama, T. Shishido, and I. Kubota. 2005. Serum heart-type fatty acid binding protein predicts cardiac events in elderly patients with chronic heart failure. J. Cardiol. 46: 9–15.

Niizeki, T., Y. Takeishi, T. Arimoto, N. Nozaki, O. Hirono, T. Watanabe, J. Nitobe, T. Miyashita, T. Miyamoto, Y. Koyama, T. Kitahara, S. Suzuki, T. Sasaki, and I. Kubota. 2008. Persistently increased serum concentration of heart-type fatty acid-binding protein predicts adverse clinical outcomes in patients with chronic heart failure. Circ. J. 72: 109–114.

Norozi, K., R. Buchhorn, A. Yasin, S. Geyer, L. Binder, J.A. Seabrook, and A. Wessel. 2011. Growth differentiation factor 15: an additional diagnostic tool for the risk stratification of developing heart failure in patients with operated congenital heart defects? Am. Heart J. 162: 131–135.

Nunez, J., J. Sanchis, V. Bodi, G.C. Fonarow, E. Nunez, V. Bertomeu-Gonzalez, G. Minana, L. Consuegra, M.J. Bosch, A. Carratala, F.J. Chorro, and A. Llacer. 2010. Improvement in risk stratification with the combination of the tumour marker antigen carbohydrate 125 and brain natriuretic peptide in patients with acute heart failure. Eur. Heart J. 31: 1752–1763.

O'Connor, C.M., M. Fiuzat, C. Lombardi, K. Fujita, G. Jia, B.A. Davison, J. Cleland, D. Bloomfield, H.C. Dittrich, P. Delucca, M.M. Givertz, G. Mansoor, P. Ponikowski, J.R. Teerlink, A.A. Voors, B.M. Massie, G. Cotter, and M. Metra. 2011. The Impact of Serial Troponin Release on Outcomes in Patients with Acute Heart Failure: Analysis from the PROTECT Pilot Study. Circ Heart Fail.

Ohmae, M. 2011. Endothelin-1 levels in chronic congestive heart failure. Wien Klin. Wochenschr 123: 714–717.

Ohtsuka, T., K. Inoue, Y. Hara, N. Morioka, K. Ohshima, J. Suzuki, A. Ogimoto, Y. Shigematsu, and J. Higaki. 2005. Serum markers of angiogenesis and myocardial ultrasonic tissue characterization in patients with dilated cardiomyopathy. Eur. J. Heart Fail. 7: 689–695.

Oikonomou, E., D. Tousoulis, G. Siasos, M. Zaromitidou, A. G. Papavassiliou, and C. Stefanadis. 2011. The role of inflammation in heart failure: new therapeutic approaches. Hellenic J. Cardiol. 52: 30–40.

Olivetti, G., G. Giordano, D. Corradi, M. Melissari, C. Lagrasta, S.R. Gambert, and P. Anversa. 1995. Gender differences and aging: effects on the human heart. J. Am. Coll. Cardiol. 26: 1068–1079.

Palmer, B.R., A.P. Pilbrow, C.M. Frampton, T.G. Yandle, L. Skelton, M.G. Nicholls, and A.M. Richards. 2008. Plasma aldosterone levels during hospitalization are predictive of survival post-myocardial infarction. Eur. Heart J. 29: 2489–2496.

Papageorgiou, N., D. Tousoulis, E. Androulakis, G. Siasos, A. Briasoulis, G. Vogiatzi, A.M. Kampoli, E. Tsiamis, C. Tentolouris, and C. Stefanadis. 2012. The Role of microRNAs in Cardiovascular Disease. Curr. Med. Chem. 19: 2605–2610.

Peacock, W.F.t., T. De Marco, G.C. Fonarow, D. Diercks, J. Wynne, F.S. Apple, and A.H. Wu. 2008. Cardiac troponin and outcome in acute heart failure. N. Engl. J. Med. 358: 2117–2126.

Peng, Y., J. Shan, X. Qi, H. Xue, C. Rong, D. Yao, Z. Guo, and S. Zheng. 2003. Effects of catecholamine-beta-adrenoceptor-cAMP system on severe patients with heart failure. Chin. Med. J. (Engl.) 116: 1459–1463.

Persson, H., H. Erntell, B. Eriksson, G. Johansson, K. Swedberg, and U. Dahlstrom. 2010. Improved pharmacological therapy of chronic heart failure in primary care: a randomized Study of NT-proBNP Guided Management of Heart Failure—SIGNAL-HF (Swedish Intervention study—Guidelines and NT-proBNP AnaLysis in Heart Failure). Eur. J. Heart Fail. 12: 1300–1308.

Pfisterer, M., P. Buser, H. Rickli, M. Gutmann, P. Erne, P. Rickenbacher, A. Vuillomenet, U. Jeker, P. Dubach, H. Beer, S.I. Yoon, T. Suter, H.H. Osterhues, M.M. Schieber, P. Hilti, R. Schindler, and H.P. Brunner-La Rocca. 2009. BNP-guided vs symptom-guided heart failure therapy: the Trial of Intensified vs Standard Medical Therapy in Elderly Patients With Congestive Heart Failure (TIME-CHF) randomized trial. JAMA 301: 383–392.

Pitt, B., F. Zannad, W.J. Remme, R. Cody, A. Castaigne, A. Perez, J. Palensky, and J. Wittes. 1999. The effect of spironolactone on morbidity and mortality in patients with severe heart failure. Randomized Aldactone Evaluation Study Investigators. N. Engl. J. Med. 341: 709–717.

Potocki, M., T. Breidthardt, T. Reichlin, S. Hartwiger, N.G. Morgenthaler, A. Bergmann, M. Noveanu, H. Freidank, A.B. Taegtmeyer, K. Wetzel, T. Boldanova, C. Stelzig, R. Bingisser, M. Christ, and C. Mueller. 2010. Comparison of midregional pro-atrial natriuretic peptide with N-terminal pro-B-type natriuretic peptide in the diagnosis of heart failure. J. Intern. Med. 267: 119–129.

Reichlin, T., T. Socrates, P. Egli, M. Potocki, T. Breidthardt, N. Arenja, J. Meissner, M. Noveanu, M. Reiter, R. Twerenbold, N. Schaub, A. Buser, and C. Mueller. 2010. Use of myeloperoxidase for risk stratification in acute heart failure. Clin. Chem. 56: 944–951.

Rosjo, H., C. Husberg, M.B. Dahl, M. Stridsberg, I. Sjaastad, A.V. Finsen, C.R. Carlson, E. Oie, T. Omland, and G. Christensen. 2010a. Chromogranin B in heart failure: a putative cardiac biomarker expressed in the failing myocardium. Circ. Heart Fail. 3: 503–511.

Rosjo, H., S. Masson, R. Latini, A. Flyvbjerg, V. Milani, M.T. La Rovere, M. Revera, A. Mezzani, G. Tognoni, L. Tavazzi, and T. Omland. 2010b. Prognostic value of chromogranin A in chronic heart failure: data from the GISSI-Heart Failure trial. Eur. J. Heart Fail. 12: 549–556.

Rouleau, J.L., M. Packer, L. Moye, J. de Champlain, D. Bichet, M. Klein, J.R. Rouleau, B. Sussex, J.M. Arnold, F. Sestier et al. 1994. Prognostic value of neurohumoral activation in patients with an acute myocardial infarction: effect of captopril. J. Am. Coll. Cardiol. 24: 583–591.

Sawicki, G. and B.I. Jugdutt. 2004. Detection of regional changes in protein levels in the *in vivo* canine model of acute heart failure following ischemia-reperfusion injury: functional proteomics studies. Proteomics 4: 2195–2202.

Shah, M.R., R.M. Califf, A. Nohria, M. Bhapkar, M. Bowers, D.M. Mancini, M. Fiuzat, L.W. Stevenson, and C.M. O'Connor. 2011. The STARBRITE trial: a randomized, pilot study of B-type natriuretic peptide-guided therapy in patients with advanced heart failure. J. Card. Fail. 17: 613–621.

Shah, R.V., A.A. Chen-Tournoux, M.H. Picard, R.R. van Kimmenade, and J.L. Januzzi. 2010. Galectin-3, cardiac structure and function, and long-term mortality in patients with acutely decompensated heart failure. Eur. J. Heart Fail. 12: 826–832.

Shah, R.V., Q.A. Truong, H.K. Gaggin, J. Pfannkuche, O. Hartmann, and J.L. Januzzi, Jr. 2012. Mid-regional pro-atrial natriuretic peptide and pro-adrenomedullin testing for the diagnostic and prognostic evaluation of patients with acute dyspnoea. Eur. Heart J.

Shlipak, M.G., R. Katz, L.F. Fried, N. S. Jenny, C. O. Stehman-Breen, A.B. Newman, D. Siscovick, B.M. Psaty, and M.J. Sarnak. 2005. Cystatin-C and mortality in elderly persons with heart failure. J. Am. Coll. Cardiol. 45: 268–271.

Siasos, G., D. Tousoulis, S. Michalea, E. Oikonomou, C. Kolia, S. Kioufis, A. Synetos, K. Vlasis, A.G. Papavassiliou, and C. Stefanadis. 2012. Biomarkers determining cardiovascular risk in patients with kidney disease. Curr. Med. Chem. 19: 2555–2571.

Smilde, T.D., K. Damman, P. van der Harst, G. Navis, B.D. Westenbrink, A.A. Voors, F. Boomsma, D.J. van Veldhuisen, and H.L. Hillege. 2009. Differential associations between renal function and "modifiable" risk factors in patients with chronic heart failure. Clin. Res. Cardiol. 98: 121–129.

Sodian, R., M. Loebe, C. Schmitt, E.V. Potapov, H. Siniawski, J. Muller, H. Hausmann, H.R. Zurbruegg, Y. Weng, and R. Hetzer. 2001. Decreased plasma concentration of brain natriuretic peptide as a potential indicator of cardiac recovery in patients supported by mechanical circulatory assist systems. J. Am. Coll. Cardiol. 38: 1942–1949.

Suzuki, S., Y. Takeishi, Y. Niizeki, Y. Koyama, T. Kitahara, T. Sasaki, M. Sagara and I. Kubota. 2008. Pentraxin 3, a new marker for vascular inflammation, predicts adverse clinical outcomes in patients with heart failure. Am. Heart J. 155: 75–81.

Takiyyuddin, M.A., J.H. Cervenka, P.A. Sullivan, M.R. Pandian, R.J. Parmer, J.A. Barbosa, and D.T. O'Connor. 1990. Is physiologic sympathoadrenal catecholamine release exocytotic in humans? Circulation 81: 185–195.

Tanaka, T., Y. Hirota, K. Sohmiya, S. Nishimura, and K. Kawamura. 1991. Serum and urinary human heart fatty acid-binding protein in acute myocardial infarction. Clin. Biochem. 24: 195–201.

Tavazzi, L., A.P. Maggioni, R. Marchioli, S. Barlera, M.G. Franzosi, R. Latini, D. Lucci, G.L. Nicolosi, M. Porcu, and G. Tognoni. 2008. Effect of rosuvastatin in patients with chronic heart failure (the GISSI-HF trial): a randomised, double-blind, placebo-controlled trial. Lancet 372: 1231–1239.

Thum, T., P. Galuppo, C. Wolf, J. Fiedler, S. Kneitz, L.W. van Laake, P.A. Doevendans, C. L. Mummery, J. Borlak, A. Haverich, C. Gross, S. Engelhardt, G. Ertl, and J. Bauersachs. 2007. MicroRNAs in the human heart: a clue to fetal gene reprogramming in heart failure. Circulation 116: 258–267.

Thygesen, K., J.S. Alpert, H.D. White, A.S. Jaffe, F.S. Apple, M. Galvani, H.A. Katus, L.K. Newby, J. Ravkilde, B. Chaitman, P.M. Clemmensen, M. Dellborg, H. Hod, P. Porela, R. Underwood, J.J. Bax, G.A. Beller, R. Bonow, E.E. Van der Wall, J.P. Bassand, W. Wijns, T.B. Ferguson, P.G. Steg, B.F. Uretsky, D.O. Williams, P.W. Armstrong, E.M. Antman, K.A. Fox, C.W. Hamm, E.M. Ohman, M.L. Simoons, P.A. Poole-Wilson, E.P. Gurfinkel, J.L. Lopez-Sendon, P. Pais, S. Mendis, J.R. Zhu, L.C. Wallentin, F. Fernandez-Aviles, K.M. Fox, A.N. Parkhomenko, S.G. Priori, M. Tendera, L.M. Voipio-Pulkki, A. Vahanian, A.J. Camm, R. De Caterina, V. Dean, K. Dickstein, G. Filippatos, C. Funck-Brentano, I. Hellemans, S.D. Kristensen, K. McGregor, U. Sechtem, S. Silber, P. Widimsky, J.L. Zamorano, J. Morais, S. Brener, R. Harrington, D. Morrow, M. Lim, M.A. Martinez-Rios, S. Steinhubl, G.N. Levine, W.B. Gibler, D. Goff, M. Tubaro, D. Dudek, and N. Al-Attar. 2007. Universal definition of myocardial infarction. Circulation 116: 2634–2653.

Tijsen, A.J., E.E. Creemers, P.D. Moerland, L.J. de Windt, A.C. van der Wal, W.E. Kok, and Y.M. Pinto. 2010. MiR423-5p as a circulating biomarker for heart failure. Circ. Res. 106: 1035–1039.

Torre-Amione, G., S.D. Anker, R.C. Bourge, W.S. Colucci, B.H. Greenberg, P. Hildebrandt, A. Keren, M. Motro, L.A. Moye, J.E. Otterstad, C.M. Pratt, P. Ponikowski, J.L. Rouleau, F. Sestier, B.R. Winkelmann, and J.B. Young. 2008. Results of a non-specific immunomodulation therapy in chronic heart failure (ACCLAIM trial): a placebo-controlled randomised trial. Lancet 371: 228–236.

Torre-Amione, G., C.M. Orrego, N. Khalil, C. Kottner-Assad, C. Leveque, R. Celis, K.A. Youker, and J.D. Estep. 2010. Therapeutic plasma exchange a potential strategy for patients with advanced heart failure. J. Clin. Apher. 25: 323–330.

Tousoulis, D., M. Charakida, and C. Stefanadis. 2005. Inflammation and endothelial dysfunction as therapeutic targets in patients with heart failure. Int. J. Cardiol. 100: 347–353.

Tousoulis, D., C. Antoniades, A. Nikolopoulou, K. Koniari, C. Vasiliadou, K. Marinou, N. Koumallos, N. Papageorgiou, E. Stefanadi, G. Siasos, and C. Stefanadis. 2007a. Interaction between cytokines and sCD40L in patients with stable and unstable coronary syndromes. Eur. J. Clin. Invest. 37: 623–628.

Tousoulis, D., C. Antoniades, C. Vasiliadou, P. Kourtellaris, K. Koniari, K. Marinou, M. Charakida, I. Ntarladimas, G. Siasos, and C. Stefanadis. 2007b. Effects of atorvastatin and vitamin C on forearm hyperaemic blood flow, asymmentrical dimethylarginine levels and the inflammatory process in patients with type 2 diabetes mellitus. Heart 93: 244–246.

Tousoulis, D., I. Andreou, M. Tsiatas, A. Miliou, C. Tentolouris, G. Siasos, N. Papageorgiou, C.A. Papadimitriou, M.A. Dimopoulos, and C. Stefanadis. 2011. Effects of rosuvastatin and allopurinol on circulating endothelial progenitor cells in patients with congestive heart failure: the impact of inflammatory process and oxidative stress. Atherosclerosis 214: 151–157.

Tousoulis, D., E. Oikonomou, G. Siasos, C. Chrysohoou, M. Charakida, Z. Siasou, M. Limperi, E.D. Papadimitriou, A.G. Papavassiliou, and C. Stefanadis. 2012. Predictive value of biomarkers in patients with heart failure. Curr. Med. Chem. 19: 2534–2547.

Tsutamoto, T., H. Sakai, T. Tanaka, M. Fujii, T. Yamamoto, A. Wada, M. Ohnishi, and M. Horie. 2007. Comparison of active renin concentration and plasma renin activity as a prognostic predictor in patients with heart failure. Circ. J. 71: 915–921.

Ueland, T., C.P. Dahl, J. Kjekshus, J. Hulthe, M. Bohm, F. Mach, A. Goudev, M. Lindberg, J. Wikstrand, P. Aukrust, and L. Gullestad. 2011. Osteoprotegerin predicts progression of chronic heart failure: results from CORONA. Circ. Heart Fail. 4: 145–152.

van Kimmenade, R.R., J.L. Januzzi, Jr., P.T. Ellinor, U.C. Sharma, J.A. Bakker, A.F. Low, A. Martinez, H.J. Crijns, C.A. MacRae, P.P. Menheere, and Y.M. Pinto. 2006. Utility of amino-terminal pro-brain natriuretic peptide, galectin-3, and apelin for the evaluation of patients with acute heart failure. J. Am. Coll. Cardiol. 48: 1217–1224.

Vasan, R.S., L.M. Sullivan, R. Roubenoff, C.A. Dinarello, T. Harris, E.J. Benjamin, D.B. Sawyer, D. Levy, P.W. Wilson, and R.B. D'Agostino. 2003. Inflammatory markers and risk of heart failure in elderly subjects without prior myocardial infarction: the Framingham Heart Study. Circulation 107: 1486–1491.

von Haehling, S., E.A. Jankowska, N.G. Morgenthaler, C. Vassanelli, L. Zanolla, P. Rozentryt, G.S. Filippatos, W. Doehner, F. Koehler, J. Papassotiriou, D.T. Kremastinos, W. Banasiak, J. Struck, P. Ponikowski, A. Bergmann, and S.D. Anker. 2007. Comparison of midregional pro-atrial natriuretic peptide with N-terminal pro-B-type natriuretic peptide in predicting survival in patients with chronic heart failure. J. Am. Coll. Cardiol. 50: 1973–1980.

von Haehling, S., G.S. Filippatos, J. Papassotiriou, M. Cicoira, E.A. Jankowska, W. Doehner, P. Rozentryt, C. Vassanelli, J. Struck, W. Banasiak, P. Ponikowski, D. Kremastinos, A. Bergmann, N.G. Morgenthaler, and S.D. Anker. 2010. Mid-regional pro-adrenomedullin as a novel predictor of mortality in patients with chronic heart failure. Eur. J. Heart Fail. 12: 484–491.

White, M., A. Ducharme, R. Ibrahim, L. Whittom, J. Lavoie, M.C. Guertin, N. Racine, Y. He, G. Yao, J.L. Rouleau, E.L. Schiffrin, and R.M. Touyz. 2006. Increased systemic inflammation and oxidative stress in patients with worsening congestive heart failure: improvement after short-term inotropic support. Clin. Sci. (Lond.) 110: 483–489.

Yan, R.T., M. White, A.T. Yan, S. Yusuf, J.L. Rouleau, A.P. Maggioni, C. Hall, R. Latini, R. Afzal, J. Floras, S. Masson, and R.S. McKelvie. 2005. Usefulness of temporal changes in neurohormones as markers of ventricular remodeling and prognosis in patients with left ventricular systolic dysfunction and heart failure receiving either candesartan or enalapril or both. Am. J. Cardiol. 96: 698–704.

Yancy, C.W., M.T. Saltzberg, R.L. Berkowitz, B. Bertolet, K. Vijayaraghavan, K. Burnham, R.M. Oren, K. Walker, D.P. Horton, and M.A. Silver. 2004. Safety and feasibility of using serial infusions of nesiritide for heart failure in an outpatient setting (from the FUSION I trial). Am. J. Cardiol. 94: 595–601.

Yin, W.H., J.W. Chen, H.L. Jen, M.C. Chiang, W.P. Huang, A.N. Feng, M.S. Young, and S.J. Lin. 2004. Independent prognostic value of elevated high-sensitivity C-reactive protein in chronic heart failure. Am. Heart J. 147: 931–938.

Zannad, F., J.J. McMurray, H. Krum, D.J. van Veldhuisen, K. Swedberg, H. Shi, J. Vincent, S.J. Pocock, and B. Pitt. 2011. Eplerenone in patients with systolic heart failure and mild symptoms. N. Engl. J. Med. 364: 11–21.

# CHAPTER 15

# The Role of Genetic Traits in Cardiovascular Disease

Alexios S. Antonopoulos,[1,2,*] Regent Lee,[2] Dimitris Tousoulis[1] and Charalambos Antoniades[1,2]

## Introduction

Cardiovascular disease (CVD), the leading cause of mortality worldwide, is a heterogeneous trait that results from the complex interplay between genetic, environmental, and clinical risk factors (Kathiresan and Srivastava 2012). Despite the recent advances in the understanding of the molecular mechanisms of atherosclerosis, structural heart disease, and heart rhythm disorders, the heritability of CVD is still poorly understood.

In the last decades, genetic research has attempted to address the missing link between genotype and phenotype. Until recently, genetic studies were "hypothesis-driven" and based mainly on the examination of a few candidate genes in genetic association or linkage studies (Zeller et al. 2011). For certain forms of CVD, genetic research has successfully identified the link between phenotypic traits and the causal genes and/or genomic variants (Lehrman et al. 1985; Watkins et al. 1993; Curran et al. 1995; Watkins et al. 1995; Abifadel et al. 2003). The completion of the International Haplotype Mapping project in 2007 and the recent technological advances in genotyping techniques have revolutionized genetic research by rendering feasible the rapid and low cost genotyping of the whole genome (C4D Genetics Consortium 2011a). Genome-wide association studies (GWAS),

[1]1st Cardiology Department, Hippokration Hospital, Athens Medical School, Athens, Greece.
[2]Department of Cardiovascular Medicine, University of Oxford, Oxford, UK.
*Corresponding author: alexios.antonopoulos@cardiov.ox.ac.uk

i.e., studies scanning genetic variants spread across the whole genome, now provide the ability to search for associations between genetic loci and CVD traits in large populations (Wellcome Trust Case Control Consortium 2007).

The genetic information brought to light by GWAS is expected to deepen our understanding of cardiovascular physiology and the molecular basis of the various disease phenotypes. This will ultimately contribute to the development of risk stratification models, as well as to novel prevention and therapeutic strategies. In the present chapter we report the recent progress in genetic research in the field of CVD and explain how genetics could be applied in clinical cardiology. The recent findings of GWAS studies in CVD are summarized, followed by discussions of their value in risk stratification models and pharmacogenomics development. Existing limitations, as well as the expectations from the next generation genetic research are also discussed.

## Genetic Research in the pre-GWAS Era

Genetic research in the pre-GWAS era managed to characterize the genetic basis of specific forms of CVD that followed the Mendelian pattern of inheritance, with a single causal gene mutation(s) being responsible for the phenotypic trait. This was done either by linkage analysis or genetic association studies. In genetic association studies, candidates' target genes are selected as plausible causes of certain phenotypes and then the association of the genotype with the phenotype is examined in a population of cases and control subjects (Kathiresan and Srivastava 2012). On the other hand, linkage association studies rely on recruitment of families/ siblings with unique, extreme phenotypes and then try to isolate the causal mutation in the responsible chromosomal segment. A prime example of this type of genetic research is the discovery of the mutation in the low density lipoprotein (LDL) cholesterol receptor gene (*LDLR*) as the cause of familial hypercholesterolemia (Lehrman et al. 1985). Other important findings uncovered by linkage analysis and genetic association studies included the identification of the causal gene mutations of familial hypercholesterolemia (Abifadel et al. 2003), hypertrophic cardiomyopathy (Watkins et al. 1993; Watkins et al. 1995), long QT syndrome (Curran et al. 1995), Marfan's syndrome (Dietz et al. 1991), or of specific forms of congenital heart disease such as of atrial or ventricular septal defects (Garg et al. 2003).

Such genetic studies have yielded important discoveries on phenotypic traits caused by rare genetic variants which confer large effects on the phenotype. However, the majority of CVD traits do not follow a Mendelian pattern of inheritance; rather, they are dependent on a complex interplay between multiple genetic variants conferring small size effects on the phenotype and non-genetic effects (Arking and Chakravarti 2009). Such heritability patterns are rendered even more complex by genetic phenomena such as pleiotropy, penetrance and expressivity: certain genetic variants can be associated with multiple phenotypic traits (pleiotropy); a certain genotype can be phenotypically expressed only in a certain proportion of the carriers (penetrance) or be expressed at different degrees of the same phenotype (expressivity) (Kathiresan and Srivastava 2012). This implies that even for Mendelian diseases, the discovery of the causal genotype associated with a certain phenotypic CVD trait can be a challenging issue.

## The Design of Genome-wide Association Studies

GWAS involve case/control or cohort studies that genotype thousands of individuals on a whole-genome scale for identification of single nucleotide polymorphisms (SNP) associated with a specific trait or disease. GWAS are based on the linkage disequilibrium (LD) between genetic loci and disease-predisposing variants. GWAS usually test for common genetic variants, i.e., genetic variants with a frequency>1:20 in the general population (Kathiresan and Srivastava 2012). The power of GWAS to detect a significant association between a genetic variant and a trait depends on the sample size of the study population, the minor-allele frequency (MAF) of the variant, the strength of LD between the SNP and the causal variants, and the effect sizes of the alleles. The effect sizes uncovered by GWAS are rather small, usually between 0.7 and 1.3 folds (Arking and Chakravarti 2009). The 'hits' of GWAS, i.e., the SNPs significantly associated with the phenotypic trait are usually then tested in larger replicated cohorts for confirmation.

Importantly, to avoid false positive associations, stringent statistical testing criteria are required for GWAS. Instead of the usually accepted $P<0.05$ level of statistical significance, a $P$ value of less than $5 \times 10^{-8}$ is used, which corresponds to genome-wide significance level of 0.05 after Bonferroni correction for ~1 million independent tests (Arking and Chakravarti 2009). It should be noted that the power of GWAS studies to detect an association between a genetic locus and a given trait is heavily influenced by the phenotypic characterization of the sample population. In other words how we perceive CVD or the methods we use to phenotype the sample population for a given trait/disease is critical for the success of a GWAS. One reason why GWAS may be susceptible to spurious results is that CVD is an heterogeneous mixture of risk phenotypes (Zeller et al. 2011). Generic cumulative outcome measures such as acute myocardial infarction (AMI), stroke, or sudden cardiac death creates an immense heterogeneicity of phenotypes in the sample population that cannot be adequately controlled. Investigators should ideally focus on specific risk factors (e.g., type 2 diabetes, plasma lipids etc.) that can be easily phenotyped, thus limiting heterogeneicity (Zeller et al. 2011). Another approach is the use of super-controls or of patients at high risk but without even early evidence of disease (Pearson and Manolio 2008); for example if type 2 diabetes is the trait under investigation, elderly patients with normal plasma glucose levels or type 2 diabetics without microalbuminuria can be used as the control group.

## Findings from Recent GWAS in Cardiovascular Disease

### *Coronary artery disease*

Genetic variation is thought to influence the risk of coronary artery disease (CAD) both directly and through effects on known CAD risk factors such as hypertension, diabetes, and hypercholesterolemia. GWAS have mapped several loci that affect susceptibility to CAD/AMI (Zeller et al. 2011). In 2007, three landmark studies simultaneously reported the identification of variants on chromosome 9p21.3 associated with susceptibility to myocardial infarction and CAD (Wellcome Trust Case Control Consortium 2007; Helgadottir et al. 2007; McPherson et al. 2007). Importantly, most of these genetic loci confer small-size effects in the range of 1.06–1.12 per risk allele (Kathiresan and Srivastava 2012). Among them some are associated with established cardiovascular risk factors, such as changes in plasma LDL cholesterol (e.g., at chromosome 6q25 and 6q25 near the *LPA* gene encoding lipoprotein (a) and at 9p13 near *LDLR* gene encoding LDL receptor), which stresses the pathophysiological importance of this risk factor. Other genetic loci such as the 12q24 and 10q24 loci are associated with blood pressure levels (Musunuru and Kathiresan 2010). Recently, a novel susceptibility locus for CAD on chromosome 6p21.3 in the major histocompatibility complex (MHC) gene was discovered (Davies et al. 2012). Since MHC genes regulate inflammation and T-cell responses, the genetic variant could be associated with the initiation and propagation of atherosclerosis (Davies et al. 2012). Nevertheless most identified genetic loci that increase susceptibility to CAD act through yet unknown mechanisms. Some of these loci appear to have pleiotropic effects, showing strong association with various other human diseases or traits (Schunkert et al. 2010). All mapped genetic loci that have been associated with susceptibility to CAD in GWAS up to present are summarized in Table 1.

Among these genetic loci known to be associated with CAD susceptibility, the 9p21.3 locus has drawn much attention, as it is the locus most strongly and consistently associated with CAD susceptibility up to present (p-value: $2.70 \times 10^{-44}$, $1.62 \times 10^{-25}$) (Kathiresan et al. 2009; Schunkert et al. 2011). The haplotype defined by the SNPs was associated with a 15 percent –20 percent increased risk in heterozygotes and a 30 percent –40

**Table 1.** Mapped genetic loci associated with coronary artery disease susceptibility in genome-wide association studies.

| Locus | SNP | Related gene(s) | p-value | Odds ratio for risk allele (95% CI) | Reference |
|---|---|---|---|---|---|
| 9p21 | rs4977574 | CDKN2A,CDKN2B, ANRIL | $2.70 \times 10^{-44}$ | 1.29 (1.25–1.34) | (Kathiresan et al. 2009) |
|  |  |  | $1.35 \times 10^{-22}$ | 1.29 (1.23–1.36) | (Schunkert et al. 2011) |
|  |  |  | $1.62 \times 10^{-25}$ | 1.20 (1.16–1.25) | (C4D Genetics Consortium 2011a) |
| 1p32 | rs17114036 | PPAP2B | $3.81 \times 10^{-19}$ | 1.17 (1.13–1.22) | (Schunkert et al. 2011) |
| 7q32 | rs11556924 | ZC3HC1 | $9.18 \times 10^{-18}$ | 1.09 (1.07–1.12) | (Schunkert et al. 2011) |
| 11q23 | rs964184 | ZNF259, APOA5, APOA1 | $1.02 \times 10^{-17}$ | 1.13 (1.10–1.16) | (Schunkert et al. 2011) |
| 6q26 | rs10455872 | LPA | $3.00 \times 10^{-15}$ | 1.70 (1.49–1.95) | (Clarke et al. 2009) |
| 9q34 | rs579459 | ABO | $4.08 \times 10^{-14}$ | 1.10 (1.07–1.13) | (Schunkert et al. 2011) |
| 6p24 | rs6903956 | c6orf105 | $2.55 \times 10^{-13}$ | 1.65 (1.44–1.90) | (Wang et al. 2011) |
| 15q25 | rs3825807 | ADAMTS7 | $4.98 \times 10^{-13}$ | 1.19 (1.13–1.24) | (Reilly et al. 2011) |
|  |  |  | $1.07 \times 10^{-12}$ | 1.08 (1.06–1.10) | (Schunkert et al. 2011) |
| 3q22 | rs9818870 | MRAS | $7.44 \times 10^{-13}$ | 1.15 (1.11–1.19) | (Erdmann et al. 2009) |
|  |  |  | $3.34 \times 10^{-8}$ | 1.12 (1.07–1.16) | (Schunkert et al. 2011) |
| 6q23 | rs12190287 | TCF21 | $1.07 \times 10^{-12}$ | 1.08 (1.06–1.10) | (Schunkert et al. 2011) |
| 6q25 | rs3798220 | LPA | $3.00 \times 10^{-11}$ | 1.51 (1.33–1.70) | (Schunkert et al. 2011) |
|  |  |  | $9.00 \times 10^{-7}$ | 1.92 (1.48–2.49) | (Clarke et al. 2009) |
| 21q22 | rs9982601 | SLC5A3, MRPS6, KCNE2 | $6.40 \times 10^{-11}$ | 1.20 (1.14–1.27) | (Kathiresan et al. 2009) |
|  |  |  | $4.22 \times 10^{-10}$ | 1.18 (1.12–1.24) | (Schunkert et al. 2011) |
| 19p13 | rs6511720 | LDLR | $9.73 \times 10^{-10}$ | 1.14 (1.09–1.18) | (Schunkert et al. 2011) |
|  |  |  |  | 1.14 (1.09–1.19) | (Kathiresan et al. 2009) |
| 17p11 | rs12936587 | RASD1, PEMT, RAI1 | $4.45 \times 10^{-10}$ | 1.07 (1.05–1.09) | (Schunkert et al. 2011) |
| 10q11 | rs1746048 | CXCL12 | $2.93 \times 10^{-10}$ | 1.33 (1.20–1.48) | (Samani et al. 2007) |
|  |  |  |  | 1.09 (1.07–1.13) | (Schunkert et al. 2011) |
| 14q32 | rs2895811 | HHIPL1 | $1.14 \times 10^{-10}$ | 1.07 (1.05–1.10) | (Schunkert et al. 2011) |
| 1p13 | rs599839 | CELSR2, PSRC1,SORT1 | $6.05 \times 10^{-10}$ | 1.14 (1.09–1.19) | (C4D Genetics Consortium 2011a) |
|  |  |  | $4.05 \times 10^{-9}$ | 1.29 (1.18–1.40) | (Samani et al. 2007) |
|  |  |  | $2.89 \times 10^{-10}$ | 1.11 (1.08–1.15) | (Schunkert et al. 2011) |

*Table 1. contd....*

*...Table 1. contd.*

| Locus | SNP | Related gene(s) | p-value | Odds ratio for risk allele (95% CI) | Reference |
|---|---|---|---|---|---|
| 13q34 | rs4773144 | COL4A1, COL4A2 | $3.84 \times 10^{-9}$ | 1.07 (1.05–1.09) | (Schunkert et al. 2011) |
| 6p24 | rs12526453 | PHACTR1 | $1.30 \times 10^{-9}$ $1.15 \times 10^{-9}$ $5.82 \times 10^{-8}$ | 1.13 (1.09–1.17) 1.10 (1.06–1.13) 1.11 (1.07–1.15) | (Kathiresan et al. 2009) (Schunkert et al. 2011) (C4D Genetics Consortium 2011a) |
| 17p13 | rs216172 | SMG6 | $1.15 \times 10^{-9}$ | 1.07 (1.05–1.09) | (Schunkert et al. 2011) |
| 6p21 | rs3869109 | HCG27, HLA-C | $1.12 \times 10^{-9}$ | 1.14 (NR) | (Davies et al. 2012) |
| 10q24 | rs12413409 | CYP17A1,CNNM2, NT5C2 | $1.03 \times 10^{-9}$ | 1.12 (1.08–1.16) | (Schunkert et al. 2011) |
| 19q13 | rs2075650 | APOE | $3.20 \times 10^{-8}$ | 1.14 (1.09–1.19) | (IBC 50K CAD Consortium 2011b) |
| 2q33 | rs6725887 | WDR12, NBEAL1 | $1.13 \times 10^{-8}$ $1.12 \times 10^{-9}$ | 1.16 (1.10–1.22) 1.14 (1.09–1.19) | (Kathiresan et al. 2009) (Schunkert et al. 2011) |
| 1p32 | rs11206510 | PCSK9 | $9.10 \times 10^{-8}$ | 1.08 (1.05–1.11) 1.15 (1.10–1.21) | (Schunkert et al. 2011) (Kathiresan et al. 2009) |
| 10p11 | rs2505083 | KIAA1462 | $3.87 \times 10^{-8}$ | 1.07 (1.04–1.09) | (C4D Genetics Consortium 2011a) |
| 7q22 | rs10953541 | BCAP29, DUS4L | $3.12 \times 10^{-8}$ | 1.08 (1.05–1.11) | (C4D Genetics Consortium 2011a) |
| 17q21 | rs46522 | UBE2Z, GIP, ATP5G1, SNF8 | $1.81 \times 10^{-8}$ | 1.06 (1.04–1.08) | (Schunkert et al. 2011) |
| 6p21 | rs17609940 | ANKS1A | $1.36 \times 10^{-8}$ | 1.07 (1.05–1.10) | (Schunkert et al. 2011) |
| 8q24 | rs17321515 | TRIB1 | $6.50 \times 10^{-7}$ | 1.06 (1.03–1.10) | (IBC 50K CAD Consortium 2011b) |
| 1q41 | rs17465637 | MIA3 | $1.27 \times 10^{-6}$ $1.36 \times 10^{-8}$ | 1.20 (1.12–1.30) 1.14 (1.09–1.20) | (Samani et al. 2007) (Schunkert et al. 2011) |
| 12q24 | rs3184504 | SH2B3 | $6.35 \times 10^{-6}$ | 1.07 (1.04–1.10) | (Schunkert et al. 2011) |
| 2p21 | rs4299376 | ABCG5, ABCG8 | $7.40 \times 10^{-5}$ | 1.07 (1.04–1.11) | (IBC 50K CAD Consortium 2011b) |
| 10q23 | rs1412444 | LIPA | $1.00 \times 10^{-5}$ | 1.08 (1.05–1.12) | (C4D Genetics Consortium 2011a) |
| 11q22 | rs974819 | PDGFD | $1.50 \times 10^{-5}$ | 1.09 (1.05–1.13) | (C4D Genetics Consortium 2011a) |

The genetic loci are listed according to the level of statistical significance. NR: non reported, SNP: single nucleotide polymorphism.

percent increased risk in homozygous individuals. Interestingly 9p21.3 locus has been also associated with aortic and intracranial aneurysms (Yasuno et al. 2010), peripheral arterial disease (Murabito et al. 2012), type 2 diabetes mellitus (Shea et al. 2011), glioma (Wrensch et al. 2009) and malignant melanoma (Bishop et al. 2009), suggesting that the genetic variant might affect pathways common to disease initiation. Interestingly, the 9p21.3 locus is in the 'gene-desert' area of the chromosome, which contains no annotated genes (Zeller et al. 2011). The nearest gene to 9p21.3 is located near to the INK4/ARF locus that contains coding sequences for cyclin dependent kinase inhibitors, *CDKN2A* (encoding p16INK4a), *CDKN2B* (encoding p15INK4b) and *CDKN2B-AS* or *ANRIL* (antisense noncoding RNA in the INK4 locus), which encodes a non-protein coding RNA (Liu et al. 2009). Both *CDKN2A* and *CDKN2B* are implicated in the regulation of cell cycle (Wrensch et al. 2009). The genetic variants of 9p21.3 are not associated with any of the traditional risk factors and the way the locus confers its biological effects still remains obscure.

Several plausible explanations have been suggested to explain 9p21.3 biological effects. Genetic variants in the locus could affect the noncoding RNA *ANRIL* sequence and its function. Alternatively, the causal variant could lie in a regulatory element, affecting the transcription of genes close to the locus, i.e., *CDKN2A*, *CDKN2B*, *ANRIL* and *ARF* or even distant ones. Identified SNPs associated with CAD risk seem to influence the 9p21 locus transcripts and mainly *ANRIL* expression, suggesting a prominent role of *ANRIL* in the mechanism via which 9p21.3 locus mediates its biological effects (Liu et al. 2009). Interestingly, evidence suggests that transforming growth factor-β (TGF-β), a signaling molecule with a well-identified role in atherosclerosis development, induces *CDKN2B* expression (Hannon and Beach 1994). Despite the existing evidence, more studies are needed to fully elucidate the biological mechanisms via which 9p21 locus could affect cardiovascular risk.

## *Plasma lipid concentrations*

The Diabetes Genetics Initiative (Saxena et al. 2007) was one of the first GWAS identifying a susceptibility locus for plasma lipid concentrations. In a total of 3,000 subjects of European descent, the study identified an index SNP in one locus for each of the three lipid traits, i.e., LDL cholesterol, HDL cholesterol, and triglycerides (Saxena et al. 2007). The identified loci were near genes that are well-known regulators of plasma lipids, such as *APOE* (apolipoprotein E) for LDL cholesterol, *CETP* (cholesterol-ester transfer protein) for HDL cholesterol and *GCKR* (glucokinase regulatory protein) for triglycerides (Saxena et al. 2007). Other studies using a GWA approach have confirmed these first findings and extended the list of genetic loci associated with plasma lipids (Sanna et al. 2011). Further to the Diabetes Genetics Initiative findings, other genetic loci harboring well-known lipid regulators have been identified in GWAS (Sanna et al. 2011). These related genes include *APOA1* (encoding apolipoprotein A1), *APOB* (encoding apolipoprotein B), *LDLR, LPL* (encoding lipoprotein lipase), *PCSK9* (encoding proprotein convertase subtilisin/kexin type 9) and *HMGCR* (encoding hydroxyl-methylglutaryl-CoA reductase) (Sanna et al. 2011). A number of these loci were known Mendelian causes of familial hypercholesterolemia (e.g., *APOB, LDLR,* and *PCSK*) and were rediscovered in GWAS. Importantly some of these loci are harboring genes that are targets of lipid-modifying treatment: the transcripts of *HMGCR, NPC1L1* and *CETP* code for the enzymes inhibited by statins, ezetimibe, and anacetrapib/dalcetrapib respectively. Interestingly, when associations were sought between GWAS 'hits' and total cholesterol as the CVD trait instead of each independent class of plasma lipoproteins, the associations for some loci became even more significant (Zeller et al. 2011). In total, up to now more than 95 genetic loci have been associated with effects on plasma lipid levels in GWAS involving more than 100,000 subjects (Kathiresan and Srivastava 2012).

## *Arterial hypertension*

Arterial hypertension is the most widespread form of CVD with an estimated prevalence of 25 percent –30 percent in the overall population. It has been argued that genetic factors could predispose to increased blood pressure levels in up to 50 percent of hypertensives (Newton-Cheh et al. 2009). However, genetic effects on blood pressure are conferred by multiple small-effect genetic variants rather than a few ones with large effects. Determining the genetic basis of arterial hypertension development has therefore been particularly challenging. In the Global BPgen consortium, 2.5 million genotyped and imputed SNPs were tested for association with systolic and diastolic blood pressure in 34,433 subjects of European ancestry (Newton-Cheh et al. 2009). Eight genetic loci were significantly associated with blood pressure levels; these were near the *CYP17A1, CYP1A2, FGF5, SH2B3, MTHFR, c10orf107, ZNF652* and *PLCD3* genes (Newton-Cheh et al. 2009). Another study, by using an extreme case-control strategy (i.e., examining cases and controls at the extremes of blood pressure distribution) identified a new genetic variant associated on a genome-wide level with blood pressure. This is an SNP upstream of the uromodulin gene (*UMOD*) and suggests a putative role of this variant in hypertension development through an effect on sodium homeostasis. A better understanding of the role of uromodulin in sodium homeostasis could potentially lead to identification of novel therapeutic targets (Padmanabhan et al. 2010). More recently, a GWAS using a multi-stage design in more than 200,000 individuals of European descent reported 29 genetic loci associated with blood pressure regulation, among which 16 were identified for first time (Ehret et al. 2011). Six of these novel loci contain genes previously known or suspected to regulate blood pressure (*GUCY1A3-GUCY1B3; NPR3-C5orf23; ADM; FURIN-FES; GOSR2; GNAS-EDN3*), while the other 10 provided new insights into the physiology of blood pressure regulation. Among the highlighted loci, only *CYP17A1* is known to harbor rare variants that have large effects on blood pressure (Ehret et al. 2011). The estimated effect of most genetic variants on blood pressure levels is approximately 1 mm Hg per allele for systolic blood pressure or 0.5 mm Hg per allele for diastolic blood pressure (Padmanabhan et al. 2010). Even though the conferred effects on blood pressure by genetic variants are rather small to be measured on an individual basis, their identification is important since even small increments in blood pressure are known to be associated with increased risk of cardiovascular events (Newton-Cheh et al. 2009).

The genetic loci discovered so far are able to describe only 1 percent –2 percent of blood pressure variability; it is believed therefore that much more remains to be discovered about the heritability of arterial hypertension. Genetic variants affecting blood pressure levels could have been missed by conducted GWAS, simply because they confer such small effects that a larger sample size is required to identify them (Zeller et al. 2011). Another major issue with GWAS focused on blood pressure levels is that of phenotyping; single blood pressure measurements in large population cohorts are often unreliable, while the effects of concomitant treatment with β-blockers or angiotensin converting enzyme inhibitors for other indications (e.g., left ventricular dysfunction) also influences blood pressure levels (Zeller et al. 2011).

## Arrhythmias

Atrial fibrillation (AF) is one of the commonest heart rhythm disorders, with a lifetime risk of one in four people above the age of 40 (Lloyd-Jones et al. 2004). Several genetic loci have been associated with atrial fibrillation risk. Gudbjartsson et al. have found that the genetic loci 16q22 and 4q25 are associated with AF risk (Gudbjartsson et al. 2009). The 16q22 locus harbors the *ZFHX3* gene which regulates myogenic and neuronal differentiation (Gudbjartsson et al. 2009). In 4q25, two sequence variants are near the *PITX2* gene, which is involved in left-right asymmetry of the heart (Gudbjartsson et al. 2007). The variants increase AF risk by 1.72 and 1.39 per copy; approximately 35 percent of individuals of European descent and 75 percent of Chinese population carry one of the variants. Common variants in 1q21 locus have been associated with lone AF (Ellinor et al. 2010).

Variants at the 1q21 locus cluster at the *KCNN3* gene which codes for a member of a family of voltage-independent, calcium-activated potassium channels (Ellinor et al. 2010). In a recent meta-analysis by the same group (Ellinor et al. 2012), six new susceptibility loci for AF were identified that harbor genes encoding transcription factors related to cardiopulmonary development, cardiac-expressed ion channels, and cell signaling molecules (Ellinor et al. 2012). Interestingly, a genetic variant (rs2824292) in 21q21 locus has been associated with susceptibility to ventricular fibrillation in myocardial infarction patients (Bezzina et al. 2010). The closest gene to the rs2824292 is *CXADR*, which has recently been identified as a modulator of cardiac conduction (Bezzina et al. 2010). This could have a major impact on the clinical management of the patients at risk; this knowledge has not been introduced though in clinical practice yet. Finally, GWAS have identified various loci affecting ECG intervals (PR, QT, RR interval and QRS duration) (Milan et al. 2010). These include variants harboring genes coding for sodium or potassium voltage gated channels, phospholamban, G-coupled protein receptors, caveolins, nitric oxide synthase 1 adaptor protein and others with a less clarified role in cardiac physiology, such as Wingless-type MMTV integration site family member 11 (*WNT11*) (Milan et al. 2010).

## *Cardiomyopathies*

Idiopathic dilated cardiomyopathy (DCM) is characterized by left ventricular dilatation and left ventricular systolic dysfunction in the absence of an obvious etiology. This form of DCM is considered to have a strong genetic component, with 20 percent –35 percent of DCM cases having an affected first-degree relative (Jefferies and Towbin 2010). Even though more than 30 genes have been implicated in monogenic forms of DCM, the results of candidate genes association studies have been inconsistent (Villard et al. 2011). Recently, a GWAS identified two loci that are associated with DCM-related heart failure (Villard et al. 2011); the 1p36.13 locus encompasses several genes among which HSPB7 has been formerly highlighted as a cause of DCM; the second locus at 10q26 harbors the BAG3 gene and the study demonstrated that rare mutations in BAG3 lead to monogenic forms of DCM, while common variant(s) of BAG3 gene have been associated with sporadic forms of DCM (Villard et al. 2011).

## Lessons Learned from GWAS

The findings of GWAS conducted so far have confirmed and enriched the knowledge acquired by genetic association studies of Mendelian diseases.

Firstly, GWAS confirmed that the complex hereditary model of CVD is dependent on the interactions between environmental factors and multiple common genetic variants conferring small size effects on the phenotype. According to the published 'hits' of GWAS in CVD, it has been estimated that the typical risk allele variant increases CVD risk by ~1.24 for qualitative traits or ~0.11 standard deviation units for quantitative traits (Arking and Chakravarti 2009). Even though this effect is modest, it is rather smaller than originally assumed. This means that in order to identify genetic variants implicated in CVD traits, GWAS have to use large sample sizes.

Secondly, among the genetic variants associated with CVD some are located within genes suspected to be causally related with CVD (e.g., those affecting established cardiovascular risk factors), while others are harboring genes not previously thought to participate in CVD pathogenesis. For example genetic variants of 9p21.3 locus are neither associated with any of the classical risk factors nor with established surrogate markers of vascular disease, such as carotid intima-media thickness nor endothelial function (Samani et al. 2008). Interestingly some genetic variants seem to have pleiotropic effects, being associated not only with CVD but with various disease traits, which could suggest common pathways being activated in disease pathogenesis. Despite the advances in the understanding of the mechanisms of atherosclerosis and heart disease, GWAS findings highlighted the need for further exploration into the mechanisms of CVD development.

Another important observation is that the therapeutic or biological value of a target gene may have little to do with the effects conferred by the genetic variant on the phenotype. This is perfectly exemplified by *HMCGR*, the gene encoding HMG-CoA reductase which is the enzyme in the metabolic pathway of cholesterol biosynthesis inhibited by statins. GWAS demonstrated that genetic variants in *HMGCR* confer changes in plasma LDL cholesterol of approximately 3mg/dl (Musunuru and Kathiresan 2010). Even though this is a rather small effect on the phenotype, statins are one of the most powerful therapeutic strategies available today for reducing cardiovascular risk. Of course statins had been discovered long before the findings of GWAS were brought to light. Nevertheless, this observation stressed the value of GWAS in the discovery of novel therapeutic targets by looking into the mechanisms via which genetic variants identified by GWAS lead to certain phenotypic traits.

## Exploring Causality with Mendelian Randomization Strategies

Despite the important insights provided by GWAS, their findings are only observational. This means that a causal relationship between a variant and a disease trait cannot be inferred by GWAS. However, Mendelian randomization studies are able to provide an unbiased answer to the issue of causality (Smith and Ebrahim 2004). Mendelian randomization studies explore causality by testing the effects of a genetic variant on an intermediate phenotype that the variant is known to affect, e.g., a variant in the promoter of a gene changing its expression. If the intermediate phenotype is indeed causally linked with the disease, then the variant should also be associated with the disease to the extent predicted by the size of the effect of the variant on the phenotype and the size of the effect of the phenotype on the disease (Musunuru and Kathiresan 2010). Despite their limitations (Smith and Ebrahim 2004), Mendelian randomization studies are powerful tools, not confounded by environmental factors or reversal causation, and are considered of similar value to randomized clinical trials (RCT). In Mendelian randomization studies, the randomization is considered to take place at the moment of conception. Therefore Mendelian randomization not only saves on the large costs and long follow up period of an RCT, but can be also used to test hypotheses, confirm the findings of existing RCTs, and revisit issues where conductingan RCT would be unethical.

Mendelian randomization has been widely used to assess the causal relationship between plasma homocysteine and CVD risk, by looking into the effects of C677T polymorphism on *MTHFR*, which is known to affect plasma homocysteine levels (Yang et al. 2012). Mendelian randomization studies based on GWAS findings are now conducted to assess the causal relationships between biomarkers and CVD, by using genetic variants highlighted in GWAS. The PROCARDIS Consortium (Clarke et al. 2009) demonstrated that two *LPA* variants were strongly associated with both an increased level of Lp(a) lipoprotein and an increased risk of coronary disease, supporting a causal role of Lp(a) lipoprotein in coronary heart disease pathogenesis. On the contrary, three published Mendelian randomization studies failed to find any association between genetic variants affecting C-reactive protein (CRP) levels and CVD risk (Zacho et al. 2008; Elliott et al. 2009; Wensley et al. 2011). Despite the putative role of CRP as a cardiovascular risk factor, a hypothesis substantially strengthened by the results of the JUPITER trial (Ridker et al. 2008), the negative findings of these studies argue against the causative role of CRP in atherosclerosis development (Zacho et al. 2008; Elliott et al. 2009; Wensley et al. 2011). Another recent Mendelian randomization study demonstrated that genetic variants affecting HDL cholesterol levels are not associated with myocardial infarction risk. The study agrees with previous RCTs findings; e.g., in the AIM-HIGH study, addition of long-acting niacin to simvastatin increased HDL cholesterol and lowered triglycerides but did not lower risk of cardiovascular events (Boden et al. 2011). This finding challenges the concept that raising of plasma HDL cholesterol should uniformly translate into reductions in risk of myocardial infarction (Voight et al. 2012). These examples demonstrate that Mendelian randomization studies can provide valuable information on the causal relationship between a given biomarker and a disease and explore their potential value as therapeutic targets.

## Existing Limitations and Future Perspectives

GWAS have been mainly focused on common genetic variants (i.e., those with a MAF> 5 percent in the population). The effects of low frequency or rare genetic variants remain largely unexplored. However, the genetic effect size decreases logarithmically as a function of the MAF of the variant, with more rare genetic variants conferring a greater effect on the phenotype (Arking and Chakravarti 2009). It has been estimated that the 13 novel loci mapped in the CARDIoGRAM consortium along with 10 ones previously discovered to be associated with CAD, could explain only 10 percent of additive genetic heritability of the disease (Schunkert et al. 2011). The still missing heritability of CVD could lie in gene-gene (GxG) and gene-environment (GxE) interactions, structural copy number variants (i.e., deletions or duplications of large regions of the genome) and rare variants, all of which have been largely overlooked by GWAS (Thanassoulis and Vasan 2010). Poor penetrance of selected genetic variants and imprecise phenotyping of the sample population are additional reasons why GWAS have managed to account for only a small proportion of the genetic variance of CVD (Thanassoulis and Vasan 2010). Moreover, GWAS have been mainly focused up to the present on populations of European descent. More GWAS are needed in populations with a different genetic background such as of African or Asian descent (Zeller et al. 2011). In addition, despite the stringent levels of statistical significance used in GWAS, the massive number of statistical tests performed always carries the risk of yielding false-positive findings. Therefore replication studies in larger cohorts are needed for all newly identified genetic loci associated with a given trait (Pearson and Manolio 2008).

It is interesting that only 12 percent of SNPs discovered by GWAS are located near protein-coding regions of genes. Around 40 percent of the identified SNPs are located in intergenic regions and another 40 percent in introns. This has stirred interest in the biological role of intergenic and intronic regions in the regulation of gene expression (Ndiaye et al. 2011). For most of the identified genetic loci, the causal genetic variant or even the causal gene has not been identified. It is time now for GWAS to move forward and study the unexplained genetic risk hidden in the low frequency (0.5%<MAF<5%) and rare genetic variants (MAF<0.5%). This would help discover the causes of all the monogenic Mendelian diseases and further contribute to our understanding of the genetics of CVD (Kathiresan and Srivastava 2012). Ideally, for each locus we need to understand the mechanisms by which the causal genetic variant affects the expression of the causal gene and of course, the mechanism by which the gene affects the phenotype (Kathiresan and Srivastava 2012). Furthermore, identification of expression quantitative trait loci (eQTL), i.e., genetic variants that are associated with the transcript level of a gene, will help us understand the causal gene and the mechanisms of disease (Raychaudhuri 2011). It has been observed that eQTL overlap with trait-associated variants, suggesting that many common disease variants act by changing transcript levels (Nicolae et al. 2010). The majority of identified eQTL so far are *cis*-effects acting on nearby genes, by affecting promoter/enhancer activity or mRNA stability. This is already known for some mapped genetic variants; for example, the genetic variant in the *PDGFD* locus associated with susceptibility to CAD showed tissue-specific *cis*-eQTL effects (C4D Genetics Consortium 2011a). Nevertheless, eQTL could also act by *trans*-effects but this remains largely unexplored (Raychaudhuri 2011).

The ultimate expectation would be to introduce GWAS findings into clinical practice for risk stratification purposes or for guidance of treatment. Genetic variants could also help determine a patient's response to a drug. For example SNPs determining the inhibition of platelets' activity by clopidogrel (e.g., 3435C>T SNP in *ABCB1* gene) could have a major impact on clinical practice (Mega et al. 2010). Similarly, attempts have been made to create a pharmacogenetic algorithm for estimating warfarin dosage (Klein et al. 2009). Furthermore, a genetic risk score system for coronary artery disease could be developed based on the genetic loci mapped by GWAS. Even though several attempts have been made to develop such genetic score systems, there was minimal or no evidence at all for their clinical utility (Thanassoulis and Vasan 2010). A recent attempt, using a genetic risk score model updated with the latest GWAS findings, only modestly improved risk reclassification for incident coronary heart disease (Thanassoulis et al. 2012). Even though these approaches have not been introduced in standard clinical practice, they seem promising for the future of pharmacogenomics and risk prediction based on genetic risk score systems.

Finally, next generation sequencing with whole-exome approaches (i.e., scan of that part of the genome coding portions of genes that are expressed) and modeling of human genetic disease in human reprogrammed cells ('disease in a dish') are thought to deepen our understanding of the mechanisms of human disease and allow cellular manipulation of genetic diseases in search of a treatment (Kathiresan and Srivastava 2012).

## Conclusion

Genome-wide association studies (GWAS) have revolutionized the field of genetic research over the last years. Mapping of susceptibility loci for cardiovascular disease has confirmed the knowledge derived from genetic association studies of Mendelian diseases and highlighted genes previously not thought to be implicated in cardiovascular physiology. Identification of expression quantitative trait loci will further enhance our understanding of the molecular pathways involved in disease initiation and progression. The findings from GWAS should be integrated in Mendelian randomization studies, which provide a unique platform for investigating causal relationships between genetic variants and disease traits. The powerful tools of whole genome/exome sequencing now available and the uncovering of cardiovascular disease genetics by GWAS are believed to ultimately lead to risk stratification strategies, the discovery of novel drugs, rapid development of pharmacogenomics and to an individualized treatment in clinical cardiology based on genetic biomarkers.

## References

Abifadel, M., M. Varret, J.P. Rabes, D. Allard, K. Ouguerram, M. Devillers, C. Cruaud, S. Benjannet, L. Wickham, D. Erlich, A. Derre, L. Villeger, M. Farnier, I. Beucler, E. Bruckert, J. Chambaz, B. Chanu, J.M. Lecerf, G. Luc, P. Moulin, J. Weissenbach, A. Prat, M. Krempf, C. Junien, N.G. Seidah, and C. Boileau. 2003. Mutations in PCSK9 cause autosomal dominant hypercholesterolemia. Nat. Genet. 34: 154–156.

Arking, D.E. and A. Chakravarti. 2009. Understanding cardiovascular disease through the lens of genome-wide association studies. Trends Genet. 25: 387–394.

Bezzina, C.R., R. Pazoki, A. Bardai, R.F. Marsman, J.S. de Jong, M.T. Blom, B.P. Scicluna, J.W. Jukema, N.R. Bindraban, P. Lichtner, A. Pfeufer, N. H. Bishopric, D.M. Roden, T. Meitinger, S.S. Chugh, R.J. Myerburg, X. Jouven, S. Kaab, L.R. Dekker, H.L. Tan, M.W. Tanck, and A.A. Wilde. 2010. Genome-wide association study identifies a susceptibility locus at 21q21 for ventricular fibrillation in acute myocardial infarction. Nat. Genet. 42: 688–691.

Bishop, D.T., F. Demenais, M.M. Iles, M. Harland, J.C. Taylor, E. Corda, J. Randerson-Moor, J.F. Aitken, M.F. Avril, E. Azizi, B. Bakker, G. Bianchi-Scarra, B. Bressac-de Paillerets, D. Calista, L.A. Cannon-Albright, A.W.T. Chin, T. Debniak, G. Galore-Haskel, P. Ghiorzo, I. Gut, J. Hansson, M. Hocevar, V. Hoiom, J.L. Hopper, C. Ingvar, P.A. Kanetsky, R.F. Kefford, M.T. Landi, J. Lang, J. Lubinski, R. Mackie, J. Malvehy, G.J. Mann, N.G. Martin, G.W. Montgomery, F.A. van Nieuwpoort, S. Novakovic, H. Olsson, S. Puig, M. Weiss, W. van Workum, D. Zelenika, K.M. Brown, A.M. Goldstein, E.M. Gillanders, A. Boland, P. Galan, D.E. Elder, N.A. Gruis, N.K. Hayward, G.M. Lathrop, J.H. Barrett, and J.A. Bishop. 2009. Genome-wide association study identifies three loci associated with melanoma risk. Nat. Genet. 41: 920–925.

Boden, W.E., J.L. Probstfield, T. Anderson, B.R. Chaitman, P. Desvignes-Nickens, K. Koprowicz, R. McBride, K. Teo, and W. Weintraub. 2011. Niacin in patients with low HDL cholesterol levels receiving intensive statin therapy. N. Engl. J. Med. 365: 2255–2267.

C4D Genetics Consortium. 2011a. A genome-wide association study in Europeans and South Asians identifies five new loci for coronary artery disease. Nat. Genet. 43: 339–344.

Clarke, R., J.F. Peden, J.C. Hopewell, T. Kyriakou, A. Goel, S.C. Heath, S. Parish, S. Barlera, M.G. Franzosi, S. Rust, D. Bennett, A. Silveira, A. Malarstig, F.R. Green, M. Lathrop, B. Gigante, K. Leander, U. de Faire, U. Seedorf, A. Hamsten, R. Collins, H. Watkins, and M. Farrall. 2009. Genetic variants associated with Lp(a) lipoprotein level and coronary disease. N. Engl. J. Med. 361: 2518–2528.

Curran, M.E., I. Splawski, K.W. Timothy, G.M. Vincent, E.D. Green, and M.T. Keating. 1995. A molecular basis for cardiac arrhythmia: HERG mutations cause long QT syndrome. Cell 80: 795–803.

Davies, R.W., G.A. Wells, A.F. Stewart, J. Erdmann, S.H. Shah, J.F. Ferguson, A.S. Hall, S.S. Anand, M.S. Burnett, S.E. Epstein, S. Dandona, L. Chen, J. Nahrstaedt, C. Loley, I.R. Konig, W.E. Kraus, C.B. Granger, J.C. Engert, C. Hengstenberg, H.E. Wichmann, S. Schreiber, W.H. Tang, S.G. Ellis, D.J. Rader, S.L. Hazen, M.P. Reilly, N.J. Samani, H. Schunkert, R. Roberts, and R. McPherson. 2012. A genome-wide association study for coronary artery disease identifies a novel susceptibility locus in the major histocompatibility complex. Circ. Cardiovasc. Genet. 5: 217–225.

Dietz, H.C., G.R. Cutting, R.E. Pyeritz, C.L. Maslen, L.Y. Sakai, G.M. Corson, E.G. Puffenberger, A. Hamosh, E.J. Nanthakumar, S.M. Curristin et al. 1991. Marfan syndrome caused by a recurrent de novo missense mutation in the fibrillin gene. Nature 352: 337–339.

Ehret, G.B., P.B. Munroe, K.M. Rice, M. Bochud, A.D. Johnson, D.I. Chasman, A.V. Smith, M.D. Tobin, G.C. Verwoert, S.J. Hwang, V. Pihur, P. Vollenweider, P.F. O'Reilly, N. Amin, J.L. Bragg-Gresham, A. Teumer, N.L. Glazer, L. Launer, J.H. Zhao, Y. Aulchenko, S. Heath, S. Sober, A. Parsa, J. Luan, P. Arora, A. Dehghan, F. Zhang, G. Lucas, A.A. Hicks, A.U. Jackson, J.F. Peden, T. Tanaka, S.H. Wild, I. Rudan, W. Igl, Y. Milaneschi, A.N. Parker, C. Fava, J.C. Chambers, E.R. Fox, M. Kumari, M.J. Go, P. van der Harst, W.H. Kao, M. Sjogren, D.G. Vinay, M. Alexander, Y. Tabara, S. Shaw-Hawkins, P.H. Whincup, Y. Liu, G. Shi, J. Kuusisto, B. Tayo, M. Seielstad, X. Sim, K.D. Nguyen, T. Lehtimaki, G. Matullo, Y. Wu, T.R. Gaunt, N.C. Onland-Moret, M.N. Cooper, C.G. Platou, E. Org, R. Hardy, S. Dahgam, J. Palmen, V. Vitart, P.S. Braund, T. Kuznetsova, C.S. Uiterwaal, A. Adeyemo, W. Palmas, H. Campbell, B. Ludwig, M. Tomaszewski, I. Tzoulaki, N.D. Palmer, T. Aspelund, M. Garcia, Y.P. Chang, J.R. O'Connell, N.I. Steinle, D.E. Grobbee, D.E. Arking, S.L. Kardia, A.C. Morrison, D. Hernandez, S. Najjar, W.L. McArdle, D. Hadley, M.J. Brown, J.M. Connell, A.D. Hingorani, I.N. Day, D.A. Lawlor, J.P. Beilby, R.W. Lawrence, R. Clarke, J.C. Hopewell, H. Ongen, A.W. Dreisbach, Y. Li, J.H. Young, J.C. Bis, M. Kahonen, J. Viikari, L.S. Adair, N.R. Lee, M.H. Chen, M. Olden, C. Pattaro, J.A. Bolton, A. Kottgen, S. Bergmann, V. Mooser, N. Chaturvedi, T.M. Frayling, M. Islam, T.H. Jafar, J. Erdmann, S.R. Kulkarni, S.R. Bornstein, J. Grassler, L. Groop, B.F. Voight, J. Kettunen, P. Howard, A. Taylor, S. Guarrera, F. Ricceri, V. Emilsson, A. Plump, I. Barroso, K.T. Khaw, A.B. Weder, S.C. Hunt, Y.V. Sun, R.N. Bergman, F.S. Collins, L.L. Bonnycastle, L.J. Scott, H.M. Stringham, L. Peltonen, M. Perola, E. Vartiainen, S.M. Brand, J.A. Staessen, T.J. Wang, P.R. Burton, M. Soler Artigas, Y. Dong, H. Snieder, X. Wang, H. Zhu, K.K. Lohman, M.E. Rudock, S.R. Heckbert, N.L. Smith, K.L. Wiggins, A. Doumatey, D. Shriner, G. Veldre, M. Viigimaa, S. Kinra, D. Prabhakaran, V. Tripathy, C.D. Langefeld, A. Rosengren, D.S. Thelle, A.M. Corsi, A. Singleton, T. Forrester, G. Hilton, C.A. McKenzie, T. Salako, N. Iwai, Y. Kita, T. Ogihara, T. Ohkubo, T. Okamura, H. Ueshima, S. Umemura, S. Eyheramendy, T. Meitinger, H.E. Wichmann, Y.S. Cho, H.L. Kim, J.Y. Lee, J. Scott, J.S. Sehmi, W. Zhang, B. Hedblad, P. Nilsson, G.D. Smith, A. Wong, N. Narisu, A. Stancakova, L.J. Raffel, J. Yao, S. Kathiresan, C.J. O'Donnell, S.M. Schwartz, M.A. Ikram, W.T. Longstreth, Jr., T.H. Mosley, S. Seshadri, N.R. Shrine, L.V. Wain, M.A. Morken, A.J. Swift, J. Laitinen, I. Prokopenko,

P. Zitting, J.A. Cooper, S.E. Humphries, J. Danesh, A. Rasheed, A. Goel, A. Hamsten, H. Watkins, S.J. Bakker, W.H. van Gilst, C.S. Janipalli, K.R. Mani, C.S. Yajnik, A. Hofman, F.U. Mattace-Raso, B.A. Oostra, A. Demirkan, A. Isaacs, F. Rivadeneira, E.G. Lakatta, M. Orru, A. Scuteri, M. Ala-Korpela, A.J. Kangas, L.P. Lyytikainen, P. Soininen, T. Tukiainen, P. Wurtz, R.T. Ong, M. Dorr, H.K. Kroemer, U. Volker, H. Volzke, P. Galan, S. Hercberg, M. Lathrop, D. Zelenika, P. Deloukas, M. Mangino, T.D. Spector, G. Zhai, J.F. Meschia, M.A. Nalls, P. Sharma, J. Terzic, M.V. Kumar, M. Denniff, E. Zukowska-Szczechowska, L.E. Wagenknecht, F.G. Fowkes, F.J. Charchar, P.E. Schwarz, C. Hayward, X. Guo, C. Rotimi, M.L. Bots, E. Brand, N.J. Samani, O. Polasek, P.J. Talmud, F. Nyberg, D. Kuh, M. Laan, K. Hveem, L.J. Palmer, Y.T. van der Schouw, J.P. Casas, K.L. Mohlke, P. Vineis, O. Raitakari, S.K. Ganesh, T.Y. Wong, E.S. Tai, R.S. Cooper, M. Laakso, D.C. Rao, T.B. Harris, R.W. Morris, A.F. Dominiczak, M. Kivimaki, M.G. Marmot, T. Miki, D. Saleheen, G.R. Chandak, J. Coresh, G. Navis, V. Salomaa, B.G. Han, X. Zhu, J.S. Kooner, O. Melander, P.M. Ridker, S. Bandinelli, U.B. Gyllensten, A.F. Wright, J.F. Wilson, L. Ferrucci, M. Farrall, J. Tuomilehto, P.P. Pramstaller, R. Elosua, N. Soranzo, E.J. Sijbrands, D. Altshuler, R.J. Loos, A.R. Shuldiner, C. Gieger, P. Meneton, A.G. Uitterlinden, N.J. Wareham, V. Gudnason, J.I. Rotter, R. Rettig, M. Uda, D.P. Strachan, J.C. Witteman, A.L. Hartikainen, J.S. Beckmann, E. Boerwinkle, R.S. Vasan, M. Boehnke, M.G. Larson, M.R. Jarvelin, B.M. Psaty, G.R. Abecasis, A. Chakravarti, P. Elliott, C.M. van Duijn, C. Newton-Cheh, D. Levy, M.J. Caulfield, and T. Johnson. 2011. Genetic variants in novel pathways influence blood pressure and cardiovascular disease risk. Nature 478: 103–109.

Ellinor, P.T., K.L. Lunetta, C.M. Albert, N.L. Glazer, M D. Ritchie, A.V. Smith, D.E. Arking, M. Muller-Nurasyid, B.P. Krijthe, S.A. Lubitz, J.C. Bis, M.K. Chung, M. Dorr, K. Ozaki, J.D. Roberts, J.G. Smith, A. Pfeufer, M.F. Sinner, K. Lohman, J. Ding, N.L. Smith, J.D. Smith, M. Rienstra, K.M. Rice, D.R. Van Wagoner, J.W. Magnani, R. Wakili, S. Clauss, J.I. Rotter, G. Steinbeck, L.J. Launer, R.W. Davies, M. Borkovich, T.B. Harris, H. Lin, U. Volker, H. Volzke, D.J. Milan, A. Hofman, E. Boerwinkle, L.Y. Chen, E.Z. Soliman, B.F. Voight, G. Li, A. Chakravarti, M. Kubo, U.B. Tedrow, L.M. Rose, P.M. Ridker, D. Conen, T. Tsunoda, T. Furukawa, N. Sotoodehnia, S. Xu, N. Kamatani, D. Levy, Y. Nakamura, B. Parvez, S. Mahida, K.L. Furie, J. Rosand, R. Muhammad, B.M. Psaty, T. Meitinger, S. Perz, H.E. Wichmann, J.C. Witteman, W.H. Kao, S. Kathiresan, D.M. Roden, A.G. Uitterlinden, F. Rivadeneira, B. McKnight, M. Sjogren, A.B. Newman, Y. Liu, M.H. Gollob, O. Melander, T. Tanaka, B.H. Stricker, S. Felix, A. Alonso, D. Darbar, J. Barnard, D.I. Chasman, S.R. Heckbert, E.J. Benjamin, V. Gudnason, and S. Kaab. 2012. Meta-analysis identifies six new susceptibility loci for atrial fibrillation. Nat. Genet. 44: 670–675.

Ellinor, P.T., K.L. Lunetta, N.L. Glazer, A. Pfeufer, A. Alonso, M.K. Chung, M.F. Sinner, P.I. de Bakker, M. Mueller, S.A. Lubitz, E. Fox, D. Darbar, N.L. Smith, J.D. Smith, R.B. Schnabel, E.Z. Soliman, K.M. Rice, D.R. Van Wagoner, B.M. Beckmann, C. van Noord, K. Wang, G.B. Ehret, J.I. Rotter, S.L. Hazen, G. Steinbeck, A.V. Smith, L.J. Launer, T.B. Harris, S. Makino, M. Nelis, D.J. Milan, S. Perz, T. Esko, A. Kottgen, S. Moebus, C. Newton-Cheh, M. Li, S. Mohlenkamp, T.J. Wang, W.H. Kao, R.S. Vasan, M.M. Nothen, C.A. MacRae, B.H. Stricker, A. Hofman, A.G. Uitterlinden, D. Levy, E. Boerwinkle, A. Metspalu, E.J. Topol, A. Chakravarti, V. Gudnason, B.M. Psaty, D.M. Roden, T. Meitinger, H.E. Wichmann, J.C. Witteman, J. Barnard, D.E. Arking, E.J. Benjamin, S.R. Heckbert, and S. Kaab. 2010. Common variants in KCNN3 are associated with lone atrial fibrillation. Nat. Genet. 42: 240–244.

Elliott, P., J.C. Chambers, W. Zhang, R. Clarke, J.C. Hopewell, J.F. Peden, J. Erdmann, P. Braund, J.C. Engert, D. Bennett, L. Coin, D. Ashby, I. Tzoulaki, I.J. Brown, S. Mt-Isa, M.I. McCarthy, L. Peltonen, N.B. Freimer, M. Farrall, A. Ruokonen, A. Hamsten, N. Lim, P. Froguel, D.M. Waterworth, P. Vollenweider, G. Waeber, M.R. Jarvelin, V. Mooser, J. Scott, A.S. Hall, H. Schunkert, S.S. Anand, R. Collins, N.J. Samani, H. Watkins, and J.S. Kooner. 2009. Genetic Loci associated with C-reactive protein levels and risk of coronary heart disease. JAMA 302: 37–48.

Erdmann, J., A. Grosshennig, P.S. Braund, I.R. Konig, C. Hengstenberg, A.S. Hall, P. Linsel-Nitschke, S. Kathiresan, B. Wright, D.A. Tregouet, F. Cambien, P. Bruse, Z. Aherrahrou, A.K. Wagner, K. Stark, S.M. Schwartz, V. Salomaa, R. Elosua, O. Melander, B.F. Voight, C.J. O'Donnell, L. Peltonen, D.S. Siscovick, D. Altshuler, P.A. Merlini, F. Peyvandi, L. Bernardinelli, D. Ardissino, A. Schillert, S. Blankenberg, T. Zeller, P. Wild, D.F. Schwarz, L. Tiret, C. Perret, S. Schreiber, N.E. El Mokhtari, A. Schafer, W. Marz, W. Renner, P. Bugert, H. Kluter, J. Schrezenmeir, D. Rubin, S.G. Ball, A.J. Balmforth, H.E. Wichmann, T. Meitinger, M. Fischer, C. Meisinger, J. Baumert, A. Peters, W.H. Ouwehand, P. Deloukas, J.R. Thompson, A. Ziegler, N.J. Samani, and H. Schunkert. 2009. New susceptibility locus for coronary artery disease on chromosome 3q22.3. Nat. Genet. 41: 280–282.

Garg, V., I.S. Kathiriya, R. Barnes, M.K. Schluterman, I.N. King, C.A. Butler, C.R. Rothrock, R.S. Eapen, K. Hirayama-Yamada, K. Joo, R. Matsuoka, J.C. Cohen, and D. Srivastava. 2003. GATA4 mutations cause human congenital heart defects and reveal an interaction with TBX5. Nature 424: 443–447.

Gudbjartsson, D.F., D.O. Arnar, A. Helgadottir, S. Gretarsdottir, H. Holm, A. Sigurdsson, A. Jonasdottir, A. Baker, G. Thorleifsson, K. Kristjansson, A. Palsson, T. Blondal, P. Sulem, V.M. Backman, G.A. Hardarson, E. Palsdottir, A. Helgason, R. Sigurjonsdottir, J.T. Sverrisson, K. Kostulas, M.C. Ng, L. Baum, W.Y. So, K.S. Wong, J.C. Chan, K.L. Furie, S.M. Greenberg, M. Sale, P. Kelly, C.A. MacRae, E.E. Smith, J. Rosand, J. Hillert, R.C. Ma, P.T. Ellinor, G. Thorgeirsson, J.R. Gulcher, A. Kong, U. Thorsteinsdottir, and K. Stefansson. 2007. Variants conferring risk of atrial fibrillation on chromosome 4q25. Nature 448: 353–357.

Gudbjartsson, D.F., H. Holm, S. Gretarsdottir, G. Thorleifsson, G.B. Walters, G. Thorgeirsson, J. Gulcher, E.B. Mathiesen, I. Njolstad, A. Nyrnes, T. Wilsgaard, E.M. Hald, K. Hveem, C. Stoltenberg, G. Kucera, T. Stubblefield, S. Carter, D. Roden, M.C. Ng, L. Baum, W.Y. So, K.S. Wong, J.C. Chan, C. Gieger, H.E. Wichmann, A. Gschwendtner, M. Dichgans, G. Kuhlenbaumer, K. Berger, E.B. Ringelstein, S. Bevan, H.S. Markus, K. Kostulas, J. Hillert, S. Sveinbjornsdottir, E.M. Valdimarsson, M.L. Lochen, R.C. Ma, D. Darbar, A. Kong, D.O. Arnar, U. Thorsteinsdottir, and K. Stefansson. 2009. A sequence variant in ZFHX3 on 16q22 associates with atrial fibrillation and ischemic stroke. Nat. Genet. 41: 876–878.

Hannon, G.J. and D. Beach. 1994. p15INK4B is a potential effector of TGF-beta-induced cell cycle arrest. Nature 371: 257–261.

Helgadottir, A., G. Thorleifsson, A. Manolescu, S. Gretarsdottir, T. Blondal, A. Jonasdottir, A. Sigurdsson, A. Baker, A. Palsson, G. Masson, D.F. Gudbjartsson, K.P. Magnusson, K. Andersen, A.I. Levey, V.M. Backman, S. Matthiasdottir, T. Jonsdottir, S. Palsson, H. Einarsdottir, S. Gunnarsdottir, A. Gylfason, V. Vaccarino, W.C. Hooper, M.P. Reilly, C.B. Granger, H. Austin, D.J. Rader, S.H. Shah, A.A. Quyyumi, J.R. Gulcher, G. Thorgeirsson, U. Thorsteinsdottir, A. Kong, and K. Stefansson. 2007. A common variant on chromosome 9p21 affects the risk of myocardial infarction. Science 316: 1491–1493.

IBC 50K CAD Consortium. 2011b. Large-scale gene-centric analysis identifies novel variants for coronary artery disease. PLoS Genet. 7: e1002260.

Jefferies, J.L. and J.A. Towbin. 2010. Dilated cardiomyopathy. Lancet 375: 752–762.

Kathiresan, S. and D. Srivastava. 2012. Genetics of human cardiovascular disease. Cell 148: 1242–1257.

Kathiresan, S., B.F. Voight, S. Purcell, K. Musunuru, D. Ardissino, P.M. Mannucci, S. Anand, J.C. Engert, N.J. Samani, H. Schunkert, J. Erdmann, M.P. Reilly, D.J. Rader, T. Morgan, J.A. Spertus, M. Stoll, D. Girelli, P.P. McKeown, C.C. Patterson, D.S. Siscovick, C.J. O'Donnell, R. Elosua, L. Peltonen, V. Salomaa, S.M. Schwartz, O. Melander, D. Altshuler, P.A. Merlini, C. Berzuini, L. Bernardinelli, F. Peyvandi, M. Tubaro, P. Celli, M. Ferrario, R. Fetiveau, N. Marziliano, G. Casari, M. Galli, F. Ribichini, M. Rossi, F. Bernardi, P. Zonzin, A. Piazza, J. Yee, Y. Friedlander, J. Marrugat, G. Lucas, I. Subirana, J. Sala, R. Ramos, J.B. Meigs, G. Williams, D.M. Nathan, C.A. MacRae, A.S. Havulinna, G. Berglund, J.N. Hirschhorn, R. Asselta, S. Duga, M. Spreafico, M.J. Daly, J. Nemesh, J.M. Korn, S.A. McCarroll, A. Surti, C. Guiducci, L. Gianniny, D. Mirel, M. Parkin,

N. Burtt, S.B. Gabriel, J.R. Thompson, P.S. Braund, B.J. Wright, A.J. Balmforth, S.G. Ball, A.S. Hall, P. Linsel-Nitschke, W. Lieb, A. Ziegler, I. Konig, C. Hengstenberg, M. Fischer, K. Stark, A. Grosshennig, M. Preuss, H.E. Wichmann, S. Schreiber, W. Ouwehand, P. Deloukas, M. Scholz, F. Cambien, M. Li, Z. Chen, R. Wilensky, W. Matthai, A. Qasim, H.H. Hakonarson, J. Devaney, M.S. Burnett, A.D. Pichard, K.M. Kent, L. Satler, J.M. Lindsay, R. Waksman, C.W. Knouff, D.M. Waterworth, M.C. Walker, V. Mooser, S.E. Epstein, T. Scheffold, K. Berger, A. Huge, N. Martinelli, O. Olivieri, R. Corrocher, P. McKeown, E. Erdmann, I.R. Konig, H. Holm, G. Thorleifsson, U. Thorsteinsdottir, K. Stefansson, R. Do, C. Xie, and D. Siscovick. 2009. Genome-wide association of early-onset myocardial infarction with single nucleotide polymorphisms and copy number variants. Nat. Genet. 41: 334–341.

Klein, T.E., R.B. Altman, N. Eriksson, B.F. Gage, S.E. Kimmel, M.T. Lee, N.A. Limdi, D. Page, D.M. Roden, M.J. Wagner, M.D. Caldwell, and J.A. Johnson. 2009. Estimation of the warfarin dose with clinical and pharmacogenetic data. N. Engl. J. Med. 360: 753–764.

Lehrman, M.A., W.J. Schneider, T.C. Sudhof, M.S. Brown, J.L. Goldstein, and D.W. Russell. 1985. Mutation in LDL receptor: Alu-Alu recombination deletes exons encoding transmembrane and cytoplasmic domains. Science 227: 140–146.

Liu, Y., H.K. Sanoff, H. Cho, C.E. Burd, C. Torrice, K.L. Mohlke, J.G. Ibrahim, N.E. Thomas, and N.E. Sharpless. 2009. INK4/ARF transcript expression is associated with chromosome 9p21 variants linked to atherosclerosis. PLoS One 4: e5027.

Lloyd-Jones, D.M., T.J. Wang, E.P. Leip, M.G. Larson, D. Levy, R.S. Vasan, R.B. D'Agostino, J.M. Massaro, A. Beiser, P.A. Wolf, and E.J. Benjamin. 2004. Lifetime risk for development of atrial fibrillation: the Framingham Heart Study. Circulation 110: 1042–1046.

McPherson, R., A. Pertsemlidis, N. Kavaslar, A. Stewart, R. Roberts, D.R. Cox, D.A. Hinds, L.A. Pennacchio, A. Tybjaerg-Hansen, A.R. Folsom, E. Boerwinkle, H.H. Hobbs, and J.C. Cohen. 2007. A common allele on chromosome 9 associated with coronary heart disease. Science 316: 1488–1491.

Mega, J.L., S.L. Close, S.D. Wiviott, L. Shen, J.R. Walker, T. Simon, E.M. Antman, E. Braunwald, and M.S. Sabatine. 2010. Genetic variants in ABCB1 and CYP2C19 and cardiovascular outcomes after treatment with clopidogrel and prasugrel in the TRITON-TIMI 38 trial: a pharmacogenetic analysis. Lancet 376: 1312–1319.

Milan, D.J., S.A. Lubitz, S. Kaab, and P.T. Ellinor. 2010. Genome-wide association studies in cardiac electrophysiology: recent discoveries and implications for clinical practice. Heart Rhythm 7: 1141–1148.

Murabito, J.M., C.C. White, M. Kavousi, Y.V. Sun, M.F. Feitosa, V. Nambi, C. Lamina, A. Schillert, S. Coassin, J.C. Bis, L. Broer, D.C. Crawford, N. Franceschini, R. Frikke-Schmidt, M. Haun, S. Holewijn, J.E. Huffman, S.J. Hwang, S. Kiechl, B. Kollerits, M.E. Montasser, I.M. Nolte, M.E. Rudock, A. Senft, A. Teumer, P. van der Harst, V. Vitart, L.L. Waite, A.R. Wood, C.L. Wassel, D.M. Absher, M.A. Allison, N. Amin, A. Arnold, F.W. Asselbergs, Y. Aulchenko, S. Bandinelli, M. Barbalic, M. Boban, K. Brown-Gentry, D.J. Couper, M.H. Criqui, A. Dehghan, M. den Heijer, B. Dieplinger, J. Ding, M. Dorr, C. Espinola-Klein, S.B. Felix, L. Ferrucci, A.R. Folsom, G. Fraedrich, Q. Gibson, R. Goodloe, G. Gunjaca, M. Haltmayer, G. Heiss, A. Hofman, A. Kieback, L.A. Kiemeney, I. Kolcic, I.J. Kullo, S.B. Kritchevsky, K.J. Lackner, X. Li, W. Lieb, K. Lohman, C. Meisinger, D. Melzer, E.R. Mohler, 3rd, I. Mudnic, T. Mueller, G. Navis, F. Oberhollenzer, J.W. Olin, J. O'Connell, C.J. O'Donnell, W. Palmas, B.W. Penninx, A. Petersmann, O. Polasek, B.M. Psaty, B. Rantner, K. Rice, F. Rivadeneira, J.I. Rotter, A. Seldenrijk, M. Stadler, M. Summerer, T. Tanaka, A. Tybjaerg-Hansen, A.G. Uitterlinden, W.H. van Gilst, S.H. Vermeulen, S.H. Wild, P.S. Wild, J. Willeit, T. Zeller, T. Zemunik, L. Zgaga, T.L. Assimes, S. Blankenberg, E. Boerwinkle, H. Campbell, J.P. Cooke, J. de Graaf, D. Herrington, S.L. Kardia, B.D. Mitchell, A. Murray, T. Munzel, A.B. Newman, B.A. Oostra, I. Rudan, A.R. Shuldiner, H. Snieder, C.M. van Duijn, U. Volker, A.F. Wright, H.E. Wichmann, J.F. Wilson, J.C. Witteman, Y. Liu, C. Hayward, I.B. Borecki, A. Ziegler, K.E. North, L.A. Cupples, and F. Kronenberg. 2012. Association between chromosome 9p21 variants and the ankle-brachial index identified by a meta-analysis of 21 genome-wide association studies. Circ. Cardiovasc. Genet. 5: 100–112.

Musunuru, K. and S. Kathiresan. 2010. Genetics of coronary artery disease. Annu. Rev. Genomics Hum. Genet. 11: 91–108.

Ndiaye, N.C., M. Azimi Nehzad, S. El Shamieh, M.G. Stathopoulou, and S. Visvikis-Siest. 2011. Cardiovascular diseases and genome-wide association studies. Clin. Chim. Acta 412: 1697–1701.

Newton-Cheh, C., T. Johnson, V. Gateva, M.D. Tobin, M. Bochud, L. Coin, S.S. Najjar, J.H. Zhao, S.C. Heath, S. Eyheramendy, K. Papadakis, B.F. Voight, L.J. Scott, F. Zhang, M. Farrall, T. Tanaka, C. Wallace, J.C. Chambers, K.T. Khaw, P. Nilsson, P. van der Harst, S. Polidoro, D.E. Grobbee, N.C. Onland-Moret, M.L. Bots, L.V. Wain, K.S. Elliott, A. Teumer, J. Luan, G. Lucas, J. Kuusisto, P.R. Burton, D. Hadley, W.L. McArdle, M. Brown, A. Dominiczak, S.J. Newhouse, N.J. Samani, J. Webster, E. Zeggini, J.S. Beckmann, S. Bergmann, N. Lim, K. Song, P. Vollenweider, G. Waeber, D.M. Waterworth, X. Yuan, L. Groop, M. Orho-Melander, A. Allione, A. Di Gregorio, S. Guarrera, S. Panico, F. Ricceri, V. Romanazzi, C. Sacerdote, P. Vineis, I. Barroso, M.S. Sandhu, R.N. Luben, G.J. Crawford, P. Jousilahti, M. Perola, M. Boehnke, L.L. Bonnycastle, F.S. Collins, A.U. Jackson, K.L. Mohlke, H.M. Stringham, T.T. Valle, C.J. Willer, R.N. Bergman, M.A. Morken, A. Doring, C. Gieger, T. Illig, T. Meitinger, E. Org, A. Pfeufer, H.E. Wichmann, S. Kathiresan, J. Marrugat, C.J. O'Donnell, S.M. Schwartz, D.S. Siscovick, I. Subirana, N.B. Freimer, A.L. Hartikainen, M.I. McCarthy, P.F. O'Reilly, L. Peltonen, A. Pouta, P.E. de Jong, H. Snieder, W.H. van Gilst, R. Clarke, A. Goel, A. Hamsten, J.F. Peden, U. Seedorf, A.C. Syvanen, G. Tognoni, E.G. Lakatta, S. Sanna, P. Scheet, D. Schlessinger, A. Scuteri, M. Dorr, F. Ernst, S.B. Felix, G. Homuth, R. Lorbeer, T. Reffelmann, R. Rettig, U. Volker, P. Galan, I.G. Gut, S. Hercberg, G.M. Lathrop, D. Zelenika, P. Deloukas, N. Soranzo, F.M. Williams, G. Zhai, V. Salomaa, M. Laakso, R. Elosua, N.G. Forouhi, H. Volzke, C.S. Uiterwaal, Y.T. van der Schouw, M.E. Numans, G. Matullo, G. Navis, G. Berglund, S.A. Bingham, J.S. Kooner, J.M. Connell, S. Bandinelli, L. Ferrucci, H. Watkins, T.D. Spector, J. Tuomilehto, D. Altshuler, D.P. Strachan, M. Laan, P. Meneton, N.J. Wareham, M. Uda, M.R. Jarvelin, V. Mooser, O. Melander, R.J. Loos, P. Elliott, G.R. Abecasis, M. Caulfield, and P.B. Munroe. 2009. Genome-wide association study identifies eight loci associated with blood pressure. Nat. Genet. 41: 666–676.

Nicolae, D.L., E. Gamazon, W. Zhang, S. Duan, M.E. Dolan, and N.J. Cox. 2010. Trait-associated SNPs are more likely to be eQTLs: annotation to enhance discovery from GWAS. PLoS Genet. 6: e1000888.

Padmanabhan, S., O. Melander, T. Johnson, A.M. Di Blasio, W.K. Lee, D. Gentilini, C.E. Hastie, C. Menni, M.C. Monti, C. Delles, S. Laing, B. Corso, G. Navis, A.J. Kwakernaak, P. van der Harst, M. Bochud, M. Maillard, M. Burnier, T. Hedner, S. Kjeldsen, B. Wahlstrand, M. Sjogren, C. Fava, M. Montagnana, E. Danese, O. Torffvit, B. Hedblad, H. Snieder, J.M. Connell, M. Brown, N.J. Samani, M. Farrall, G. Cesana, G. Mancia, S. Signorini, G. Grassi, S. Eyheramendy, H.E. Wichmann, M. Laan, D.P. Strachan, P. Sever, D.C. Shields, A. Stanton, P. Vollenweider, A. Teumer, H. Volzke, R. Rettig, C. Newton-Cheh, P. Arora, F. Zhang, N. Soranzo, T.D. Spector, G. Lucas, S. Kathiresan, D.S. Siscovick, J. Luan, R.J. Loos, N.J. Wareham, B.W. Penninx, I.M. Nolte, M. McBride, W.H. Miller, S.A. Nicklin, A.H. Baker, D. Graham, R.A. McDonald, J.P. Pell, N. Sattar, P. Welsh, P. Munroe, M.J. Caulfield, A. Zanchetti, and A.F. Dominiczak. 2010. Genome-wide association study of blood pressure extremes identifies variant near UMOD associated with hypertension. PLoS Genet. 6: e1001177.

Pearson, T.A. and T.A. Manolio. 2008. How to interpret a genome-wide association study. JAMA 299: 1335–1344.

Raychaudhuri, S. 2011. Mapping rare and common causal alleles for complex human diseases. Cell 147: 57–69.

Reilly, M.P., M. Li, J. He, J.F. Ferguson, I.M. Stylianou, N.N. Mehta, M.S. Burnett, J.M. Devaney, C.W. Knouff, J.R. Thompson, B.D. Horne, A.F. Stewart, T.L. Assimes, P.S. Wild, H.

Allayee, P.L. Nitschke, R.S. Patel, N. Martinelli, D. Girelli, A.A. Quyyumi, J.L. Anderson, J. Erdmann, A.S. Hall, H. Schunkert, T. Quertermous, S. Blankenberg, S.L. Hazen, R. Roberts, S. Kathiresan, N.J. Samani, S.E. Epstein, and D.J. Rader. 2011. Identification of ADAMTS7 as a novel locus for coronary atherosclerosis and association of ABO with myocardial infarction in the presence of coronary atherosclerosis: two genome-wide association studies. Lancet 377: 383–392.

Ridker, P.M., E. Danielson, F.A. Fonseca, J. Genest, A.M. Gotto, Jr., J.J. Kastelein, W. Koenig, P. Libby, A.J. Lorenzatti, J.G. MacFadyen, B.G. Nordestgaard, J. Shepherd, J.T. Willerson, and R. J. Glynn. 2008. Rosuvastatin to prevent vascular events in men and women with elevated C-reactive protein. N. Engl. J. Med. 359: 2195–2207.

Samani, N.J., J. Erdmann, A.S. Hall, C. Hengstenberg, M. Mangino, B. Mayer, R.J. Dixon, T. Meitinger, P. Braund, H.E. Wichmann, J.H. Barrett, I.R. Konig, S.E. Stevens, S. Szymczak, D.A. Tregouet, M.M. Iles, F. Pahlke, H. Pollard, W. Lieb, F. Cambien, M. Fischer, W. Ouwehand, S. Blankenberg, A.J. Balmforth, A. Baessler, S.G. Ball, T.M. Strom, I. Braenne, C. Gieger, P. Deloukas, M.D. Tobin, A. Ziegler, J.R. Thompson, and H. Schunkert. 2007. Genomewide association analysis of coronary artery disease. N. Engl. J. Med. 357: 443–453.

Samani, N.J., O.T. Raitakari, K. Sipila, M.D. Tobin, H. Schunkert, M. Juonala, P.S. Braund, J. Erdmann, J. Viikari, L. Moilanen, L. Taittonen, A. Jula, E. Jokinen, T. Laitinen, N. Hutri-Kahonen, M.S. Nieminen, Y.A. Kesaniemi, A.S. Hall, J. Hulkkonen, M. Kahonen, and T. Lehtimaki. 2008. Coronary artery disease-associated locus on chromosome 9p21 and early markers of atherosclerosis. Arterioscler. Thromb. Vasc. Biol. 28: 1679–1683.

Sanna, S., B. Li, A. Mulas, C. Sidore, H.M. Kang, A.U. Jackson, M.G. Piras, G. Usala, G. Maninchedda, A. Sassu, F. Serra, M.A. Palmas, W.H. Wood, 3rd, I. Njolstad, M. Laakso, K. Hveem, J. Tuomilehto, T.A. Lakka, R. Rauramaa, M. Boehnke, F. Cucca, M. Uda, D. Schlessinger, R. Nagaraja, and G.R. Abecasis. 2011. Fine mapping of five loci associated with low-density lipoprotein cholesterol detects variants that double the explained heritability. PLoS Genet. 7: e1002198.

Saxena, R., B.F. Voight, V. Lyssenko, N.P. Burtt, P.I. de Bakker, H. Chen, J.J. Roix, S. Kathiresan, J.N. Hirschhorn, M.J. Daly, T.E. Hughes, L. Groop, D. Altshuler, P. Almgren, J.C. Florez, J. Meyer, K. Ardlie, K. Bengtsson Bostrom, B. Isomaa, G. Lettre, U. Lindblad, H.N. Lyon, O. Melander, C. Newton-Cheh, P. Nilsson, M. Orho-Melander, L. Rastam, E.K. Speliotes, M.R. Taskinen, T. Tuomi, C. Guiducci, A. Berglund, J. Carlson, L. Gianniny, R. Hackett, L. Hall, J. Holmkvist, E. Laurila, M. Sjogren, M. Sterner, A. Surti, M. Svensson, R. Tewhey, B. Blumenstiel, M. Parkin, M. Defelice, R. Barry, W. Brodeur, J. Camarata, N. Chia, M. Fava, J. Gibbons, B. Handsaker, C. Healy, K. Nguyen, C. Gates, C. Sougnez, D. Gage, M. Nizzari, S.B. Gabriel, G.W. Chirn, Q. Ma, H. Parikh, D. Richardson, D. Ricke, and S. Purcell. 2007. Genome-wide association analysis identifies loci for type 2 diabetes and triglyceride levels. Science 316: 1331–1336.

Schunkert, H., I.R. Konig, S. Kathiresan, M.P. Reilly, T.L. Assimes, H. Holm, M. Preuss, A.F. Stewart, M. Barbalic, C. Gieger, D. Absher, Z. Aherrahrou, H. Allayee, D. Altshuler, S.S. Anand, K. Andersen, J.L. Anderson, D. Ardissino, S.G. Ball, A.J. Balmforth, T.A. Barnes, D.M. Becker, L.C. Becker, K. Berger, J.C. Bis, S.M. Boekholdt, E. Boerwinkle, P.S. Braund, M.J. Brown, M.S. Burnett, I. Buysschaert, J.F. Carlquist, L. Chen, S. Cichon, V. Codd, R.W. Davies, G. Dedoussis, A. Dehghan, S. Demissie, J.M. Devaney, P. Diemert, R. Do, A. Doering, S. Eifert, N.E. Mokhtari, S.G. Ellis, R. Elosua, J.C. Engert, S.E. Epstein, U. de Faire, M. Fischer, A.R. Folsom, J. Freyer, B. Gigante, D. Girelli, S. Gretarsdottir, V. Gudnason, J.R. Gulcher, E. Halperin, N. Hammond, S.L. Hazen, A. Hofman, B.D. Horne, T. Illig, C. Iribarren, G.T. Jones, J.W. Jukema, M.A. Kaiser, L.M. Kaplan, J.J. Kastelein, K.T. Khaw, J.W. Knowles, G. Kolovou, A. Kong, R. Laaksonen, D. Lambrechts, K. Leander, G. Lettre, M. Li, W. Lieb, C. Loley, A.J. Lotery, P.M. Mannucci, S. Maouche, N. Martinelli, P.P. McKeown, C. Meisinger, T. Meitinger, O. Melander, P.A. Merlini, V. Mooser, T. Morgan, T.W. Muhleisen, J.B. Muhlestein, T. Munzel, K. Musunuru, J. Nahrstaedt, C.P. Nelson, M.M. Nothen, O. Olivieri, R.S. Patel, C.C. Patterson, A. Peters, F. Peyvandi,

L. Qu, A.A. Quyyumi, D.J. Rader, L.S. Rallidis, C. Rice, F.R. Rosendaal, D. Rubin, V. Salomaa, M.L. Sampietro, M.S. Sandhu, E. Schadt, A. Schafer, A. Schillert, S. Schreiber, J. Schrezenmeir, S.M. Schwartz, D.S. Siscovick, M. Sivananthan, S. Sivapalaratnam, A. Smith, T.B. Smith, J.D. Snoep, N. Soranzo, J.A. Spertus, K. Stark, K. Stirrups, M. Stoll, W.H. Tang, S. Tennstedt, G. Thorgeirsson, G. Thorleifsson, M. Tomaszewski, A.G. Uitterlinden, A.M. van Rij, B.F. Voight, N.J. Wareham, G.A. Wells, H.E. Wichmann, P.S. Wild, C. Willenborg, J.C. Witteman, B.J. Wright, S. Ye, T. Zeller, A. Ziegler, F. Cambien, A.H. Goodall, L.A. Cupples, T. Quertermous, W. Marz, C. Hengstenberg, S. Blankenberg, W.H. Ouwehand, A.S. Hall, P. Deloukas, J.R. Thompson, K. Stefansson, R. Roberts, U. Thorsteinsdottir, C.J. O'Donnell, R. McPherson, J. Erdmann, and N.J. Samani. 2010. Large-scale association analysis identifies 13 new susceptibility loci for coronary artery disease. Nat. Genet. 43: 333–338.

Schunkert, H., I.R. Konig, S. Kathiresan, M.P. Reilly, T.L. Assimes, H. Holm, M. Preuss, A.F. Stewart, M. Barbalic, C. Gieger, D. Absher, Z. Aherrahrou, H. Allayee, D. Altshuler, S.S. Anand, K. Andersen, J.L. Anderson, D. Ardissino, S.G. Ball, A.J. Balmforth, T.A. Barnes, D.M. Becker, L.C. Becker, K. Berger, J.C. Bis, S.M. Boekholdt, E. Boerwinkle, P.S. Braund, M.J. Brown, M.S. Burnett, I. Buysschaert, J.F. Carlquist, L. Chen, S. Cichon, V. Codd, R.W. Davies, G. Dedoussis, A. Dehghan, S. Demissie, J.M. Devaney, P. Diemert, R. Do, A. Doering, S. Eifert, N.E. Mokhtari, S.G. Ellis, R. Elosua, J.C. Engert, S.E. Epstein, U. de Faire, M. Fischer, A.R. Folsom, J. Freyer, B. Gigante, D. Girelli, S. Gretarsdottir, V. Gudnason, J.R. Gulcher, E. Halperin, N. Hammond, S.L. Hazen, A. Hofman, B.D. Horne, T. Illig, C. Iribarren, G.T. Jones, J.W. Jukema, M.A. Kaiser, L.M. Kaplan, J.J. Kastelein, K.T. Khaw, J.W. Knowles, G. Kolovou, A. Kong, R. Laaksonen, D. Lambrechts, K. Leander, G. Lettre, M. Li, W. Lieb, C. Loley, A.J. Lotery, P.M. Mannucci, S. Maouche, N. Martinelli, P.P. McKeown, C. Meisinger, T. Meitinger, O. Melander, P.A. Merlini, V. Mooser, T. Morgan, T.W. Muhleisen, J.B. Muhlestein, K. Munzel, K. Musunuru, J. Nahrstaedt, C.P. Nelson, M.M. Nothen, O. Olivieri, R.S. Patel, C.C. Patterson, A. Peters, F. Peyvandi, L. Qu, A.A. Quyyumi, D.J. Rader, L.S. Rallidis, C. Rice, F.R. Rosendaal, D. Rubin, V. Salomaa, M.L. Sampietro, M.S. Sandhu, E. Schadt, A. Schafer, A. Schillert, S. Schreiber, J. Schrezenmeir, S.M. Schwartz, D.S. Siscovick, M. Sivananthan, S. Sivapalaratnam, A. Smith, T.B. Smith, J.D. Snoep, N. Soranzo, J.A. Spertus, K. Stark, K. Stirrups, M. Stoll, W.H. Tang, S. Tennstedt, G. Thorgeirsson, G. Thorleifsson, M. Tomaszewski, A.G. Uitterlinden, A.M. van Rij, B.F. Voight, N.J. Wareham, G.A. Wells, H.E. Wichmann, P.S. Wild, C. Willenborg, J.C. Witteman, B.J. Wright, S. Ye, T. Zeller, A. Ziegler, F. Cambien, A.H. Goodall, L.A. Cupples, T. Quertermous, W. Marz, C. Hengstenberg, S. Blankenberg, W.H. Ouwehand, A.S. Hall, P. Deloukas, J.R. Thompson, K. Stefansson, R. Roberts, U. Thorsteinsdottir, C.J. O'Donnell, R. McPherson, J. Erdmann, and N.J. Samani. 2011. Large-scale association analysis identifies 13 new susceptibility loci for coronary artery disease. Nat. Genet. 43: 333–338.

Shea, J., V. Agarwala, A.A. Philippakis, J. Maguire, E. Banks, M. Depristo, B. Thomson, C. Guiducci, R.C. Onofrio, S. Kathiresan, S. Gabriel, N.P. Burtt, M.J. Daly, L. Groop, and D. Altshuler. 2011. Comparing strategies to fine-map the association of common SNPs at chromosome 9p21 with type 2 diabetes and myocardial infarction. Nat. Genet. 43: 801–805.

Smith, G.D. and S. Ebrahim. 2004. Mendelian randomization: prospects, potentials, and limitations. Int. J. Epidemiol. 33: 30–42.

Thanassoulis, G. and R.S. Vasan. 2010. Genetic cardiovascular risk prediction: will we get there? Circulation 122: 2323–2334.

Thanassoulis, G., G.M. Peloso, M.J. Pencina, U. Hoffmann, C.S. Fox, L.A. Cupples, D. Levy, R.B. D'Agostino, S.J. Hwang, and C.J. O'Donnell. 2012. A genetic risk score is associated with incident cardiovascular disease and coronary artery calcium: the Framingham Heart Study. Circ. Cardiovasc. Genet. 5: 113–121.

Villard, E., C. Perret, F. Gary, C. Proust, G. Dilanian, C. Hengstenberg, V. Ruppert, E. Arbustini, T. Wichter, M. Germain, O. Dubourg, L. Tavazzi, M. C. Aumont,

P. DeGroote, L. Fauchier, J.N. Trochu, P. Gibelin, J.F. Aupetit, K. Stark, J. Erdmann, R. Hetzer, A.M. Roberts, P.J. Barton, V. Regitz-Zagrosek, U. Aslam, L. Duboscq-Bidot, M. Meyborg, B. Maisch, H. Madeira, A. Waldenstrom, E. Galve, J.G. Cleland, R. Dorent, G. Roizes, T. Zeller, S. Blankenberg, A.H. Goodall, S. Cook, D.A. Tregouet, L. Tiret, R. Isnard, M. Komajda, P. Charron, and F. Cambien. 2011. A genome-wide association study identifies two loci associated with heart failure due to dilated cardiomyopathy. Eur. Heart. J. 32: 1065–1076.

Voight, B.F., G.M. Peloso, M. Orho-Melander, R. Frikke-Schmidt, M. Barbalic, M.K. Jensen, G. Hindy, H. Holm, E.L. Ding, T. Johnson, H. Schunkert, N.J. Samani, R. Clarke, J.C. Hopewell, J.F. Thompson, M. Li, G. Thorleifsson, C. Newton-Cheh, K. Musunuru, J.P. Pirruccello, D. Saleheen, L. Chen, A. Stewart, A. Schillert, U. Thorsteinsdottir, G. Thorgeirsson, S. Anand, J.C. Engert, T. Morgan, J. Spertus, M. Stoll, K. Berger, N. Martinelli, D. Girelli, P.P. McKeown, C.C. Patterson, S.E. Epstein, J. Devaney, M.S. Burnett, V. Mooser, S. Ripatti, I. Surakka, M.S. Nieminen, J. Sinisalo, M.L. Lokki, M. Perola, A. Havulinna, U. de Faire, B. Gigante, E. Ingelsson, T. Zeller, P. Wild, P.I. de Bakker, O.H. Klungel, A.H. Maitland-van der Zee, B.J. Peters, A. de Boer, D.E. Grobbee, P.W. Kamphuisen, V.H. Deneer, C.C. Elbers, N.C. Onland-Moret, M.H. Hofker, C. Wijmenga, W.M. Verschuren, J.M. Boer, Y.T. van der Schouw, A. Rasheed, P. Frossard, S. Demissie, C. Willer, R. Do, J.M. Ordovas, G.R. Abecasis, M. Boehnke, K.L. Mohlke, M.J. Daly, C. Guiducci, N.P. Burtt, A. Surti, E. Gonzalez, S. Purcell, S. Gabriel, J. Marrugat, J. Peden, J. Erdmann, P. Diemert, C. Willenborg, I.R. Konig, M. Fischer, C. Hengstenberg, A. Ziegler, I. Buysschaert, D. Lambrechts, F. Van de Werf, K.A. Fox, N.E. El Mokhtari, D. Rubin, J. Schrezenmeir, S. Schreiber, A. Schafer, J. Danesh, S. Blankenberg, R. Roberts, R. McPherson, H. Watkins, A.S. Hall, K. Overvad, E. Rimm, E. Boerwinkle, A. Tybjaerg-Hansen, L.A. Cupples, M.P. Reilly, O. Melander, P.M. Mannucci, D. Ardissino, D. Siscovick, R. Elosua, K. Stefansson, C.J. O'Donnell, V. Salomaa, D.J. Rader, L. Peltonen, S.M. Schwartz, D. Altshuler, and S. Kathiresan. 2012. Plasma HDL cholesterol and risk of myocardial infarction: a mendelian randomisation study. Lancet 380: 572–580.

Wang, F., C.Q. Xu, Q. He, J.P. Cai, X.C. Li, D. Wang, X. Xiong, Y.H. Liao, Q.T. Zeng, Y.Z. Yang, X. Cheng, C. Li, R. Yang, C.C. Wang, G. Wu, Q.L. Lu, Y. Bai, Y.F. Huang, D. Yin, Q. Yang, X.J. Wang, D.P. Dai, R.F. Zhang, J. Wan, J.H. Ren, S.S. Li, Y.Y. Zhao, F.F. Fu, Y. Huang, Q.X. Li, S.W. Shi, N. Lin, Z.W. Pan, Y. Li, B. Yu, Y.X. Wu, Y.H. Ke, J. Lei, N. Wang, C.Y. Luo, L.Y. Ji, L.J. Gao, L. Li, H. Liu, E.W. Huang, J. Cui, N. Jia, X. Ren, H. Li, T. Ke, X.Q. Zhang, J.Y. Liu, M.G. Liu, H. Xia, B. Yang, L.S. Shi, Y.L. Xia, X. Tu, and Q.K. Wang. 2011. Genome-wide association identifies a susceptibility locus for coronary artery disease in the Chinese Han population. Nat. Genet. 43: 345–349.

Watkins, H., C. MacRae, L. Thierfelder, Y.H. Chou, M. Frenneaux, W. McKenna, J.G. Seidman, and C.E. Seidman. 1993. A disease locus for familial hypertrophic cardiomyopathy maps to chromosome 1q3. Nat. Genet. 3: 333–337.

Watkins, H., D. Conner, L. Thierfelder, J.A. Jarcho, C. MacRae, W.J. McKenna, B.J. Maron, J.G. Seidman, and C.E. Seidman. 1995. Mutations in the cardiac myosin binding protein-C gene on chromosome 11 cause familial hypertrophic cardiomyopathy. Nat. Genet. 11: 434–437.

Wellcome Trust Case Control Consortium. 2007. Genome-wide association study of 14,000 cases of seven common diseases and 3,000 shared controls. Nature 447: 661–678.

Wensley, F., P. Gao, S. Burgess, S. Kaptoge, E. Di Angelantonio, T. Shah, J.C. Engert, R. Clarke, G. Davey-Smith, B.G. Nordestgaard, D. Saleheen, N.J. Samani, M. Sandhu, S. Anand, M.B. Pepys, L. Smeeth, J. Whittaker, J.P. Casas, S.G. Thompson, A.D. Hingorani, and J. Danesh. 2011. Association between C reactive protein and coronary heart disease: mendelian randomisation analysis based on individual participant data. BMJ 342: d548.

Wrensch, M., R.B. Jenkins, J.S. Chang, R.F. Yeh, Y. Xiao, P.A. Decker, K.V. Ballman, M. Berger, J.C. Buckner, S. Chang, C. Giannini, C. Halder, T.M. Kollmeyer, M.L. Kosel, D.H. LaChance, L. McCoy, B.P. O'Neill, J. Patoka, A.R. Pico, M. Prados, C. Quesenberry,

T. Rice, A.L. Rynearson, I. Smirnov, T. Tihan, J. Wiemels, P. Yang, and J.K. Wiencke. 2009. Variants in the CDKN2B and RTEL1 regions are associated with high-grade glioma susceptibility. Nat. Genet. 41: 905–908.

Yang, Q., L. Bailey, R. Clarke, W.D. Flanders, T. Liu, A. Yesupriya, M.J. Khoury, and J.M. Friedman. 2012. Prospective study of methylenetetrahydrofolate reductase (MTHFR) variant C677T and risk of all-cause and cardiovascular disease mortality among 6000 US adults. Am. J. Clin. Nutr. 95: 1245–1253.

Yasuno, K., K. Bilguvar, P. Bijlenga, S.K. Low, B. Krischek, G. Auburger, M. Simon, D. Krex, Z. Arlier, N. Nayak, Y.M. Ruigrok, M. Niemela, A. Tajima, M. von und zu Fraunberg, T. Doczi, F. Wirjatijasa, A. Hata, J. Blasco, A. Oszvald, H. Kasuya, G. Zilani, B. Schoch, P. Singh, C. Stuer, R. Risselada, J. Beck, T. Sola, F. Ricciardi, A. Aromaa, T. Illig, S. Schreiber, C.M. van Duijn, L.H. van den Berg, C. Perret, C. Proust, C. Roder, A.K. Ozturk, E. Gaal, D. Berg, C. Geisen, C.M. Friedrich, P. Summers, A.F. Frangi, M.W. State, H.E. Wichmann, M.M. Breteler, C. Wijmenga, S. Mane, L. Peltonen, V. Elio, M.C. Sturkenboom, P. Lawford, J. Byrne, J. Macho, E.I. Sandalcioglu, B. Meyer, A. Raabe, H. Steinmetz, D. Rufenacht, J.E. Jaaskelainen, J. Hernesniemi, G.J. Rinkel, H. Zembutsu, I. Inoue, A. Palotie, F. Cambien, Y. Nakamura, R.P. Lifton, and M. Gunel. 2010. Genome-wide association study of intracranial aneurysm identifies three new risk loci. Nat. Genet. 42: 420–425.

Zacho, J., A. Tybjaerg-Hansen, J.S. Jensen, P. Grande, H. Sillesen, and B.G. Nordestgaard. 2008. Genetically elevated C-reactive protein and ischemic vascular disease. N. Engl. J. Med. 359: 1897–1908.

Zeller, T., S. Blankenberg, and P. Diemert. 2011. Genomewide association studies in cardiovascular disease—an update. 2011. Clin. Chem. 58: 92–103.

# CHAPTER 16

# MicroRNAs: A Novel Cardiac Biomarker

Nikolaos Papageorgiou,* Dimitris Tousoulis, Anna Kontogeorgou, Emmanuel Androulakis and Christodoulos Stefanadis

## Introduction

Coronary artery disease (CAD) along with its clinical manifestations represents the world's leading cause of morbidity/mortality. However, its diagnosis in the early stages ensures timely initiation of an effective therapy and subsequently reduction of mortality rates (Tousoulis et al. 2012; Briasoulis et al. 2012; Charakida et al. 2006).

During the last few years it has been well established that genetic components and alterations of gene families are frequently implicated in cardiovascular disease (CVD) (Tousoulis et al. 2008a). microRNAs are a class of approximately 22 nucleotide noncoding RNAs and are currently considered as major regulators of physiological processes including differentiation, proliferation, and apoptosis (Bartel et al. 2004; Ambros et al. 2004; Tousoulis et al. 2011a; Tousoulis et al. 2010). Primary longer RNAs (pri-miRNAs) are cleaved in the nucleus by RNase III enzyme Drosha and form pre-miRNAs, which are subsequently transported to the cytoplasm and produce mature miRNAs (Lee et al. 2003; Gregory et al. 2004; Papageorgiou et al. 2012; Shyu et al. 2008). There are observations regarding their role in the pathogenesis of cardiac diseases that have triggered the conduction of numerous clinical and experimental studies (Thum et al. 2008). Research studies suggest that miRNAs are implicated

1st Department of Cardiology, 'Hippokration' Hospital, University of Athens Medical School, Athens, Greece.
*Corresponding author: drnpapageorgiou@yahoo.com

in CVD including atherosclerosis and plaque rupture, heart failure, cardiac arrhythmias, and hypertrophy (Pan et al. 2010; Ono et al. 2011). Additionally, there are either established or under research biochemical markers that may be used for diagnostic purposes in CAD patients (Tousoulis et al. 2008b) and miRNAs belong to this category.

In the present chapter, we will review the role of microRNAs in cardiovascular disease and their potential to perform as novel biomarkers.

## Biological Aspects of miRNAs: Structure and Function

miRNAs represent a novel class of single-stranded RNA of approximately 22 nucleotides. They bind to complementary sequences located on the 3ft untranslated region (UTR) of the target genes and thus regulate gene expression (Zhang et al. 2009). They are organized in clusters with common transcriptional region and are located both in introns and exons of coding as well as noncoding genes (Wang et al. 2009). Lin-4 was the first miR discovered and since then over 700 human miRs have been registered (Lee et al. 1993; Wightman et al. 1993).

Regarding the maturation of microRNAs, primary miRNAs (pri-miRNAs) are transcribed by RNA polymerase II or III (Pol II or Pol III) to large pri-miRNA which are thousands of nucleotides long and incorporate the mature miRNAs (Condorelli et al. 2010; Borchert et al. 2006). Afterwards, RNase III enzyme and DGCR8/Pasha cleave intranuclear pri-miRNAs to yield pre-miRNAs of approximately 70 nucleotides (Catalucci et al. 2009; Zeng 2006). Accordingly, these molecules are transported by RAN GTP and exportin 5 into the cytoplasm where Dicer promotes the generation of a transient oligonucleotide duplex by the pre-miRNA (Metias et al. 2009; Bohnsack et al. 2004; Lund et al. 2004). This molecule is incorporated into the RNA induced silencing complex (RISC) which is an accumulation of proteins that remove the loop portion of the pre-miRNA (Maniataki et al. 2005). Finally, a strand from the duplex is detached and thus yields mature miRNAs (Suarez et al. 2009; Zeng 2006). Of note, recent studies have identified an alternative pathway for miRs biogenesis which bypass the Drosha processing step to produce miRNA by Dicer (Ruby et al. 2007). Furthermore, mature miRNA binds to complementary regions in the target gene and thus regulates its expression (Zhang et al. 2008a; Zhang et al. 2008b; Hutvagner et al. 2002; Davison et al. 2006). Subsequently, regulation depends on the degree of compliance between the miRNA and the respective gene. Perfect compliance can potentially lead to target mRNA cleavage while it has been reported that even imperfect binding of miRNA to its target region can downregulate the abundance of mRNA. It is also worth mentioning that miRNAs are in abundance and are considered to regulate a large proportion of protein-coding genes (Zeng et al. 2003; Lim et al. 2005). Additionally, it is important to mention that miRNA expression profile may vary depending on the tissue. Specific miRNAs such as miR-1, miR-133, let-7, and miR-126-3p are abundantly expressed in cardiac muscles while their expression may have completely different pattern in other cell types, such as endothelial cells and smooth muscle cells (Wang et al. 2008).

## Role of miRNAs in the Cardiovascular System

### *Expression of miRNAs*

Numerous studies have attempted to identify miRNA expression. Regulation concerns multiple steps during RNA biogenesis, yet the corresponding mechanisms are not fully clarified. The most important mechanism seems to be transcriptional regulation while some miRNAs are controlled at the post-transcriptional level. Current data indicates that the great majority of miRNAs are under the control of developmental, tissue specific signaling or both (Kim et al. 2006; Chen et al. 2005b; Baskerville et al. 2005; Krichevsky et al. 2003).

Several techniques have been applied to quantify miRNAs in the laboratory; however, technical problems that mainly derive from miRNA short length remain. More specifically, the most common technique is based on microarrays (Liu et al. 2004; Babak et al. 2004). Although microarrays is a widely accepted method, the small size of miRNAs makes the procedure challenging for conventional techniques. Recently, microarrays that detect mature miRNA have been developed; however, the issue of cross-hybridization of related miRNAs is far from being overcome (Monticelli et al. 2005; Nelson et al. 2004). Additionally, real time polymerase chain reaction (RT-PCR) predominates in sensitivity but is difficult to use when the number of miRNAs exceeds 300. Innovations are constantly being developed; however much improvement is essential (Chen et al. 2005a; Lu et al. 2005).

The expression profile of miRNAs differs according to the tissue/cell type. More specifically, cardiac muscle cells express several miRNAs, such as miR-1, miR-133, miR-126-3p, let-7, miR-26a, though not exclusively (Wang et al. 2008). These miRNAs may have completely different expression in endothelial cells, smooth muscle cells, and fibroblasts. The aforementioned scenario suggests that miRNAs have tissue-specific functions and thereby provide important data for future studies (Zhang et al. 2009).

The effects of various miRs in cardiovascular system have been thoroughly documented both *in vitro* and *in vivo* (Tasuguchi et al. 2007; Dong et al. 2010; van Rooij et al. 2007; Thum et al. 2007). Actually, miR-1 seems to regulate the balance between proliferation and differentiation during cardiogenesis. Also, miR-21 was examined experimentally where it exerted anti-apoptotic and proproliferative properties on vascular smooth muscle cells (VSMCs) (Ji et al. 2007).

## *Release of miRNAs in circulation*

Findings suggest that circulating miRNAs are in stable form protected from RNase-dependent degradation. Of note, current knowledge claims that miRNAs are actively secreted in microvesicles or exosomes (Gilad et al. 2008; Hunter et al. 2008), but the underlying mechanisms are still not clear.

## *Myocardial development*

It is widely accepted that the great majority of miRNAs exhibit tissue/cell specific expression (Lagos-Quintana et al. 2002). Since they regulate the expression of mRNAs that direct cell proliferation, differentiation, and apoptosis, miRNAs are indirectly implicated in tissue differentiation and organ development (Garofalo et al. 2008). Interestingly, Dicer-lacking mice that have lost miRNAs manifest various developmental abnormalities in their cardiovascular system (Yang et al. 2005). Several studies have examined the role of different miRNAs in cardiovascular system regulation (Zhao et al. 2005; Foshay et al. 2007; Callis et al. 2007). Hundreds of miRNAs have been identified so far, yet miR-1, miR-133, and miR-208 are muscle specific and are mainly expressed in cardiac and skeletal muscles (Wang et al. 2008). Particularly, current data indicate that miR-1 is expressed in cardiac and skeletal muscle of embryonic mice (Yang et al. 2005; Foshay et al. 2007) and it is thought to participate in mechanisms of proliferation and differentiation during cardiogenesis (Kwon et al. 2005) as well as in asymmetrical cell division. Actually, overexpression of miR-1 thickens ventricle walls as a result of premature differentiation and early withdrawal of cardiac cells from the cell cycle (Ono et al. 2011).

## Role of miRNAs in Atherosclerosis

It is well established that miRNAs contribute to the pathogenesis of cardiovascular disease including atherosclerosis, myocardial infarction, cardiac arrhythmias, heart failure and hypertrophy (van Rooil et al. 2008; Baudhuin et al. 2009). Particularly, it has been suggested that miRs regulate endothelial nitric oxide synthase (eNOS) function, inflammation, apoptosis and angiogenesis and thus contribute to atherosclerotic plaque formation and rupture (Haver et al. 2010).

## Association with endothelial nitric oxide synthase

Nitric oxide is a vasculoprotective molecule that is produced by eNOS and exhibits anti-atherosclerotic properties. Damage of NO homeostasis is a key mechanism that predisposes to atherosclerotic lesions (Santovito et al. 2012). miRs exhibit either a favorable or suppressive effect on eNOS activity. The exposure of endothelial cells (ECs) to shear stress results in extended upregulation of various miRs, especially miR 21. Actually, a recent study demonstrated that shear stress upregulates miR-21 which in turn stimulates nitric oxide synthase (NOS) and attenuates endothelial cell apoptosis (Weber et al. 2010). On the other hand, overexpression of miR 221 and miR222 results in diminished eNOS activity and consequently impaired NO bioavailability (Fleissner et al. 2010). Additionally, miR 214 is strongly correlated to eNOS activity regulation (Chan et al. 2009) and is likely to suppress eNOS gene expression via complementary mechanisms (Jamaluddin et al. 2011).

## Association with vascular smooth muscle cells

Vascular smooth muscle cell is a main component of the vessel wall and participates in atherosclerotic processes through proliferation, apoptosis, migration, and differentiation (Ji et al. 2007). Normally, VSMC rarely proliferate whereas inflammation and injury of the vessel could potentially trigger VSMC proliferation (Ji et al. 2007; Parmacek et al. 2009). Several studies concur that certain miRs regulate the aforementioned processes by transforming VSMC biology and function (Liu et al. 2009; Zhang et al. 2009; Elia et al. 2009). miR143 and miR145 are of particular interest as they are highly expressed in VSMC (Cordes et al. 2009) and affect the dominance of either contractile or proliferative phenotype (Cordes et al. 2009; Cheng et al. 2007). Of note, miRs143 and 145 are typically suppressed in impaired vasculature (Cordes et al. 2009; Xin et al. 2009). Actually, overexpression of either miR143 or miR145 downregulates VSMC proliferation (Elia et al. 2009; Kawai-Kowase et al. 2007) as well as favors differentiation (Cheng et al. 2007). Recent clinical studies (Elia et al. 2009; Cordes et al. 2009; Boettger et al. 2009) examined the effect of miR 143/145 mutation in VSMC properties and noticed that the proliferative phenotype dominated at the expense of the contractile phenotype. Furthermore, recent experimental data suggests that miR-143 and miR-145 regulate VSMCs homeostasis (Elia et al. 2009; Cordes et al. 2009). According to a recent study, platelet derived growth factor (PDGF), which stimulates VSMC migration, can attenuate miR143/145 expression which is in agreement with the aforementioned scenario. Additionally, PDGF also stimulates mir-221 and -222 expression which are probably involved in VSMC proliferation and neointimal lesion formation (Davis et al. 2009). Particularly, both *in vivo* (Martin et al. 2011) and *in vitro* (Qin et al. 2012) it has been found that miR221/222 knockdown reduces VSMC proliferation.

## Association with angiogenesis

Various studies have attempted to clarify the role of miRs in angiogenesis. It is widely accepted that many of the so far known miRs have a favorable effect on angiogenesis whereas others suppress angiogenic mechanisms (Urbich et al. 2008). Dicer is an essential enzyme for miR maturation. A recent study observed that mutation of Dicer suppresses angiogenesis (Suarez et al. 2008) which potentially implicates miRs in angiogenic processes. Hypoxia, inflammation and shear stress probably mobilize miRs expression (Urbich et al. 2008) and thus indirectly regulate angiogenic factors (Poliseno et al. 2006). miR130 (Chen et al. 2008), miR27b (Kuehbacher et al. 2007), miR378 (Lee et al. 2007), and Let-7f are thought to have angiogenic properties. Actually, the current data suggests that miRNAs, such as miR-92a, could either stimulate or inhibit angiogenesis (Haver et al. 2010; Urbich et al. 2008). Also, miRNA-210, enhanced by vascular endothelium growth factor (VEGF) inhibits hypoxia-induced angiogenesis (Urbich et al. 2008). However, miR221/222 attenuates angiogenesis acting in a post transcriptional level (Kuehbacher et al. 2007). Of note, miR221/222 also diminishes NO bioavailability and thus indirectly affects angiogenesis (Suarez et al. 2009; Murohara et al. 1998).

## Association with inflammation

Systemic inflammation has been proved essential for the establishment of atherosclerotic lesions. It is characterized by extended upregulation of adhesive molecules and inflammatory mediators. As a result leukocytes, macrophages and VSMC adhere to the injured endothelium and migrate into the vessel wall to build up the atheroma (Ross et al. 1999; Silvestre et al. 2008). miRs mainly affect the circulation of inflammatory and adhesive molecules. Particularly, recent experimental data indicates that in human vein endothelial cells, miR-126 inhibits vascular cell adhesion molecule-1 (VCAM-1) which mediate leukocyte adherence to the endothelium (Harris et al. 2008). Moreover, miR155 inhibits T cell adhesion to ECs and downregulates ECs migration (Zhu et al. 2011). miR221/222 belongs to the same family and potentially regulates inflammatory molecules. Actually, recent clinical studies observed that miR155 was remarkably reduced in atherosclerotic patients (Fichtlscherer et al. 2010) while miR221/222 was found upregulated (Minami et al. 2009). miR125a/b is frequently implicated in inflammatory processes and is so far connected to the downregulation of endothelin-1 (ET-1) (Papageorgiou et al. 2011) and inflammatory cytokines (interleukin-6/-2, tumor necrosis factor-a) (Chen et al. 2009). Additionally, other miRNAs regulate the expression of E-selectin or exhibit proatherogenic properties in a tumor necrosis factor-a (TNF-a) mediated process (Yoshizaki et al. 2008; Suarez et al. 2010). Actually, miR31 and miR17-3p have anti-inflammatory properties and target E-selectin and intracellular adhesion molecule-1 (ICAM-1) respectively (Suarez et al. 2010). The aforementioned miRNAs functions leave space for further research and potentially suggest new therapeutic approaches (Fasanaro et al. 2009; O'Sullivan et al. 2011).

## Role of microRNAs in myocardial infarction (MI)

Cardiac troponin is a dominant and widely accepted marker of myocardial injury. However, there are still cases in which their utility is limited and thus is it necessary to broaden our diagnostic and therapeutic horizons and focus on novel biomarkers such as DNA polymorphisms and miRNAs (Margulies et al. 2009; Zacharowski et al. 2006) (Table 1).

Regarding miRNAs, mir-208 has shown interesting properties since it is related to troponin I; it is also detectable in circulation after MI and reflects its extent (Ji et al. 2009; Wang et al. 2010) (Table 2). Additionally, miR-208 is superior to cardiac troponins in the detection of MI in patients with renal dysfunction since troponin may be notionally increased (Xu et al. 2012). Furthermore, a recent study demonstrated that the administration of the antagomir miR-92a after MI contributed to vessel growth and recovery of damaged tissues. Thus, this miR could serve as a therapeutic target for ischemic disease (Bonauer et al. 2009). Also, a male Sprague-Dawley rat model suggests that miR-1 could as well be a useful biomarker for MI (Cheng et al. 2010). More specifically, serum miR-1 levels were significantly and quickly elevated within the first six hours and three days; after the MI they returned to basal level. MI also stimulates various microRNAs including miR-1, -133a, -133b and -499, both in humans and mice, thus proposing prospective novel biomarkers of IM (D' Alesandra et al. 2010). The hypothesis that miRNAs induced by ischemic preconditioning play an important role in protection against myocardial injury has also been tested. In this model, miRNA-21 caused significant increases in miRNA-1, miRNA-21 and miRNA-24 levels.

**Table 1.** Role of microRNAs in atherogenesis.

| microRNAs | Effects |
|---|---|
| miR 17-3p<br>(Qin et al. 2012) | Negative effect on inflammation (ICAM-1) |
| miR 155<br>(Quin et al. 2012) | Negative effect on NO synthase expression<br>Negative effect on immune cell adhesion |
| miR 143/145<br>(Xin et al. 2009; Elia et al. 2009;<br>Cordes et al. 2009) | Negative effect on VSMC proliferation<br>Positive effect on VSMC differentiation |
| miR 214<br>(Chan et al. 2009) | Negative effect on NO synthase expression |
| miR 130<br>(Urbich at al. 2008; Suarez et al. 2008) | Positive effect on angiogenesis |
| miR 27b<br>(Urbich at al. 2008; Suarez et al. 2008) | Positive effect on angiogenesis |
| miR 378<br>(Urbich et al. 2008; Suarez et al. 2008;<br>Lee et al. 2007) | Positive effect on angiogenesis |
| miR 125a/b<br>(Quin et al. 2012; Chen et al. 2009) | Negative effect on inflammation<br>(ET-1, cytokines) |
| miR 31<br>(Quin et al. 2012; Yoshizaki et al. 2008) | Negative effect on inflammation<br>(E-selectin) |
| miR 126<br>(Harris et al. 2008) | Negative effect on inflammation<br>(VCAM-1, adhesion molecules) |
| miR 21<br>(Weber et al. 2010; Fleissner et al. 2010) | Positive effect on VSMC proliferation<br>Positive effect on NO synthase expression |
| miR 221/222<br>(Liu et al. 2009; Boettger et al. 2009;<br>Davis et al. 2009) | Negative effect on NO synthase expression<br>Positive effect on VSMC proliferation<br>Positive effect on angiogenesis<br>Positive effect on inflammation |

**Abbreviations:** NO: nitric oxide, VCAM-1: vascular cell adhesion molecule 1, ICAM-1: intracellular adhesion molecule 1, VSMC: vascular smooth muscle cell, ET-1: endothelin 1.

## Role of microRNAs in Heart Failure

Cardiac hypertrophy often leads to heart failure characterized by multiple abnormalities both structural and functional and contributes significantly to mortality and morbidity (Tousoulis et al. 2011b). Emerging evidence suggests for the first time that miRNAs are expressed in hypertrophic hearts (Cheng et al. 2007; Song et al. 2010; Chen et al. 2008). In a recent clinical study, miRNAs profiles of human dilated cardiomyopathy, ischemic cardiomyopathy, and pressure overload hypertrophy were compared with that of normal hearts. The profiles have been demonstrated to be distinguishable between these groups (Ikeda et al. 2007). With data collated from several other reports, it could be assumed that miRNAs expression patterns could vary in different forms of cardiac injury, thereby providing diagnostic value with regard to

**Table 2.** Effects of miRNAs on the myocardium.

| Specific microRNAs | Activity | Comments |
|---|---|---|
| miR 198 (Margulies et al. 2009; Xu et al. 2012) | (+) | Risk for CAD or ACS (positive effect) |
| miR 134 (Margulies et al. 2009; Xu et al. 2012) | (+) | Risk for CAD or ACS (positive effect) |
| miR 208 (Ji et al. 2009) | (+) | Association with troponin I (positive effect) |
| miR 270 (Margulies et al. 2009; Xu et al. 2012) | (+) | Risk for CAD or ACS (positive effect) |
| miR 21 (Margulies et al. 2009; Xu et al. 2012) | (+) | Protection against myocardial injury (positive effect) |
| miR 24 (Margulies et al. 2009; Xu et al. 2012) | (+) | Protection against myocardial injury (positive effect) |
| miR 92a (Bonauer et al. 2009) | (−) | Vessel growth (positive effect) Recovery of damaged tissue (positive effect) |
| miR 133 (Care et al. 2007; Luo et al. 2008) | (+) | Heart adaptation after ischaemic stress (positive effect) |
| miR 1 (Cheng et al. 2010; Ikeda et al. 2009; Yang et al. 2007; D' Alessandra et al. 2010) | (+) | Infarct size (negative effect) MI (negative effect) Heart adaptation after ischaemic stress (positive effect) Protection against myocardial injury (positive effect) |
| miR 499 (Wang et al. 2010; D' Alessandra et al. 2010) | (+) | MI (positive effect) |

**Abbreviations**: MI: myocardial infarction; ACS: acute coronary syndrome; CAD: coronary artery disease; UA: unstable angina, + or − indicate upregulation or downregulation respectively

discrimination between ischemic and nonischemic heart failure. (Sucharov et al. 2008; Naga Prasad et al. 2009). Of note, among these expression profiles, some consistent changes (i.e., miRNA-21) were observed, a fact potentially indicating that miRNAs may be involved in common pathways mediating the hypertrophic response (Wang et al. 2009). Particularly, it has been indicated that miRNA-1 expression may be inversely correlated with cardiac hypertrophy (Ikeda et al. 2009). Also, miRNA-1 and miRNA-133 were attenuated in hypertrophic cardiomyopathy, atrial dilatation, as well as cardiac hypertrophy (Care et al. 2007). Moreover, studies have investigated whether miRNAs expression profiles that were regulated in heart failure could be normalized by left ventricular assist device (LVAD) therapy (Matkovich et al. 2009). The assay used a microarray containing probes for 467 miRNAs in cardiomyopathic hearts with or without treatment with LVADs. Interestingly, more than 70% of the upregulated miRNAs were fully normalized in hearts with LVAD treatment, while the others showed a tendency for reduction. In agreement, it has been demonstrated that LVAD treatment also succeeded in normalizing abnormal miRNAs expression, albeit with a more pronounced effect in states of ischemic cardiomyopathy (Schipper et al. 2008).

## Role of microRNAs in Cardiac Arrhythmias

Cardiac arrhythmias are characterized by important electrophysiological changes and remain a major health problem due to their unpredictable nature. Yang et al. attempted to investigate the role of miRNA-1 in arrhythmiogenesis in a rat model of MI and in human hearts with CAD (Yang et al. 2007). According to their findings, the nucleotide miRNA-1 was significantly upregulated in ischemic heart tissue, and its administration not only exacerbated arrhythmiogenesis, but also, its lack induced the opposite effect in heart tissues. Moreover, it has been shown that mice lacking miRNA-1-2 exhibited a high incidence of electrophysiological abnormalities often resulting in sudden death (Zhao et al. 2007). miRNA-133 is also implicated in arrythmiogenesis, as exogenous administration into diabetic rabbit myocytes resulted in suppression of expression of ERG, a long-QT syndrome gene (encoding a K+ channel). Specifically, ERG protein level was downregulated without altering its transcript level, which resulted in QT prolongation and the associated arrhythmias in diabetic hearts (Krutzfeldt et al. 2006). Moreover, it has been indicated that downregulation of miRNA-1 and miRNA-133 has been associated with an increase in protein levels of HCN2 and HCN4, which are two important cardiac pacemaker channel proteins enhancing automaticity and development of arrhythmia (Luo et al. 2008). Those miRNAs have also been implicated in expression changes of KCNQ1 and KCNE1, proteins that form a channel complex when assembled (Xu et al. 2007). Additional evidence suggests the potential involvement of miRNA-133 in the regulation of L-type calcium channel expression (Lu et al. 2009).

## Conclusion

A great amount of evidence indicates the importance of genetic components and altered expression of genes in cardiac function and pathophysiology. MicroRNAs are currently considered to play an important role in regulation of target genes in various cell types. These noncoding RNAs are implicated in several processes, such as proliferation, differentiation, and apoptosis, thus contributing to tissue differentiation and organ development. Specifically, the participation of miRNAs in cardiac development has been well documented in several studies. Additionally, these molecules seem to be highly sensitive early biomarkers for cardiovascular diseases which may represent a new revolution in modern cardiology. Also, measurable circulating miRNAs have raised the hypothesis that the miRNA signature could discriminate between high and low-risk patients. Finally, miRNA-based strategies and manipulation of these nucleotides may be proved an important novel therapeutic option. However, many more large scale studies are required to evaluate the role of microRNAs in cardiovascular disease and their potential to perform as novel biomarkers or even therapeutic targets.

# References

Ambros, V. 2004. The functions of animal microRNAs. Nature 431: 350–355.

Babak, T., W. Zhang, Q. Morris, B.J. Blencowe, and T.R. Hughes. 2004. Probing microRNAs with microarrays: tissue specificity and functional inference. RNA 10: 1813–1819.

Bartel, D.P. 2004 MicroRNAs: genomics, biogenesis, mechanism, and function. Cell 116: 281–297.

Baskerville, S. and D.P. Bartel. 2005. Microarray profiling of microRNAs reveals frequent coexpression with neighboring miRNAs and host genes. RNA 11: 241–247.

Baudhuin, L.M. 2009. Genetic markers for coronary artery disease. Clin. Lab. Sci. 22: 226.

Boettger, T., N. Beetz, S. Kostin, J. Schneider, M. Krüger, L. Hein, and T. Braun. 2009. Acquisition of the contractile phenotype by murine arterial smooth muscle cells depends on the miR143/145 gene cluster. J. Clin. Invest. 119: 2634–2647.

Bohnsack, M.T., K. Czaplinski, and D. Gorlich. 2004. Exportin 5 is a RanGTP-dependent dsRNA-binding protein that mediates nuclear export of pre-miRNAs. RNA 10: 185–191.

Bonauer, A., G. Carmona, M. Iwasaki, M. Mione, M. Koyanagi, A. Fischer, J. Burchfield, H. Fox, C. Doebele, K. Ohtani, E. Chavakis, M. Potente, M. Tjwa, C. Urbich, A.M. Zeiher, and S. Dimmeler. 2009. MicroRNA-92a controls angiogenesis an and functional recovery of ischemic tissues in mice. Science 324: 1710–1713.

Borchert, G.M., W. Lanier, and B.L. Davidson. 2006. RNA polymerase III transcribes human microRNAs. Nat. Struct. Mol. Biol. 13: 1097–101.

Briasoulis, A., D. Tousoulis, E.S. Androulakis, N. Papageorgiou, G. Latsios, and C. Stefanadis. 2012. Endothelial dysfunction and atherosclerosis: focus on novel therapeutic approaches. Recent Pat Cardiovasc. Drug Discov. 7: 21–32.

Callis, T.E., J.F. Chen, and D.Z. Wang. 2007. MicroRNAs in skeletal and cardiac muscle development. DNA Cell Biol. 26: 219–225.

Care, A., D. Catalucci, F. Felicetti, D. Bonci, A. Addario, P. Gallo, M.L. Bang, P. Segnalini, Y. Gu, N.D. Dalton, L. Elia, M.V. Latronico, M. Hoydal, C. Autore, M.A. Russo, G.W. II Dorn, O. Ellingsen, P. Ruiz-Lozano, K.L. Peterson, C.M. Croce, C. Peschle, and G. Condorelli. 2007. MicroRNA-133 controls cardiac hypertrophy. Nat. Med. 13: 613–618.

Catalucci, D., P. Gallo, and G. Condorelli. 2009 MicroRNAs in cardiovascular biology and heart disease. Circ. Cardiovasc. Genet. 2: 402–408.

Chan, L.S., P.Y. Yue, N.K. Mak, and R.N. Wong. 2009. Role of microRNA-214 in ginsenoside-Rg1-induced angiogenesis. Eur. J. Pharm. Sci. 38: 370–377.

Charakida, M., D. Tousoulis, and C. Stefanadis. 2006. Early atherosclerosis in childhood: diagnostic approaches and therapeutic strategies. Int. J. Cardiol. 109: 152–159.

Chen, C., D.A. Ridzon, A.J. Broomer, Z. Zhou, D.H. Lee, J.T. Nguyen, M. Barbisin, N.L. Xu, V.R. Mahuvakar, M.R. Andersen, K.Q. Lao, K.J. Livak, and K.J. Guegler. 2005a. Real-time quantification of microRNAs by stemloop RTPCR. Nucleic Acids Res. 33: e179.

Chen, P.Y., H. Manninga, K. Slanchev, M. Chien, J.J. Russo, J. Ju, R. Sheridan, B. John, D.S. Marks, D. Gaidatzis, C. Sander, M. Zavolan, and T. Tuschl. 2005b. The developmental miRNA profiles of zebrafish as determined by small RNA cloning. Genes Dev. 19: 1288–1293.

Chen, J.F., E.P. Murchison, R. Tang, T.E. Callis, M. Tatsuguchi, Z. Deng, M. Rojas, S.M. Hammond, M.D. Schneider, C.H. Selzman, G. Meissner, C. Patterson, G.J. Hannon, and D.Z. Wang. 2008. Targeted deletion of Dicer in the heart leads to dilated cardiomyopathy and heart failure. Proc. Natl. Acad. Sci. USA 105: 2111–2116.

Chen, T., Z. Huang, L. Wang, Y. Wang, F. Wu, S. Meng, and C. Wang. 2009. MicroRNA-125a-5p partly regulates the inflammatory response, lipid uptake, and ORP9 expression in oxLDL-stimulated monocyte/macrophages. Cardiovasc Res. 83: 131–139

Chen, Y. and D.H. Gorski. 2008. Regulation of angiogenesis through a microRNA (miR-130a) that downregulates antiangiogenic homeobox genes GAX and HOXA5. Blood 111: 1217–1226.

Cheng, Y., N. Tan, J. Yang, X. Liu, X. Cao, P. He, X. Dong, S. Qin, and C. Zhang. 2010. A translational study of circulating cell-free microRNA-1 in acute myocardial infarction. Clin. Sci. (Lond.) 119: 87–95

Cheng, Y., R. Ji, J. Yue, J. Yang, X. Liu, H. Chen, D.B. Dean, and C. Zhang. 2007. MicroRNAs are aberrantly expressed in hypertrophic heart: do they play a role in cardiac hypertrophy? Am. J. Pathol. 170: 1831–1840.

Condorelli, G., M.V. Latronico, and G.W. Dorn 2nd. 2010. microRNAs in heart disease: putative novel therapeutic targets? Eur. Heart J. 31: 649–658.

Cordes, K.R., N.T. Sheehy, M.P. White, E.C. Berry, S.U. Morton, A.N. Muth T.H. Lee, J.M. Miano, K.N. Ivey, and D. Srivastava. 2009. miR-145 and miR-143 regulate smooth muscle cell fate and plasticity. Nature 460: 705–710.

D'Alessandra, Y., P. Devanna, F. Limana, S. Straino, A. Di Carlo, P.G. Brambilla, M. Rubino, M.C. Carena, L. Spazzafumo, M. De Simone, B. Micheli, P. Biglioli, F. Achilli, F. Martelli, S. Maggiolini, G. Marenzi, G. Pompilio, and M.C. Capogrossi. 2010. Circulating microRNAs are new and sensitive biomarkers of myocardial infarction. Eur. Heart J. 31: 2765–2773.

Davis, B.N., A.C. Hilyard, P.H. Nguyen, G. Lagna, and A. Hata. 2009. Induction of microRNA-221 by platelet-derived growth factor signaling is critical for modulation of vascular smooth muscle phenotype. J. Biol. Chem. 284: 3728–3738.

Davison, T.S., C.D. Johnson, and B.F. Andruss. 2006. Analyzing micro-RNA expression using microarrays. Methods Enzymol. 411: 14–34.

Dong, D.L., C. Chen, R. Huo, N. Wang, Z. Li, Y.J. Tu, J.T. Hu, X. Chu, W. Huang, and B.F. Yang. 2010. Reciprocal repression between microRNA-133 and calcineurin regulates cardiac hypertrophy: a novel mechanism for progressive cardiac hypertrophy. Hypertension 55: 946–952.

Elia, L., M. Quintavalle, J. Zhang, R. Contu, L. Cossu, M.V. Lantronico, K.L. Peterson, C. Indolfi, D. Catalucci, J. Chen, S.A. Courtnteidge, and G. Condorelli. 2009. The knockout of miR-143 and -145 alters smooth muscle cell maintenance and vascular homeostasis in mice: correlates with human disease. Cell Death Differ. 16: 1590–1598.

Fasanaro, P., S. Greco, M. Lorenzi, M. Pescatori, M. Brioschi, R. Kulshreshtha, C. Banfi, A. Stubbs, G.A. Calin, M. Ivan, M.C. Capogrossi, and F. Martelli. 2009. An integrated approach for experimental target identification of hypoxia-induced miR-210. JBC 284: 35134–35143.

Fichtlscherer, S., S. De Rosa, H. Fox, T. Schwietz, A. Fischer, C. Liebetrau, M. Weber, C.W. Hamm, T. Röxe, M. Müller-Ardogan, A. Bonauer, A.M. Zeiher, and S. Dimmeler. 2010. Circulating microRNAs in patients with coronary artery disease. Circ. Res. 107: 677–684.

Fleissner, F., V. Jazbutyte, J. Fiedler, S.K. Gupta, X. Yin, Q. Xu, P. Galuppo, S. Kneitz, M. Mayr, G. Ertl, J. Bauersachs, and T. Thum. 2010. Short communication: asymmetric dimethylarginine impairs angiogenic progenitor cell function in patients with coronary artery disease through a microRNA 21-dependentmechanism. Circ. Res. 107: 138–143.

Foshay, K.M. and G.I. Gallicano. 2007. Small RNAs, big potential: the role of MicroRNAs in stem cell function. Curr. Stem Cell Res. Ther. 2: 264–71.

Garofalo, M., G. Condorelli, and C.M. Croce. 2008. MicroRNAs in diseases and drug response. Curr. Opin. Pharmacol. 8: 661–667.

Gilad, S., E. Meiri, Y. Yogev, S. Benjamin, D. Lebanony, N. Yerushalmi, H. Benjamin, M. Kushnir, H. Cholakh, N. Melamed, Z. Bentwich, M. Hod, Y. Goren, and A. Chajut. 2008. Serum microRNAs are promising novel biomarkers. PLoS ONE 3: e3148.

Gregory, R.I., K. Yan, G. Amuthan, T. Chendrimada, B. Doratotaj, N. Cooch, and R. Shiekhattar. 2004. The microprocessor complex mediates the genesis of microRNAs. Nature 432: 235–240.

Harris, T.A., M. Yamakuchi, M. Ferlito, J.T. Mendell, and C.J. Lowenstein. 2008. MicroRNA-126 regulates endothelial expression of vascular cell adhesion molecule 1. PNAS 105: 1516–1521.

Haver, V.G., R.H. Slart, C.L. Zeebregts, M.P. Peppelenbosch, and R.A. Tio. 2010. Rupture of vulnerable atherosclerotic plaques: MicroRNAs conducting the orchestra? Trends Cardiovasc. Med. 20: 65–71.

Hunter, M.P., N. Ismail, X. Zhang, B.D. Aguda, E.J. Lee, L. Yu, T. Xiao, J. Schafer, M.L. Lee, T.D. Schmittgen, S.P. Nana-Sinkam, D. Jarjoura, and C.B. Marsh. 2008. Detection of microRNA expression in human peripheral blood microvesicles. PLoS ONE 3: pe3694.

Hutvagner, G. and P.D. Zamore. 2002. A microRNA in a multiple-turnover RNAi enzyme complex. Science 297: 2056–2060.

Ikeda, S., A. He, S.W. Kong, J. Lu, R. Bejar, N. Bodyak, N.K.H. Lee, Q. Ma, P.M. Kang, T.R. Golub, and W.T. Pu. 2009. MicroRNA-1 negatively regulates expression of the hypertrophy-associated genes calmodulin and Mef2a. Mol. Cell Biol. 29: 2193–2204.

Ikeda, S., S.W. Kong, J. Lu, E. Bisping, H. Zhang, P.D. Allen, R.D. Golub, B. Pieske, and W.T. Pu. 2007. Altered microRNA expression in human heart disease. Physiol. Genomics 31: 367–373.

Jamaluddin, S., S. Weakley, L. Zhang, P. Kougias, P.H. Lin, Q. Yao, C. Chen. 2011. miRNAs: roles and clinical applications in vascular disease, Expert. Rev. Mol. Diagn. 11: 79–89.

Ji, R., Y. Cheng, J. Yue, J. Yang, X. Liu, H. Chen, D.B. Dean, and C. Zhang. 2007. MicroRNA expression signature and antisense-mediated depletion reveal an essential role of microRNA in vascular neointimal lesion formation. Circ. Res. 100: 1579–1588.

Ji, X., R. Takahashi, Y. Hiura, G. Hirokawa, Y. Fukushima, N. Iwai. 2009. Plasma miR-208 as a biomarker of myocardial injury. Clin. Chem. 55: 1944–1949.

Kawai-Kowase, K. and G.K. Owens. 2007. Multiple repressor pathways contribute to phenotypic switching of vascular smooth muscle cells. Am. J. Physiol. Cell Physiol. 292: C59–C69

Kim, V.N. and J.W. Nam. 2006. Genomics of microRNA. Trends Genet. 22: 165–173.

Krichevsky, A.M., K.S. King, C.P. Donahue, K. Khrapko, and K.S. Kosik. 2003. A microRNA array reveals extensive regulation of microRNAs during brain development. RNA 9: 1274–1281.

Krutzfeldt, J., M.N. Poy, and M. Stoffel. 2006. Strategies to determine the biological function of microRNAs. Nat. Genet. 38: S14–S19.

Kuehbacher, A., C. Urbich, A.M. Zeiher, and S. Dimmeler. 2007. Role of Dicer and Drosha for endothelial microRNA expression and angiogenesis. Circ. Res. 101: 59–68.

Kwon, C., Z. Han, E.N. Olson, and D. Srivastava. 2005. MicroRNA1 influences cardiac differentiation in Drosophila and regulates Notch signaling. Proc. Natl. Acad. Sci. USA 102: 18986–18991.

Lagos-Quintana, M., R. Rauhut, A. Yalcin, J. Meyer, W. Lendeckel, and T. Tuschl. 2002. Identification of tissue-specific microRNAs from mouse. Curr. Biol. 12: 735–739.

Lee, D.Y., Z. Deng, C.H. Wang, and B.B. Yang. 2007. MicroRNA-378 promotes cell survival, tumor growth, and angiogenesis by targeting SuFu and Fus-1 expression. Proc. Natl. Acad. Sci. USA 104: 20350–20355.

Lee, R.C., R.L. Feinbaum, and V. Ambros. 1993. The C. elegans heterochronic gene lin-4 encodes small RNAs with antisense complementarity to lin-14. Cell 75: 843–854.

Lee, Y., C. Ahn, J. Han, H. Choi, J. Kim, J. Yim, and J. Lee, P. Provost, O. Rådmark, S. Kim, V.N. Kim. 2003. The nuclear RNase III Drosha initiates microRNA processing. Nature 425: 415–419.

Lim, L.P., N.C. Lau, P. Garrett-Engele, A. Grimson, J.M. Schelter, J. Castle, J, D.P. Bartel, P.S. Linsley, and J.M Johnson. 2005. Microarray analysis shows that some microRNAs downregulate large numbers of target mRNAs. Nature 433: 769–773.

Liu, C.G., G.A. Calin, B. Meloon, N. Gamliel, C. Sevignani, M. Ferracin, C.D. Dumitru, M. Shimizu, S. Zupo, M. Dono, H. Alder, F. Bullrich, M. Negrini, and C.M. Croce. 2004. An oligonucleotide microchip for genome-wide microRNA profiling in human and mouse tissues. Proc. Natl. Acad. Sci. USA 101: 9740–9744.

Liu, X., Y. Cheng, S. Zhang, Y. Lin, J. Yang, and C. Zhang. 2009. A necessary role of miR-222 and miR-221 in vascular smooth muscle cell proliferation and neointimal hyperplasia. Circ. Res. 104: 476–487.

Lu, J., G. Getz, E.A. Miska, E. Alvarez-Saavedra, J. Lamb, D. Peck, A. Sweet-Cordero, B.L. Ebert, R.H. Mak, A.A. Ferrando, J.R. Downing, T. Jacks, H.R. Horvitz, and T.R. Golub. 2005 MicroRNA expression profiles classify human cancers. Nature 435: 834–838.

Lu, Y., J. Xiao, H. Lin, Y. Bai, X. Luo, Z. Wang, and B.A. Yang. 2009. Single antimicroRNA antisense oligodeoxyribonucleotide (AMO) targeting multiple microRNAs offers an improved approach for microRNA interference. Nucleic Acids Res. 37: e24.

Lund, E., S. Guttinger, A. Calado, J.E. Dahlberg, and U. Kutay. 2004. Nuclear export of microRNA precursors. Science 303: 95–98.

Luo, X., H. Lin, Z. Pan, J. Xiao, Y. Zhang, Y. Lu, B. Yang, and Z. Wang. 2008. Down-regulation of miR-1/miR-133 contributes to re-expression of pacemaker channel genes HCN2 and HCN4 in hypertrophic heart. J. Biol. Chem. 283: 20045–20052.

Maniataki, E. and Z. Mourelatos. 2005. A human, ATP-independent, RISC assembly machine fueled by pre-miRNA. Genes Dev. 19: 2979–2990.

Margulies, K.B. 2009. MicroRNAs as novel myocardial biomarkers. Cin. Chem. 55: 1897–1899.

Martin, K., J. O'Sullivan, and N. Caplice. 2011. New therapeutic potential of microRNA treatment to target vulnerable atherosclerotic lesions and plaque rupture. Cur. Opin. Cardiol. 26: 569–575.

Matkovich, S.J., D.J. Van Booven, K.A. Youker, G. Torre-Amione, A. Diwan, W.H. Eschenbacher, L.E. Dorn, M.A. Watson, K.B. Margulies, and G.W. Dorn. 2009. Reciprocal regulation of myocardial microRNAs and messenger RNA in human cardiomyopathy and reversal of the microRNA signature by biomechanical support. Circulation 119: 1263–1271.

Metias, S.M., E. Lianidou, and G.M. Yousef. 2009. MicroRNAs in clinical oncology: at the crossroads between promises and problems. J. Clin. Pathol. 62: 771–776.

Minami, Y., M. Satoh, C. Maesawa, Y. Takahashi, T. Tabuchi, T. Itoh, and M. Nakamura. 2009. Effect of atorvastatin on microRNA 221/222 expression in endothelial progenitor cells obtained from patients with coronary artery disease. Eur. J. Clin. Invest. 39: 359–367.

Monticelli, S., K.M. Ansel, C. Xiao, N.D. Socci, A.M. Krichevsky, T.H. Thai, N. Rajewsky, D.S. Marks, C. Sander, K. Rajewsky, A. Rao, and K.S. Kosik. 2005. MicroRNA profiling of the murine hematopoietic system. Genome Biol. 6: R71.

Murohara, T., T. Asahara, M. Silver, C. Bauters, H. Masuda, C. Kalka, M. Kearne, Y.D. Chen, J.F. Symes, M.C. Fishman, P.L. Huang, and J.M. Isner. 1998. Nitric oxide synthase modulates angiogenesis in response to tissue ischemia. J. Clin. Invest. 101: 2567–2578.

Naga Prasad, S.V., Z.H. Duan, M.K. Gupta, V.S. Surampudi, S. Volinia, G.A. Calin, C.G. Liu, A. Kotwal, C.S. Moravec, R.C. Starling, D.M. Perez, S. Sen, Q. Wu, E.F. Plow, C.M. Croce, and S. Karnik. 2009. Unique microRNA profile in end-stage heart failure indicates alterations in specific cardiovascular signaling networks. J. Biol. Chem. 284: 27487–27499.

Nelson, P.T., D.A. Baldwin, L.M. Scearce, J.C. Oberholtzer, J.W. Tobias, and Z. Mourelatos. 2004. Microarray-based, high-throughput gene expression profiling of microRNAs. Nat. Methods 1: 155–161.

O'Sullivan, J.F., K. Martin, and N.M. Caplice. 2011. Microribonucleic acids for prevention of plaque rupture and in-stent restenosis: "A finger in the dam". J. Am. Coll. Cardiol. 57: 383–9.

Ono, K., Y. Kuwabara, and J. Han. 2011. MicroRNAs and cardiovascular diseases. FEBS J. 278: 1619–1633.

Pan, Z.W., Y.J. Lu, and B.F. Yang. 2010. MicroRNAs: a novel class of potential therapeutic targets for cardiovascular diseases. Acta Pharmacol. Sin. 31: 1–9.

Papageorgiou, N., D. Tousoulis, E. Androulakis, A. Giotakis, G. Siasos, G. Latsios, and C. Stefanadis. 2011. Lifestyle Factors and Endothelial Function. Curr. Vasc. Pharmacol. 10: 94–106.

Papageorgiou, N., D. Tousoulis, E. Androulakis, G. Siasos, A. Briasoulis, G. Vogiatzi, A.M. Kampoli, E. Tsiamis, C. Tentolouris, and C. Stefanadis. 2012. The role of microRNAs in cardiovascular disease. Curr. Med. Chem. 19: 2605–2610.

Parmacek, M.S. 2009. MicroRNA-modulated targeting of vascular smooth muscle cells. J. Clin. Invest. 119: 2526–2528.

Poliseno, L., A. Tuccoli, L. Mariani, M. Evangelista, L. Citti, K. Woods, A. Mercatanti, S. Hammond, and G. Rainaldi. 2006. MicroRNAs modulate the angiogenic properties of HUVECs. Blood 108: 3068–3071.

Qin, B., H. Yang, and B. Xiao. 2012. Role of microRNAs in endothelial inflammation and senescence. Mol. Biol. Rep. 39: 4509–4518.

Ross, R. 1999. Atherosclerosis–an inflammatory disease. N. Engl. J. Med. 340: 115–126.

Ruby, J.G., C.H. Jan, and D.P. Bartel. 2007. Intronic microRNA precursors that bypass Drosha processing. Nature 448: 83–86.

Santovito, D., A. Mezzetti, and F. Cipollone. 2012. MicroRNAs and atherosclerosis: New actors for an old movie. Nutr. Metab. Cardiovasc. Dis. [Epub ahead of print].

Schipper, M.E., J. van Kuik, N. de Jonge, H.F. Dullens, and R.A. de Weger. 2008. Changes in regulatory microRNA expression in myocardium of heart failure patients on left ventricular assist device support. J. Heart Lung Transplant. 27: 1282–1285.

Shyu, AB., M.F. Wilkinson, and A. van Hoof. 2008. Messenger RNA regulation: to translate or to degrade. EMBO J. 27: 471–481.

Silvestre, J.S., Z. Mallat, A. Tedgui, and B.I. Levy. 2008. Post-ischaemic neovascularisation and inflammation. Cardiovasc. Res. 78: 242–249.

Song, X.W., Q. Li, L. Lin, X.C. Wang, D.F. Li, G.K. Wang, A.J. Ren, Y.R. Wang, Y.W. Qin, W.J. Yuan, and Q. Jing. 2010. MicroRNAs are dynamically regulated in hypertrophic hearts, and miR-199a is essential for the maintenance of cell size in cardiomyocytes. J. Cell Physiol. 225: 437–443.

Suarez, Y. and W.C. Sessa. 2009. MicroRNAs as novel regulators of angiogenesis. Circ. Res. 104: 442–454.

Suarez, Y., C. Fernandez-Hernando, J. Yu, S.A. Gerber, K.D. Harrison, J.S. Pober, M.L. Iruela-Arispe, M. Merkenschlager, and W.C. Sessa. 2008. Dicer-dependent endothelial microRNAs are necessary for postnatal angiogenesis. Proc. Natl. Acad. Sci. USA 105: 14082–14087.

Suarez, Y., C. Wang, T.D. Manes, and J.S. Pober. 2010. Cutting edge: TNF-induced micro- RNAs regulate TNF-induced expression of E-selectin and intercellular adhesion molecule-1 on human endothelial cells: feedback control of inflammation. J. Imm. 184: 21–25.

Sucharov, C., M.R. Bristow, and J.D. Port. 2008. miRNA expression in the failing human heart: functional correlates. J. Mol. Cell Cardiol. 45: 185–192.

Tatsuguchi, M., H.Y. Seok, T.E. Callis, J.M. Thomson, J.F. Chen, M. Newman, M. Rojas, S.M. Hammond, and D.Z. Wang. 2007. Expression of microRNAs is dynamically regulated during cardiomyocyte hypertrophy. J. Mol. Cell. Cardiol. 42: 1137–1141.

Thum, T., D. Catalucci, and J. Bauersachs. 2008. MicroRNAs: novel regulators in cardiac development and disease. Cardiovasc. Res. 79: 562–570.

Thum, T., P. Galuppo, C. Wolf, J. Fiedler, S. Kneitz, L.W. van Laake, P.A. Doevendans, C.L. Mummery, J. Borlak, A. Haverich, C. Gross, S. Engelhardt, G. Ertl, and J. Bauersachs. 2007. MicroRNAs in the human heart: a clue to fetal gene reprogramming in heart failure. Circulation 116: 258–267.

Tousoulis, D., A. Briasoulis, N. Papageorgiou, C. Antoniades, and C. Stefanadis. 2008a. Candidate gene polymorphisms and the 9p21 locus in acute coronary syndromes. Trends Mol. Med. 14: 441–449.

Tousoulis, D., A.M. Kampoli, E. Stefanadi, C. Antoniades; G. Siasos; A.G. Papavassiliou, and C. Stefanadis. 2008b. New biochemical markers in acute coronary syndromes. Curr. Med. Chem. 15: 1288–1296.

Tousoulis, D., M. Koutsogiannis, N. Papageorgiou, G. Siasos, C. Antoniades, E. Tsiamis, and C. Stefanadis. 2010. Endothelial dysfunction: potential clinical implications. Minerva Med. 101: 271–284.

Tousoulis, D., A.M. Kampoli, N. Papageorgiou, E. Androulakis, C. Antoniades, K. Toutouzas, and C. Stefanadis. 2011a. Pathophysiology of atherosclerosis: the role of inflammation. Curr. Pharm. Des. 17: 4089–110.

Tousoulis, D., N. Papageorgiou, A. Briasoulis, E. Androulakis, M. Charakida, E. Tsiamis, and C. Stefanadis. 2011b. Conflicting effects of nitric oxide and oxidative stress in chronic heart failure: potential therapeutic strategies. Heart Fail. Rev.

Tousoulis, D., G. Hatzis, N. Papageorgiou, E. Androulakis, G. Bouras, A. Giolis, C. Bakogiannis, G. Siasos, G. Latsios, C. Antoniades, and C. Stefanadis. 2012. Assessment of acute coronary syndromes: focus on novel biomarkers. Curr. Med. Chem. 19: 2572–2587.

Urbich, C., A. Kuehbacher, and S. Dimmeler. 2008. Role of microRNAs in vascular diseases, inflammation, and angiogenesis, Cardiovasc. Res. 79: 581–588.

van Rooij, E., L.B. Sutherland, J.E. Thatcher, J.M. DiMaio, R.H. Naseem, W.S. Marshall, J.A. Hill, and E.N. Olson. 2008. Dysregulation of microRNAs after myocardial infarction reveals a role of miR-29 in cardiac fibrosis. Proc. Natl. Acad. Sci. USA 105: 13027–13032.

van Rooij, E., L.B. Sutherland, X. Qi, J.A. Richardson, J. Hill, and E.N. Olson. 2007. Control of stressdependent cardiac growth and gene expression by a microRNA. Science 316: 575–579.

Wang, G.K., J.Q. Zhu, J.T. Zhang, Q. Li, Y. Li, J. He, Y.W. Qin, and Q. Jing. 2010. Circulating microRNA: a novel potential biomarker for early diagnosis of acute myocardial infarction in humans. Eur. Heart J. 31: 659–666.

Wang, K., S. Zhang, B. Marzolf, P. Troisch, A. Brightman, Z. Hu, L.E. Hood, and D.J. Galas. 2009. Circulating microRNAs, potential biomarkers for drug-induced liver injury. Proc. Natl. Acad. Sci. USA 106: 4402–4440.

Wang, N., Z. Zhou, X. Liao, and T. Zhang. 2009. Role of microRNAs in cardiac hypertrophy and heart failure. IUBMB Life 61: 566–571.

Wang, Z., X. Luo, Y. Lu, and B. Yang. 2008. miRNAs at the heart of the matter. J. Mol. Med. (Berl.) 86: 771–783.

Weber, M., M.B. Baker, J.P. Moore, and C.D. Searles. 2010. MiR-21 is induced in endothelial cells by shear stress and modulates apoptosis and eNOS activity. Biochem. Biophys. Res. Commun. 393: 643–648.

Wightman, B., I. Ha, and G. Ruvkun. 1993. Posttranscriptional regulation of the heterochronic gene lin-14 by lin-4 mediates temporal pattern formation in C. elegans. Cell 75: 855–862.

Xin, M., E.M. Small, L.B. Sutherland, X. Qi, J. McAnally, C.F. Plato , J.A. Richardson, R. Bassel-Duby, and E.N. Olson. 2009. MicroRNAs miR-143 and miR-145 modulate cytoskeletal dynamics and responsiveness of smoothmuscle cells to injury. Genes Dev. 23: 2166–2178.

Xu, C., Y. Lu, Z. Pan, W. Chu, X. Luo, H. Lin, J. Xiao, H. Shan, Z. Wang, and B. Yang. 2007. The muscle-specific microRNAs miR-1 and miR-133 produce opposing effects on apoptosis by targeting HSP60, HSP70 and caspase-9 in cardiomyocytes. J. Cell Sci. 120: 3045–3052.

Xu, J., J. Zhao, G. Evan, C. Xiao, Y. Cheng, and J. Xiao. 2012. Circulating microRNAs: novel biomarkers for cardiovascular diseases. J. Mol. Med. (Berl.) 90: 865–875.

Yang, B., H. Lin, J. Xiao, Y. Lu, X. Luo, B. Li, Y. Zhang, C. Xu, Y. Bai, H. Wang, G. Chen, and Z. Wang. 2007. The muscle-specific microRNA miR-1 regulates cardiac arrhythmogenic potential by targeting GJA1 and KCNJ2. Nat. Med. 13: 486–491.

Yang, W.J., D.D. Yang, S. Na, G.E. Sandusky, Q. Zhang, and G. Zhao. 2005. Dicer is required for embryonic angiogenesis during mouse development. J. Biol. Chem. 280: 9330–9335.

Yoshizaki, K., H. Wakita, K. Takeda, and K. Takahashi. 2008. Conditional expression of microRNA against E-selectin inhibits leukocyte-endothelial adhesive interaction under inflammatory condition. Biochem. Biophys. Res. Comm. 371: 747–751.

Zacharowski, K., P. Zacharowski, S. Reingruber, and P. Petzelbauer. 2006. Fibrin(ogen) and its fragments in the pathophysiology and treatment of myocardial infarction. J. Mol. Med. (Berl) 84: 469–477.

Zeng, Y. 2006. Principles of micro-RNA production and maturation. Oncogene 25: 6156–6162.

Zeng, Y. and B.R. Cullen. 2003. Sequence requirements for micro RNA processing and function in human cells. RNA 9: 112–123.

Zhang, C. 2008a. MicroRNAs: role in cardiovascular biology and disease. Clin. Sci. (Lond). 114: 699–706.

Zhang, C. 2008b. MicroRNomics: a newly emerging approach for disease biology. Physiol. Genomics. 33: 139–147.

Zhang, R. and B. Su. 2009. Small but influential: the role of microRNAs on gene regulatory network and 3′UTR evolution. J. Genet. Genomics 36: 1–6.

Zhao, Y., E. Samal, and D. Srivastava. 2005. Serum response factor regulates a musclespecific microRNA that targets Hand2 during cardiogenesis. Nature 436: 214–220.

Zhao, Y., J.F. Ransom, A. Li, V. Vedantham, M. von Drehle, A.N. Muth, T. Tsuchihashi, M.T. McManus, R.J. Schwartz, and D. Srivastava. 2007. Dysregulation of cardiogenesis, cardiac conduction, and cell cycle in mice lacking miRNA-1-2. Cell 129: 303–317.

Zhu, N., D. Zhang, S. Chen, X. Liu, L. Lin, X. Huang, Z. Guo, J. Liu , Y. Wang, W. Yuan, and Y. Qin. 2011. Endothelial enriched microRNAs regulate angiotensin II-induced endothelial inflammation and migration. Atherosclerosis 215: 286–293.

# Index

Milton Keynes UK
Ingram Content Group UK Ltd.
UKHW021848071024
449327UK00021B/1552